String-Math 2012

Proceedings of Symposia in
PURE MATHEMATICS

Volume 90

String-Math 2012

July 16–21, 2012
Universität Bonn, Bonn, Germany

Ron Donagi
Sheldon Katz
Albrecht Klemm
David R. Morrison
Editors

American Mathematical Society
Providence, Rhode Island

2010 *Mathematics Subject Classification.* Primary 11G55, 14D21, 14F05, 14J28, 14M30, 32G15, 53D18, 57M27, 81T40. 83E30.

Library of Congress Cataloging-in-Publication Data

String-Math (Conference) (2012 : Bonn, Germany)
String-Math 2012 : July 16-21, 2012, Universität Bonn, Bonn, Germany/Ron Donagi, Sheldon Katz, Albrecht Klemm, David R. Morrison, editors.
pages cm. – (Proceedings of symposia in pure mathematics; volume 90)
Includes bibliographical references.
ISBN 978-0-8218-9495-8 (alk. paper)
1. Geometry, Algebraic–Congresses. 2. Quantum theory– Mathematics–Congresses. I. Donagi, Ron, editor. II. Katz, Sheldon, 1956- editor. III. Klemm, Albrecht, 1960- editor. IV. Morrison, David R., 1955- editor. V. Title.

QA564.S77 2012
516.3′5–dc23 2015017523

DOI: http://dx.doi.org/10.1090/pspum/090

Color graphic policy. Any graphics created in color will be rendered in grayscale for the printed version unless color printing is authorized by the Publisher. In general, color graphics will appear in color in the online version.

∞ The paper used in this book is acid-free and falls within the guidelines
established to ensure permanence and durability.
Visit the AMS home page at http://www.ams.org/

10 9 8 7 6 5 4 3 2 1 20 19 18 17 16 15

Contents

Preface

The conference 'String-Math 2012' was held on July 16–21, 2012 at the Hausdorff Center for Mathematics, Universität Bonn. This was the second in a series of large meetings exploring the interface of mathematics and string theory. This volume presents the proceedings of that conference.

The nature of the interactions between mathematicians and physicists has been thoroughly transformed in recent years. String theory and quantum field theory have contributed a series of profound ideas which gave rise to entirely new mathematical fields and revitalized older ones. The influence flows in both directions, with mathematical techniques and ideas contributing crucially to major advances in string theory. By now there is a large and rapidly growing number of both mathematicians and physicists working at the string-theoretic interface between the two academic fields.

For mathematics, string theory has been a source of many significant inspirations, ranging from Seiberg-Witten theory in four-manifolds, to enumerative geometry and Gromov-Witten theory in algebraic geometry, to work on the Jones polynomial in knot theory, to recent progress in the geometric Langlands program and the development of derived algebraic geometry and n-category theory. In the other direction, mathematics has provided physicists with powerful tools, ranging from powerful differential geometric techniques for solving or analyzing key partial differential equations, to toric geometry, to K-theory and derived categories in D-branes, to the geometry of special holonomy manifolds as string compactifications, to the use of modular forms and other arithmetic techniques. The depth, power and novelty of the results obtained in both fields thanks to their interaction is truly mind-boggling.

The annual String-Math conferences are becoming the central venue for these profound and wide-ranging interactions. They bring together leading mathematicians and mathematically minded physicists working in this interface. These meetings promote and publicize such interactions, giving attendees greater opportunities to cross cultural boundaries, learn aspects of other fields relevant for their research, and advertise important developments to audiences that might not otherwise hear of them or appreciate their importance.

The 2012 conference was organized by Sergei Gukov, Daniel Huybrechts, Hans Jockers, Albrecht Klemm, Wolfgang Lück, Hans-Peter Nilles, Catharina Stroppel, Peter Teichner, and Don Zagier. The Steering Committee consisted of Dan Freed, Nigel Hitchin, Maxim Kontsevich, David Morrison, Karen Uhlenbeck, Edward Witten, and Shing-Tung Yau.

The meeting covered a wide array of topics at the interface of mathematics and high energy physics, including

- Topological field and string theory in various dimensions
- Homological mirror symmetry
- String topology
- Arithmetic of strings
- Gromov-Witten theory and enumerative geometry
- BPS state counting and Wall crossing formulas
- Geometric Langlands program
- A-twisted Landau-Ginzburg models
- Compactifications, special holonomy and special structure manifolds
- Heterotic strings, gauge bundle construction and (2,0) mirror symmetry
- Elliptic cohomology
- Large N dualities and integrability
- Non-perturbative dualities, F-theory
- Topological T-duality
- String measures
- Chiral de Rham complexes
- Noncommutative geometry

Altogether, this conference brought together nearly 200 mathematicians and physicists. There were 34 invited plenary talks given by leaders in both fields. Additionally, there were 39 contributed talks given in parallel sessions on the Wednesday of the meeting. All the talks are available at the conference website: http://www.hcm.uni-bonn.de/events/eventpages/2012/string-math-2012/. The conference also included a public lecture on 'Quo Vadis LHC?' by Christophe Grojean of CERN. The conference was preceeded by the Bethe Forum 'Lecture Series on Mathematical String Theory,' intended as a preparation to String-Math 2012 for graduate students and researchers alike.

Friedrich Hirzebruch, the founder and longtime director of the Max Planck Institute for Mathematics who encouraged and developed the interaction between mathematics and string theory over many years, passed away less than two months before the conference. Hirzebruch was renowned for his early work on the Riemann-Roch theorem, which became crucial for the understanding of many physical anomalies. In the early nineties, along with T. Höfer, he explained the relation between the Euler number of orbifolds, as introduced by string physicists, and the formulas of his student Lothar Göttsche for the Betti numbers of the Hilbert schemes of points on algebraic surfaces. The latter became decisive in the microscopic interpretation of black hole entropy by Strominger and Vafa. In the last years of his life, Hirzebruch explored elliptic genera and their connections with physics. He wrote a beautiful book on the subject with Thomas Berger and Rainer Jung. At the end of the interview that he gave for the Simons Foundation, one can hear him say that he "would not mind to be an expert in string theory." The venue of the conference was the main lecture hall of mathematics, where Hirzebruch had directed the famous Arbeitstagung for many years. His absence in the audience of eminent researchers in this field, so close to his heart, was keenly felt. A number of the papers in this volume are dedicated to his memory.

The string/math collaboration is clearly here to stay, and we expect this conference series to continue as long as the subject remains active and exciting. The venues and years of the first seven conferences of the String-Math series are:

- String-Math 2011, Philadelphia (Penn), June 6–11, 2011
- String-Math 2012, Bonn (Hausdorff Center for Mathematics), July 16–21, 2012
- String-Math 2013, Stony Brook (Simons Center for Geometry and Physics), June 17–21, 2013
- String-Math 2014, Edmonton (U. of Alberta), June 9–13, 2014
- String-Math 2015, Sanya (Tsinghua Sanya International Mathematics Forum), Dec. 31, 2015 – Jan. 5, 2016
- String-Math 2016, Paris (Institut Poincaré), June 27–July 2, 2016
- String-Math 2017, Hamburg

We gratefully acknowledge support obtained from the following sources: The Bethe Center for Theoretical Physics, the Hausdorff Center for Mathematics, the Max-Planck-Institut for Mathematics and the Sonderforschungsbereich TR45 "Periods, moduli spaces and arithmetic of algebraic varieties." We are also very grateful to Sergei Gelfand and Chris Thivierge of the AMS for their help in preparing this volume.

The editors of String-Math 2012:

Ron Donagi
Sheldon Katz
Albrecht Klemm
David R. Morrison

List of Participants

Ido Adam
IFT-UNESP

Nezhla Aghaei
Physikalisches Institut der University of Bonn

Xiaohua AI
École Polytechnique

Tavanfar Alireza
CERN

Lara Anderson
Harvard University

David Andriot
ASC LMU Munich

Lilia Anguelova
Perimeter Institute for Theoretical Physics

Nima Arkani-Hamed
IAS

Elena Mirela Babalic
IFIN-HH Horia Hulubei National Institute of Physics and Nuclear Engineering

Francesco Benini
Simons Center, Stonybrook U.

Marco Bertolini
Duke University

Nana Geraldine Cabo Bizet
BCTP, University of Bonn

Michael Blaszczyk
BCTP, University of Bonn

Nikolay Bobev
Simons Center for Geometry and Physics, SUNY- Stony Brook

Giulio Bonelli
SISSA

Vincent Bouchard
University of Alberta, Department of Mathematics

Tom Bridgeland
University of Oxford, All Souls College

Alexandr Buryak
University of Amsterdam

David Bücher
Universität Hamburg

Ana Ros Camacho
Universität Hamburg

Nils Carqueville
LMU Munich

Sergio Cecotti
SISSA

Chi-Ming Chang
Harvard University

Athanasios Chatzistavrakidis
BCTP, University of Bonn

Adrian Clingher
University of Missouri - St. Louis

Ioana-Alexandra Coman
Horia-Hulubei National Institute of Physics and Nuclear Engineering

Andrei Constantin
Oxford University

Clay Cordova
Harvard University

Rhys Davies
Mathematical Institute, University of Oxford

Andreas Deser
Max-Planck-Institute for Physics

Anindya Dey
University of Texas at Austin

Tudor Dimofte
Institute for Advanced Study

Ron Donagi
University of Pennsylvania

Will Donovan
University of Edinburgh

Philippe Durand
Département de Mathématiques (Equipe M2N)

Valeriy Dvoeglazov
Universidad de Zacatecas

Chris Elliott
Northwestern University

Magnus Engenhorst
Mathematisches Institut

Ahmad Reza Estakhr
Shiraz

Jens Fjelstadt
Nanjing University

Omar Foda
University of Melbourne

Daniel Labardini Fragoso
Universität Bonn, Mathematisches Institut

Edward Frenkel
University of California, Berkeley

Hiroyuki Fuji
Nagoya University/Faculty of Science

Maxime Gabella
IPhT CEA/Saclay

Matthias Gaberdiel
ETH Zurich

Navaneeth Krishna Gaddam
BCTP, University of Bonn

Sergey Grigorian
Simons Center for Geometry and Physics

Dima Grigoriev
Université de Lille

Jie Gu
BCTP, University of Bonn

Sam Gunningham
Northwestern University

Babak Haghighat
Harvard University

Daniel Halpern-Leistner
UC Berkeley

Jeff Harvey
Enrico Fermi Institute, University of Chicago

Mans Henningson
Fundamental Physics, Chalmers University of Technology

Nigel Hitchin
Oxford

Stefan Hohenegger
MPI Munich

Daigo Honda
University of Tokyo

Zheng Hua
Max Planck Institute for Mathematics

Minxin Huang
Kavli IPMU, University of Tokyo

Giovanni Cerulli Irelli
University of Bonn

Zbigniew Jaskolski
Wroclaw University

Hans Jockers
BCTP, University of Bonn

Larisa Jonke
Physikalisches Institut der University of Bonn

Benjamin Jurke
Northeastern University

Joel Kamnitzer
University of Toronto

Dila Kandel
Golden Gate International College

Anna Karlsson
Chalmers University of Technology

Amir-Kian Kashani-Poor
École Normale Supérieure

Sheldon Katz
University of Illinois

Christoph Keller
Caltech

Bilal Khadija
Ibn Tofail

Imran Parvez Khan
Comsatis Institute of Information Technology, Islamabad

Tae-Su Kim
Seoul National University

Axel Kleinschmidt
Max Planck Institute for Gravitational Physics

Albrecht Klemm
Physikalisches Institut, Universität Bonn

Denis Klevers
University of Pennsylvania

Johanna Knapp
Kavli IPMU

Maxim Kontsevich
IHES

Peter Koroteev
University of Minnesota

Christian Gueha Koundjo
I.E.P.D Cameroun

Can Kozcaz
CERN

Daniel Krefl
University of California, Berkeley

Sven Krippendorf
BCTP, University of Bonn

Stefan Krämer
BCTP, University of Bonn

Mohammed Labbi
Max Planck Institute for Mathematics

Joshua Lapan
McGill University

Calin Iuliu Lazaroiu
Horia Hulubei National Institute of Physics and Nuclear Engineering, Bucarest

Wolfgang Lerche
CERN

Wei Li
MPI for Gravitational Physics

Oscar Loaiza-Brito
University of Guanajuato

Daniel Vieira Lopes
BCTP, University of Bonn

Farhang Loran
Isfahan University of Technology

Christoph Lüdeling
BCTP

Andrew Macpherson
Imperial College London

Jean-Pierre Magnot
Université Blaise Pascal

Andreas Malmendier
Colby College

Jan Manschot
BCTP, MPI for Mathematics

Kishore Marathe
CUNY Brooklyn College

Kazunobu Maruyoshi
SISSA

Lionel Mason
The Mathematical Institute, University of Oxford

Jock McOrist
University of Cambridge

Noppadol Mekareeya
Max Planck Institute for Physics

Ilarion Melnikov
Max Planck Institute for Gravitational Physics

Stefan Mendez-Diez
University of Alberta

Hartmut Monien
University of Bonn

Samuel Monnier
Lab. de Physique Théorique, ENS

Gregory Moore
Rutgers University

Partha Mukhopadhyay
The Institute of Mathematical Sciences

Motohico Mulase
University of California

Han Muxin
Centre de Physique Théorique

Hans Peter Nilles
Physikalisches Institut, Universität Bonn

Sebastian Novak
Universität Hamburg

Paul-Konstantin Oehlmann
BCTP, University of Bonn

Ryo Ohkawa
Kyoto University

Ashraf Oiws
Cairo University

Tadashi Okazaki
Osaka University

Andrei Okounkov
Columbia University

Takuya Okuda
University of Tokyo

Hirosi Ooguri
California Institute of Technology

Domenico Orlando
CERN

Rahul Pandharipande
ETH Zurich

Seo-Ree Park
Seoul National University

Sara Pasquetti
Queen Mary University

Mattia Pedrini
SISSA

Damian Kaloni Mayorga Pena
BCTP, University of Bonn

Daniel Persson
Chalmers University of Technology

Jochen Peschutter
BCTP

Vasily Pestun
IAS, Princeton

Franco Pezzella
INFN and Naples University

Boris Pioline
CERN

David Plencner
LMU Munich

Maximilian Poretschkin
BCTP, University of Bonn

Leonardo Rastelli
Yang Institute for Theoretical Physics

Susanne Reffert
CERN

Jonas Reuter
Physikalisches Institut der Universität Bonn

Patricia Ritter
CECs

Daniel Roggenkamp
Rutgers University

Andy Royston
Rutgers University

Fabio Ferrari Ruffino
Universidade de Sao Paulo

Francesco Sala
Heriot-Watt University

Karim Salehi
Zakho University

Osvaldo Pablo Santillán
Universidad de Buenos Aires

Raffaele Savelli
Max-Planck-Institute for Physics

Emanuel Scheidegger
Mathematisches Institut, Universität Freiburg

Ricardo Schiappa
Instituto Superior Tecnico

Marc Schiereck
BCTP, University of Bonn

Cornelius Schmidt-Colinet
IPMU, Tokyo

Matthias Schmitz
BCTP, University of Bonn

Domenico Seminara
Physics Department, Florence University

Ashoke Sen
Harish-Chandra Research Institute

Eric Sharpe
Virginia Tech

Samson Shatashvili
Trinity College of Dublin

Vivek Shende
MIT

Nick Sheridan
MIT

Artan Sheshmani
University of British Columbia

Bernd Siebert
Universität Hamburg, FB Mathematik

Yan Soibelman
Kansas State University

Masoud Soroush
BCTP, University of Bonn

Pawel Sosna
University of Hamburg

Stephan Stieberger
MPI Physik Munich

Catharina Stroppel
University of Chicago

Paulina Suchanek
DESY, Theory Group

Piotr Sulkowski
University of Amsterdam and Caltech

Rui Sun
AEI

Roman Sverdlov
Institute of Mathematical Sciences

Balazs Szendroi
Mathematical Institute, University of Oxford

Meng-Chwan Tan
National University of Singapore

Alessandro Tanzini
SISSA

Jamie Tattersall
BCTP, University of Bonn

Washington Taylor
MIT

Jörg Teschner
DESY

Richard Thomas
Imperial College London

Maike Tormählen
Leibniz Universität Hannover

Alessandro Torrielli
University of Surrey

Hagen Triendl
IPhT/CEA Saclay

Efstratios Tsatis
University of Patras

Grigory Vartanov
DESY

Janu Verma
Kansas State University

Jena Vinod
MATS University

Anastasia Volovich
Brown University

Roberto Volpato
Max Planck Institut Potsdam

Marcel Vonk
University of Santiago de Compostela

Johannes Walcher
McGill University

Konrad Waldorf
Universität Hamburg

James Wallbridge
Harvard University

Katrin Wendland
Mathematisches Institut, Freiburg University

Clemens Wieck
BCTP, University of Bonn

Martijn Wijnholt
LMU Munich

Matthias Wilhelm
Humboldt University Berlin/University of Bonn

Simon Wood
The University of Tokyo, IPMU

Thomas Wotschke
BCTP, University of Bonn

Junya Yagi
University of Hamburg

Hyun Seok Yang
Center for Quantum Spacetime, Sogang University

Yi Yang
National Chiao Tung University

Shing Tung Yau
Harvard University

Don Zagier
MPI für Mathematik, Bonn

Jose Miguel Zapata-Rolon
Cologne

Jie Zhou
Department of Mathematics, Harvard University

Plenary talks

Proceedings of Symposia in Pure Mathematics
Volume **90**, 2015
http://dx.doi.org/10.1090/pspum/090/01516

The Quiver Approach to the BPS Spectrum of a $4d$ $\mathcal{N}=2$ Gauge Theory

Sergio Cecotti

Dedicated to the Memory of Professor Friedrich Hirzebruch

ABSTRACT. We present a survey of the computation of the BPS spectrum of a general four–dimensional $\mathcal{N}=2$ supersymmetric gauge theory in terms of the Representation Theory of quivers with superpotential. We focus on SYM with a general gauge group G coupled to standard matter in arbitrary representations of G (consistent with a non–positive beta–function). The situation is particularly tricky and interesting when the matter consists of an odd number of *half*–hypermultiplets: we describe in detail $SU(6)$ SYM coupled to a $\frac{1}{2}\,\mathbf{20}$, $SO(12)$ SYM coupled to a $\frac{1}{2}\,\mathbf{32}$, and E_7 SYM coupled to a $\frac{1}{2}\,\mathbf{56}$.

1. Introduction

In the last few years many new powerful methods were introduced to compute the exact BPS spectrum of a four–dimensional $\mathcal{N}=2$ supersymmetric QFT. We may divide the methods in two broad classes: *i)* geometric methods [**1**–**5**] and *ii)* algebraic methods [**6**–**14**]. The geometric methods give a deep understanding of the non–perturbative physics, while the algebraic ones are quite convenient for actual computations. In the algebraic approach the problem of computing the BPS spectrum is mapped to a *canonical* problem in the Representation Theory (RT) of (basic) associative algebras. A lot of classical results in RT have a direct physical interpretation and may be used to make the BPS spectral problem 'easy' for interesting classes of $\mathcal{N}=2$ theories. Besides, by comparing RT and physics a lot of interesting structures emerge which shed light on both subjects.

1.1. From $\mathcal{N}=2$ QFT to quiver representations. To fix the notation, we recall how the BPS states are related to quiver representations, referring to [**10**] for more details. The conserved charges of the theory (electric, magnetic, and flavor) are integrally quantized, and hence take value in a lattice $\Gamma=\oplus_v\,\mathbb{Z}e_v$. On Γ we have a skew–symmetric integral pairing, $\langle\gamma,\gamma'\rangle_{\text{Dirac}}\in\mathbb{Z}$, given by the Dirac electro–magnetic pairing; the flavor charges then correspond to the zero–eigenvectors of the matrix $B_{uv}\equiv\langle e_u,e_v\rangle_{\text{Dirac}}\in\mathbb{Z}$.

Following [**7**] we say that our $\mathcal{N}=2$ model has the *quiver property* if we may find a set of generators $\{e_v\}$ of Γ such that the charge vectors $\gamma\in\Gamma$ of all the BPS

2010 *Mathematics Subject Classification.* Primary 81T60.

particles satisfy

$$\gamma \in \Gamma_+ \quad \text{or} \quad -\gamma \in \Gamma_+, \tag{1}$$

where $\Gamma_+ \equiv \oplus_v \mathbb{Z}_+ e_v$ is the *positive cone* in Γ. Given a $\mathcal{N}=2$ theory with the quiver property, we associate a 2–acyclic quiver Q to the data $(\Gamma_+, \langle\cdot,\cdot\rangle_{\text{Dirac}})$: to each positive generator e_v of Γ_+ we associate a node v of Q and we connect the nodes u v with B_{uv} arrows $u \to v$ (a negative number meaning arrows in the opposite direction). The positive cone $\Gamma_+ \subset \Gamma$ is then identified with the cone of dimension vectors of the representations X of Q through $\mathbf{dim}\, X \equiv \sum_v \dim X_v\, e_v$.

The emergence of the quiver Q may be understood as follows. Fix a particle with charge $\gamma = \sum_v N_v\, e_v \in \Gamma_+$; on its word–line we have a one dimensional supersymmetric theory with 4 supercharges, and the BPS particles correspond to states which are SUSY vacua of this 1d theory. The 1d theory turns out to be a quiver theory in the sense that its Kähler target space is the representation space of Q of dimension $\sum_v N_v\, e_v$

$$\prod_{\substack{\text{arrows}\\ u\to v}} \mathbb{C}^{N_u N_v} \Big/\!\!\Big/ \prod_{\substack{\text{nodes}\\ v}} GL(N_v, \mathbb{C}) \qquad \text{(symplectic quotient)}. \tag{2}$$

To completely define the 1d theory we need to specify a $\prod_{\text{nodes}} GL(N_v,\mathbb{C})$–invariant superpotential $\mathcal{W}$ (and the FI terms implicit in (2)); gauge invariance requires $\mathcal{W}$ to be a function of the traces of the products of the bi–fundamental Higgs fields along the closed oriented loops in Q. It turns out that this function must be linear (a sum of single–trace operators) and thus canonically identified with a linear combination (with complex coefficients) of the oriented cycles in Q. Thus $\mathcal{W}$ is a potential for the quiver Q in the sense of DWZ [**15**]. One shows [**10**] that a 1d configuration is a classical SUSY vacuum if and only if the bi–fundamental Higgs fields associated to the arrows of Q form a *stable* module X of the Jacobian algebra[1]

$$\mathscr{J}(Q,\mathcal{W}) := \mathbb{C}Q/(\partial\mathcal{W}), \tag{3}$$

and two field configurations are physically equivalent iff the corresponding modules are isomorphic. *Stability* is defined in terms of the central charge Z of the $\mathcal{N}=2$ SUSY algebra. Being conserved, Z is a linear combinations of the various charges; hence may be seen as a linear map $Z\colon \Gamma \to \mathbb{C}$. We assume $\operatorname{Im} Z(\Gamma_+) \geq 0$, so that we have a well–defined function $\arg Z\colon \Gamma_+ \to [0,\pi]$. Then $X \in \mathsf{mod}\,\mathscr{J}(Q,\mathcal{W})$ is *stable* (with respect to the given central charge Z) iff, for all proper non–zero submodules Y, $\arg Z(Y) < \arg Z(X)$. In particular, *X is stable $\Rightarrow$ X is a brick,* (a module X of an associative algebra is called a *brick* if $\operatorname{End} X = \mathbb{C}$). The isoclasses of stable modules of given dimension γ typically form a family parameterized by a Kähler manifold $\mathcal{M}_\gamma$; from the viewpoint of the 1d theory the space $\mathcal{M}_\gamma$ corresponds to zero–modes which should be quantized producing $SU(2)_{\text{spin}} \times SU(2)_R$ quantum numbers. In particular, a d–dimensional family corresponds (at least) to a BPS supermultiplet with spin content $(0,\frac{1}{2}) \otimes \frac{d}{2}$ (thus rigid modules corresponds to hypermultiplets, $\mathbb{P}^1$–families to vector supermultiplets, and so on). Notice that the full dependence of the BPS spectrum from the parameters of the theory is encoded

[1] A module $X \in \mathsf{mod}\,\mathscr{J}(Q,\mathcal{W})$ of dimension $\sum_v N_v e_v$ is specified by giving, for each arrow $u \xrightarrow{\alpha} v$, an $N_v \times N_u$ matrix X_α such that the matrices $\{X_\alpha\}$ satisfy the relations $\partial_{X_\beta}\mathcal{W}(X_\alpha) = 0$ for all arrows β in Q. Two such representations are *isomorphic* if they are related by a $\prod_v GL(N_v,\mathbb{C})$ transformation.

in the central charge Z, which depends on these parameters as specified by the Seiberg–Witten geometry.

For a given $\mathcal{N} = 2$ theory $(Q, \mathcal{W})$ is *not* unique; indeed there may be several sets of generators $\{e_v\}$ with the above properties. Two allowed $(Q, \mathcal{W})$ are related by a Seiberg duality, which precisely coincides with the mutations of a quiver with potential in the sense of cluster algebras [**15**] (this, in particular, requires $\mathcal{W}$ to be non–degenerate in that sense). Therefore, to a QFT we associate a full *mutation class* of quivers. If the mutation class is finite we say that the corresponding $\mathcal{N} = 2$ QFT is *complete* [**7**] which, in particular, implies that no BPS state has spin larger than 1.

T_2**–duality.** The Seiberg duality/DWZ mutation is not the only source of quiver non–uniqueness. The quiver mutations preserve both the number of nodes and 2–acyclicity. There are more general dualties which do not share these properties. As an example consider the Gaiotto theory corresponding to the A_1 $(2, 0)$ 6d theory on a sphere with 3 regular punctures (the T_2 theory) [**16**]. T_2 consists of 4 free hypermultiplets, carrying 4 flavor charges, which corresponds to a disconnected quiver with 4 nodes and no arrows. On the other hand, we may associate to it a quiver with only *three* nodes, each pair of nodes being connected by a pair of opposite arrows $\rightleftarrows$ [**10**]. We refer to the equivalence of the two quivers as 'T_2–duality'.

2. The $(Q, \mathcal{W})$ class associated to an $\mathcal{N} = 2$ theory

The BPS states correspond to the stable bricks of the Jacobian algebra. This reduces our problem to a standard problem in Representation Theory *provided* we know which $(Q, \mathcal{W})$ mutation class is associated to our $\mathcal{N} = 2$ theory. Determining the mutation class for several interesting gauge theories is the main focus of the present note.

For $\mathcal{N} = 2$ models having a corner in their parameter space with a weakly coupled Lagrangian description, we have a very physical criterion to check whether a candidate pair $(Q, \mathcal{W})$ is correct. Simply use the category $\mathsf{mod}\,\mathscr{J}(Q, \mathcal{W})$ to compute the would–be BPS spectrum in the limit of vanishing YM coupling $g_{\text{YM}} \to 0$ and compare the result with the prediction of perturbation theory. The weakly coupled spectrum should consist of

- finitely many mutually–local states with bounded masses as $g_{\text{YM}} \to 0$:
  (1) vector multiplets making one copy of the adjoint representation of the gauge group G (photons and W–bosons);
  (2) hypermultiplets making definite (quaternionic) representations R_a of G (quarks);
- particles non–local relatively to the W–bosons with masses $O(1/g^2_{\text{YM}})$ (heavy dyons).

We ask which pairs $(Q, \mathcal{W})$ have such a property (the Ringel property [**11**]).

2.1. Magnetic charge and weak coupling regime. Consider a quiver $\mathcal{N} = 2$ gauge theory having a weak coupling description with gauge group G (of rank r). We pick a particular pair $(Q, \mathcal{W})$ in the corresponding Seiberg mutation–class which is appropriate for the weak coupling regime (along the Coulomb branch).

$\mathsf{mod}\,\mathscr{J}(Q,\mathcal{W})$ should contain, in particular, one–parameter families of representations corresponding to the massive W–boson vector–multiplets which are in one–to–one correspondence with the positive roots of G. We write δ_a $(a = 1, 2, \ldots, r)$ for the charge (*i.e.* dimension) vector of the W–boson associated to the *simple–root* α_a of G.

At a generic point in the Coulomb branch we have an unbroken $U(1)^r$ symmetry. The $U(1)^r$ electric charges, properly normalized so that they are integral for all states, are given by the fundamental coroots[2] $\alpha_a^\vee \in \mathfrak{h}$ $(a = 1, 2, \ldots, r)$. The a–th electric charge of the W–boson associated to b–th simple root α_b then is

$$q_a = \alpha_a(\alpha_b^\vee) = C_{ab}, \qquad \text{(the Cartan matrix of } G). \tag{4}$$

Therefore the vector in $\Gamma \otimes \mathbb{Q}$ corresponding to the a–th unit electric charge is

$$\mathfrak{q}_a = (C^{-1})_{ab}\, \delta_b. \tag{5}$$

Then the magnetic weights (charges) of a representation X are given by

$$m_a(X) \equiv \langle \dim X, \mathfrak{q}_a \rangle_{\text{Dirac}} = (C^{-1})_{ab}\, B_{ij}\, (\dim X)_i\, (\delta_b)_j. \tag{6}$$

Dirac quantization requires the r linear forms $m_a(\cdot)$ to be *integral* [**11**]. This integrality condition is quite a strong constraint on the quiver Q, and is our main tool to determine it.

At weak coupling, $g_{\text{YM}} \to 0$, the central charge takes the classical form [**11**]

$$Z(X) = -\frac{1}{g^2_{\text{YM}}} \sum_i C_a\, m_a(X) + O(1), \tag{7}$$

where $C_a = -i\langle \varphi_a \rangle > 0$ in the region of the Coulomb branch covered by the quiver Q. It is convenient to define the light category, $\mathscr{L}(Q,\mathcal{W})$, as the subcategory of the modules $X \in \mathsf{mod}\,\mathscr{J}(Q,\mathcal{W})$ with $m_a(X) = 0$ for all a such that all their submodules have $m_a(Y) \leq 0$. Comparing with the definition of stability in §. 1.1, we see that all BPS states with bounded mass in the limit $g_{\text{YM}} \to 0$ correspond to modules in $\mathscr{L}(Q,\mathcal{W})$, and, in facts, for a $\mathcal{N} = 2$ theory which has a weakly coupled Lagrangian description the stable objects of $\mathscr{L}(Q,\mathcal{W})$ precisely match the perturbative states. They are just the gauge bosons, making one copy of the adjoint of G, together with finitely many hypermultiplets transforming in definite representations of G. The detailed structure of $\mathscr{L}(Q,\mathcal{W})$ is described in [**11**].

Remarks and Properties

(1) $\mathsf{mod}\,\mathscr{J}(Q,\mathcal{W})$ contains *many* ligh subcategories, one for each weakly coupled corner. *E.g.* $SU(2)$ $N_f = 4$ has a $SL(2,\mathbb{Z})$ orbit of such subcategories;
(2) $m(\Gamma_+) \ngeq 0 \Rightarrow$ the light category is *not* the restriction to a subquiver, and its quiver is *not* necessarily 2–acyclic (as in the T_2 case [**10, 11**]);
(3) the category $\mathscr{L}(Q,\mathcal{W})$ is *tame* (physically: no light BPS state of spin > 1);
(4) *universality of the SYM sector*: for given gauge group G

$$\mathscr{L}(Q_{\text{SYM}}, \mathcal{W}_{\text{SYM}}) \subset \mathscr{L}(Q,\mathcal{W})$$

where $(Q_{\text{SYM}}, \mathcal{W}_{\text{SYM}})$ is the pair for pure G SYM. Only finitely many bricks $X \in \mathscr{L}(Q,\mathcal{W})$ and $X \notin \mathscr{L}(Q_{\text{SYM}}, \mathcal{W}_{\text{SYM}})$, they correspond to 'quarks'.

[2] $\mathfrak{h}$ stands for the Cartan subalgebra of the complexified Lie algebra of the gauge group G.

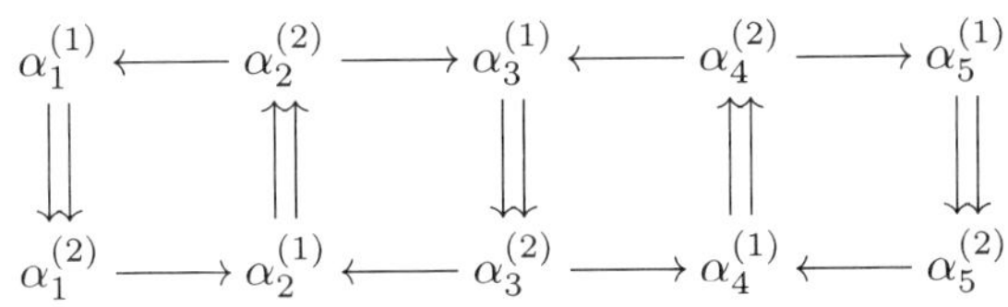

FIGURE 1. The *square* form of the quiver for pure $SU(6)$ SYM

3. First examples

As a warm–up we consider four classes of (simple) examples.

3.1. Example 1: $SU(2)$ SQCD with $N_f \leq 4$. These examples are discussed in detail in [**7**, **10**, **11**]; here we limit ourselves to a description of the resulting categories. One shows [**11**] that the category $\mathsf{mod}\mathscr{J}(Q,\mathcal{W})$ is Seiberg–duality equivalent to the Abelian category $\mathrm{Coh}(\mathbb{P}^1_{N_f})$ of coherent sheaves on $\mathbb{P}^1_{N_f}$ which is $\mathbb{P}^1$ with N_f 'double points', that is, the variety in the weighted projective space $W\mathbb{P}(2,2,\ldots,2,1,1)$ of equations

$$X_i^2 - \lambda_i X_{N_f+1} - \mu_i X_{N_f+2} = 0, \qquad i = 1,2,\ldots,N_f, \quad (\lambda_i : \mu_i) \in \mathbb{P}^1. \tag{8}$$

In $\mathrm{Coh}(\mathbb{P}^1_{N_f})$ we have two quantum numbers, *degree* and *rank*

$$\text{rank} = \text{magnetic charge}, \qquad \text{degree} = 2\times \text{ electric charge}. \tag{9}$$

The light subcategory $\mathrm{Coh}(\mathbb{P}^1_{N_f}) \supset \mathscr{L} = \{\text{sheaves of finite length}\}$ *a.k.a.* 'skyscrapers', while the dyons correspond to line bundles of various degree.

For $N_f = 4$ the curve $\mathbb{P}^1_4$ is Calabi–Yau, hence an elliptic curve E. The moduli space of the degree 1 skyscrapers, which is the curve E itself, is isomorphic to its Jacobian $J(E)$ which parameterizes the line bundles of fixed degree. Quantization of $J(E)$ then produces magnetic charged vector–multiplets. Of course, $E \sim J(E)$ reflects the S–duality of the theory. See [**11**] for more details.

3.2. Example 2: SYM with a simply–laced gauge group G. The quiver exchange matrix B is fixed by the Dirac charge quantization [**11**] (cfr. §. 2.1). The standard quiver (the square form) corresponds to

$$B_\Box = C \otimes S, \qquad \text{where} \quad \begin{cases} C \text{ is the Cartan matrix of } G, \\ S \text{ is the modular } S\text{–matrix.} \end{cases} \tag{10}$$

The square quiver is represented (for $G = SU(6)$) in Figure 1; it is supplemented by a quartic superpotential $\mathcal{W}$ [**10**, **11**]. The charge vector of the a–th simple root W–boson is equal to $\delta_a \equiv \alpha_a^{(1)} + \alpha_a^{(2)}$, *i.e.* the a–th simple–root W bosons corresponds to the $\mathbb{P}^1$–family of bricks associated with the minimal imaginary root of the a–th $\widehat{A}(1,1)$ affine subquiver $\upuparrows_a$. The a–th magnetic charge (weight) is (cfr. eqn.(6))

$$m_a(X) = \dim X_{\alpha_a^{(1)}} - \dim X_{\alpha_a^{(2)}}. \tag{11}$$

From the discussion around eqn.(7), the light subcategory $\mathscr{L}^{\mathrm{YM}}(G)$ containing the perturbative BPS spectrum is then given by the modules $X \in \mathsf{mod}\mathscr{J}(Q,\mathcal{W})$ with $m_a(X) = 0$ such that all their submodules Y satisfy $m_a(Y) \leq 0$, $\forall a$.

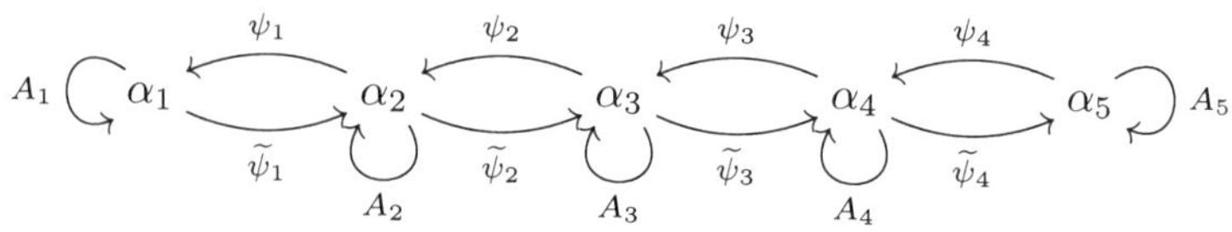

FIGURE 2. The reduced quiver Q' for $SU(6)$ pure SYM.

We may break $G \to SU(2)_a \times U(1)^{r-1}$ at weak coupling and describe the Higgs mechanism perturbatively; that is, the gauge breaking should respect the light *sub*category. Mathematically, this gives the following result at the level of Abelian categories of modules

$$X \in \mathscr{L}^{\mathrm{YM}}(G) \;\Rightarrow\; X\big|_{\upuparrows_a} \in \mathscr{L}^{\mathrm{YM}}(SU(2)) \quad \forall\, a, \tag{12}$$

which may be checked directly. Then, if X is indecomposable, in each Kronecker subquiver $\upuparrows_a$ we may set one of the arrows to 1 with the result that the category $\mathscr{L}^{\mathrm{YM}}(G)$ gets identified with the category of modules of a Jacobian algebra

$$\mathscr{L}^{\mathrm{YM}}(G) = \mathsf{mod}\,\mathscr{J}(Q', \mathcal{W}') \tag{13}$$

where the *reduced* quiver Q' is the *double*[3] of the Dynkin graph[4] G with loops A_v attached at the nodes (i.e. the '$\mathcal{N}=2$ quiver' of G), see Figure 2 for the $SU(6)$ example. The reduced quiver Q' is equipped with the superpotential

$$\mathcal{W}' = \sum_{\overrightarrow{a\colon \mathrm{edges}\in G}} \mathrm{tr}\big(\widetilde{\psi}_a A_{t(a)} \psi_a - \psi_a A_{h(a)} \widetilde{\psi}_a\big). \tag{14}$$

Given a module $X \in \mathsf{mod}\,\mathscr{J}(Q', \mathcal{W}')$, consider the linear map

$$\ell\colon (X_{\alpha_1}, X_{\alpha_2}, \cdots, X_{\alpha_r}) \mapsto (A_1 X_{\alpha_1}, A_2 X_{\alpha_2}, \cdots, A_r X_{\alpha_r}). \tag{15}$$

It is easy to check that $\ell \in \mathrm{End}\, X$, hence X a brick $\Rightarrow A_i = \lambda \in \mathbb{C}$ for all i (in fact, $\lambda \in \mathbb{P}^1$). Fixing $\lambda \in \mathbb{P}^1$, the brick X is identified with a brick of the double $\overline{G}$ of the Dynkin graph[5]

(16) $\overline{A_5}$

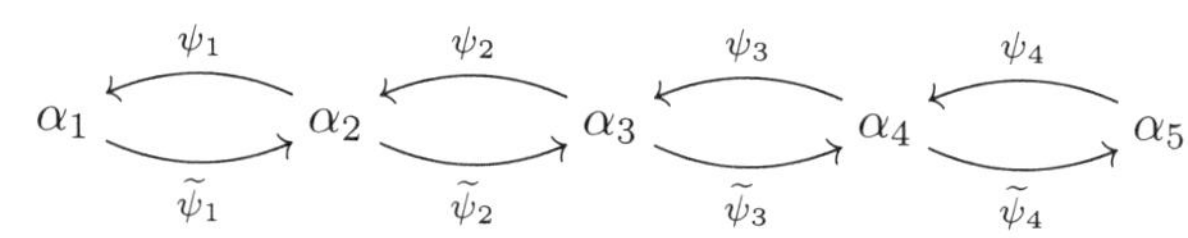

subjected to relations

$$\sum_{t(a)=v} \psi_a \widetilde{\psi}_a - \sum_{h(a)=v} \widetilde{\psi}_a \psi_a = 0. \tag{17}$$

[3] Given an unoriented graph L, its *double quiver* $\overline{L}$ is obtained by replacing each edge a of L by a pair of opposite arrows $\bullet \underset{\widetilde{\psi}_a}{\overset{\psi_a}{\rightleftarrows}} \bullet$. To write eqn.(14) we have picked an arbitrary orientation of G, the algebra $\mathscr{J}(Q', \mathcal{W}')$ being independent of choices, up to isomorphism.

[4] By abuse of notation, we use the same symbol G for the gauge group and its Dynkin graph.

[5] The reduced quiver is not 2–acyclic: this is related to the fact that it describes a subset of states which are *all* mutually local, hence have trivial Dirac pairing. At the level of the quiver this means that the *net* number of arrows from node i to node j must vanish (while we need to have arrows since the perturbative sector is not a free theory).

The algebra defined by the double quiver $\overline{G}$ with the relations (17) is known as the Gelfand–Ponomarev *preprojective algebra* of the graph G, written $\mathcal{P}(G)$ [**17**]. There are three basic results on the preprojective algebra of a graph L:

- Gelfand and Ponomarev [**17**]: $\dim \mathcal{P}(L) < \infty$ if and only if L is an ADE Dynkin graph;
- Crawley–Boevey [**18**]: Let $C_L = 2 - I_L$ be the Cartan matrix of the graph L. Then for all $X \in \mathsf{mod}\,\mathcal{P}(L)$

$$2 \dim \operatorname{End} X = (\mathbf{dim}\, X)^t\, C_L(\mathbf{dim}\, X) + \dim \operatorname{Ext}^1(X, X) \tag{18}$$

- Lusztig [**19**]: Let X be an indecomposable module of $\mathcal{P}(L)$ belonging to a family of non–isomorphic ones parameterized by the (Kähler) moduli space $\mathcal{M}(X)$. Then

$$\dim \mathcal{M}(X) = \tfrac{1}{2} \dim \operatorname{Ext}^1(X, X). \tag{19}$$

If L is an ADE graph G, the integral quadratic form $v^t\, C_G\, v$ is positive–definite and even; then $X \neq 0$ implies $(\mathbf{dim}\, X)^t\, C_L(\mathbf{dim}\, X) \geq 2$ with equality if and only if $\mathbf{dim}\, X$ is a positive root of G. From eqns.(18)(19) it follows that if X is a brick of $\mathcal{P}(G)$ it must be rigid with $\mathbf{dim}\, X$ a positive root of G. Going back to $\mathscr{L}^{\mathrm{YM}}(G)$, we see that a module in the light category is a brick iff $\mathbf{dim}\, X$ is a positive root of G and $\mathcal{M}(X) = \mathbb{P}^1$. By the dictionary between physics and Representation Theory, this means that the BPS states which are stable and have bounded mass as $g_{\mathrm{YM}} \to 0$ are *vector–multiplets in the adjoint of* the gauge group G. In fact, a more detailed analysis shows [**11**] that there is precisely one copy of the adjoint in each weakly coupled BPS chamber. This is, clearly, the result expected for pure SYM at weak coupling; in particular, is shows that the identification [**6**] of $(Q, \mathcal{W})$ is correct.

3.3. Example 3: SQCD with G simply–laced and N_a quarks in the a–th fundamental representation. We consider $\mathcal{N} = 2$ SQCD with a simply–laced gauge group $G = ADE$ coupled to N_a *full* hypermultipletss in the representation F_a with Dynkin label $[0, \cdots, 0, 1, 0, \cdots, 0]$ (1 in the a–th position, $a = 1, 2, \ldots, r$). The prescription for the quiver is simple [**10**]: one replaces the a–th Kronecker subquiver $\downdownarrows_a$ of the pure G SYM quiver (cfr. §. 3.2) as follows

(20)

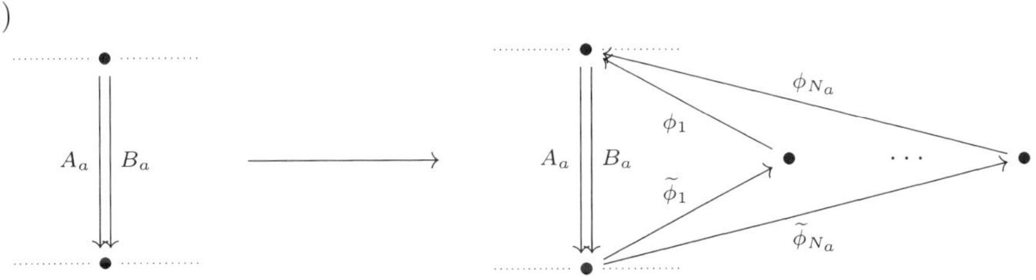

and replaces the pure SYM superpotential $\mathcal{W}_{\mathrm{SYM}}$ with

$$\mathcal{W} \longrightarrow \mathcal{W}_{\mathrm{SYM}} + \sum_{i=1}^{N_i} \operatorname{tr}\big[(\alpha_i\, A_a - \beta_i\, B_a)\phi_i\, \widetilde{\phi}_i\big], \tag{21}$$

$$(\alpha_i : \beta_i) \equiv \lambda_a \in \mathbb{P}^1 \text{ pairwise distinct.} \tag{22}$$

The exchange matrix of the resulting quiver, B, has N_i zero eigenvalues corresponding to the N_a flavor charges carried by the quarks. Formally [**10**], we may extend this construction to the case in which we have quarks in several distinct

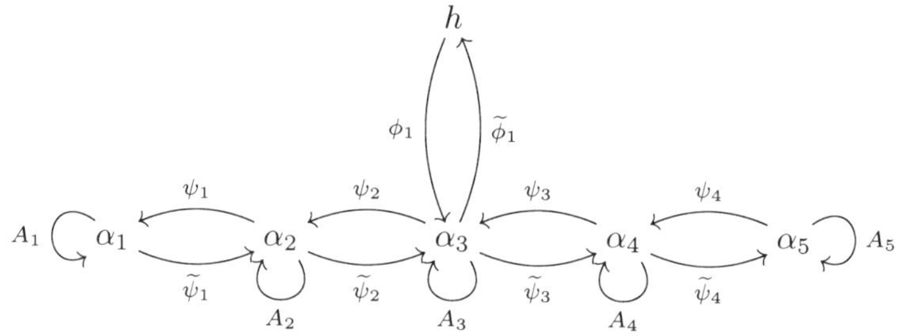

FIGURE 3. The *reduced* quiver $\overline{A_5[3,1]}$ for the light category of $G = SU(6)$ SYM coupled to one hypermultiplet h in the 3–rd fundamental rep. (*i.e.* the **20**).

fundamental representations, just be applying the substitutions (20)(21) to all the corresponding Kronecker subquivers of the (square) pure SYM quiver.

Going through the same steps as in §. 3.2, one sees that the light category $\mathscr{L} = \mathsf{mod}\,\mathscr{J}(Q',\mathcal{W}')$ with Q' the double of the graph $G[a, N_a]$ obtained by adding N_a extra nodes to the Dynkin graph G connected with a single hedge to the a–th node of G and having loops only at all 'old' nodes of G [**11**] (see Figure 3 for a typical example) and superpotential

$$\mathcal{W}' = \mathcal{W}'_{\text{SYM}} + \sum\nolimits_i \operatorname{tr}\big[(\alpha_i\, A_a - \beta_i)\phi_i\,\widetilde{\phi}_i\big]. \tag{23}$$

As in §. 3.2, X is a brick $\Rightarrow A_i = \lambda \in \mathbb{P}^1$. Now we have two distinct cases:

(1) λ is generic (*i.e.* $\lambda \neq \lambda_i$, $i = 1, 2, \dots, N_a$): the Higgs fields $\phi_i, \widetilde{\phi}_i$ are *massive* and may be integrated out. Then X is a brick of $\mathcal{P}(G)$ and its charge vector $\mathbf{dim}\, X$ is a positive root of G. These are the same representations as for the light category of pure SYM and they correspond to W–bosons in the adjoint of G;

(2) $\lambda = \lambda_a$, then X is a brick of the preprojective algebra $\mathcal{P}(G[i,1])$. Right properties (finitely many, rigid, in right reprs. of G) if and only if $G[i,1]$ is also a Dynkin graph.

By comparison one gets the following [**11**]:

Theorem. *(1) Consider $\mathcal{N} = 2$ SYM with simple simply–laced gauge group G coupled to a hyper in a representation of the form $F_a = [0, \cdots, 0, 1, 0, \cdots, 0]$. The resulting QFT is* Asymptotically Free *if and only if the augmented graph $G[a,1]$ obtained by adding to the Dynkin graph of G an extra node connected by a single edge to the a–th node of G is also an ADE Dynkin graph. (2) The model has a Type IIB engineering iff, in addition, the extra node is an extension node in the extended (affine) augmented Dynkin graph $\widehat{G[a,1]}$.*

See Figure 4 for the full list of asymptotically free theories of this class. Note that in case (2) the light category automatically contains hypermultiplets in the right representation of G since, if a is an extension node in $\widehat{G[a,1]}$ we have

$$\text{Ad}(G[a,1]) = \text{Ad}(G) \oplus [0, \cdots, 0, 1, 0, \cdots, 0] \oplus \overline{[0, \cdots, 0, 1, 0, \cdots, 0]} \oplus \text{singlets}. \tag{24}$$

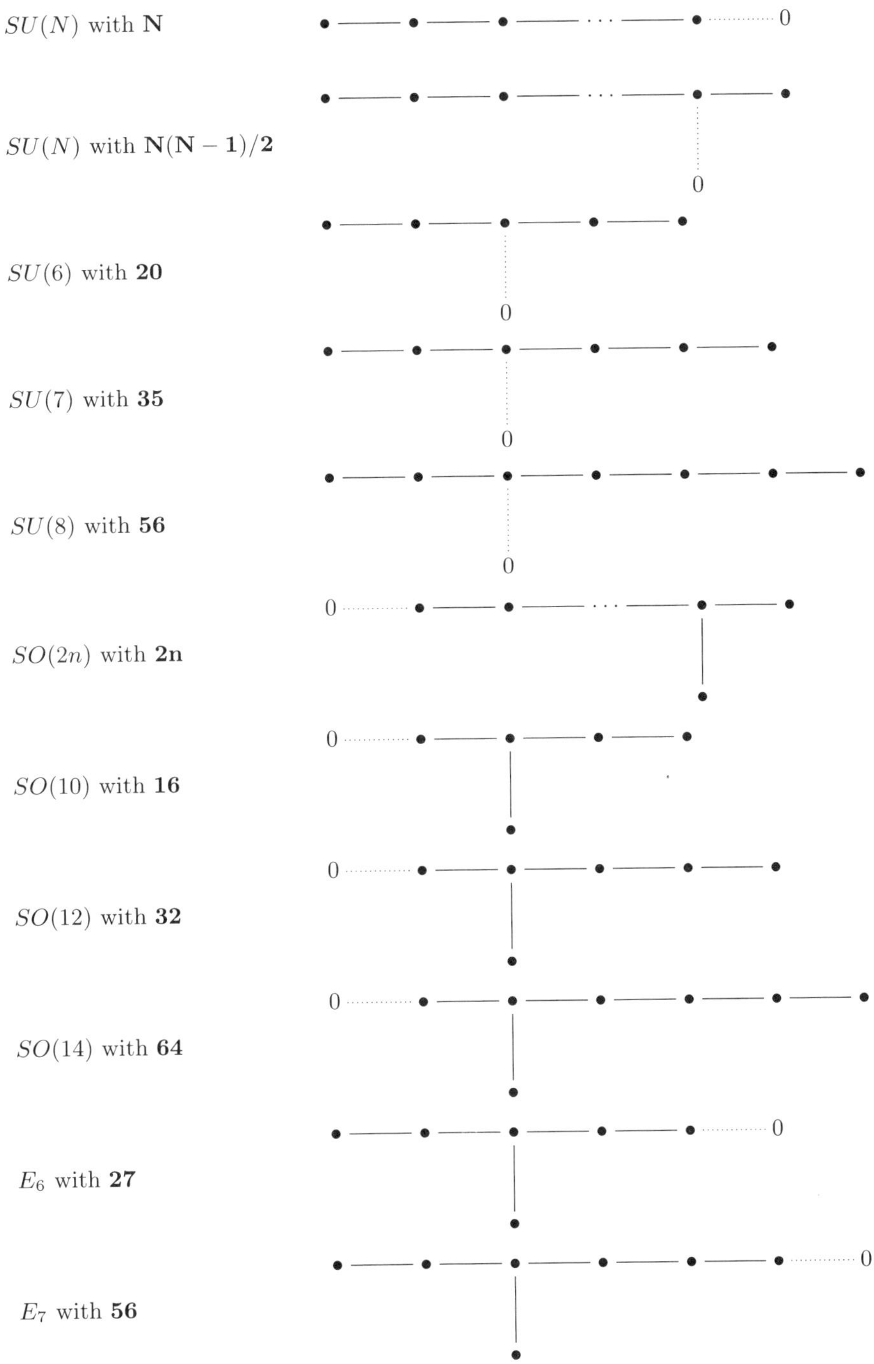

FIGURE 4. The augmented graphs $G[a, 1]$ corresponding to pairs of gauge group $G = ADE$ and fundamental representation which give an asymptotically free $\mathcal{N} = 2$ gauge theory.

Besides those in Figure 4 there is another asymptotically free pair (group, representation), namely $SU(N)$ with the two–index *symmetric* representation (which is not *fundamental*) whose augmented graph is identified with the non–simply–laced Dynkin graph of type B_N [**13**].

3.4. Example 4: G non–simply–laced. The Dynkin graph of a non–simply laced Lie group G arises by folding a parent simply–laced Dynkin graph G_{parent} along an automorphism group U. Specifically, the $G_{\text{parent}} \to G$ foldings are

$$\begin{aligned} &D_{n+1} \longrightarrow B_n \quad A_{2n-1} \longrightarrow C_n \\ &D_4 \longrightarrow G_2 \qquad E_6 \longrightarrow F_4. \end{aligned} \tag{25}$$

$U = \mathbb{Z}_2$ in all cases except for $D_4 \to G_2$ where it is $\mathbb{Z}_3$. To each node of the folded Dynkin diagrams there is attached an integer d_a, namely the number of nodes of the parent graph which were folded into it. This number corresponds to one–half the length–square of the corresponding simple co–root $\alpha_a^\vee$

$$d_a = \frac{1}{2}(\alpha_a^\vee, \alpha_a^\vee) \equiv \frac{2}{(\alpha_a, \alpha_a)} \qquad a = 1, 2, \dots, r. \tag{26}$$

In general, the light category of a (quiver) $\mathcal{N} = 2$ gauge theory with group G has the structure

$$\mathscr{L} = \bigvee_{\lambda \in \mathbb{P}^1/U} \mathscr{L}_\lambda \tag{27}$$

with U acting on the category $\mathscr{L}_\lambda$ through monodromy functors $\mathscr{M}_u$ [**13**]

$$\mathscr{L}_{u\cdot\lambda} = \mathscr{M}_u(\mathscr{L}_\lambda) \qquad u \in U. \tag{28}$$

Since the cylinder $\mathbb{C}^* \subset \mathbb{P}^1$ is identified with the Gaiotto plumbing cylinder associated to the gauge group G, this monodromical construction is equivalent to the geometric realization of the non–simply–laced gauge groups in the Gaiotto framework [**20**] or in F–theory [**21**]. In the simply–laced case the light category was described in terms of the preprojective algebra of G; likewise, to each gauge group $G = BCFG$ we may associate a generalized 'preprojective' algebra of the form $\mathscr{J}(Q', \mathcal{W}')$. Q' is the same reduced quiver as in the A_r case (see Figure 2 for the $r = 5$ example) while the reduced superpotential is

$$\mathcal{W} = \sum_{a \xrightarrow{\alpha} b} \Big(\alpha\, A_{s(\alpha)}^{n(\alpha)}\, \alpha^* - \alpha^*\, A_{t(\alpha)}^{m(\alpha)}\, \alpha\Big), \tag{29}$$

where the sum is over the edges $a \overset{\alpha}{\text{———}} b$ of A_r and

$$(n(\alpha), m(\alpha)) = \left(\frac{d_a}{(d_a, d_b)}, \frac{d_b}{(d_a, d_b)}\right). \tag{30}$$

One checks [**13**] that $\mathsf{mod}\,\mathscr{J}(Q', \mathcal{W}')$ has the monodromic property (28) and the dimension vectors of its bricks are the positive roots of G, so that the light category corresponds to vector multiplets forming a single copy of the adjoint of G, as required for pure SYM. From the light subcategory $\mathsf{mod}\,\mathscr{J}(Q', \mathcal{W}')$ one reconstructs the full *non–perturbative* Abelian category $\mathsf{mod}\,\mathscr{J}(Q, \mathcal{W})$, which describes the model in all physical regimes, by using the Dirac integrality conditions described in §. 2.1. See [**13**] for details.

4. Half–hypers

4.1. Coupling *full* hypermutliplets to SYM. The construction of the pairs $(Q_{N_f}, \mathcal{W}_{N_f})$ for $G = ADE$ SQCD coupled to N_f fundamental *full* hypermultiplets of refs.[**10, 11**] was relatively easy: each hypermultiplet has a gauge invariant mass m_i, and taking the *decoupling limit* $m_i \to \infty$ we make $N_f \to N_f - 1$. At the level of modules categories this decoupling processes insets

$$\mathsf{mod}\,\mathscr{J}(Q_{N_f-1}, \mathcal{W}_{N_f-1}) \xrightarrow{\subset} \mathsf{mod}\,\mathscr{J}(Q_{N_f}, \mathcal{W}_{N_f}) \tag{31}$$

as an extension–closed, exact, full, *controlled* Abelian subcategory [**11**]. In general, a *control function* is a linear map $\eta\colon \Gamma \to \mathbb{Z}$, and the controlled subcategory is the full subcategory over the objects X such that $\eta(X) = 0$ while for all their subobjects $\eta(Y) \leq 0$. The light subcategory is an example of controlled one with control function the magnetic charge. All decoupling limits of QFT correspond to controlled subcategories in the RT language.

For the decoupling limit $m_i \to \infty$ the control function $f_i\colon \Gamma \to \mathbb{Z}$ corresponds to the flavor charge dual to m_i. Choosing f_i so that $f_i(\Gamma_+) \geq 0$, we realize Q_{N_f-1} as a full subquiver of Q_{N_f} missing one node, the functor $\mathsf{mod}\,\mathscr{J}(Q_{N_f-1}, \mathcal{W}_{N_f-1}) \to \mathsf{mod}\,\mathscr{J}(Q_{N_f}, \mathcal{W}_{N_f})$ being the restriction. This gives a recursion relation in N_f of the form

(32)

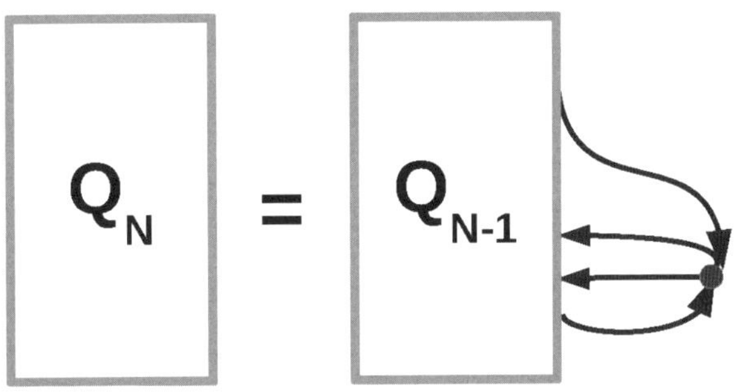

where the blue node in the right corresponds to the controlling flavor charge f_i. By repeated use of this relation, we eventually get to pure G SYM whose quiver is known, see §. 3.2. The decoupling process may be easily inverted to get a recursive map $Q_{N_f-1} \to Q_{N_f}$. Indeed, to define such a map we have only to determine the *red* arrows in eqn.(32) which connect Q_{N_f-1} to the extra (blue) node in the RHS of (32) which corresponds to an additional massive quark. Given the electric weight (*i.e.* the G–representation) of the added quark, ω, the red arrows are uniquely determined by the Dirac pairing of ω with the charges associated with the nodes of Q_{N_f-1}.

This strategy does not work for SYM coupled to *half*–hypermultiplets: they carry no flavor symmetry, have no mass parameter. They are tricky theories, always on the verge of inconsistency: most of them are indeed quantum inconsistent, but there are a few consistent models which owe their existence to peculiar *'miracles'*. The typical example being $G = E_7$ SYM coupled to half a **56**.

4.2. Coupling *half* hypermutliplets. We use yet another decoupling limit: *extreme Higgs*. Given a $\mathcal{N} = 2$ gauge theory with group G_r, of rank r, we take a v.e.v. of the adjoint field $\langle\Phi\rangle \in \mathfrak{h}$ such that

$$\alpha_b(\langle\Phi\rangle) = \begin{cases} t\, e^{i\phi}, & t \to +\infty, \quad b = a \\ O(1) & \text{otherwise} \end{cases} \tag{33}$$

States having electric weight ρ such that $\rho(\langle\Phi\rangle) = O(t)$ decouple, and we remain with a gauge theory with a gauge group G_{r-1} whose Dynkin diagram is obtained by deleting the a–th node from that of G_r (coupled to specific matter). *E.g.* starting from $G_7 = E_7$ coupled to $\frac{1}{2}\,\mathbf{56}$ and choosing $a = 1$ we get $G_6 = \mathrm{Spin}(12)$ coupled to $\frac{1}{2}\,\mathbf{32}$ corresponding to deleting the black node in the Dynkin graph

(34)

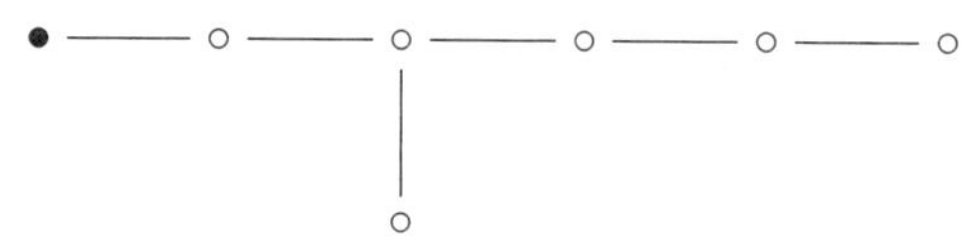

Again, the decoupling limit should correspond to a controlled Abelian *sub*category of the representations of $(Q_{G_r}, \mathcal{W}_{G_r})$. One can choose $(Q_{G_r}, \mathcal{W}_{G_r})$ in its mutation–class and the phase ϕ in (33) so that the control function $\lambda(\cdot)$ is *non–negative* on the positive–cone Γ_+. Then $Q_{G_{r-1}}$ is a full subquiver of Q_{G_r} and $\mathcal{W}_{G_{r-1}}$ is just the restriction of $\mathcal{W}_{G_r}$. It is easy to see that the complementary full subquiver is a two–nodes Kronecker one $\upuparrows$ [**12**]. Putting everything together, we get a recursion of the quiver with respect to the rank r of G_r of the form

(35)

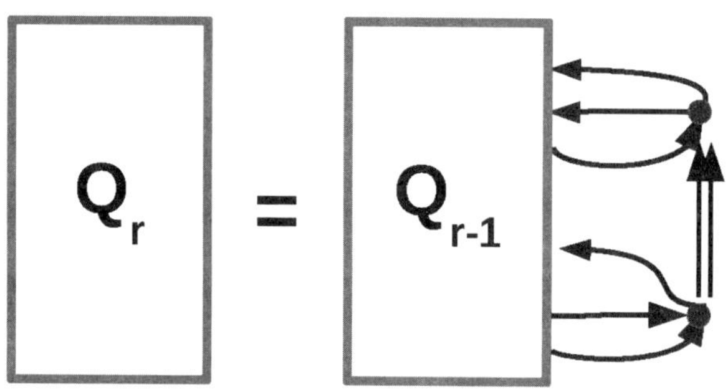

If we know the simpler quiver $Q_{G_{r-1}}$, to get Q_{G_r} we need just the fix the red arrows connecting the Kronecker to $Q_{G_{r-1}}$ in the above figure. Just as in §. 4.1, the red arrows are uniquely fixed by Dirac charge quantization. Indeed, by the recursion assumption, we know the representations X_{α_a} associated to all simple–root W–bosons of G_r; under the maximal torus $U(1)^r \subset G$ the simple–root W–bosons have charges $q_a(X_{\alpha_b}) = C_{ab}$ (Cartan matrix), while the dual magnetic charges are given by eqn.(6) which explicitly depends on the red arrows. It turns out [**12**] that $m_a(X) \in {}^L\Gamma_{\mathrm{root}}$ for all X for a *unique* choice of the arrows which are then fixed. Then Q_{G_r} is uniquely determined if we know $Q_{G_{r-1}}$. $\mathcal{W}_{G_r}$ is also essentially determined, up to some higher–order ambiguity [**12**].

Taking a suitable chain of such Higgs decouplings/symmetry breakings

$$G_r \to G_{r-1} \to G_{r-2} \to \cdots\cdots \to G_k, \tag{36}$$

we eventually end up with a *complete* $\mathcal{N} = 2$ with gauge group $G_k = SU(2)^k$. The complete $\mathcal{N} = 2$ quivers are known by classification [**7**]. Inverting the Higgs procedure, we may construct the pair $(Q_{G_r}, \mathcal{W}_{G_r})$ for the theory of interest by 'pulling back' through the chain (36) the pair $(Q_{\text{max comp}}, \mathcal{W}_{\text{max comp}})$ of their maximal

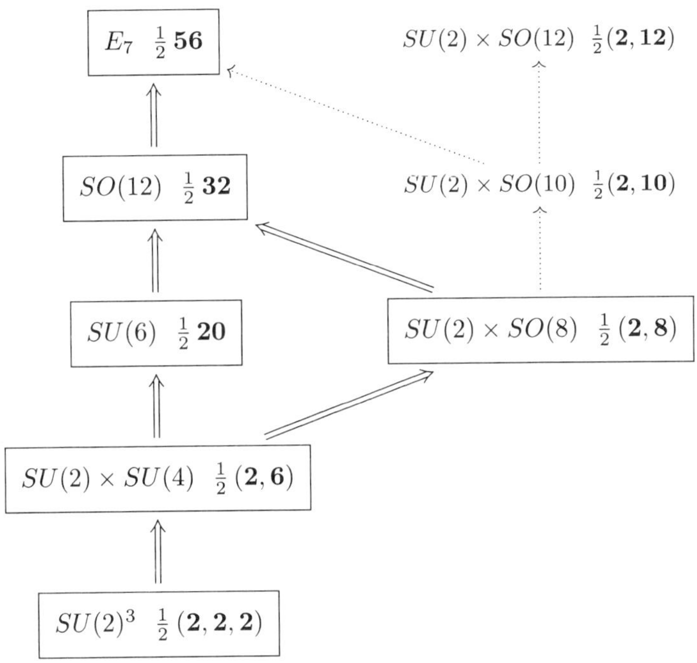

FIGURE 5. The Higgs breaking chain for various SYM models coupled to half hypers.

complete subsector. For the models of interest the *'pull back'* chain is presented in figure 5. The bottom model $SU(2)^3$ with $\frac{1}{2}(\mathbf{2},\mathbf{2},\mathbf{2})$ is complete [**7**, **11**].

The pair $(Q_{E_7}, \mathcal{W}_{E_7})$ for the model $G = E_7$ coupled to $\frac{1}{2}$ **56** is given in Figure 6; the other models in Figure 5 correspond to the restriction to suitable subquivers of $(Q_{E_7}, \mathcal{W}_{E_7})$ [**12**]. The light category deduced from these pairs contains light vectors forming one copy of the adjoint of G plus light hypermultiplets in the G–representation $\frac{1}{2}\,\mathbf{R}$, with $\mathbf{R}$ irreducible *quaternionic* [**12**]. Indeed, the light category has again the form $\mathsf{mod}\,\mathscr{J}(Q', \mathcal{W}')$ for a reduced pair $(Q', \mathcal{W}')$. See Figure 7 for the the reduced pair for $G = E_7$ coupled to $\frac{1}{2}$**56**; the other models are obtained by restriction of this one. Note that Q'_{E_7} (and hence all reduced quivers Q'_{G_r} in the Higgs chain) contains as a full subquiver the quiver of the Gaiotto A_1 theory on S^2 with 3 punctures (the T_2 theory) described in [**10**]. Hence for all these models the 'T_2–duality' of §. 1.1 is operative; this duality is crucial — together with special properties of the relevant Dynkin graphs — to check the above claims on the BPS spectrum at weak coupling. Details may be found in [**12**].

E_7 coupled to $\frac{1}{2}\,\mathbf{56}$

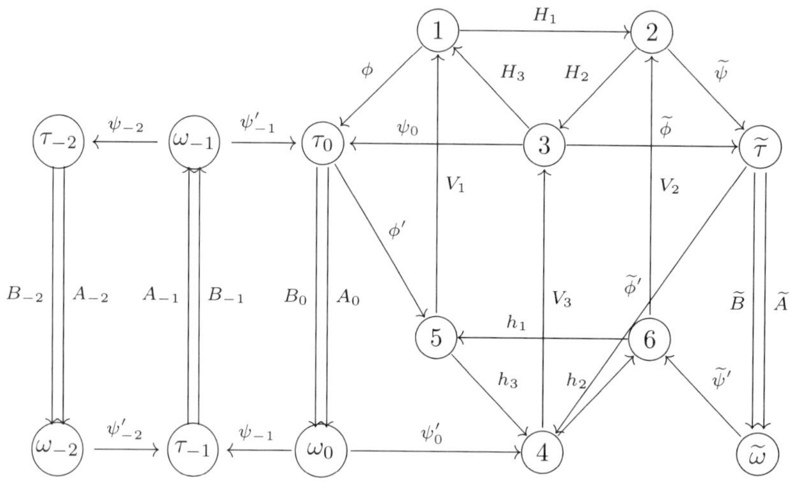

$$
\begin{aligned}
\mathcal{W}_{E_7} = {} & H_1H_3H_2 + h_3h_1h_2 + A\psi V_3\psi' + B\psi H_2V_2h_2\psi' + \phi V_1\phi' + \psi V_3h_3\phi' + \\
& \phi H_3V_3\psi' B + \widetilde{A}\widetilde{\psi}V_2\widetilde{\psi}' + \widetilde{B}\widetilde{\psi}H_1V_1h_1\widetilde{\psi}' + \widetilde{\phi}V_3\widetilde{\phi}' + \widetilde{\psi}V_2h_2\widetilde{\phi}' + \widetilde{\phi}H_2V_2\widetilde{\psi}'\widetilde{B} + \\
& + A_0\psi'_{-1}B_{-1}\psi_{-1} - B_0\psi'_{-1}A_{-1}\psi_{-1} + A_{-1}\psi'_{-2}B_{-2}\psi_{-2} - B_{-1}\psi'_{-2}A_{-2}\psi_{-2}
\end{aligned}
$$

FIGURE 6. Quiver and superpotential for the $\mathcal{N} = 2$ E_7 SYM coupled to $\frac{1}{2}\mathbf{56}$ quark.

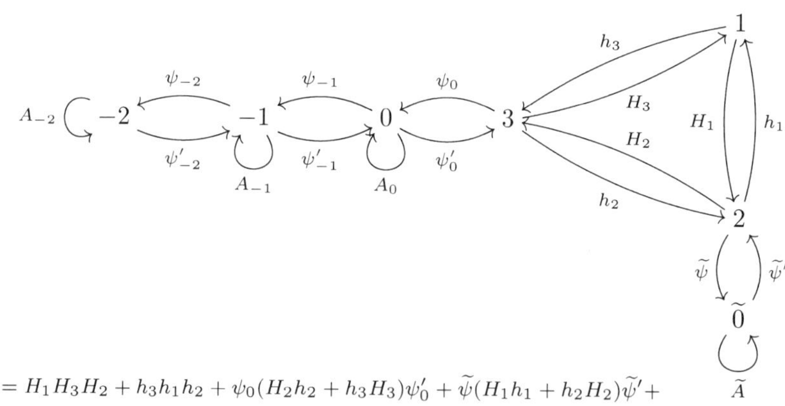

$$
\begin{aligned}
\mathcal{W}'_{E_7} = {} & H_1H_3H_2 + h_3h_1h_2 + \psi_0(H_2h_2 + h_3H_3)\psi'_0 + \widetilde{\psi}(H_1h_1 + h_2H_2)\widetilde{\psi}' + \\
& + A_0\psi_0\psi'_0 + A_0\psi'_{-1}\psi_{-1} - A_{-1}\psi_{-1}\psi'_{-1} + A_{-1}\psi'_{-2}\psi_{-2} - A_{-2}\psi_{-2}\psi'_{-2} + \widetilde{A}\widetilde{\psi}\widetilde{\psi}'.
\end{aligned}
$$

FIGURE 7. Reduced pair of the light category $\mathscr{L}_{E_7} \equiv \mathsf{mod}\,\mathscr{J}(Q', \mathcal{W}')$ for E_7 SYM with $\frac{1}{2}\mathbf{56}$ quark.

References

[1] D. Gaiotto, G. W. Moore, and A. Neitzke, *Wall-crossing, Hitchin systems, and the WKB approximation*, Adv. Math. **234** (2013), 239–403, DOI 10.1016/j.aim.2012.09.027. MR3003931

[2] D. Gaiotto, G. W. Moore, and A. Neitzke, *Framed BPS states*, Adv. Theor. Math. Phys. **17** (2013), no. 2, 241–397. MR3250763

[3] D. Gaiotto, G. W. Moore, and A. Neitzke, *Wall-crossing in coupled 2d-4d systems*, J. High Energy Phys. **12** (2012), 082, front matter + 166. MR3045271

[4] D. Gaiotto, G. W. Moore, and A. Neitzke, *Spectral networks*, Ann. Henri Poincaré **14** (2013), no. 7, 1643–1731, DOI 10.1007/s00023-013-0239-7. MR3115984

[5] D. Gaiotto, G. W. Moore, and A. Neitzke, *Spectral networks and snakes*, Ann. Henri Poincaré **15** (2014), no. 1, 61–141, DOI 10.1007/s00023-013-0238-8. MR3147409

[6] S. Cecotti, A. Neitzke, and C. Vafa, "R-Twisting and 4d/2d Correspondences," `arXiv:1006.3435 [hep-th]`.

[7] S. Cecotti and C. Vafa, *Classification of complete $\mathcal{N} = 2$ supersymmetric theories in 4 dimensions*, Surveys in differential geometry. Geometry and topology, Surv. Differ. Geom., vol. 18, Int. Press, Somerville, MA, 2013, pp. 19–101, DOI 10.4310/SDG.2013.v18.n1.a2. MR3087917

[8] S. Cecotti and M. Del Zotto, *On Arnold's 14 'exceptional' $\mathcal{N} = 2$ superconformal gauge theories*, J. High Energy Phys. **10** (2011), 099, 53, DOI 10.1007/JHEP10(2011)099. MR2909694

[9] M. Alim, S. Cecotti, C. Córdova, S. Espahbodi, A. Rastogi, and C. Vafa, *BPS quivers and spectra of complete $\mathcal{N} = 2$ quantum field theories*, Comm. Math. Phys. **323** (2013), no. 3, 1185–1227, DOI 10.1007/s00220-013-1789-8. MR3106506

[10] M. Alim, S. Cecotti, C. Córdova, S. Espahbodi, A. Rastogi, and C. Vafa, *$\mathcal{N} = 2$ quantum field theories and their BPS quivers*, Adv. Theor. Math. Phys. **18** (2014), no. 1, 27–127. MR3268234

[11] S. Cecotti, *Categorical tinkertoys for $\mathcal{N} = 2$ gauge theories*, Internat. J. Modern Phys. A **28** (2013), no. 5-6, 1330006, 124, DOI 10.1142/S0217751X13300068. MR3036676

[12] S. Cecotti and M. Del Zotto, *Half-hypers and quivers*, J. High Energy Phys. **9** (2012), 135, front matter+49. MR3044914

[13] S. Cecotti and M. Del Zotto, *4d $\mathcal{N} = 2$ gauge theories and quivers: the non-simply laced case*, J. High Energy Phys. **10** (2012), 190, front matter + 34. MR3033809

[14] S. Cecotti and M. Del Zotto, *Infinitely many $\mathcal{N} = 2$ SCFT with ADE flavor symmetry*, J. High Energy Phys. **1** (2013), 191, front matter + 21. MR3045469

[15] H. Derksen, J. Weyman, and A. Zelevinsky, *Quivers with potentials and their representations. I. Mutations*, Selecta Math. (N.S.) **14** (2008), no. 1, 59–119, DOI 10.1007/s00029-008-0057-9. MR2480710 (2010b:16021)

[16] D. Gaiotto, *$N = 2$ dualities*, J. High Energy Phys. **8** (2012), 034, front matter + 57. MR3006961

[17] I. M. Gel′fand and V. A. Ponomarev, *Model algebras and representations of graphs* (Russian), Funktsional. Anal. i Prilozhen. **13** (1979), no. 3, 1–12. MR545362 (82a:16030)

[18] W. Crawley-Boevey, *On the exceptional fibres of Kleinian singularities*, Amer. J. Math. **122** (2000), no. 5, 1027–1037. MR1781930 (2001f:14009)

[19] G. Lusztig, *Quivers, perverse sheaves, and quantized enveloping algebras*, J. Amer. Math. Soc. **4** (1991), no. 2, 365–421, DOI 10.2307/2939279. MR1088333 (91m:17018)

[20] Y. Tachikawa, *$\mathcal{N} = 2$ S-duality via outer-automorphism twists*, J. Phys. A **44** (2011), no. 18, 182001, 7, DOI 10.1088/1751-8113/44/18/182001. MR2788718

[21] M. Bershadsky, K. Intriligator, S. Kachru, D. R. Morrison, V. Sadov, and C. Vafa, *Geometric singularities and enhanced gauge symmetries*, Nuclear Phys. B **481** (1996), no. 1-2, 215–252, DOI 10.1016/S0550-3213(96)90131-5. MR1423982 (97j:81240)

Scuola Internazionale di Studi Avanzati, via Bonomea 265, I-34100 Trieste, Italy
E-mail address: `cecotti@sissa.it`

Proceedings of Symposia in Pure Mathematics
Volume **90**, 2015
http://dx.doi.org/10.1090/pspum/090/01525

Supermoduli Space is Not Projected

Ron Donagi and Edward Witten

ABSTRACT. We prove that for genus $g \geq 5$, the moduli space of super Riemann surfaces is not projected (and in particular is not split): it cannot be holomorphically projected to its underlying reduced manifold. Physically, this means that certain approaches to superstring perturbation theory that are very powerful in low orders have no close analog in higher orders. Mathematically, it means that the moduli space of super Riemann surfaces cannot be constructed in an elementary way starting with the moduli space of ordinary Riemann surfaces. It has a life of its own.

CONTENTS

2010 *Mathematics Subject Classification.* Primary 14H10, 14H81, 81T30, 83E30.

1. Introduction

Ordinary geometry has a generalization in $\mathbb{Z}_2$-graded supergeometry, which is the arena for supersymmetric theories of physics. In this generalization, ordinary manifolds are replaced by supermanifolds, which are endowed with $\mathbb{Z}_2$-graded rings of functions. In addition to a vast physics literature, supermanifolds have also been much studied mathematically; for example, see [**1**–**6**]. A basic example of a supermanifold is a super Riemann surface, which for our purposes is a complex supermanifold of dimension $(1|1)$ with a superconformal structure, a notion that we explain in section 3. See for example [**7**–**12**] (and see [**13**] for the generalization to super Riemann surfaces of dimension $(1|n)$, $n > 1$).

Mathematically, the theory of super Riemann surfaces and their moduli spaces generalizes the theory of ordinary Riemann surfaces and their moduli spaces in a strikingly rich way. Physically, the main importance of super Riemann surfaces is their role in superstring perturbation theory. Perturbative calculations in superstring theory are carried out by integration over the moduli space $\mathfrak{M}_g$ of super Riemann surfaces, and its analogs for super Riemann surfaces with punctures. See for example [**14**, **15**]. We will write $\mathfrak{M}_{g,n}$ for the moduli space of genus g super Riemann surfaces with n marked points or Neveu-Schwarz punctures (we do not consider in this paper the more general moduli spaces of super Riemann surfaces with Ramond punctures).

Both physically and mathematically, one of the most basic questions about $\mathfrak{M}_g$ is whether it can be projected holomorphically to its reduced space, which is the moduli space $\mathcal{SM}_g$ that parametrizes ordinary Riemann surfaces with a spin structure. (By such a projection, one means a holomorphic map that is left inverse to the natural inclusion of $\mathcal{SM}_g$ in $\mathfrak{M}_g$.) A complex supermanifold that can be projected holomorphically to its reduced space is said to be projected; if the fibers of the projection are linear in a certain sense that will be described later, the supermanifold is said to be split.

Mathematically, if $\mathfrak{M}_g$ is split, this means that it can be reconstructed from the purely bosonic moduli space $\mathcal{SM}_g$ in an elementary fashion and, in a sense, need not be studied independently. Physically, if $\mathfrak{M}_g$ is split – or at least projected – then a possible strategy in superstring perturbation theory is to integrate over $\mathfrak{M}_g$ by first integrating over the fibers of its projection to $\mathcal{SM}_g$. Indeed, practical calculations in superstring perturbation theory – such as the $g = 2$ calculations that are surveyed in [**16**] – are usually done in this way.

However, there has been no evidence that $\mathfrak{M}_g$ is split, or even projected, in general. The validity of superstring perturbation theory does not depend on a projection, so the existence of string theory gives no hint that $\mathfrak{M}_g$ is projected. The existence of holomorphic projections for small g follows from the cohomological nature of the obstructions to splitting and the nature of the reduced spaces $\mathcal{SM}_g$ for small g, and gives little indication of what happens for larger g. The goal of the present paper is to show that actually $\mathfrak{M}_g$ is not projected or split in general.

In fact, we show the following:

THEOREM 1.1. *The supermanifold $\mathfrak{M}_g$ is non-projected, and in particular non-split, for $g \geq 5$.*

(We suspect that this result may hold for $g \geq 3$.)

Our second main result is:

THEOREM 1.2. *The supermanifold $\mathfrak{M}_{g,1}$ is non-projected for $g \geq 2$, for the case of an even spin structure.*

Once this is established, a simple argument gives

THEOREM 1.3. *The supermanifold $\mathfrak{M}_{g,n}$ is non-projected for $g \geq 2$ and $g-1 \geq n \geq 1$. This holds for both even and odd spin structures if g is odd and for even spin structure if g is even.*

We do not resolve the question of projectedness of $\mathfrak{M}_{g,n}$ for the case of even g, odd spin structure, and $n \geq 1$.

REMARK 1.4. Though $\mathfrak{M}_g$ and $\mathfrak{M}_{g,n}$, like their bosonic counterparts, have natural Deligne-Mumford compactifications, we do not consider the compactifications in the present paper. Our assertion is that $\mathfrak{M}_g$ and $\mathfrak{M}_{g,n}$ are non-projected without any consideration of the compactification. Asking for the compactification to be projected would only be a stronger condition.

REMARK 1.5. Although the upper bound on n in Theorem 1.3 is probably not optimal, one should expect to require some sort of upper bound precisely because we do not consider the compactification of $\mathfrak{M}_{g,n}$ and thus we require the n marked points to be distinct. For $n >> g$, requiring the n points to be distinct very likely kills the obstructions to splitness, though we would expect the compactification of $\mathfrak{M}_{g,n}$ to be non-split.

This paper is organized as follows. In section 2, we review the basic notions of supermanifold theory. We start with several constructions of new supermanifolds from old: as submanifolds, coverings, branched coverings, blowups and blowdowns. We explain the concepts of projection and splitting and the cohomology classes ω_i that obstruct a splitting. A useful result is the Compatibility Lemma 2.11, comparing the leading obstruction ω_2 for a supermanifold S and a submanifold S'. In fact, the part of this discussion that involves ω_2 is considerably simpler than the full story, and is the only part that will be used in the rest of the paper. The reader interested only in the main results of this work may safely skip our analysis of the higher obstructions.

In section 3, we introduce our main objects of study, namely super Riemann surfaces and their moduli spaces. In sections 3.1 and 3.2 we review some of the basics of super Riemann surfaces and examine in particular the effects on them of the branched covering, blowup and blowdown constructions considered previously for supermanifolds. In section 3.3, we exhibit an explicit and basic example of a non-split supermanifold X_η. This has dimension $(1|2)$, and can be thought of as a family of super Riemann surfaces parametrized by a single odd parameter.

The main theorems are proved in sections 4 and 5. Let us outline some of the ideas of the proofs.

The above-mentioned non-split supermanifold X_η embeds naturally into $\mathfrak{M}_{g,1}$. Using the Compatibility Lemma and some standard algebraic geometry, we use this embedding to prove that the leading obstruction to a projection of $\mathfrak{M}_{g,1,\mathrm{even}}$ is non-vanishing, proving Theorem 1.2. The argument fails for odd spin structure. However, by considering unramified covers of X_η, it is then possible to deduce Theorem 1.3.

To prove Theorem 1.1, we follow a similar path. We describe a covering space $\widetilde{\mathfrak{M}}_{g_0,1} \to \mathfrak{M}_{g_0,1}$ parametrizing super Riemann sufaces of genus g_0 together with a particular type of branched covering, and we find an explicit embedding of super-moduli spaces, mapping $\widetilde{\mathfrak{M}}_{g_0,1}$ into $\mathfrak{M}_g$. In fact, we can find such an embedding for every $g \geq 5$. The normal bundle sequence for this embedding splits, so non-splitness of $\mathfrak{M}_g$ follows from that of $\widetilde{\mathfrak{M}}_{g_0,1,\mathrm{even}}$. In fact, we see that $\widetilde{\mathfrak{M}}_{g_0,1,\mathrm{even}}$ is itself reducible. Conveniently, its two components go to the two components of $\mathfrak{M}_g$, so we are able to deduce non-splitness for both of the latter.

In the appendix we discuss in more detail the simplest instance of our family of branched coverings, namely the case $g_0 = 2$, $g = 5$. We give an elementary construction of the families involved, check that both parities on $\mathcal{SM}_5$ arise from even parity on $\mathcal{SM}_{2,1}$, and analyze the parameter spaces in $\mathcal{SM}_5$ and in the ordinary (non-spin) moduli space $\mathcal{M}_5$. Somewhat surprisingly, they both turn out to be curves of genus 0.

In a sequel to this work [**23**], we will provide some further interpretations of the first non-trivial class $\omega := \omega_2$. We show that this class is, in a certain sense, a superanalog of what in ordinary algebraic geometry is the Atiyah class of the tangent bundle. In the case of the moduli space of super Riemann surfaces, we give a concrete description of ω in terms of sheaves on $C \times C$, where C is an ordinary Riemann surface, and use this to give an alternative proof of Theorem 1.2.

2. Supermanifolds

A *supermanifold*, like an ordinary manifold, is a locally ringed space which is locally isomorphic to a certain local model. A locally ringed space is a pair $(M, \mathcal{O})$ consisting of a topological space M and a sheaf of algebras $\mathcal{O} = \mathcal{O}_M$ on it, whose stalk $\mathcal{O}_x$ at each point $x \in M$ is a *local ring*. One type of example is affine space $\mathbb{A}^m$, for which M is an m-dimensional vector space, while $\mathcal{O}$ can be the sheaf of functions on M that are continuous, differentiable, or analytic. (Here and elsewhere, we will always work in characteristic 0, in fact over the real or complex numbers.) A manifold is then a commutative locally ringed space which is locally isomorphic to one of these local models. In the complex case, we also have the option of taking the functions in $\mathcal{O}$ to be algebraic, but we must then allow a larger collection of local models, namely all non-singular affine varieties, i.e. closed non-singular algebraic subvarieties of affine space.[1]

A $\mathbb{Z}/2$-graded sheaf of algebras $A = A_0 \oplus A_1$ is *supercommutative* if it is commutative up to the usual sign rule: for $f \in A_i, g \in A_j$, the rule is $fg = (-1)^{ij} gf$. Given a manifold M and a vector bundle V over it, we define the supercommutative locally ringed space $S := S(M, V)$ to be the pair $(M, \mathcal{O}_S)$, where $\mathcal{O}_S$ is the sheaf of $\mathcal{O}_M$-valued sections of the exterior algebra $\wedge^\bullet V^\vee$ on the dual bundle $V^\vee$. (If we interpret V as a locally free sheaf of $\mathcal{O}_M$-modules, then $\mathcal{O}_S$ is simply $\wedge^\bullet V^\vee$.) This sheaf $\mathcal{O}_S$ is $\mathbb{Z}/2$-graded and supercommutative, and its stalks are local rings. The simplest example is *affine superspace* $\mathbb{A}^{m|n} = (\mathbb{A}^m, \mathcal{O}_{\mathbb{A}^{m|n}}) = S(\mathbb{A}^m, {\mathcal{O}_{\mathbb{A}^m}}^{\oplus n})$, where M is ordinary affine m-space and $V = \mathcal{O}_M^{\oplus n}$ is the trivial rank n bundle on it. A *supermanifold* is then a supercommutative locally ringed space which is locally isomorphic to some local model $S(M, V)$. [2] It is *split* if it is globally isomorphic to some $S(M, V)$.

It is important to note that the isomorphisms above are isomorphisms of $\mathbb{Z}/2$-graded algebras (over the real or complex numbers). They need not preserve the $\mathbb{Z}$-grading of $\wedge^\bullet V^\vee$. So if z is a function on M and θ_i are fiber coordinates on V, then z cannot go to $z + \theta_1$, but it can go to $z + \theta_1\theta_2$.

The definition of a supermanifold endows the sheaf $\mathcal{O}_S$ with the structure of a sheaf of (real or complex) algebras, and also with a surjective homomorphism to $\mathcal{O}_M$ (more on this below). It does not endow the sheaf $\mathcal{O}_S$ with the structure of a sheaf of $\mathcal{O}_M$-modules: multiplication by $\mathcal{O}_M$ need not commute with the gluing isomorphisms. We say that the supermanifold S is *projected* when the sheaf $\mathcal{O}_S$ can be given the structure of a sheaf of $\mathcal{O}_M$-algebras commuting with the given projection to $\mathcal{O}_M$. (It may be possible to do this in more than one way.) Clearly, every split supermanifold is projected, but we will see that not every supermanifold is projected, and likewise there are obstructions for a projected supermanifold to be split.

The structure sheaf $\mathcal{O}_S$ of a supermanifold S contains the ideal J consisting of all nilpotents. It is the ideal generated by all odd functions. Given a supermanifold S, we can recover its underlying manifold $(M, \mathcal{O}_M)$, which we call the reduced space S_{red}, as well as a bundle V on it: $\mathcal{O}_M$ is recovered as $\mathcal{O}_S/J$, while the dual $V^\vee$ is recovered as J/J^2. In fact, the supermanifold S determines the split supermanifold $\text{Gr}(S)$ whose reduced space is M and whose structure sheaf is $\text{Gr}_J(\mathcal{O}_S) := \bigoplus_{i=0}^\infty (J^i/J^{i+1})$, and the latter determines and is uniquely determined by the pair (M, V):

$$S \Longrightarrow \text{Gr}(S) \Longleftrightarrow (M, V).$$

We say that a supermanifold S with $\text{Gr}(S) \cong S(M, V)$ is modeled on M, V, where $S(M, V)$ is the split supermanifold determined by the pair M, V as above. We will

---

[1] In each of these cases, the stalk $\mathcal{O}_x$ is the ring of germs of functions, i.e. functions defined on any neighborhood of x, where two functions are identified if their restrictions to some open subset coincide. This is indeed a local ring: the germs of functions that vanish at x form the unique maximal ideal.

[2] In the continuous, differentiable or analytic settings, we may as well restrict our local models to the affine superspaces $\mathbb{A}^{m|n}$. In the algebraic setting we must allow all $S(M, V)$ with affine M, as in the bosonic algebraic case.

discuss in section 2.2 how to characterize all supermanifolds S that are modeled on M, V.

The definition of a supermanifold leads more or less immediately to various standard notions: a morphism of supermanifolds, a submanifold (a term we use instead of the clunky 'sub-supermanifold'), immersion and submersion, a product, a fiber product (or pullback), a fiber. (The fiber of a submersion of supermanifolds is a supermanifold. The fiber of a general morphism of supermanifolds is a possibly singular locally ringed superspace. In the algebraic world, it is a superscheme.) By a family of supermanifolds parametrized by a supermanifold P, we mean a submersion $\pi : S \to P$. If $0 \in P_{\mathrm{red}}$ is a point, the family is interpreted as a deformation of the fiber $S_0 := \pi^{-1}(0)$. It is also straightforward to define vector bundles, sheaves of $\mathcal{O}_S$-modules, and so on.

We have noted that a supermanifold S determines its reduced space M. Moreover, M has a natural embedding in S; this corresponds to the existence of a projection from $\mathcal{O}_S$ to $\mathcal{O}_M = \mathcal{O}_S/J$. In the other direction, a projection $S \to M$ would be equivalent to an embedding of $\mathcal{O}_M$ in $\mathcal{O}_S$, and also to endowing $\mathcal{O}_S$ with an $\mathcal{O}_M$-algebra structure. We emphasize that in general these structures do not exist: a general supermanifold may not be projected.

We note that the product of two split supermanifolds is split, and for any morphism $f : M' \to M$, if S is a split or projected supermanifold with reduced space M, then $S' = f^*S$ is well-defined and it too is split or projected. Some of the simplest examples of supermanifolds are coverings and submanifolds, discussed below in section 2.1. If $\widetilde{S} \to S$ is a finite covering map, it is not too hard to see (Corollary 2.8) that $\widetilde{S}$ is split if and only if so is S (and projected if S is). However, a submanifold S' of a split supermanifold S need not in general be split as a supermanifold.

A sheaf $\mathcal{F}$ on a supermanifold S can be viewed simply as a sheaf on the reduced space M, and sheaf cohomology on S is defined to be the cohomology of the corresponding sheaf on the reduced space: $H^*(S, \mathcal{F}) := H^*(M, \mathcal{F})$. This is the same definition that one uses for sheaf cohomology on a non-reduced scheme in ordinary algebraic geometry, and as in that case, it is usually more illuminating to denote this cohomology as $H^*(S, \mathcal{F})$ rather than $H^*(M, \mathcal{F})$. For example, it is much more natural to think of a sheaf of $\mathcal{O}_S$-modules as a sheaf on S rather than as a sheaf on M.

The tangent bundle TS of the supermanifold S is an example of such a sheaf of $\mathcal{O}_S$-modules. It can be defined in terms of derivations: $TS := \mathrm{Der}(\mathcal{O}_S)$. It is a $\mathbb{Z}_2$-graded vector bundle, or locally free sheaf of $\mathcal{O}_S$-modules. When $S = \mathbb{A}^{m|n}$, TS is the free $\mathcal{O}_S$-module generated by even tangent vectors $\partial/\partial x_i, i = 1, \ldots, m$ and odd tangent vectors $\partial/\partial\theta_j, j = 1, \ldots, n$. In general, TS need not have such distinguished complementary even and odd subbundles: the even and odd parts are not sheaves of $\mathcal{O}_S$-modules. But the restriction $TS_{|M}$ of TS to the reduced space $M = S_{\mathrm{red}}$ does split. (By restriction to $M = S_{\mathrm{red}}$ we mean pullback, under the natural inclusion $M \to S$, of a sheaf of $\mathcal{O}_S$ modules to a sheaf of $\mathcal{O}_M$ modules. In this case, this is accomplished by setting all odd functions to 0.) We refer to the graded pieces of $TS_{|M}$ as the even and odd tangent bundles of S. Explicitly, these are given by:

$$T_+S := TM, \qquad T_-S := V.$$

S is then modeled on the pair M, V, though of course it may not be isomorphic to the split model $S(M, V)$. By definition, the *dimension* of S is the pair $(m|n)$, where m, n are the ranks of $T_{\pm}S$: $m := \dim(M)$ and $n := \text{rank}(V)$. We note that specifying a map $f_v : \mathbb{C}^{0|1} \to S$, where $\mathbb{C}^{0|1}$ is the $(0|1)$ dimensional affine superspace, is equivalent to specifying an odd tangent vector $v \in T_-S$, i.e. a point $p \in M$ and an odd tangent vector $v \in T_{-,p}S$ at that point.

2.1. Examples of supermanifolds. The simplest supermanifolds are the affine superspaces $\mathbb{A}^{m|n}$, defined above. In the algebraic case, their global function ring is the polynomial ring in m commuting (even) variables $x_i, i = 1, \ldots, m$, and n anticommuting (odd) variables $\theta_j, j = 1, \ldots, n$. In the other cases, the even part is extended to allow continuous, differentiable or analytic functions of the x_i, but the odd part is unchanged. In these cases, it is often convenient to illustrate arguments about supermanifolds by referring to these local coordinates x_i, θ_j. As noted above, in the algebraic case a supermanifold is usually not locally isomorphic to affine superspace, but we can still use analytic local coordinates.

A less trivial example of a complex supermanifold is complex projective superspace $\mathbb{P}^{m|n}$, for $m, n \geq 0$. We can think of it globally, as a quotient, or locally, as pieced together from affine charts. The global description involves a quotient by the purely even group $\mathbb{C}^{\times}$, so we use homogeneous coordinates $x_0 \ldots x_m|\theta_1 \ldots \theta_n$, subject to an overall scaling of all x's and θ's by the same nonzero even complex parameter $\lambda \in \mathbb{C}^{\times}$, and with a requirement that not all the bosonic coordinates x_α are allowed to vanish simultaneously. The local description specifies $\mathbb{P}^{m|n}$ as the union of its open subsets U_α, for $\alpha = 0, \ldots, m$, defined by the condition $x_\alpha \neq 0$. Each U_α can be identified with affine superspace $\mathbb{A}^{m|n}$ by the ratios x_β/x_α, $\beta \neq \alpha$, and θ_j/x_α, for $j = 1, \ldots, n$. The gluing relations are the obvious ones. Note that for $m = 0$ there is a unique open set U_α, so $\mathbb{P}^{0|n}$ is the same thing as $\mathbb{A}^{0|n}$.

Next we describe three ways of constructing supermanifolds: as submanifolds, blowups, and branched covers.

2.1.1. *Submanifolds of supermanifolds.* One way to construct new supermanifolds from a given supermanifold is by imposing one (or more) equations. E.g. in $\mathbb{P}^{m|n}$ impose: $P(z^0 \ldots z^n|\theta^1 \ldots \theta^n) = 0$, where P is a homogeneous polynomial in the homogeneous coordinates of $\mathbb{CP}^{m|n}$ that is either even or odd. If P is even and sufficiently generic, this will give a complex supermanifold of dimension $m - 1|n$. For suitable odd P, it gives a complex supermanifold of dimension $m|n - 1$. We reserve the name *divisor* to the case of codimension $(1|0)$, i.e. when the defining polynomial P is even.

2.1.2. *Coverings of supermanifolds.* Given a supermanifold $S = (M, \mathcal{O}_S)$ it is straightforward to lift a finite unramified covering $f : \widetilde{M} \to M$ of the underlying reduced space M to a finite unramified covering $F : \widetilde{S} \to S$. The reduced space is of course $\widetilde{M}$, and the sheaf $\mathcal{O}_{\widetilde{S}} \to \widetilde{M}$ is the pullback $f^{-1}(\mathcal{O}_S)$ of the sheaf $\mathcal{O}_S$ over M. This pullback has the local structure of the sheaf of functions on a supermanifold since the covering map f is a local isomorphism.

2.1.3. *Branched coverings of supermanifolds.* A variant of the above allows us to construct branched coverings as well. Start with:

- a supermanifold $S = (M, \mathcal{O}_S)$,
- a branched covering $f : \widetilde{M} \to M$ of the underlying reduced space M, with smooth branch divisor $B \subset M$, and

- a divisor $D \subset S$ whose intersection with M is B.

We construct a supermanifold $\widetilde{S}$ and a morphism $F : \widetilde{S} \to S$ whose branch divisor is D and whose reduced version is $F_{\rm red} = f$. For the moment, assume that S has global coordinates $(z_1, \ldots, z_m|\theta_1, \ldots, \theta_n)$, where $(z_1, \ldots, z_m)$ are coordinates on M, the branch divisor B is given by $z_1 = 0$, and the corresponding coordinates on $\widetilde{M}$ are $(w_1, z_2, \ldots, z_m)$, so that pulling back by the branched covering $f : \widetilde{M} \to M$ sends z_1 to $(w_1)^k$ for some k. (We can always achieve this after restricting to sufficiently small open subsets.) Now the divisor D is given by the vanishing of some even function $z' = z'(z, \theta)$ whose image modulo the θ's is z_1. This implies that the coordinate ring $(z_1, \ldots, z_m|\theta_1, \ldots, \theta_n)$ is also generated by $(z', z_2, \ldots, z_m|\theta_1, \ldots, \theta_n)$. We can therefore construct $\mathcal{O}_{\widetilde{S}}$ as the sheaf of $\mathcal{O}_S$ algebras generated by w', which is defined to be the k-th root of z'. The reduced space of the resulting $\widetilde{S}$ is naturally identified with $\widetilde{M}$: just send w' to w, z_i to themselves and the θ_j to 0. Now this sheaf is unique up to an isomorphism which itself is unique up to a k-th root of unity (=a deck transformation of the covering). This allows us to patch the open pieces to obtain the desired global branched covering $F : \widetilde{S} \to S$.

The above construction extends to families of supermanifolds. In fact, a branched covering of a family of supermanifolds is a special case of a branched covering of a single supermanifold: given a family $\pi : S \to B$ and the above data, we momentarily forget π, so we get the branched covering $F : \widetilde{S} \to S$ of the total space of the family, and then we remember that S (and hence also $\widetilde{S}$) are families over B.

2.1.4. *Blowups of supermanifolds.* Starting with a supermanifold X and its codimension $(k|l)$ submanifold Y, we construct the blowup $\widetilde{X}$ of X along Y. Let $y_1, \ldots, y_{m-k}$ and $\eta_1, \ldots, \eta_{n-l}$ be coordinates on an open set $W \subset Y$, while $x_1, \ldots, x_k$ and $\theta_1, \ldots, \theta_l$ are normal coordinates to Y in X, so that x, y, θ, η together form coordinates on an open $U \subset X$. We cover U by affine open subsets U_α, $\alpha = 1, \ldots, k$, given by $x_\alpha \neq 0$, and replace these by new affines $\widetilde{U}_\alpha$ with coordinates:

$$y_1, \ldots, y_{m-k}, \quad \eta_1, \ldots, \eta_{n-l}, \quad x_\beta/x_\alpha, \beta \neq \alpha, \quad x_\alpha, \quad \theta_j/x_\alpha, j = 1, \ldots, l.$$

The $\widetilde{U}_\alpha$ are glued in the obvious way to give the blowup $\widetilde{X}$. (As in ordinary algebraic geometry, one can describe the blowup more intrinsically using the Proj construction, by means of which one can more generally blow up an arbitrary sheaf of ideals.) As in ordinary algebraic geometry, $\widetilde{X}$ comes with an *exceptional divisor* E and a map $\pi : \widetilde{X} \to X$ such that E becomes a bundle over Y with fiber $\mathbb{P}^{(k-1)|l}$ while $\widetilde{X} \backslash E$ maps isomorphically to $X - Y$. Particularly interesting is the case $k = 1$, in which the underlying reduced manifold remains unchanged by the blowup, and only the odd directions are modified. We will encounter this in section 3.4.2.

2.2. Obstructions to splitting. The obstructions to splitting of a supermanifold have been analyzed by a number of authors, including Green, Berezin, Manin, Vaintrob, Rothstein, Onishchik and others, cf. [**1**–**6**]. In this section we will describe the space of all supermanifolds modeled on a given M, V in terms of the cohomology of a certain sheaf G of non-abelian groups, and explain how the condition for a supermanifold S to be split is equivalent to vanishing of a certain sequence of abelian cohomology classes. We go into much more detail in this section than is needed for our later applications, which only depend on the leading obstruction ω_2.

2.2.1. *Green's cohomological description and the obstruction classes.* Let V be a rank n vector bundle on a manifold M, and let $S(M,V)$ be the corresponding split supermanifold. In great generality, the set of all objects of some kind that are locally isomorphic to some model object is given as the first cohomology of the sheaf of automorphism groups of the model. When these automorphism groups are non abelian, the first cohomology is not a group, only a pointed set: the "point" corresponds to the model object itself. We start by applying this principle to all supermanifolds with given reduced space M and given odd dimension; all of these are locally isomorphic to $S(M,V)$, since any two vector bundles of the same rank on M are locally isomorphic. We then restrict to obtain cohomological descriptions of (1) isomorphism classes of pairs consisting of a supermanifold together with an isomorphism of its graded version with $S(M,V)$, and (2) isomorphism classes of supermanifolds whose graded version is globally isomorphic to the given $S(M,V)$ (Green's theorem).

By definition, a supermanifold S with reduced space M and odd dimension n is a sheaf of $\mathbb{Z}_2$-graded algebras on M that is locally isomorphic to $\mathcal{O}_{S(M,V)} = \wedge^\bullet V^\vee$. Consider the sheaf of these local isomorphisms, namely the sheaf $\mathrm{Isom}(S(M,V),S)$ on M whose sections on an open $U \subset M$ are the isomorphisms between $S(M,V)_{|U}$ and $S_{|U}$. In case $S = S(M,V)$, this becomes the sheaf of non-abelian groups:

$$\mathrm{Isom}(S(M,V),S(M,V)) = \mathrm{Aut}(\mathcal{O}_{S(M,V)}) = \mathrm{Aut}(\wedge^\bullet V^\vee),$$

where $\mathrm{Aut}(\wedge^\bullet V^\vee)$ denotes the sheaf of automorphisms of the $\mathbb{Z}_2$-graded sheaf of algebras $\wedge^\bullet V^\vee$. These automorphisms send J to itself, so they act on $\mathcal{O}_M = \mathcal{O}_{S(M,V)}/J$, but this action is trivial since these are automorphisms of sheaves on M. We will describe the structure of $\mathrm{Aut}(\wedge^\bullet V^\vee)$ below: by (2.1) it maps onto $\mathrm{Aut}(V)$ with a kernel G, which in turn is filtered by subgroups G^i with graded pieces given (cf. (2.3)) by: $G^i/G^{i+1} \cong T_{(-)^i}M \otimes \wedge^i V^\vee$.

In general, since S is locally isomorphic to $S(M,V)$, $\mathrm{Isom}(S(M,V),S)$ is locally isomorphic to $\mathrm{Aut}(\wedge^\bullet V^\vee)$, i.e. it is a torsor (=principal homogeneous space) over $\mathrm{Aut}(\wedge^\bullet V^\vee)$. Conversely, every such torsor determines a corresponding supermanifold. The set of isomorphism classes of supermanifolds S with a given reduced space M and given odd dimension n is therefore given by the first cohomology set

$$H^1(M,\mathrm{Aut}(\mathcal{O}_{S(M,V)})) = H^1(M,\mathrm{Aut}(\wedge^\bullet V^\vee)).$$

As noted above, since the group involved is non-abelian, this cohomology is not a group but only a set with a base point. The base point corresponds to $S(M,V)$. (So far, we could have used a different rank n bundle V'; this would have yielded another description of the same set of isomorphism classes of supermanifolds S with the given reduced space M and given odd dimension n, the only difference being that the base point would now be $S(M,V')$.)

Let $G = G_V$ be the kernel of the map that sends an automorphism of $\wedge^\bullet V^\vee$ to the induced automorphism of $V^\vee = \mathrm{Gr}^1(\wedge^\bullet V^\vee)$, or equivalently to the (transpose inverse) automorphism of of V:

$$1 \to G \to \mathrm{Aut}(\wedge^\bullet V^\vee) \to \mathrm{Aut}(V) \to 1. \qquad (2.1)$$

Equivalently, G is the group of those automorphisms of the split model $S(M,V)$ that preserve both M and V.

Consider a supermanifold $S=(M,\mathcal{O}_S)$ with an isomorphism $\rho : V \cong T_{-}S = \mathrm{Gr}^1(\mathcal{O}_S)^{\vee}$, or equivalently an isomorphism of $\mathbb{Z}$-graded sheaves of algebras:

$$\rho : \wedge^{\bullet}V^{\vee} \cong \mathrm{Gr}(\mathcal{O}_S). \tag{2.2}$$

We can compare the pair (S,ρ) to $S(M,V)$, which comes with the natural isomorphism $\mathrm{Id} : V \cong T_{-}S(M,V)$. We find that the sheaf

$$Isom(\ (S(M,V),\mathrm{Id}),\ (S,\rho)\ ),$$

consisting of those local isomorphisms that send Id to ρ, is a torsor over G. The terms of the long exact sequence of cohomology sets of (2.1):

$$\begin{aligned} 1 \to H^0(M,G) &\to H^0(M,\mathrm{Aut}(\wedge^{\bullet}V^{\vee})) \to H^0(M,\mathrm{Aut}(V)) \\ \to H^1(M,G) &\to H^1(M,\mathrm{Aut}(\wedge^{\bullet}V^{\vee})) \to H^1(M,\mathrm{Aut}(V)) \end{aligned}$$

therefore have the following interpretations:

- $H^1(M,\mathrm{Aut}(V)$ is the set of isomorphism classes of rank n bundles on M, with the base point corresponding to V.
- As noted above, $H^1(M,\mathrm{Aut}(\wedge^{\bullet}V^{\vee}))$ is the set of isomorphism classes of supermanifolds S with reduced space M and odd dimension n. The map to $H^1(M,\mathrm{Aut}(V))$ sends S to $T_{-}S$.
- $H^1(M,G)$ is the set of isomorphism classes of pairs (S,ρ) where S is a supermanifold with reduced space M and ρ is an isomorphism $V \cong T_{-}S$.
- The set of isomorphism classes of supermanifolds S modeled on M,V (the isomorphism is required to be the identity on the reduced space M, but can act on the odd directions V) is therefore identified with the quotient of $H^1(M,G)$ by $H^0(M,\mathrm{Aut}(V))$. The base point corresponds to $S(M,V)$. This is the main result of [**1**]. In the present paper we will use only the previous identification of $H^1(M,G)$ itself.

The group G has a descending filtration by normal subgroups $G^i \quad (i=2,3,\ldots)$. We give three descriptions of these subgroups: algebraic, geometric, and analytic. We then use these groups to describe an obstruction theory for the splitting of a supermanifold.

Algebraically, we define:

$$G^i = \{ g \in G \quad | \quad g(x) - x \in J^i \quad \forall x \in \wedge^{\bullet}V^{\vee} \}.$$

One has $G^2 = G$, while G^i is trivial if i exceeds the odd dimension n of S. Modulo higher order terms, each $g \in G$ is a $\wedge^i V^{\vee}$-valued, even derivation. In other words, there is a natural isomorphism for $i \geq 2$:

$$G^i/G^{i+1} \cong T_{(-)^i}M \otimes \wedge^i V^{\vee}. \tag{2.3}$$

On the right hand side, $T_{(-)^i}M \otimes \wedge^i V^{\vee}$ is understood simply as a sheaf of abelian groups under addition. This isomorphism is easiest to see from the geometric or analytic descrptions of the G^i, which we give next.

Geometrically, we interpret the G^i in terms of a filtration of S itself. Given a supermanifold $S=(M,\mathcal{O}_S)$ with nilpotent subsheaf $J \subset \mathcal{O}_S$, it is convenient to introduce

$$S^{(i)} := (M, \mathcal{O}_{S^{(i)}} = \mathcal{O}_S/J^{i+1}). \tag{2.4}$$

These $S^{(i)}$ are locally ringed subspaces of S, though they are not supermanifolds, except for the extremes $i = 0, n$. (They are superanalogs of non-reduced schemes in ordinary algebraic geometry.) They form an increasing filtration of S:

$$S_{red} = S^{(0)} \subset S^{(1)} \subset \cdots \subset S^{(i-1)} \subset S^{(i)} \subset \cdots \subset S^{(n)} = S$$

Recall that automorphisms of the exterior algebra preserve J, hence they preserveJ^i so they preserve the filtration of S by the $S^{(i)}$. In the above, we can in particular take S to be the split model $S(M, V)$. An equivalent definition of the G^i is as those automorphisms of $S(M, V)$ that act as the identity on $S(M, V)^{(i-1)}$.

Analytically, it is natural to interpret these groups in terms of vector fields on the split model $S(M, V)$. Concretely, the (Lie algebra $\mathfrak{g}$ of the) group G is generated by vector fields on $S(M, V)$ that are schematically of the form $\theta^{2k}\partial_x$ or $\theta^{2k+1}\partial_\theta$, $k \geq 1$. These expressions are shorthand for vector fields on $S(M, V)$ that in local coordinates $x_1, \ldots, x_m|\theta_1, \ldots, \theta_n$ take the form

$$\sum_{a_1,\ldots,a_{2k}=1}^{n} \sum_{b=1}^{m} f_{a_1,\ldots,a_{2k};b}(x_1, \ldots, x_m)\theta_{a_1} \ldots \theta_{a_{2k}} \frac{\partial}{\partial x^b} \tag{2.5}$$

or

$$\sum_{a_1,\ldots,a_{2k+1}=1}^{n} \sum_{s=1}^{n} f_{a_1,\ldots,a_{2k+1};s}(x_1, \ldots, x_m)\theta_{a_1} \ldots \theta_{a_{2k+1}} \frac{\partial}{\partial \theta^s}, \tag{2.6}$$

respectively. In these terms, the (Lie algebra $\mathfrak{g}^i$ of the) subgroup G^i is generated by vector fields on $S(M, V)$ that are schematically of the form $\theta^j\partial_x$ or $\theta^j\partial_\theta$, depending on the parity of j, for $j \geq i$. The abelian $G^i/G^{i+1} \cong T_{(-)^i}M \otimes \wedge^i V^\vee$ can be identified with its Lie algebra. In agreement with (2.3), it is just the sheaf of vector fields on $S(M, V)$ that are schematically $\theta^i\partial_x$ or $\theta^i\partial_\theta$, depending on the parity of i. Since $\mathfrak{g}$ is nilpotent, the exponential map $\exp : \mathfrak{g} \to G$ is a bijection, inducing bijections on global sections: $\exp : H^0(\mathfrak{g}) \to H^0(G)$ and $\exp : H^0(\mathfrak{g}^i) \to H^0(G^i)$. But since exp does not respect the group structure of the two sheaves $\mathfrak{g}, G$, the bijection on H^0's need not be an isomorphism of groups, and there is no induced bijection on H^1's.

Using either of these equivalent descriptions of the G^i, we obtain natural interpretations for their cohomologies H^0 and H^1. by a *splitting* of the supermanifold S we mean an isomorphism from the split supermanifold $S(M, V)$ to S that induces the identity on both the underlying reduced space M and the odd tangent bundle V. The family of all splittings of S is parametrized by

$$\text{Splittings}(S) := \text{Isom}_{M,V}(S(M, V), S).$$

A bit more generally, we have the notion of a splitting of the superspace $S^{(i)}$, i.e. an isomorphism from $S(M, V)^{(i)}$ to $S^{(i)}$ that induces the identity on both M and V, and the parameter space Splittings($S^{(i)}$) of all such splittings. For the split $S = S(M, V)$, we have identifications Splittings(S) $\cong H^0(G)$ and Splittings($S^{(i-1)}$) $\cong H^0(G/G^i)$. For a general S, we get instead that Splittings($S^{(i-1)}$) is an $H^0(G/G^i)$-torsor, which is non-empty if and only if S is split. Not all splittings of $S^{(i-1)}$ lift to splittings of S. The variety Splittings$(S^{(i-1)})^S$ of those splittings that do lift to S is (a torsor over) $H^0(G)/H^0(G^i) \subset H^0(G/G^i)$. Similarly, $H^0(G^i)$ itself parametrzes those splittings of $S = S(M, V)$ that induce the identity splitting of $S^{(i-1)}$. In full generality, we may consider splittings of $S^{(j-1)}$ that lift to $S^{(k-1)}$ and induce the identity splitting of $S^{(i-1)}$, whenever $2 \leq i \leq j \leq k$. In terms of the G^i, this is:

$$\text{Splittings}(S^{(j-1)})_{S^{(i-1)}}^{S^{(k-1)}} \quad \cong \quad H^0(G^i/G^k) \quad / \quad H^0(G^j/G^k). \tag{2.7}$$

Another useful case is when $j = i + 1$ and $k = \infty$ or $k = j$: the splittings of $S^{(i)}$ that lift to S and are trivial on $S^{(i-1)}$ are parametrized by $H^0(G^i)/H^0(G^{i+1})$, while all splittings of $S^{(i)}$ that are trivial on $S^{(i-1)}$ are parametrized by $H^0(G^i/G^{i+1}) \cong H^0(M, T_{(-)^i}M \otimes \wedge^i V^\vee)$. The latter is a vector space, and we will see below (in the proof of Corollary 2.5) that the former is actually a linear subspace.

The obstruction theory for splitting of the supermanifold S is based on filtering $H^1(M, G)$ by the images of the $H^1(G^i)$. The geometric interpretation of $H^1(G^i)$ is as the set of isomorphism classes of pairs $\varphi_{i-1} = (S, \rho_{i-1})$, where S is a supermanifold with reduced space M, and $\rho_{i-1} \in \text{Splittings}(S^{(i-1)})$ is an isomorphism between $S(M,V)^{(i-1)}$ and $S^{(i-1)}$. In order for a class $\varphi = (S, \rho)$ in $H^1(M, G)$ to vanish, it is necessary and sufficient that, for each $i \geq 2$, this class should be the image of some $\varphi_{i-1} \in H^1(M, G^i)$. This is clear, since for sufficiently high i the G^i vanish and the $S^{(i)} = S$.

There is nothing to check for $i = 2$, since $G^2 = G$, $\rho_1 = \rho$ and $\varphi_1 = \varphi$. If a given class (S, ρ) is in the image for some i, then to decide if it is in the image for $i + 1$, we look at the exact sequence

$$H^1(M, G^{i+1}) \to H^1(M, G^i) \xrightarrow{\omega} H^1(M, T_{(-)^i}M \otimes \wedge^i V^\vee). \tag{2.8}$$

The obstruction for a class $\varphi_{i-1} \in H^1(M, G^i)$ to come from some $\varphi_i \in H^1(M, G^{i+1})$ is that its image

$$\omega_i = \omega(\varphi_{i-1}) \in H^1(M, T_{(-)^i}M \otimes \wedge^i V^\vee) \tag{2.9}$$

must vanish. This class ω_i is called the i-th obstruction class for splitting of S, and we refer to $\varphi_i \in H^1(M, G^{i+1})$ as a level i trivialization (or level i splitting) of S. The condition for lifting a given φ_{i-1} to an isomorphism $\varphi_i : S^{(i)} \to S(M,V)^{(i)}$ is that $\omega_i = \omega(\varphi_{i-1}) = 0$. This was the basis for the original definition of the classes ω_i in [**2, 3**]; for the interpretation we have described above via cohomology of the sheaf of non-abelian groups G, see [**6**].

Note that we have defined ω_i only if the ω_j vanish, for all $2 \leq j < i$, and only after a level $i-1$ trivialization φ_{i-1} has been chosen. We discuss these issues below. The splitness of S is equivalent to the existence of a level i trivialization φ_i for some $i \geq n$, and hence can be investigated recursively using the obstruction classes ω_i.

We summarize the limitations in the definition of the higher obstructions ω_i as follows:

(1) As noted above, ω_i is defined only if the ω_j vanish, for all $2 \leq j < i$.
(2) Even then, $\omega_i = \omega(\varphi_{i-1})$ may depend not only on the supermanifold S but also on the choice of trivialization φ_{i-1}. In section 2.2.2 below we give an example showing that even for the split model $S(M,V)$, a non-standard choice of trivialization φ_2 may lead to $\omega_3 \neq 0$, while of course the standard choice of trivialization φ_2 gives $\omega_3 = 0$. So the non-vanishing of some higher ω_i for a particular choice of φ_{i-1} is not sufficient to deduce that a supermanifold is non-split.
(3) On the other hand, we will show, also in section 2.2.2, that for even i, while ω_i may depend on φ_{i-2}, it is independent of how φ_{i-2} is lifted to φ_{i-1}.

(4) In section 2.2.3 we will describe analogous classes ω_i^-, for even i, that obstruct the existence of a projection $\varphi_i^- : S^{(i)} \to S^{(0)}$. The ω_i^- for odd i vanish identically: the projection φ_{2k-2}^- lifts uniquely to φ_{2k-1}^-. The obstruction ω_{2k}^- depends only on φ_{2k-2}^-. When φ_{2k-2}^- is taken as the image of φ_{2k-2} (i.e. it is the projection determined by the level i trivialization φ_{2k-2}) and ω_{2k} is defined, it equals ω_{2k}^-. In this sense, ω_{2k} can be made to depend only on the projection data.

(5) These obstructions to projection also depend on previous choices. For instance, Proposition 4.9.5 of [**2**] shows that when $\omega_2 = 0$, so that $\varphi_2 = \varphi_2^- = \varphi_3^-$ can be chosen, but $\omega_3 \neq 0$, the next class ω_4^- depends linearly and non trivially on the choice of φ_2.

None of these points affect the first obstruction class ω_2, which is an invariant of any supermanifold S and obstructs a splitting or projection. Our proofs that various moduli spaces are non-split and in fact non-projected will boil down to showing that ω_2 is nonzero.

2.2.2. *Illustrations.* Consider for example a split supermanifold $S = S(M, V)$ of dimension $1|3$. The filtration is

$$1 = G^4 \subset G^3 \subset G^2 = G,$$

and we have a short exact sequence

$$1 \to G^3 \to G \to G^2/G^3 \to 1 \tag{2.10}$$

with

$$G^3 = G^3/G^4 = Hom(\wedge^3 T-, T-) = T_- \otimes \wedge^3 T_-^*$$
$$G^2/G^3 = Hom(\wedge^2 T-, T+) = T_+ \otimes \wedge^2 T_-^*.$$

The trivial class $1 \in H^1(G)$ has the standard trivialization $\varphi_2 = 1 \in H^1(G^3)$, whose obstruction $\omega_3(1)$ vanishes. Since $G^4 = 1$, the map ω in sequence (2.8) for $i = 3$ is injective. So any exotic lift $\varphi_2 \neq 1$ of the trivial class $1 \in H^1(G)$ must be obstructed; it cannot be extended to a splitting of S. We will show that the coboundary map:

$$H^0(Hom(\wedge^2 T_-, T_+)) \to H^1(Hom(\wedge^3 T_-, T_-)) \tag{2.11}$$

can be non-zero. Any non-trivial $\varphi_2 \neq 1$ in its image would then be an obstructed, exotic level 2 trivialization of $S(M, V)$, with $\omega(\varphi_2) \neq 1$.

For a split supermanifold $S = S(M, V)$ of arbitrary dimension $m|n$, the Lie algebra $\mathfrak{g}$ is the subalgebra of $\mathrm{Der}(\wedge^\bullet V)$ consisting of even derivations sending J^i to J^{i+2} for each i. The action of each $g \in \mathfrak{g}$ gives a first-order differential operator $V^\vee \to \wedge^3 V^\vee$, so we have a sheaf map

$$\mathfrak{g} \to D^1(V^\vee, \wedge^3 V^\vee). \tag{2.12}$$

Several simplifications occur when $n = 3$: $\mathfrak{g}$ becomes abelian, the exponential map $\exp : \mathfrak{g} \to G$ sends $x \mapsto 1 + x$ and induces an isomorphism of groups. In particular it induces a bijection $\exp : H^1(\mathfrak{g}) \to H^1(G)$. The map (2.12) becomes an isomorphism, which in fact takes the Lie algebra variant of short exact sequence (2.10) to the symbol sequence:

$$0 \to \wedge^2 V^\vee \to D^1(V^\vee, \wedge^3 V^\vee) \xrightarrow{\sigma} T_+ \otimes \wedge^2 V^\vee \to 0. \tag{2.13}$$

The coboundary map in (2.11) can therefore be identified as cup product with the extension class of the symbol sequence (2.13), which is induced from the Atiyah class of V.

For simplicity, consider the case that V is the direct sum of three line bundles L_i. Our coboundary decouples as the sum of three maps:

$$c_1(L_i) : H^0(T_+ \otimes L_j^\vee \otimes L_k^\vee) \to H^1(L_j^\vee \otimes L_k^\vee),$$

where $\{i, j, k\}$ is a permutation of $\{1, 2, 3\}$, and we have identified the Atiyah class of a line bundle with its first Chern class. This is clearly non-zero for general choices. For example, this is the case for the super line $\mathbb{P}^{(1|3)}$, where $T_+ = \mathcal{O}_{\mathbb{P}^1}(2)$ and the L_i are $\mathcal{O}_{\mathbb{P}^1}(1)$; or for a superelliptic curve where two of the L_i are trivial, as is T_+, while the third L_i has non-zero degree.

The situation is very different though for even i. In that case, ω_i depends on φ_{i-2}, but is independent of the choice of its lift φ_{i-1}. This is seen by considering the rescaling action along the fibers of V. Under this action, the θ have degree 1, ∂_θ has degree -1, and the x's and ∂_x are neutral. We see that the vector fields in equations (2.5) and (2.6) both have the same degree $2k$. The Lie algebra $\mathfrak{g}$ is therefore graded, with a two-dimensional graded piece for each even weight $2k$. It follows that the coboundary map:

$$H^0(Hom(\wedge^{2k-1}T_-, T_-)) \to H^1(Hom(\wedge^{2k}T_-, T_+))$$
$$H^0(G^{2k-1}/G^{2k}) \to H^1(G^{2k}/G^{2k+1})$$

goes between pieces of different weights, so it must vanish, and therefore the ambiguity, given by the composition

$$H^0(G^{2k-1}/G^{2k}) \to H^1(G^{2k}) \xrightarrow{\omega} H^1(Hom(\wedge^{2k}T_-, T_+))$$

vanishes as well. Actually, a stronger result holds: in an appropriate sense, the even ω_i can be chosen to be independent of φ_j for all odd $j < i$. We give a direct construction of these choice-independent ω_{2k} next.

2.2.3. *Analog for projections.* The group G has a subgroup G_- that is generated by the vector fields $\theta^{2k+1}\partial_\theta$. This is the subgroup of G that preserves a projection. Accordingly, supermanifolds S with a chosen isomorphism ρ as in eqn. (2.2) and that are projected but not necessarily split are labeled by a class in $H^1(M, G_-)$.

In general, a supermanifold S modeled on M, V is projected if and only if the class $x_S \in H^1(M, G)$ that represents it is in the image of $H^1(M, G_-)$. If the subgroup G_- of G were normal, we would have an exact sequence $H^1(M, G_-) \to H^1(M, G) \to H^1(M, G/G_-)$ and then S would be projected if and only if x_S maps to zero in $H^1(M, G/G_-)$. But G_- is not normal, and it does not appear to be possible to express the obstruction to a projection in such simple terms.

Instead, we define G^i_- to be the subgroup of G generated by G^i and G_-. This gives a sequence of subgroups

$$G := G^2_- \supset G^3_- = G^4_- \supset G^5_- = G^6_- \supset G^7_- = \dots$$

In order for a class $x_S \in H^1(M, G)$ to come from $H^1(M, G_-)$, it is necessary and sufficient that, for each (odd) $i \geq 2$, this class should be in the image of $H^1(M, G^i_-)$. Is it possible to convert this to the vanishing of a series of obstructions? For that, we need each G^i_- to be a normal subgroup of its predecessor G^{i-2}_-. This turns out to be true.

We can see the non-normality for G_- as well as the normality for G^i_- quite explicitly from the analytic description in terms of the Lie algebra $\mathfrak{g}$, which is generated by the vector fields on the split model $S(M,V)$ that are schematically of the form $\theta^{2k}\partial_x$ or $\theta^{2k+1}\partial_\theta$, $k \geq 1$, as in (2.5) and (2.6). The Lie algebra $\mathfrak{g}_-$ of G_- is the subalgebra generated by the vector fields $\theta^{2k+1}\partial_\theta$. The condition for a subgroup to be normal is that its Lie algebra should be an ideal in the ambient Lie algebra. The Lie bracket of vector fields on $S(M,V)$ schematically gives formulas such as

$$[\theta^2\partial_x, \theta^3\partial_\theta] = \theta^4\partial_x + \theta^5\partial_\theta. \tag{2.14}$$

(Recall that our schematic notation suppresses the coefficients, which depend on x.) This shows that $\mathfrak{g}_-$ is not an ideal in $\mathfrak{g}$ since the bracket of $\theta^2\partial_x \in \mathfrak{g}$ with $\theta^3\partial_\theta \in \mathfrak{g}_-$ has a non-vanishing $\theta^4\partial_x$ term, and is therefore not in $\mathfrak{g}_-$.

The Lie subalgebra $\mathfrak{g}^i_-$ corresponding to G^i_-, for odd i, is generated by all the above vector fields except $\theta^2\partial_x, \ldots, \theta^{i-1}\partial_x$. The same equation (2.14) shows that for $i \geq 5$, $\mathfrak{g}^i_-$ is not an ideal in $\mathfrak{g}$, since the bracket of $\theta^2\partial_x \in \mathfrak{g}$ with $\theta^3\partial_\theta \in \mathfrak{g}^i_-$ has a non-vanishing $\theta^4\partial_x$ term, and is therefore not in $\mathfrak{g}^i_-$. (For $i = 3$, $\mathfrak{g}^3_-$ *is* normal in $\mathfrak{g}$, as it equals $\mathfrak{g}^3$.)

On the other hand, $\mathfrak{g}^i_-$ is always an ideal in $\mathfrak{g}^{i-2}_-$. As above, $\mathfrak{g}^{i-2}_-$ is generated by all the vector fields except $\theta^2\partial_x, \ldots, \theta^{i-3}\partial_x$. The bracket of any two vector fields in $\mathfrak{g}_-$ is in $\mathfrak{g}_-$. The smallest mixed term with a ∂_x factor from $\mathfrak{g}^{i-2}_-$ and a ∂_θ factor from $\mathfrak{g}^i_-$ is $[\theta^{i-1}\partial_x, \theta^3\partial_\theta] = \theta^{i+1}\partial_x + \theta^{i+2}\partial_\theta$, which is in $\mathfrak{g}^i_-$, as is the remaining bracket $[\theta^{i-1}\partial_x, \theta^{i+1}\partial_x] = \theta^{2i}\partial_x$.

We conclude that each $G^{2k+2}_- = G^{2k+1}_-$ is a normal subgroup of its predecessor $G^{2k}_- = G^{2k-1}_-$. As in (2.8), if a given class x_S is in the image of $H^1(G^{2k}_-)$ for some k, then to decide whether it is in the image of $H^1(G^{2k+1}_-)$, we look at the exact sequence

$$H^1(M, G^{2k+1}_-) \to H^1(M, G^{2k}_-) \overset{\omega^-}{\to} H^1(M, T_+M \otimes \wedge^{2k}V^\vee). \tag{2.15}$$

The obstruction for a class $\varphi^-_{2k-1} \in H^1(M, G^{2k}_-)$ to come from $H^1(M, G^{2k+1}_-)$ is that its image

$$\omega^-_{2k} = \omega^-(\varphi^-_{2k-1}) \in H^1(M, T_+M \otimes \wedge^{2k}V^\vee)$$

must vanish. We may as well call this class ω^-_{2k} the $2k$-th obstruction class for projectedness of S. It has properties, and limitations, analogous to those of the ω_i. By construction, it depends only on the odd φ^-_{2k-1}, and these are the same as the even φ^-_{2k-2}. When both are defined, ω^-_{2k} and ω_{2k} clearly agree. This is the sense in which ω_{2k} can be made to depend only on the previous even choices.

Note in particular that $\omega^-_2 = \omega_2$, so the vanishing of ω_2 is a necessary condition for a projection. (This follows more directly from the fact that G_- is contained in G^3.) This is the only necessary criterion for projectedness that we will use in the present paper.

2.2.4. *Some immediate applications.*

COROLLARY 2.1. *Any C^∞ supermanifold S is split.*

PROOF. A C^∞ locally-free sheaf is fine, hence its H^1 vanishes. □

COROLLARY 2.2. *Any supermanifold S of dimension $(m|1)$ is split.*

PROOF. $(\wedge^i V^\vee)_0 = 0, \quad i \geq 2.$ □

COROLLARY 2.3. *A supermanifold S of dimension $(m|2)$ is determined by the triple (M, V, ω), where $\omega = \omega_2 \in H^1(M, Hom(\wedge^2 T_-, T_+))$, and any such triple arises from some S. A supermanifold of dimension $(m|2)$ is projected if and only if it is split.*

The last statement reflects the fact that ω_2, which is the only obstruction to a splitting when the odd dimension is 2, is also an obstruction to a projection.

Similarly,

COROLLARY 2.4. *For any supermanifold S, the following conditions are equivalent: $\omega_2(S) \neq 0$; $S^{(2)}$ is not split; $S^{(2)}$ is not projected. Moreover, if $\omega_2(S) \neq 0$, then S is not projected.*

Indeed, with M, V given, $S^{(2)}$ is classified up to isomorphism by $\omega_2(S)$, which obstructs both a splitting and a projection of $S^{(2)}$. A projection of S could be restricted to a projection of $S^{(2)}$, so $\omega_2(S) \neq 0$ implies that S is not projected.

One can easily construct explicit examples of non-split supermanifolds, beginning in dimension $(1|2)$. For example, a non-degenerate conic in $\mathbb{P}^{2|2}$, e.g.

$$x^2 + y^2 + z^2 + \theta^1\theta^2 = 0, \tag{2.16}$$

is non-split. The obstruction class ω_2 for this supermanifold is evaluated in [**3**]. We will see another explicit example in section 3.5.

LEMMA 2.5. *The space Splittings(S) of all splittings of a split supermanifold S has the structure of an iterated fibration by affine spaces. The action of the automorphism group Aut(S) preserves this fibration and induces an affine action on the affine fibers. (Similarly if S is replaced by $S^{(i)}$ for some i.)*

PROOF. Let M, V be the reduced space and odd tangent bundle of S. The group $\mathrm{Aut}(S)$ acts on the manifold M, and V is an $\mathrm{Aut}(S)$-equivariant vector bundle on M. Let $S_0 := S(M, V)$ be the split version of S. It inherits an action of $\mathrm{Aut}(S)$, and the distinguished splitting is $\mathrm{Aut}(S)$-invariant.

Recall that $\mathrm{Splittings}(S) := \mathrm{Isom}_{M,V}(S_0, S)$ denotes the variety of splittings of the split supermanifold S, i.e. isomorphisms of S_0 with S that act trivially on M and V. The actions of $\mathrm{Aut}(S)$ on S, S_0 induce an action of $\mathrm{Aut}(S)$ on $\mathrm{Splittings}(S)$. (This action combines the $\mathrm{Aut}(S)$ action on S with the inverse action on S_0.)

Similarly, we have the variety $\mathrm{Splittings}(S^{(i)})^S = \mathrm{Splittings}(S^{(i)})^S_{S^{(1)}}$ of those splittings of $S^{(i)}$ that lift to S (and restrict to the identity on $S^{(1)}$, as we always require.) The iterated fibration is:

$$\mathrm{Splittings}(S) \ldots \to \mathrm{Splittings}(S^{(i)})^S \to \mathrm{Splittings}(S^{(i-1)})^S \to \ldots \tag{2.17}$$

For S_0 itself, $\mathrm{Splittings}(S_0)$ is a group, isomorphic to $H^0(G)$. The distinguished splitting, corresponding to the unit element, is a fixed point of the above natural $\mathrm{Aut}(S)$-action. As a special case of (2.7), the iterated fibration (2.17) becomes:

(2.18) $$H^0(G)\ldots \to H^0(G)/H^0(G^{i+1}) \to H^0(G)/H^0(G^i) \to \ldots$$

The typical step here is a surjective group homomorphism:

$$\pi : H^0(G)/H^0(G^{i+1}) \to H^0(G)/H^0(G^i))$$

whose kernel is $H^0(G^i))/H^0(G^{i+1})$ and whose other fibers are cosets of this kernel. The sequence $0 \to G^{i+1} \to G^i \to G^i/G^{i+1} \to 0$ shows that this kernel is the image of the group homomorphism

$$h : H^0(G^i)) \to H^0(G^i/G^{i+1}).$$

This image is clearly a subgroup of the vector space

$$H^0(G^i/G^{i+1}) = H^0(M, T_{(-)^i}M \otimes \wedge^i V^\vee).$$

It is also invariant under homotheties: if $h(x) = y$, then (because the exponential map is bijective) x can be extended to a 1-parameter subgroup of $H^0(G^i))$, and h of this 1-parameter subgroup is a 1-parameter subgroup of the vector space $H^0(G^i/G^{i+1})$, i.e. it is a line. So $\ker(\pi)$, which is the image of h, is a subgroup closed under homotheties, i.e. it is a vector subspace of $H^0(G^i/G^{i+1})$, and $\mathrm{Aut}(S)$ acts on it linearly. The general fibers of π are cosets of this, i.e. affine spaces with affine $\mathrm{Aut}(S)$ action.

Compare this with S, which is split but perhaps not $\mathrm{Aut}(S)$-invariantly so. Then Splittings(S) is a Splittings(S_0)-torsor, and iterated fibration (2.17) is a torsor over iterated fibration (2.18). (This means that the objects of (2.17) are torsors over corresponding objects of (2.18), and the maps in (2.17) are torsor maps compatible with the corresponding maps of groups in (2.18).) There are induced actions of $\mathrm{Aut}(S)$ on all terms of the two iterated fibrations, and the crucial observation is that these $\mathrm{Aut}(S)$-actions are compatible with the above torsor structure. In particular, the fibers of $\mathrm{Splittings}(S^i)^S \to \mathrm{Splittings}(S^{i-1})^S$ in the iterated fibration (2.17) are affine spaces modeled on the vector space $\ker(\pi)$, the $\mathrm{Aut}(S)$ action on them is affine, and the linearization of this action is the linear $\mathrm{Aut}(S)$ action on $\ker(\pi)$. □

COROLLARY 2.6. *Let S be an algebraic supermanifold (i.e. the reduced space M is a complex algebraic variety and V is an algebraic vector bundle), and let F be a reductive subgroup of its automorphism group $\mathrm{Aut}(S)$. If S, or $S^{(i)}$ for some i, is split, then the splitting can be chosen to be F-invariant.*

PROOF. [3] An F-invariant splitting of S is a fixed point p of the F-action on Splittings(S). We find a fixed point p^i of the action of F on $\mathrm{Splittings}(S^{(i)})^S$ for each i, inductively. For sufficiently high i this will give us the desired fixed point of the action of F on Splittings(S). For $i = 1$ there is nothing to do, since $\mathrm{Splittings}(S^{(1)})^S$ is a single point. So we assume inductively that we have a fixed point $p^{i-1} \in \mathrm{Splittings}(S^{(i-1)})^S$ and we need to lift it to a fixed point $p^i \in \mathrm{Splittings}(S^{(i)})^S$ in the fiber above p^{i-1}, which was seen in Lemma 2.5 to be an affine space with an affine action of F. The inductive step, and the proof of the Corollary, now follow from the observation that an algebraic affine action of a reductive group on an affine space must have a fixed point.

If F is finite, the existence of fixed points is clear: the average of a finite set of points in an affine space (in characteristic zero) is well defined, and the average of

[3]We thank P. Deligne for help with this proof.

the points in any F-orbit is clearly F-invariant. When the vector space in question, $A := H^0(M, T_{(-)^i}M \otimes \wedge^i V^\vee)$, is finite dimensional, e.g. when M is compact, the reductive case follows similarly, by averaging with respect to an invariant measure on a compact form F_c of the group. Alternatively, the affine action of F on the vector space A extends to a linear action on the vector space $B := A \oplus C$, where $C = \mathbb{C}$, and the action on B commutes with the projection to C. (The affine action is recovered on the slice A^1 above $1 \in \mathbb{C}$, while its linearization is realized on the slice A^0 above $0 \in \mathbb{C}$.) This B contains $A = A^0$, with a 1-dimensional quotient C on which F acts trivially. In the finite dimensional case, complete reducibility of the action of our reductive F gives an F-invariant embedding of C in B, and points of its image are fixed by F.

It is true in general that any algebraic linear representation of F, such as its action on $B = A \oplus C$, is the union of its finite dimensional subrepresentations. The existence of fixed points for the affine action in the general case then follows: choose any point p in the affine fiber A^1, think of it as a point of B above $1 \in \mathbb{C}$, and find a finite dimensional subrepresentation of B containing this point. Using the finite dimensional case, we find a fixed point p^i in this subrepresentation.

In our case, with $A = H^0(M, T_{(-)^i}M \otimes \wedge^i V^\vee)$, we can use some geometry to show directly that any $\rho \in A$ is contained in some finite dimensional subrepresentation. For this, choose a compactification $\overline{M}$ of M. (This can always be done, by a well known result of Nagata [**17**], which has been modernized and improved in [**18**] by Conrad, following Deligne.) The vector bundle $T_{(-)^i}M \otimes \wedge^i V^\vee$ on M extends to a coherent sheaf F on $\overline{M}$. Our space $A = H^0(M, T_{(-)^i}M \otimes \wedge^i V^\vee)$ is the union of its finite dimensional subspaces A_j consisting of meromorphic sections of F with a pole of order $\leq j$ on the boundary of $\overline{M}$. (These A_j may not be F-invariant.) Now given any $\rho \in A$, the F-action on ρ gives a map $f : F \to A$. The inverse images $f^{-1}(A_j)$ are Zariski closed subsets of F, since each is given locally by the vanishing of a set of Laurent coefficients. They form an increasing sequence of Zariski closed subsets that cover F, so one of them must equal F. (Over an uncountable field, if a variety is the union of a countable collection of its subvarieties, then it is also the union of a finite subset of these. Topologically, this is the Baire category theorem.) It follows that the span of the F-orbit of ρ is contained in some A_j, hence this span is a finite dimensional F-invariant subspace containing ρ as desired. $\square$

COROLLARY 2.7. *Let $\pi : \widetilde{S} \to S$ be a finite covering map of supermanifolds. If S, or $S^{(i)}$ for some i, is projected or split, then likewise $\widetilde{S}$, or $\widetilde{S}^{(i)}$, is projected or split. Conversely (in characteristic zero, as always assumed in this paper), if $\widetilde{S}$ or $\widetilde{S}^{(i)}$ is split, then so is S or $S^{(i)}$.*

PROOF. The first part is an immediate consequence of the construction in section 2.1.2 of a covering map of supermanifolds. In this construction, if S or $S^{(i)}$ for some i is split or projected, then the cover $\widetilde{S}$ or $\widetilde{S}^{(i)}$ immediately acquires the same property.

For the converse part, if there is a finite group F that acts freely on $\widetilde{S}$ with quotient S, we can appeal to Corollary 2.6: the splitting of $\widetilde{S}$ can be replaced by one which is F-invariant, and the latter is a pullback from a splitting of S. In general, we can always find a covering $\overline{S} \to \widetilde{S}$, a freely acting group F, and a subgroup $G \subset F$ such that $\widetilde{S}$ is the quotient of $\overline{S}$ by the action of G, while S is the quotient of $\overline{S}$ by the action of F. To construct $\overline{S}$, we simply construct a Galois cover $\overline{M}$ of

M lying over $\widetilde{M}$ (the reduced space of $\widetilde{S}$), and let $\overline{S}$ be the corresponding cover of S. The splitness of $\widetilde{S}$ implies the splitness of $\overline{S}$ by the first part of this proof, and the splitness of S then follows from Corollary 2.6. □

We do not know whether Corollary 2.6 and the converse part of Corollary 2.7 hold with "split" replaced by "projected."

The special case of Corollary 2.7 that we will use later is the following:

COROLLARY 2.8. *Let $\pi : \widetilde{S} \to S$ be a finite covering map of supermanifolds (in characteristic zero). If $\omega_2(S) \neq 0$, then $\omega_2(\widetilde{S}) \neq 0$, so $\widetilde{S}$ is not projected.*

PROOF. The hypothesis that $\omega_2(S) \neq 0$ is equivalent to $S^{(2)}$ being non-split. By Corollary 2.7, therefore, $\widetilde{S}^{(2)}$ is not split. This in turn is equivalent to $\omega_2(\widetilde{S}) \neq 0$, and, as ω_2 obstructs a projection, it follows that $\widetilde{S}$ is not projected. □

One can also give a direct proof of this Corollary. (We thank one of the referees for pointing this out.) For a local isomorphism $\pi : \widetilde{S} \to S$, the structure sheaf $\mathcal{O}_{\widetilde{S}}$ is the pullback of $\mathcal{O}_S$. It follows that $\omega_2(\widetilde{S}) = \pi^*\omega_2(S)$. But the composition of π^* with the trace map is multiplication by the degree of π, so π^* is injective and the result follows.

2.2.5. *Submanifolds.* Suppose that S is a supermanifold modeled on M, V. Let M' be a submanifold of M, and let V' be a subbundle of $V|_{M'}$. In general, a submanifold S' of S with reduced space M' and odd tangent bundle V' does not exist. To analyze the obstruction to this, we start with the split supermanifold $S(M, V)$, which has $S(M', V')$ as a submanifold. The sheaf of groups G has a subsheaf G_* consisting of automorphisms of $S(M, V)$ that restrict to automorphisms of $S(M', V')$. The condition that a supermanifold S modeled on M, V contains a submanifold S' modeled on M', V' is that the class in $H^1(M, G)$ that represents S should be the image of a class in $H^1(M, G_*)$. As in the discussion of projections, there would be an easy criterion for this if G_* were normal, but this is not the case.

However, we can partially reduce to the normal case if we replace S with $S^{(2)}$ and ask whether $S^{(2)}$ contains a subspace $S'^{(2)}$ modeled on $S(M', V')^{(2)}$. For this, we replace G by the sheaf of abelian groups $G/G^3 = Hom(\wedge^2 T_-M, T_+M)$ and G_* by G_*/G_*^3, which is the subsheaf of $Hom(\wedge^2 T_-M, T_+M)$ consisting of homomorphisms that, when restricted to M', map $\wedge^2 T_-M'$ to T_+M'. We call this sheaf $Hom^*(\wedge^2 T_-M, T_+M)$. The quotient of the two is the sheaf $Hom(\wedge^2 T_-M', N)$ supported on M', where N is the normal bundle to M' in M. So we get an exact sequence in cohomology

$$\begin{aligned} &H^1(M, Hom^*(\wedge^2 T_-M, T_+M)) \to H^1(M, Hom(\wedge^2 T_-M, T_+M) \\ &\qquad \xrightarrow{b} H^1(M', Hom(\wedge^2 T_-M', N), \end{aligned} \tag{2.19}$$

leading to a necessary condition for existence of S':

COROLLARY 2.9. *A necessary condition for a supermanifold S modeled on M, V to contain a submanifold S' modeled on M', V' is that, in the notation of* (2.19),

$$b(\omega_2) = 0 \in H^1(M', Hom(\wedge^2 T_-M', N)).$$

From the definition of G_* in terms of automorphisms of $S(M, V)$ that restrict to automorphisms of $S(M', V')$, there is a tautological map from G_* to the sheaf

G' of automorphisms of $S(M', V')$ that act trivially on M' and its normal bundle. Thus, G_* maps to both G and G',

$$\begin{array}{ccc} G_* & \stackrel{u}{\to} & G \\ r\downarrow & & \\ G', & & \end{array} \tag{2.20}$$

leading to corresponding maps in cohomology

$$\begin{array}{ccc} H^1(M, G_*) & \stackrel{u}{\to} & H^1(M, G) \\ r\downarrow & & \\ H^1(M', G'). & & \end{array} \tag{2.21}$$

REMARK 2.10. As already noted, if S' exists, the class $x_S \in H^1(M, G)$ that represents S is the image of some $x_S^* \in H^1(M, G_*)$. The map from G_* to G' just restricts an element of G_* to its action on $S(M', V')$, so the image of x_S^* in $H^1(M', G')$ is the class $x_{S'}$ that represents S'.

If we abelianize by replacing G_*, G, and G' by their quotients by G_*^3, G^3, and G'^3, we can complete the picture (2.20) to a commuting square:

$$\begin{array}{ccc} Hom^*(\wedge^2 T_-M, T_+M) & \stackrel{u}{\to} & Hom(\wedge^2 T_-M, T_+M) \\ r\downarrow & & \iota\downarrow \\ Hom(\wedge^2 T_-M', T_+M') & \stackrel{j}{\to} & Hom(\wedge^2 T_-M', T_+M|_{M'}) \end{array} \tag{2.22}$$

Here u and r are the linearizations of the corresponding maps in (2.20). The sheaves on the top row are sheaves on M and the sheaves on the bottom row are sheaves on M'. The vertical maps are restrictions from M to M', followed by restriction from $\wedge^2 T_-M$ to $\wedge^2 T_-M'$. Finally, j comes from the inclusion $T_+M' \subset T_+M|_{M'}$. In cohomology we get:

$$\begin{array}{ccc} H^1(M, Hom^*(\wedge^2 T_-M, T_+M)) & \stackrel{u}{\to} & H^1(M, Hom(\wedge^2 T_-M, T_+M)) \\ r\downarrow & & \iota\downarrow \\ H^1(M', Hom(\wedge^2 T_-M', T_+M')) & \stackrel{j}{\to} & H^1(M', Hom(\wedge^2 T_-M', T_+M)). \end{array} \tag{2.23}$$

The linearization of Remark 2.10 says that there is a class $\omega_2^*(S)$ in the upper left corner of the square (2.23) that maps horizontally to $\omega_2(S)$ and vertically to $\omega_2(S')$. So commutativity of the square implies that $j(\omega_2(S')) = \iota(\omega_2(S))$. This statement is our Compatibility Lemma:

COROLLARY 2.11. *If S is a supermanifold with submanifold S', then the classes $\omega_2(S')$ and $\omega_2(S)$ are compatible in the sense that $j(\omega_2(S')) = \iota(\omega_2(S))$.*

Particularly important is the case that the normal sequence decomposes:

COROLLARY 2.12. *Let S be a supermanifold, and $S' \subset S$ a submanifold, with reduced spaces $M' \subset M$, such that the normal sequence of M' decomposes: $TM|_{M'} = T_{M'} \oplus N$, where N is the (even) normal bundle. If $\omega_2(S') \neq 0$, then also $\omega_2(S) \neq 0$ and S is not projected.*

PROOF. This follows from Corollary 2.11 and the fact that, when the normal sequence decomposes, the map j is injective. □

Are there analogous results with ω_2 replaced by ω_j, for some $j > 2$? Clearly a necessary hypothesis is that $\omega_i(S) = \omega_i(S') = 0$, $i < j$, so that $\omega_j(S)$ and $\omega_j(S')$ are defined. This would imply that there are splittings $S^{(j-1)} \to M$ and $S'^{(j-1)} \to M'$. One actually needs a stronger hypothesis to make the argument. One needs to know that the class in $H^1(M, G)$ associated to S is a pullback from G_*^j, which implies that there is a splitting $S^{(j-1)} \to M$ that restricts to a splitting $S'^{(j-1)} \to M'$. Under this hypothesis, the analog of Corollary 2.11 holds, with essentially the same proof.

Rather than an embedding, consider any map of supermanifolds $f : S' \to S$, with reduced manifolds $S_{\text{red}} = M$ and $S'_{\text{red}} = M'$. The differentials:

$$d_\pm : T_\pm S' \to f^*(T_\pm S)$$

induce maps on cohomology groups:

$$\iota : H^1(M, (T_\pm S) \otimes \wedge^i (T_- S)^\vee) \to H^1(M', (T_\pm S) \otimes \wedge^i (T_- S')^\vee)$$

and

$$j : H^1(M', (T_\pm S') \otimes \wedge^i (T_- S')^\vee) \to H^1(M', (T_\pm S) \otimes \wedge^i (T_- S')^\vee).$$

Perhaps surprisingly, Corollary 2.11 holds for an arbitrary map:

COROLLARY 2.13. *For any map of supermanifolds $f : S' \to S$, the classes $\omega_2(S')$ and $\omega_2(S)$ are compatible, in the sense that:*

$$j(\omega_2(S')) = \iota(\omega_2(S)).$$

PROOF. The map $f : S' \to S$ factors through the graph embedding $\Gamma : S' \to S \times S'$ and the projection on the first factor. The result is clear for the projection and behaves well under composition, so this corollary follows from the embedding case which is Corollary 2.11. □

3. Super Riemann Surfaces

3.1. Basics. A *Super Riemann Surface*[4] is a pair $\mathcal{S} = (S, \mathcal{D})$ where $S = (C, \mathcal{O}_S)$ is a complex analytic supermanifold of dimension $(1|1)$ and $\mathcal{D}$ is an everywhere non-integrable odd distribution, $\mathcal{D} \subset TS$. Recall that the square of an odd vector field v is an even vector field $v^2 = \frac{1}{2}\{v, v\}$. The distribution $\mathcal{D}$ generated by v is said to be *integrable* if $v^2 \in \mathcal{D}$, and *everywhere non-integrable* if v^2 is everywhere independent of v. In the latter case, v and v^2 span the full tangent bundle TS, and thus the nonintegrable distribution $\mathcal{D}$ is actually part of an exact sequence

$$0 \to \mathcal{D} \to TS \to \mathcal{D}^2 \to 0. \tag{3.1}$$

The everywhere nonintegrable odd distribution $\mathcal{D}$ endows the $(1|1)$ supermanifold S with what is called a superconformal structure.

As is the case for supermanifolds in general, interesting phenomena usually concern not a single Super Riemann Surface but a family of Super Riemann Surfaces over a base that is itself a supermanifold. This means a family of complex analytic supermanifolds of dimension $(1|1)$ together with a subbundle of the relative tangent bundle, of rank $(0|1)$ and everywhere non-integrable.

[4]The term "super Riemann surface" was introduced in [**9**]. The same objects were called "superconformal manifolds" in [**7**]. In our terminology, a super Riemann surface, or SRS for short, is a $(1|1)$ complex supermanifold with a superconformal structure.

LEMMA 3.1. *Locally on a super Riemann surface S one can choose coordinates x and θ, referred to as superconformal coordinates, such that $\mathcal{D}$ is generated by the odd vector field*

$$v := \frac{\partial}{\partial\theta} + \theta\frac{\partial}{\partial x}. \tag{3.2}$$

More specifically, if x is any even function on S that reduces mod nilpotents to a local parameter on the reduced space, then there is (locally in the analytic or etale topologies) an odd function θ on S such that $x|\theta$ are superconformal coordinates.

PROOF. In general, in any local coordinate system $x|\theta$, an odd vector field is of the form $v = a(x|\theta)\partial_\theta + b(x|\theta)\partial_x$, where a is an even function and b is an odd one. (In general, a and b depend on additional odd and/or even parameters as well as on $x|\theta$.) As b is odd, it is nilpotent. The definition of an odd distribution is such that v defines an odd distribution where and only where $a \neq 0$. (On a supermanifold, to say that an even quantity is nonzero is taken to mean that it is invertible.) The condition that v and v^2 generate TS implies that if we write $b(x|\theta) = b_0(x)+\theta b_1(x)$, then $b_1 \neq 0$ everywhere. Given the conditions that a and b_1 are everywhere nonzero, it is elementary to find a change of variables that locally puts v in the form given in eqn. (3.2). The last statement in the lemma holds because in this change of variables, it is only necessary to change θ, not x. □

Applying the vector field v of (3.2) to a function $f(x)+\theta g(x)$ gives $g(x)+\theta f'(x)$, so applying it twice gives $f'(x) + \theta g'(x)$. In other words,

$$v^2 = \frac{\partial}{\partial x}. \tag{3.3}$$

Since a super Riemann surface has dimension $(1|1)$, a divisor in it has dimension $(0|1)$. One way to specify a subvariety of dimension $(0|1)$ in any complex supermanifold is to give a point p and an odd tangent vector at p. There is then a unique subvariety of dimension $(0|1)$ that passes through p with the given tangent vector. In the case of a super Riemann surface $\mathcal{S}$, let p be the point defined by equations $x = x_0$, $\theta = \theta_0$. For this to make sense, with $\theta_0 \neq 0$, we must work over a ring with odd elements; the constant θ_0 is an odd element of the ground ring. The fiber at p of the distribution $\mathcal{D}$ gives an odd tangent direction at p, and this determines a divisor D passing through p. Concretely, with $\mathcal{D}$ generated as in (3.2), D is given in parametric form by $x = x_0 + \alpha\theta$, $\theta = \theta_0 + \alpha$, where α is an odd parameter. Alternatively, D is given by the equation

$$x = x_0 - \theta_0\theta. \tag{3.4}$$

In a general supermanifold of dimension $(1|1)$, a divisor that in some local coordinate system is defined by an equation of this kind is called a minimal divisor. (Algebraically, one might call it a prime divisor.) Since the parameters (x_0, θ_0) in the equation were the coordinates of an arbitrary point $p \in \mathcal{S}$, this construction gives a natural 1-1 correspondence between points and minimal divisors on a super Riemann surface $\mathcal{S}$. On a general complex supermanifold S of dimension $(1|1)$, there is no such natural correspondence between points and minimal divisors. (Rather, to a $(1|1)$ supermanifold S, one associates a dual supermanifold $\widehat{S}$ that parametrizes minimal divisors in S; the relationship between S and $\widehat{S}$ is actually symmetrical.)

REMARK 3.2. For a concrete special case of the relationship between points and minimal divisors on a super Riemann surface, consider, in local superconformal coordinates $x|\theta$, the divisor $x = 0$. The equation $x = 0$ is of the form (3.4) for $x_0 = \theta_0 = 0$, so the distinguished point associated to the divisor $x = 0$ is given by $x = \theta = 0$. On a general $(1|1)$ supermanifold, one could make an automorphism $x \to x$, $\theta \to \theta + \alpha$, with α an odd parameter, and thus there would be no distinguished point on the divisor $x = 0$. On a super Riemann surface, there is no such automorphism preserving the superconformal structure.

The example that we have just described is typical in the sense that any minimal divisor on a super Riemann surface takes the form $x = 0$ in some system $x|\theta$ of local superconformal coordinates. This assertion follows from the last remark in Lemma 3.1, or alternatively from the fact that if $x|\theta$ are local superconformal coordinates, then so are $x - a + \eta\theta|\theta - \eta$, where here a and η are arbitrary even and odd parameters.

3.2. Moduli.

In ordinary bosonic algebraic geometry, one encounters the coarse moduli space $\mathcal{M}_g$ of genus g curves (or: Riemann surfaces) as well as the moduli stack M_g. The stack is characterized by the collection of maps to it: a map $\phi : B \to M_g$ from a variety B to the moduli stack M_g is specified by a family of genus g curves parametrized by B. Such a family determines a map $\overline{\phi}$ of B to the moduli space $\mathcal{M}_g$ too; this is just the composition of ϕ with the natural map $\pi : M_g \to \mathcal{M}_g$. But not every such map $\overline{\phi}$ arises from a family over B; it does if and only if it factors through M_g. For example, the identity map of $\mathcal{M}_g$ to itself does not arise from a family: there is no universal curve over $\mathcal{M}_g$. The problem arises from the automorphisms. Every curve of genus ≤ 2, and some curves in any genus, have non-trivial automorphisms. At the corresponding points of moduli, the map $\pi : M_g \to \mathcal{M}_g$ is not a local isomorphism: it looks locally like a quotient by the automorphism group.

For super Riemann surfaces, the problem is more severe. Any split super Riemann surface $\mathcal{S}$, e.g. any SRS over a point (as opposed to a family of SRS's), has a non-trivial automorphism which is the identity on $\mathcal{S}_{\mathrm{red}}$ and acts as -1 in the odd direction. In local coordinates, this is $x \mapsto x$, $\theta \mapsto -\theta$. So super Riemann surfaces with non-trivial automorphisms are dense in any family of super Riemann surfaces, and unlike the case of ordinary Riemann surfaces, one cannot avoid this by taking a finite cover of the moduli space (for example, by fixing a level structure). So the stacky nature of the moduli problem is more essential for super Riemann surfaces than for ordinary ones and what we will loosely call super moduli space and denote as $\mathfrak{M}_g$ must be properly understood not as a supermanifold but as the supermanifold analog of a stack. We return to this in section 3.2.1, but for now we consider local questions.

By Corollary 2.2, a $(1|1)$ supermanifold S defined over a field (as opposed to one that is defined over a ring with odd elements) is split, so it is specified by a pair (C, V) where C is an ordinary Riemann surface and V is the analytic line bundle on C underlying $\mathcal{D}$. We refer to the genus of C also as the genus of $\mathcal{S}$. The calculation (3.3) shows that $V^{\otimes 2} \sim T_C$, so V (or rather, its dual) is a spin structure on C. An ordinary Riemann surface with a choice of spin structure is often called a spin curve, and thus the reduced space C of a super Riemann surface $\mathcal{S}$ is a spin curve. It follows also that the reduced space of the moduli space $\mathfrak{M}_g$ of super Riemann surfaces of genus g is the moduli space (or rather stack) $\mathcal{SM}_g$ of spin curves. Just like $\mathcal{SM}_g$, $\mathfrak{M}_g$ has two components, corresponding to even and odd spin structures.

For a super Riemann surface constructed from a pair (C, V), we will denote V as ${T_C}^{1/2}$, and its n-th power by either $T_C^{\otimes \frac{n}{2}} = T_C^{\frac{n}{2}}$ or equivalently $K_C^{\otimes(-\frac{n}{2})} = K_C^{-\frac{n}{2}}$.

As in ordinary algebraic geometry, to understand the deformation theory of a super Riemann surface $\mathcal{S}$, we first must consider the automorphisms. Locally, the superconformal vector fields, i.e. infinitesimal automorphisms of $\mathcal{S}$, are given by vector fields that preserve the distribution $\mathcal{D}$. In superconformal coordinates, a short calculation shows that an even superconformal vector field takes the form

$$f(x)\frac{\partial}{\partial x} + \frac{f'(x)}{2}\theta\frac{\partial}{\partial \theta}, \tag{3.5}$$

while an odd one takes the form

$$-g(x)\left(\frac{\partial}{\partial \theta} - \theta\frac{\partial}{\partial x}\right). \tag{3.6}$$

We stress that f and g are functions of x only, and not θ. One defines a subsheaf W of the tangent bundle TS whose local sections are of the form (3.5) and (3.6). It is called the sheaf of superconformal vector fields. Since it is defined by the condition of preserving the superconformal structure of $\mathcal{S}$, W is a sheaf of $\mathbb{Z}_2$-graded Lie algebras. From this point of view, W is not naturally a locally-free sheaf (multiplying a superconformal vector field by a function of x and θ, or even a function of x, does not in general give a new superconformal vector field). However, forgetting its structure as a sheaf of graded Lie algebras, W can be given the structure of a locally-free sheaf. For this, we just think of W as a subsheaf of the sheaf of sections of TS, so that in view of the exact sequence (3.1), W can be projected to the sheaf of sections of $\mathcal{D}^2$. A short calculation in local superconformal coordinates shows that this projection is an isomorphism, so W can be identified with the sheaf of sections of $\mathcal{D}^2$. Indeed, in local superconformal coordinates $x|\theta$, $TS/\mathcal{D} \cong \mathcal{D}^2$ is generated by ∂_x, so a general section of $\mathcal{D}^2$ is $a(x|\theta)\partial_x$ for some function $a(x|\theta)$. So we must show that $a(x|\theta)\partial_x$ can in a unique fashion be written as a section of W modulo $\mathcal{D}$. This follows from the formula

$$(f(x)\partial_x + (f'(x)/2)\theta\partial_\theta) - g(x)\left(\partial_\theta - \theta\partial_x\right) = (f(x) + 2\theta g(x))\partial_x \bmod \mathcal{D}, \tag{3.7}$$

which shows that a general section of W is $a(x|\theta)\partial_x \bmod \mathcal{D}$, with $a(x|\theta) = f(x) + 2\theta g(x)$.

Just as in bosonic algebraic geometry, the first-order deformations of $\mathcal{S}$ are given by the first cohomology of $\mathcal{S}$ with values in the sheaf of infinitesimal automorphisms. So first-order deformations are given by $H^1(\mathcal{S}, W)$, or equivalently by $H^1(\mathcal{S}, \mathcal{D}^2)$. (Sheaf cohomology on a SRS $\mathcal{S}$ means cohomology of the same sheaf on the underlying supermanifold S, which in turn was defined to be the cohomology of the corresponding sheaf on the reduced space C.) This gives the tangent space to the moduli space $\mathfrak{M}_g$ of super Riemann surfaces at the point corresponding to a super Riemann surface $\mathcal{S}$:

$$T_{\mathcal{S}}\mathfrak{M}_g = H^1(\mathcal{S}, W) = H^1(\mathcal{S}, \mathcal{D}^2). \tag{3.8}$$

If $\mathcal{S}$ is split (by which we mean that the underlying supermanifold S is split), we can make this more explicit. For $\mathcal{S}$ split with reduced space C, the sheaf W of superconformal vector fields is the direct sum of its even and odd parts W_+ and W_-, where W_+ is the sheaf of sections of T_C and W_- is the sheaf of sections of

$T_C^{1/2}$. This enables us to identify the even and odd tangent spaces $T_\pm\mathfrak{M}_g$. Their fibers at the point in $\mathcal{SM}_g$ corresponding to C are

$$
\begin{aligned}
T_{+,\mathcal{S}}\mathfrak{M}_g &= H^1(C, T_C)\\
T_{-,\mathcal{S}}\mathfrak{M}_g &= H^1(C, T_C^{1/2}).
\end{aligned}
\tag{3.9}
$$

3.2.1. *More on the moduli stack.* A closer examination of the formula (3.9) for the odd normal bundle to $\mathcal{SM}_g$ in $\mathfrak{M}_g$ leads to a better understanding of why the "stacky" nature of the moduli problem is more central for super Riemann surfaces than for ordinary ones.

Suppose that B is an algebraic variety that parametrizes a family of curves with spin structure. We denote this family as $\pi : X \to B$ and denote a fiber of this fibration as C. By definition, since B parametrizes a family of curves with spin structure, each C comes with an isomorphism class of line bundle $\mathcal{L} = T_C^{1/2}$ with an isomorphism $\varphi : \mathcal{L}^2 \cong TC$. We define $\mathcal{T} \to X$ to be the relative tangent bundle, i.e. the tangent bundle along the fibers of $\pi : X \to B$. If there is a line bundle $\mathcal{L} \to X$ with an isomorphism $\varphi : \mathcal{L}^2 \cong \mathcal{T}$, and such that $\mathcal{L}$ restricted to each C is isomorphic to the given $T_C^{1/2}$, then we call $\mathcal{L}$ a relative spin bundle.

However, in general, the existence of a relative spin bundle is obstructed. The essential reason is that locally, after a line bundle $\mathcal{L}$ with an isomophism $\varphi : \mathcal{L}^2 \cong \mathcal{T}$ is chosen, we are still left with the group of automorphisms $\{\pm 1\}$ acting on $\mathcal{L}$ without changing φ. Thus, locally, the pair $\mathcal{L}$, φ is unique up to isomorphism but not up to a unique isomorphism, and this leads to a global obstruction. One may cover B with small open sets $\mathcal{O}_\alpha$ and choose for each of them a line bundle $\mathcal{L}_\alpha \to \pi^{-1}(\mathcal{O}_\alpha)$ with an isomorphism $\varphi_\alpha : \mathcal{L}_\alpha^2 \cong \mathcal{T}$. Since locally the pair $\mathcal{L}_\alpha$, φ_α is unique up to isomorphism, on each $\mathcal{O}_\alpha \cap \mathcal{O}_\beta$, one can pick an isomorphism $\psi_{\alpha\beta} : \mathcal{L}_\alpha \to \mathcal{L}_\beta$ such that $\varphi_\alpha = \varphi_\beta \circ (\psi_{\alpha\beta} \otimes \psi_{\alpha\beta})$. This last condition determines $\psi_{\alpha\beta}$ uniquely up to sign, but in general there is no natural way to fix the sign. In a triple intersection $\mathcal{O}_\alpha \cap \mathcal{O}_\beta \cap \mathcal{O}_\gamma$, set $\lambda_{\alpha\beta\gamma} = \psi_{\gamma\alpha}\psi_{\beta\gamma}\psi_{\alpha\beta}$. In general, $\lambda_{\alpha\beta\gamma} = \pm 1$. If the signs of the local isomorphisms $\psi_{\alpha\beta}$ can be chosen so that $\lambda_{\alpha\beta\gamma} = +1$ for all α, β, γ, then the $\mathcal{L}_\alpha$ can be glued together via the isomorphisms $\psi_{\alpha\beta}$ to make a relative spin bundle $\mathcal{L}$. In general, the $\lambda_{\alpha\beta\gamma}$ are a 2-cocycle representing an element $\varrho \in H^2(B, \mathbb{Z}_2)$. When $\varrho \neq 0$, this obstructs the existence of $\mathcal{L}$; in this situation, we can say that $\mathcal{L}$ exists not as an ordinary line bundle, but as a twisted line bundle, twisted by the $\mathbb{Z}_2$ gerbe $\widehat{\varrho}$ corresponding to ϱ. In general, such an obstruction can arise even if B is simply-connected, so it cannot be removed by replacing B by an unramified cover.

Let us consider in this light the case that B is the spin moduli space $\mathcal{SM}_g$ that parametrizes pairs consisting of a curve C with an isomorphism class of spin structure. A universal spin curve $\pi : X \to \mathcal{SM}_g$ exists if one suitably interprets $\mathcal{SM}_g$ as an orbifold or stack, to account for the possibility that a curve C may have non-trivial automorphisms that preserve its spin structure. If there were a universal relative spin bundle $\mathcal{L}$ in this situation, then we would interpret (3.9) to mean that the odd normal bundle to $\mathcal{SM}_g$ in $\mathfrak{M}_g$ is the vector bundle over $\mathcal{SM}_g$ with fiber $H^1(C, \mathcal{L}|_C)$. However, the existence of such an $\mathcal{L}$ is obstructed[5] for sufficiently large g [**19**, **20**]. (We do not know if the obstruction can be eliminated by endowing C

[5]Relatively simple families realizing the obstruction have been described by J. Ebert and O. Randal-Williams and by B. Hassett, who has also pointed out reference [**21**].

with a suitable level structure or by otherwise taking a finite cover of $\mathcal{SM}_g$.) This obstruction means that the "odd normal bundle" to $\mathcal{SM}_g$ in $\mathfrak{M}_g$ is not a vector bundle, even in the orbifold sense. It is better described as a twisted vector bundle, twisted by a $\mathbb{Z}_2$-valued gerbe.

Thus, to properly understand the moduli stack of super Riemann surfaces, one should think of a curve C with spin structure, even if C has no non-trivial geometrical automorphisms that preserve its spin structure, as having a $\mathbb{Z}_2$ group of automorphisms $\{\pm 1\}$ acting on its spin bundle. When C is such that there is a nontrivial group F of geometrical automorphisms that preserve the spin structure, the automorphism group that is relevant in $\mathfrak{M}_g$ is the double cover $\widehat{F}$ of F that acts on $T_C^{1/2}$; in general, this is a nontrivial double cover of F. (The stacky structure of the Deligne-Mumford compactification of $\mathfrak{M}_g$ is still more subtle because in general there are separate groups $\{\pm 1\}$ acting on the spin bundles of different components of C.)

To fully understand $\mathfrak{M}_g$, one should generalize the theory of supermanifolds to a theory of superstacks and understand $\mathfrak{M}_g$ in this framework. A very special case of this more general theory is as follows. Let $\widehat{\varrho}$ be a $\mathbb{Z}_2$-gerbe over an ordinary manifold (or algebraic variety) M. Then by a $\widehat{\varrho}$-twisted supermanifold S with reduced space M, we mean a $\widehat{\varrho}$-twisted sheaf $\mathcal{O}_S$ of $\mathbb{Z}_2$-graded algebras, such that the even part of $\mathcal{O}_S$ is an ordinary sheaf, the odd part is a $\widehat{\varrho}$-twisted sheaf, and $\mathcal{O}_S$ is locally isomorphic to the sheaf of sections of $\wedge^\bullet V$, where $V \to M$ is a $\widehat{\varrho}$-twisted vector bundle. In the approximation of ignoring geometrical automorphisms (or eliminating them by picking a level structure), we can view $\mathfrak{M}_g$ as a $\widehat{\varrho}$-twisted supermanifold with reduced space $\mathcal{SM}_g$, where $\widehat{\varrho}$ is the gerbe associated to the obstruction ϱ to finding a relative spin bundle. The obstruction class ω_2 to splitting of a $\widehat{\varrho}$-twisted supermanifold can be defined rather as for ordinary supermanifolds, with similar properties.

For the limited purposes of the present paper, we do not really need to go in that direction. Our considerations showing that $\mathfrak{M}_g$ is not projected involve concrete 1-parameter families of spin curves, over which a relative spin bundle will be visible. So the supermanifold framework is adequate for our purposes. We will construct explicit families of super Riemann surfaces, parametrized by a base B that in our examples generally will have dimension $(1|2)$, and show that the class $\omega_2(\mathfrak{M}_g)$ that obstructs projection or splitting of $\mathfrak{M}_g$ is non-zero by showing that it has a nonzero restriction to B – or more precisely a nonzero pullback to B, where this more precise statement accounts for the fact that some of the spin curves parametrized by B have geometrical automorphisms.

3.3. Punctures.

The notion of a puncture or a marked point on an ordinary Riemann surface has two analogs on a super Riemann surface. In string theory, they are known as Neveu-Schwarz (NS) and Ramond punctures, respectively.

An NS puncture in a super Riemann surface $\mathcal{S}$ is the obvious analog of a puncture in an ordinary Riemann surface. It is simply the choice of a point in $\mathcal{S}$, given in local coordinates $x|\theta$ by $x = x_0$, $\theta = \theta_0$, for some x_0, θ_0. As we have learned in section 3.1, such an NS puncture determines and is determined by a minimal divisor on $\mathcal{S}$. Just as in the classical case, deformation theory in the presence of an NS puncture at a point p is obtained by restricting the sheaf W of superconformal vector fields to its subsheaf W_p consisting of superconformal vector fields that leave fixed the point p. By this definition, W_p is a subsheaf of the sheaf of sections of TS, but not a locally free subsheaf. However, rather as we explained in the absence of the puncture, W_p can be given a natural structure of a locally-free sheaf; in fact, it is isomorphic to $\mathcal{D}^2(-F)$, where F is the minimal divisor that corresponds to the point p in the correspondence described in section 3.1. To see this, we use local coordinates $x|\theta$, and take p to be the point $x = \theta = 0$. The condition for a superconformal vector field to vanish at p is then that $f(0) = g(0) = 0$ in eqns. (3.5) and (3.6). This means precisely that $a(x|\theta) = f(x) + 2\theta g(x)$ vanishes at $x = 0$ in the computation described in eqn. (3.7). But the divisor F corresponding to the point p is defined by $x = 0$, as explained in Remark 3.2. So the condition that $a(0|\theta) = 0$ precisely means that $a(x|\theta)\partial_x$ is a section of $\mathcal{D}^2(-F)$.

If $\mathfrak{M}_{g,1}$ is the moduli space of super Riemann surfaces with a single NS puncture, then its reduced space is $\mathcal{SM}_{g,1}$, which parametrizes spin curves of genus g with a single puncture p. The analog of eqn. (3.9) is

$$\begin{aligned} T_{+,S}\mathfrak{M}_{g,1} &= H^1(C, T_C(-p)) \\ T_{-,S}\mathfrak{M}_{g,1} &= H^1(C, T_C^{1/2}(-p)), \end{aligned} \tag{3.10}$$

where the twisting by $\mathcal{O}(-p)$ reflects the conditions $f(0) = g(0) = 0$. Eqn. (3.10) has an obvious generalization for the moduli space $\mathfrak{M}_{g,n}$ of super Riemann surfaces with any number n of NS punctures.

We also are interested in the case of a $(1|1)$ supermanifold $\mathcal{S}$ that is endowed with a superconformal structure that degenerates along a divisor in the following way. We assume that the underlying supermanifold $S = (C, \mathcal{O}_S)$ is still smooth, but the odd distribution $\mathcal{D} \subset T_-S$ is no longer everywhere non-integrable: the local form (3.2) for a generator is replaced by

$$v := \frac{\partial}{\partial\theta} + x^k\theta\frac{\partial}{\partial x}. \tag{3.11}$$

For negative k, such v is meromorphic, and $\mathcal{D}$ fails to be a distribution along the divisor $x = 0$. (One can multiply by x^{-k} to remove the pole, but the resulting vector field $x^{-k}\partial_\theta + \theta\partial_x$ vanishes modulo nilpotents at $x = 0$ and does not define a distribution there; the notion of odd distribution is explained for instance in the proof of Lemma 3.1.) For $k \geq 1$, such a $\mathcal{D}$ is a distribution but the non-integrability fails along the divisor $x = 0$, with multiplicity k. In fact, $v^2 = x^k\partial/\partial x$ vanishes along the divisor $x = 0$ to order k. We say that $\mathcal{S} = (S, \mathcal{D})$ is a *SRS with a parabolic structure* of order k at the divisor $x = 0$ (and we use this definition also for negative k). The basic case $k = 1$ is called a Ramond puncture by string theorists. The local form is:

$$v := \frac{\partial}{\partial\theta} + x\theta\frac{\partial}{\partial x}. \tag{3.12}$$

It can be shown that in the presence of a Ramond puncture, the sheaf of superconformal vector fields is still isomorphic to $\mathcal{D}^2$, but $\mathcal{D}^2$ is no longer isomorphic to $TS/\mathcal{D}$; rather, $\mathcal{D}^2 \cong TS/\mathcal{D} \otimes \mathcal{O}(-F)$, where now F is the divisor on which the superconformal structure degenerates (thus, in the example (3.12), F is the divisor $x = 0$).

For our purposes, the importance of parabolic structures is that they arise naturally in branched coverings, as we will see in section 3.4.1. In string theory, and also in the context of the Deligne-Mumford compactification of supermoduli space, there is a close analogy between NS and Ramond punctures, and it is natural to define moduli spaces – or rather stacks – that parametrize super Riemann surfaces with a specified number of punctures of each type. However, in this paper, we limit ourselves to the moduli spaces $\mathfrak{M}_{g,n}$ of super Riemann surfaces of genus g with n NS punctures.

The moduli space $\mathfrak{M}_{g,1}$ can be interpreted as the total space of the universal genus g super Riemann surface parametrized by $\mathfrak{M}_g = \mathfrak{M}_{g,0}$. It has dimension $(3g-2|2g-1)$ and comes with a morphism to $\mathfrak{M}_g$ whose fibers are the $(1|1)$ supermanifolds S underlying the genus g super Riemann surfaces parametrized by $\mathfrak{M}_g$. In the stacky sense, it is not necessary to correct this statement for small g to take into account the generic automorphisms of a Riemann surface. In our applications, we will always be pulling back the relevant structures on moduli stacks to concrete families of Riemann surfaces or super Riemann surfaces and again we need not worry about the automorphisms.

3.4. Effects of geometric operations.

3.4.1. *Effect of a branched covering.* Let $\pi : \widetilde{C} \to C$ be a branched covering of ordinary Riemann surfaces. There are some ramification points $\widetilde{p}_j \in \widetilde{C}$ whose images in C are the branch points $p_j = \pi(\widetilde{p}_j)$, where the covering map π has local degree $k_j \geq 2$. Let $\widetilde{S} \to S$ be a branched covering of $(1|1)$-dimensional complex supermanifolds whose reduction is π, as in section 2.1.3. Corresponding to each $\widetilde{p}_j$ there is now a ramification divisor $R_j \subset \widetilde{S}$ sitting over a branch divisor $B_j \subset S$. Let $\mathcal{S} = (S, \mathcal{D})$ be a SRS with underlying supermanifold S. The structure induced on $\widetilde{S}$ is not that of a SRS, but rather a SRS with parabolic structure of (negative) order $1 - k_j$ along the R_j. Indeed, if v is given near $p_j \in C$ by (3.2), and the local coordinate w near $\widetilde{p}_j \in \widetilde{C}$ satisfies $w^{k_j} = x$, then the induced (meromorphic) vector field upstairs is $\widetilde{v} = \partial/\partial\theta + \theta\partial/\partial x = \partial/\partial\theta + \frac{1}{k_j}w^{1-k_j}\theta\partial/\partial w$.

The above remains true, of course, when $k_j = 1$, except that p_j is no longer a branch point and $\widetilde{v}$ no longer has a pole. A little more generally, if $\mathcal{S} = (S, \mathcal{D})$ is a SRS with underlying supermanifold S and parabolic structure of order m_j at minimal divisor B_j, and $\pi : \widetilde{C} \to C$ is a branched covering of ordinary Riemann surfaces with local degree $k_j \geq 1$ at $\widetilde{p}_j$, then a branched covering $\widetilde{S} \to S$ inherits the structure of a SRS with parabolic structure at each R_j of order $k_j(m_j - 1) + 1$. This follows immediately from equation (3.11) and the local form of the branched cover, as given in section 2.1.3.

3.4.2. *Effect of a blowup.* Let S be a $(1|1)$-dimensional complex supermanifold, with local coordinates $x|\theta$ defining the point, or codimension $(1|1)$ submanifold $p : \{x = \theta = 0\}$. In section 2.1.4 we described the blowup $\widehat{S}$ of S at p: it is again a $(1|1)$-dimensional complex supermanifold, with the same reduced manifold as S, but with a new (and "larger") structure sheaf. It has local coordinates $\widehat{x}|\widehat{\theta}$ such that the map $\widehat{S} \to S$ sets $\widehat{x} = x, \widehat{\theta} = \theta/x$, replacing the point p by the minimal divisor $\widehat{x} = 0$.

Now let $\mathcal{S} = (S, \mathcal{D})$ be a SRS with local superconformal coordinates $x|\theta$, and thus with the distribution $\mathcal{D}$ generated by $v = \partial_\theta + \theta\partial_x$. As in section 3.4.1, let $\pi : \widetilde{\mathcal{S}} \to \mathcal{S}$ be a branched cover of $\mathcal{S}$, with k-fold ramification along the divisor $x = 0$. Concretely, $\widetilde{S}$ is described by local coordinates $y|\theta$, where $y^k = x$, and the distribution is generated by $v = \partial_\theta + (\theta/ky^{k-1})\partial_y$, with parabolic structure of degree

$-(k-1)$. In $\mathcal{S}$, the divisor $x=0$ has the distinguished point $x=\theta=0$ (see Remark 3.2), and this can be pulled back to the point $y=\theta=0$ in $\widetilde{S}$. Then we can blow up this point, to get a new complex supermanifold $\widehat{S}$ with local coordinates $\widehat{y}=y$ and $\widehat{\theta}=\theta/y$. The generator of the distribution becomes $v=(1/y)\partial_{\widehat{\theta}}+(1/ky^{k-2})\widehat{\theta}\partial_x$. Away from $y=0$, the same distribution is generated by $\widehat{v}=yv=\partial_{\widehat{\theta}}+(1/ky^{k-3})\widehat{\theta}\partial_y$.

For $k=3$, which actually will be the basic case in our later applications, this simple blowup has eliminated the parabolic structure, and $\widehat{S}$ has an ordinary super Riemann surface structure near $y=0$. More generally, for any k, the effect of a blowup has been to increase the order of a parabolic structure by 2.

Actually, we can make a blowup to increase the order of a parabolic structure by any positive even number. This is not achieved by a series of blowups as above; the problem is that after the first blowup, there is no distinguished point to blow up inside the exceptional divisor. Instead, we need to blow up a more complicated multiple point, specified by a certain sheaf of ideals. We describe this ideal as follows. Downstairs, the distinguished point $x=\theta=0$ on the divisor $x=0$ determines the ideal I generated by x and θ. The ideal $\pi^*(I)$ is generated by y^k and θ. On the other hand, upstairs, we have the ideal J generated by y and hence also its ℓth power J^ℓ, generated by y^ℓ. We let $I_\ell=(J^\ell,\pi^*I)$ be the ideal generated by J^ℓ and π^*I. For $k\geq\ell$ it is generated, in local coordinates, by y^ℓ and θ. Blowing up along this ideal has the effect of replacing θ by $\theta':=\theta/y^\ell$ and thus increasing the order of the parabolic structure by 2ℓ, i.e. from $1-k$ to $1-k+2\ell$. If k is odd, we can thus produce a SRS with no parabolic structure. If k is even, we can reduce to the case of parabolic structure of degree 1. These are the two cases (no parabolic structure and parabolic structure of degree 1, corresponding to a Ramond puncture) that are usually considered in string theory.

In short, we can eliminate the parabolic structure caused by a branched covering of local degree k if and only if k is odd, in which case we need to blowup the sheaf I_ℓ, with $\ell=(k-1)/2$. Recalling the result in section 2.1.3, we conclude[6] :

PROPOSITION 3.3. *To a family $f:\mathcal{S}\to\mathfrak{M}$ of super Riemann surfaces parametrized by a super manifold $\mathfrak{M}$, with underlying super manifold $S\to\mathfrak{M}$, together with a branched covering map $\pi_{\mathrm{red}}:\widetilde{C}\to C$ of the reduced space $C:=S_{\mathrm{red}}$ with all its local degrees* odd, *and a divisor $D\subset S$ whose reduced manifold is the branch locus B of π_{red}, there is naturally associated a family of super Riemann surfaces $\widetilde{f}:\widetilde{\mathcal{S}}\to\mathfrak{M}$ which factors: $\widetilde{f}=f\circ\pi$ through a branched covering map $\pi:\widetilde{\mathcal{S}}\to\mathcal{S}$ whose reduced version is the given π_{red}.*

Although we do not need this in the sequel, we note that the above has an interesting converse: any minimal divisor $\widetilde{x}=0$ on a $(1|1)$-dimensional complex supermanifold $\widetilde{S}$, with local coordinates $\widetilde{x}|\widetilde{\theta}$, can be blown down to a point $p:\{x=\theta=0\}$ on a $(1|1)$-dimensional complex supermanifold S with the same reduced space, the same structure sheaf away from p, and coordinates $x=\widetilde{x},\quad \theta=x\widetilde{\theta}$ near p. This process is natural, and can be described independently of the choice of coordinates. Given a $(1|1)$ supermanifold S with minimal divisor D, we blow down D by allowing only functions that are constant along D. If more generally $S\to B$ is a family of $(1|1)$ supermanifolds with minimal divisor D, we only allow functions on S whose restriction to D is a pullback from B. In bosonic algebraic

[6] A related assertion has been made on p. 61 of [**22**].

geometry, one could make the same definition by allowing only functions that are constant along a given divisor, but in general this would not give the sheaf of functions on an algebraic variety. For a $(1|1)$ supermanifold, the computation in the local coordinates $x|\theta$ does show that the blowdown works. So for example, let $\widetilde{\mathcal{S}} = (\widetilde{S}, \widetilde{\mathcal{D}})$ be a SRS that has along the minimal divisor $\widetilde{x} = 0$ a parabolic structure of multiplicity $m+2$. As we have just seen, this divisor can be blown down, reducing the multiplicity of the parabolic structure to m. The blowdown process can be repeated, further reducing the multiplicity.

Everything that we have said in this section works naturally in families. Start with a family of SRS parametrized by some base supermanifold B with parabolic structure along a relative minimal divisor (i.e. a divisor intersecting each fiber in a minimal divisor). We can blow the divisor down, and thereby reduce the order of the family of parabolic structures by 2. Or we can blow up the relative point corresponding to the given divisor, and thereby increase the order of the family of parabolic structures by 2, or blow up a more subtle sheaf of ideals, increasing it by 2ℓ for any integer ℓ.

3.5. A non-split supermanifold.

In this section we exhibit a particular non-split supermanifold X_η. It has dimension $(1|2)$, and is fibered over the odd line $\mathbb{C}^{0|1}$. The fibers are super Riemann surfaces. In fact, we interpret $\mathbb{C}^{0|1}$ as an odd tangent line to $\mathfrak{M}_g$, and build our X_η by restricting the universal super Riemann surface $\mathfrak{M}_{g,1}$ to this line. This example will serve as a crucial ingredient in our proof of non-projectedness of $\mathfrak{M}_{g,1}$ in section 4.

Let $\mathcal{S} = (S, \mathcal{D})$ be a split SRS of genus g, $C := S_{\mathrm{red}}$ the underlying Riemann surface. Let $\eta \in H^1(C, T_C^{1/2})$ be an odd tangent vector. It determines a map $f_\eta : \mathbb{C}^{0|1} \to \mathfrak{M}_g$. By pulling back the universal super Riemann surface $\mathfrak{M}_{g.1} \to \mathfrak{M}_g$, we obtain a $(1|2)$-dimensional supermanifold $X_\eta := f_\eta^*(\mathfrak{M}_{g,1})$. By definition, X_η comes with a projection $\pi_\eta : X_\eta \to \mathbb{C}^{0|1}$.

PROPOSITION 3.4. *X_η is projected if and only if $\eta = 0$, in which case it is actually split.*

PROOF. First we remark that when $\eta = 0$, the map f_η is constant, so X_η is the product $S \times \mathbb{C}^{0|1}$ so in particular it is split. In general, we are in the situation of Corollary 2.3, so X_η is determined by its first obstruction:

$$\omega := \omega(X_\eta) \in H^1((X_\eta)_{\mathrm{red}}, \mathrm{Hom}(\wedge^2 T_- X_\eta, T_+ X_\eta)).$$

We can identify:

$$(X_\eta)_{\mathrm{red}} = C$$

$$T_+ X_\eta = TC$$

$$\wedge^2 T_- X_\eta = T_- S \otimes T_- \mathbb{C}^{0|1} = T_C^{1/2} \otimes \mathcal{O} = T_C^{1/2}.$$

So ω lives in

$$H^1((X_\eta)_{\mathrm{red}}, \mathrm{Hom}(\wedge^2 T_- X_\eta, T_+ X_\eta)) = H^1(C, T_C^{1/2}).$$

Our proposition follows from:

LEMMA 3.5. *Under the natural identifications, $\omega(X_\eta) = \eta$.*

We will give a very pedestrian explanation. Since X_η has odd dimension 2, the class in $H^1(C,G)$ associated to X_η is a 1-cocycle on the split model $S(M,V) = S \times \mathbb{C}^{0|1}$ valued in vector fields of the form

$$w(x)\eta\theta\frac{\partial}{\partial x}, \tag{3.13}$$

where x is a local coordinate on $S_{\text{red}} = C$ and θ is a local odd coordinate on S that vanishes along C. We can view this as a 1-cocycle that deforms $S\times\mathbb{C}^{0|1}$ away from being split. $S \times \mathbb{C}^{0|1}$ has other first-order deformations, but they do not affect its splitness.

On the other hand, to deform the super Riemann surface $\mathcal{S}$ in an η-dependent fashion, leaving it fixed at $\eta = 0$, we use a one-cocycle valued in odd superconformal vector fields on $\mathcal{S}$, multiplied by η. Given the form (3.6) of an odd superconformal vector field, the one-cocycle is valued in vector fields of the form

$$-g(x)\eta\left(\frac{\partial}{\partial\theta} - \theta\frac{\partial}{\partial x}\right). \tag{3.14}$$

If we forget the superconformal structure and simply view this as a one-cocycle that we use to deform the complex structure of $S \times \mathbb{C}^{0|1}$, it is a sum of two terms, namely $-g(x)\eta\partial_\theta$ and $g(x)\eta\theta\partial_x$, that can be considered separately. The first term does not affect the splitness of $S\times\mathbb{C}^{0|1}$, but the second does. Indeed if we set $w = g$, the second term coincides with the cocycle (3.13) that characterizes X_η. The value that the cocycle has in one interpretation is the same as the value that it has in the other interpretation, since either way the vector field $w(x)\eta\theta\partial_x$ or $g(x)\eta\theta\partial_x$ can be naturally identified with a section over C of $T_C^{1/2}$ and concretely the cocycles under discussion represent elements of $H^1(C, T_C^{1/2})$.

$\square$

4. Non-projectedness of $\mathfrak{M}_{g,1}$

We are now ready to prove Theorem 1.2, which says that the even spin component of the moduli space $\mathfrak{M}_{g,1}$ of super Riemann surfaces with one NS puncture is non-projected. The result follows from the more precise:

PROPOSITION 4.1. *The first obstruction to the splitting of* $\mathfrak{M}_{g,1}$*:*

$$\omega := \omega_2 \in H^1(\mathcal{SM}_{g,1}, Hom(\wedge^2 T_-, T_+))$$

does not vanish for $g \geq 2$ *(and even spin structure), so the supermanifold* $\mathfrak{M}_{g,1}$ *is non-projected.*

Here and in the rest of this section, $T_\pm$ refer to $T_\pm\mathfrak{M}_{g,1}$. Our proof here is based on some of the general results we obtained about supermanifolds and their obstructions. In the sequel to this paper [**23**] we give a different proof which relies on a cohomological interpretation of the obstruction class.

PROOF. Fix a spin curve $(C, T_C^{1/2}) \in \mathcal{SM}_g^+$ and an odd tangent vector $\eta \in H^1(C, T_C^{1/2})$. We have already seen in section 3.5 an example of a family X_η of super Riemann surfaces with a non-projected total space. We identify this X_η as a submanifold of $\mathfrak{M}_{g,1}$. In terms of the projection $\pi : \mathfrak{M}_{g,1} \to \mathfrak{M}_g$, it is $\pi^{-1}(\mathbb{C}^{(0|1)})$, where $\mathbb{C}^{(0|1)}$ is embedded in $\mathfrak{M}_g$ via the odd tangent vector η. We wish to apply

Corollary 2.11, with $S = \mathfrak{M}_{g,1}$, $S' = X_\eta$, $M = \mathcal{SM}_{g,1}$, $M' = C$. We note that T_-X_η is a rank 2 vector bundle on C, in fact it is an extension:

$$0 \to T_C^{1/2} \to T_-X_\eta \to \mathbb{C}\eta \to 0,$$

where the third term stands for the trivial bundle with fiber $\mathbb{C}\eta$. The choice of η therefore identifies $\wedge^2 T_-X_\eta$ with $T_C^{1/2}$. The maps ι, j appearing in Corollary 2.11 can therefore be written explicitly in our case:

$\iota : H^1(C, \mathrm{Hom}(\wedge^2 T_-, T_+)) \to H^1(C, \mathrm{Hom}(T_C^{1/2}, T_+))$

$j : H^1(C, \mathrm{Hom}(T_C^{1/2}, T_C)) \to H^1(C, \mathrm{Hom}(T_C^{1/2}, T_+))$.

According to Corollary 2.11, $\iota(\omega_{|C}) = j(\omega(X_\eta))$. We already know by Lemma 3.5 that $\omega(X_\eta) = \eta \neq 0$, so in order to show that $\omega \neq 0$, it suffices to show that j is injective. Unfortunately, Lemma 2.12 does not apply. Instead, we note that j fits into an exact sequence. We start with the short exact sequence of sheaves on C:

$$0 \to T_C \to i^*T_+ \to i^*\pi^*T_+\mathfrak{M}_g \to 0 \tag{4.1}$$

induced on (even) tangent spaces by the fibration $\pi : \mathfrak{M}_{g,1} \to \mathfrak{M}_g$ and the inclusion $i : C \to \mathfrak{M}_{g,1}$. Note that the third term is isomorphic to the trivial sheaf $W \otimes \mathcal{O}_C$ with W the tangent space $T_{+,C}\mathfrak{M}_g$ to $\mathfrak{M}_g$ at the point $[C]$. We apply the functor $\mathrm{Hom}(T_C^{1/2}, \cdot)$ to this sequence; the cohomology sequence of the resulting exact sequence reads in part

$$W \otimes H^0(C, K_C^{1/2}) \to H^1(C, \mathrm{Hom}(T_C^{1/2}, T_C)) \xrightarrow{j} H^1(C, \mathrm{Hom}(T_C^{1/2}, T_+)).$$

For generic choice of the even spin curve $(C, T_C^{1/2})$, we have $H^0(C, K_C^{1/2}) = 0$, so j is injective, completing the proof. □

REMARK 4.2. Our proof fails for the odd component; in fact the map j above is identically zero in this case, because $H^0(C, K_C^{1/2})$ is generically 1-dimensional rather than 0. In more detail, the vanishing of j follows from surjectivity of the above map

$$W \otimes H^0(C, K_C^{1/2}) \to H^1(C, \mathrm{Hom}(T_C^{1/2}, T_C)),$$

which can be written explicitly as

$$H^1(C, T_C) \otimes H^0(C, K_C^{1/2}) \to H^1(C, T_C^{1/2}),$$

whose surjectivity is equivalent (in the generic case when $H^0(C, K_C^{1/2})$ is 1-dimensional) to injectivity of the trasposed map

$$H^0(C, K_C^{1/2}) \otimes H^0(C, K_C^{3/2}) \to H^0(C, K_C^2),$$

but the latter map is just multiplication with a fixed non-zero section of $H^0(C, K_C^{1/2})$, which is indeed injective.

5. Compact families of curves and non-projectedness of $\mathfrak{M}_g$

To show that $\mathfrak{M}_g$ is non-projected, it suffices to show that its first obstruction $\omega := \omega_2(\mathfrak{M}_g) \in H^1(\mathcal{SM}_g, \mathrm{Hom}(\wedge^2 T_-, T_+))$ does not vanish. Here $\mathcal{SM}_g = (\mathfrak{M}_g)_{\mathrm{red}}$ is the moduli space of ordinary Riemann surfaces with a spin structure. $T_\pm$ stand respectively for the even and odd tangent bundles of $\mathfrak{M}_g$. These are vector bundles on $(\mathfrak{M}_g)_{\mathrm{red}}$; their fibers at $C \in \mathcal{SM}_g$ are $H^1(C, T_C)$ and $H^1(C, T_C^{1/2})$, respectively.

The basic idea of the proof is to construct a compact curve $Y \subset \mathcal{SM}_g$, or more precisely a family of spin curves parametrized by a compact curve Y, and to show that the pullback of the class ω to Y is nonzero. The families of genus g curves that we study are constructed in a standard fashion to parametrize a family of ramified covers of a fixed curve of genus less than g. The construction is such that we can prove that both components of $\mathfrak{M}_g$ are non-projected.

5.1. Examples of compact families of curves.

One general way to produce compact curves in $\mathcal{M}_g$ depends on the existence of a small compactification. We briefly review this approach.

The Satake compactification $\overline{\mathcal{A}_g}$ of the moduli space $\mathcal{A}_g$ of abelian varieties is a projective variety whose boundary is $\overline{\mathcal{A}_{g-1}}$, hence of codimension g. The closure $\overline{\mathcal{M}_g}$ of the Abel-Jacobi image of $\mathcal{M}_g$ does not meet this boundary transversally: it meets the boundary in the compactification $\overline{\mathcal{M}_{g-1}}$ of $\mathcal{M}_{g-1}$, which for $g \geq 3$ has codimension 3 in $\overline{\mathcal{M}_g}$. (Contrast this with the Deligne-Mumford compactification, whose boundary has codimension 1.) The difference $\overline{\mathcal{M}_g} \setminus \mathcal{M}_g$ consists of this boundary plus a locus in the interior of $\mathcal{A}_g$, namely the closure of the locus of reducible curves consisting of two components meeting at a point. The genera $g_1, g_2 > 0$ of these components add up to g. Most components of this locus also have codimension 3. But for every genus, there is one component whose codimension is 2: this happens when one of the g_i equals 1. (When both $g_1 = g_2 = 1$, the codimension is just 1; but this happens only for $g = 2$.)

We can embed the projective variety $\overline{\mathcal{A}_g}$, and hence also its subvariety $\overline{\mathcal{M}_g}$, in a large projective space $\mathbb{P}^N$. Consider the 1-dimensional intersection of $\overline{\mathcal{M}_g}$ with a generic linear subspace in $\mathbb{P}^N$ of the appropriate codimension, which is $3g-4$. The above dimension count showed that when $g \geq 3$, the complement $\overline{\mathcal{M}_g} \setminus \mathcal{M}_g$ has codimension at least 2 in $\overline{\mathcal{M}_g}$. It follows that our generic 1-dimensional intersection is contained in the interior $\mathcal{M}_g$. This provides a large but non-explicit family of compact curves in $\mathcal{M}_g$ for $g \geq 3$. (On the other hand, when $g \leq 2$, it is known that $\mathcal{M}_g$ is an affine variety; hence it can contain no compact curves.)

In addition, several explicit constructions are known. Kodaira, Atiyah and Hirzebruch constructed examples [**25**–**27**] of surfaces X with smooth maps $\pi : X \to B$ to a smooth curve B of genus g'. All the fibers of such a map are smooth curves C of some genus g. In fact, their surfaces X are certain branched covers of the product of two curves, so they can be fibered as above in two distinct ways. In their smallest example, the base genus is 129 and the fiber genus is 6, for one fibration; and the base genus is 3 and the fiber genus is 321, for the other fibration. More efficient constructions are known, producing examples of lower genera. The construction in [**28**] gives base genus 9 and fiber genus 4, for one projection, and base genus 2 and fiber genus 25, for the other.

A fibered surface X as above determines a map from B to $\mathcal{M}_g$. When this map is non-constant, the signature of X is non-zero. Conversely, if a universal curve over $\mathcal{C} \to \mathcal{M}_g$ existed, every map from B to $\mathcal{M}_g$ would determine such a fibered surface X. Actually this fails, since a universal curve over $\mathcal{M}_g$ does not exist near curves with extra automorphsms. For example, it is easy to see that the base B of a fibered X must have genus $g' \geq 2$, since the universal cover of B must map to the Siegel half space by the period map. However, we will see in the Appendix an example of a genus 0 curve in $\mathcal{M}_5$. There is no family of genus $g = 5$ curves fibered over this $\mathbb{P}^1$, but there is such a family over a certain cover B of $\mathbb{P}^1$, with genus $g' = 19$. Earlier examples of compact genus 0 curves in moduli spaces appeared in [**29**].

5.2. Covers with triple ramification. The strategy behind the explicit constructions mentioned at the end of the previous section is to start with a curve C of some lower genus g_0 and a branched cover $\pi : \widetilde{C} \to C$ having branch divisor $B \subset C$. Then keep the curve C fixed, and allow B to move in a 1-parameter family P. The crucial condition is that the points of B should never collide, i.e. the cardinality of $B = B_p$ must remain constant as p varies in P. This means that the topology of the pair (C, B) remains constant as B varies, and therefore as p varies locally in P, the cover $\pi : \widetilde{C_p} \to C$ deforms along with B. Globally there may be monodromy: there are usually many branched covers with a specified branch divisor B. So this construction produces a family of covers $\pi : \widetilde{C_p} \to C$ parametrized by points p of some cover $\widetilde{P}$ of P.

One way to guarantee compactness of our parameter space $\widetilde{P}$, or to enforce the condition that points of B should never collide, is to start with a C which admits a free action of a finite group G, and to take the branch divisors B_p to be the orbits, parametrized by the quotient $P := C/G$. This is the basic idea behind the constructions in [**25–28**], where the simplest examples use double covers. For our purpose in this work, we will avoid collisions by taking B to consist of a single point: $B_p = \{p\}$.

It is well known that a double cover cannot have just one branch point. But this *is* possible for degree ≥ 3. One way to see the impossibility in degree 2 is to note that this would produce a branched cover $\widetilde{C}$ of odd Euler characteristic. The same argument excludes a single, total branch point in all even degrees, but allows a single, total branch point in any odd degree; in any degree d, odd or even, it does not exclude a single branch point of various non-total types, e.g. the $(3, 1, \ldots, 1) = (3, 1^{d-3})$ pattern works for all $d \geq 3$, as we will see shortly. We settle then on the following version of the construction:

Fix a curve C of genus g_0 and an integer $d \geq 3$. Consider the family $\widetilde{P}$ of all branched covers $\pi : \widetilde{C} \to C$ of degree d having a *single* ramification point $\widetilde{p} \in \widetilde{C}$, of local ramification degree 3, over a branch point $p \in C$. The fiber $\pi^{-1}(p)$ consists of 3 times $\widetilde{p}$ plus $d-3$ other points $\widetilde{p}_i, \quad i = 1, \ldots, d-3$.

We can easily see that such branched covers indeed exist. Consider the fundamental group of the complement, $\pi_1(C \setminus p)$. This is generated by a standard symplectic basis $\alpha_i, \beta_i, \quad i = 1, \ldots, g_0$, plus a loop l around the point p. The only relation is $lr = 1$, where $r := \prod_{i=1}^{g_0} [\alpha_i, \beta_i]$. There is thus a short exact sequence

$$0 \to K \to \pi_1(C \setminus p) \to \pi_1(C) \to 0 \tag{5.1}$$

where the kernel K is generated by l (or equivalently, r). Now a d-sheeted cover of C which is not branched except possibly at p is specified by a subgroup $S \subset \pi_1(C \setminus p)$ of index d, and this cover is unbranched at p if and only if $K \subset S$. Every subgroup of index 2 is normal, so it is the kernel of a homomorphism $\pi_1(C \setminus p) \to \mathbb{Z}/2$ and must contain all commutators, hence must contain K. But for $d \geq 3$, subgroups of index d which do not contain K do exist. For example, when $d = 3$, we can map $\pi_1(C \setminus p)$ to the symmetric group S_3 by sending $\alpha_1 \to (23), \quad \beta_1 \to (123), \quad$ and $\alpha_i, \beta_i \to 1$ for $i \neq 1$, so $r \to (123)$. We then take S to be the inverse image of the non-normal subgroup $S_2 \subset S_3$ stabilizing one of the three permuted objects. This S has index 3 in $\pi_1(C \setminus p)$ but does not contain the element $r \in K$. More generally, for any $d \geq 3$, we can map $\pi_1(C \setminus p)$ onto the symmetric group S_d by sending $\alpha_1 \to (23)$, $\beta_1 \to (12\ldots d)$, and $\alpha_i, \beta_i \to 1$ for $i \neq 1$, so again $r \to (123)$. We then take S to be the inverse image of the non-normal subgroup $S_{d-1} \subset S_d$ fixing 1 and permuting $2, \ldots, d$. This S has index d in $\pi_1(C \setminus p)$, but the loop $l = r^{-1}$ goes to a 3-cycle, so the resulting cover $\widetilde{C} \to C$ has the desired $(3, 1^{d-3})$ pattern.

The genus of any such $\widetilde{C}$ is easily seen to be $g = d(g_0 - 1) + 2$. We may as well take $g_0 = 2$ so $g = d + 2$, and thus by varying d, we get all values $g \geq 5$. In the appendix we will give a detailed, algebro-geometric description of the $g_0 = 2, d = 3, g = 5$ case, in which the branching is total. This example is attributed in [**30**] (a few paragraphs above theorem 2.34) to Kodaira.

We saw that for each $p \in C$, we get a finite number of covers $\widetilde{C} \to C$. The parameter space $\widetilde{P}$ of our branched covers of C is therefore itself a certain cover of C. If we now allow the curve C to vary through the moduli space $\mathcal{M}_{g_0}$, we get a family of covers $\pi : \widetilde{C} \to C$ parametrized by a certain cover $\widetilde{\mathcal{M}}_{g_0,1} := \widetilde{\mathcal{M}}^d_{g_0,1}$ of the universal curve $\mathcal{M}_{g_0,1}$, and a morphism $\widetilde{\mathcal{M}}_{g_0,1} \to \mathcal{M}_g$ sending the isomorphism class of a cover $\pi : \widetilde{C} \to C$ to the isomorphism class of $\widetilde{C}$.

5.3. Maps from $\widetilde{\mathcal{SM}}_{g_0,1}$ to $\mathcal{SM}_g$. The construction in the previous section 5.2 gives a map of moduli spaces, from $\widetilde{\mathcal{M}}_{g_0,1}$ to $\mathcal{M}_g$. In the present section we extend this to a map of the spin moduli spaces, from $\widetilde{\mathcal{SM}}_{g_0,1}$ to $\mathcal{SM}_g$. In the next section we discuss the further extension to a map (in the sense of stacks) of supermoduli spaces, from $\widetilde{\mathfrak{M}}_{g_0,1}$ to $\mathfrak{M}_g$.

For fixed $d \geq 3$, let $\widetilde{\mathcal{M}}_{g_0,1} := \widetilde{\mathcal{M}}^d_{g_0,1}$ denote the moduli space of all branched covers $\pi : \widetilde{C} \to C$ as in the previous section: of degree d over a curve C of genus g_0, having a single ramification point $\widetilde{p} \in \widetilde{C}$, of local ramification degree 3, over a branch point $p \in C$. (Recall from Proposition 3.3 that in order to lift a family of branched coverings of Riemann surfaces to branched coverings of super Riemann surfaces we need all the local ramification degrees to be odd. This is satisfied in our situation, where these local degrees equal 3 or 1.)

There are several forgetful morphisms:

- $P_1 : \widetilde{\mathcal{M}}_{g_0,1} \to \mathcal{M}_g$ sends $(\pi : \widetilde{C} \to C)$ to $\widetilde{C}$.
- $P_2 : \widetilde{\mathcal{M}}_{g_0,1} \to \mathcal{M}_{g_0,1}$, sending $(\pi : \widetilde{C} \to C) \mapsto (C, p)$, is a finite covering.
- $P_3 : \widetilde{\mathcal{M}}_{g_0,1} \to \mathcal{M}_{g_0}$ sends $(\pi : \widetilde{C} \to C)$ to C.

Our goal in this section and the next is to construct super versions of the space $\widetilde{\mathcal{M}}_{g_0,1}$ and the morphism P_1. We do this in several steps, designed to match the input of Proposition 3.3:

(1) Start with a typical branched covering map $\pi : \widetilde{C} \to C$ as in section 5.2.
(2) Put it into a universal family $\Pi : \widetilde{U} \to U$, parametrized by $\widetilde{\mathcal{M}}_{g_0,1}$.
(3) Add spin: $S\Pi : S\widetilde{U} \to SU$ over $S\widetilde{\mathcal{M}}_{g_0,1}$.
(4) Construct the super space $\widetilde{\mathfrak{M}}_{g_0,1}$ and a universal genus g_0 SRS $\mathcal{U}$ over it, $\mathcal{F} : \mathcal{U} \to \widetilde{\mathfrak{M}}_{g_0,1}$, whose reduced spaces are $S\widetilde{\mathcal{M}}_{g_0,1}$ and SU, respectively.
(5) Construct the divisor $\mathcal{B} \subset \mathcal{U}$ whose reduced version is the branch divisor SB of $S\Pi$.

(6) By Proposition 3.3 we then get a family of super Riemann surfaces $\widetilde{\mathcal{F}} : \widetilde{\mathcal{U}} \to \widetilde{\mathfrak{M}}_{g_0,1}$ of genus g, along with a branched covering map $\varPi : \widetilde{\mathcal{U}} \to \mathcal{U}$ whose reduced version is $S\Pi : S\widetilde{U} \to SU$ and which satisfies $\widetilde{\mathcal{F}} = \mathcal{F} \circ \varPi$.

(7) By the universal property of moduli spaces, this gives a morphism of supermanifolds, from $\widetilde{\mathfrak{M}}_{g_0,1}$ to $\mathfrak{M}_g$, whose reduced version is the spin lift of the above forgetful map $P_1 : \widetilde{\mathcal{M}}_{g_0,1} \to \mathcal{M}_g$.

In fact, we need quite a bit less than this for the proof of our main results: we will use the construction only in the case $g_0 = 2$, and only in the vicinity of one fiber of $\widetilde{\mathfrak{M}}_{g_0,1}$ to $\mathfrak{M}_{g_0}$. In fact, for $g_0 = 2$ one can give an elementary construction of a relative spin line bundle $L = K^{1/2}$ on the universal curve over a particular double cover of $\mathcal{SM}_2^+$, the one that parametrizes triples of Weierstrass points on the (hyperelliptic) genus 2 curve. So in this case one could prove a stronger result, about a map between moduli spaces (rather than stacks). Since we will need this only in the vicinity of a single curve, we will not work out the details of this improvement.

In the remainder of this section we will fill in the details of steps (2) and (3), leading to the map of spin moduli spaces. The subsequent steps are treated in section 5.4.

(2) Over $\widetilde{\mathcal{M}}_{g_0,1}$ we construct two universal spaces $a : U \to \widetilde{\mathcal{M}}_{g_0,1}$ and $\widetilde{a} : \widetilde{U} \to \widetilde{\mathcal{M}}_{g_0,1}$, with an intertwining map $\Pi : \widetilde{U} \to U$ satisfying $a \circ \Pi = \widetilde{a}$. Here U is just the universal genus g_0 curve over $\widetilde{\mathcal{M}}_{g_0,1}$, meaning that it is the pullback of the universal curve $\mathcal{M}_{g_0,1} \to \mathcal{M}_{g_0}$ via the above forgetful map $P_3 : \widetilde{\mathcal{M}}_{g_0,1} \to \mathcal{M}_{g_0}$:

$$U = \widetilde{\mathcal{M}}_{g_0,1} \times_{\mathcal{M}_{g_0}} \mathcal{M}_{g_0,1}.$$

Similarly, $\widetilde{U}$ is the pullback of the universal genus g curve $\mathcal{M}_{g,1} \to \mathcal{M}_g$ via the forgetful map $P_1 : \widetilde{\mathcal{M}}_{g_0,1} \to \mathcal{M}_g$:

$$\widetilde{U} = \widetilde{\mathcal{M}}_{g_0,1} \times_{\mathcal{M}_g} \mathcal{M}_{g,1}.$$

Note that Π is a branched covering map. Its restriction to the fibers of $U, \widetilde{U}$ above a point of $\widetilde{\mathcal{M}}_{g_0,1}$ is just the map $\pi : \widetilde{C} \to C$ classified by that point. The branch divisor $B \subset U$ of Π is the pullback via

$$P_2 \times 1 : U = \widetilde{\mathcal{M}}_{g_0,1} \times_{\mathcal{M}_{g_0}} \mathcal{M}_{g_0,1} \to \mathcal{M}_{g_0,1} \times_{\mathcal{M}_{g_0}} \mathcal{M}_{g_0,1} =: \overline{U}$$

of the diagonal

$$\Delta \cong \mathcal{M}_{g_0,1} \subset \mathcal{M}_{g_0,1} \times_{\mathcal{M}_{g_0}} \mathcal{M}_{g_0,1} = \overline{U}.$$

(3) Next, we add a spin structure. The space $\mathcal{SM}_{g_0}$ parametrizes genus g_0 curves with a spin structure. By pulling $\mathcal{M}_{g_0,1}$ and $\widetilde{\mathcal{M}}_{g_0,1}$ back to the cover $\mathcal{SM}_{g_0} \to \mathcal{M}_{g_0}$, we get $\mathcal{SM}_{g_0,1}$, parametrizing genus g_0 curves with a spin structure and a marked point, and $\widetilde{\mathcal{SM}}_{g_0,1}$, parametrizing branched covers $\pi : \widetilde{C} \to C$ with a single branch point p as above and with a spin structure L on C, $L^2 \cong K_C$. Similarly, we let $SU, S\widetilde{U}, S\Pi$ and $S\overline{U}$ denote the pullbacks of $U, \widetilde{U}, \Pi$ and $\overline{U}$ from $\widetilde{\mathcal{M}}_{g_0,1}$ to $\widetilde{\mathcal{SM}}_{g_0,1}$ (or equivalently, from $\mathcal{M}_{g_0}$ to $\mathcal{SM}_{g_0}$). We note that $K_{\widetilde{C}} \cong \pi^* K_C(2p)$, so the spin structure L on C determines a spin structure $\widetilde{L} := \pi^* L(p)$ on $\widetilde{C}$. (This would fail if one or more of the ramification points of π had even local degree over C.)

5.4. Maps from $\widetilde{\mathfrak{M}}_{g_0,1}$ to $\mathfrak{M}_g$. Our goal here is to lift the map of spin moduli spaces constructed in the previous section to a map of supermoduli spaces. We continue filling in the details of steps (4)-(7) outlined above.

(4) Start with $\mathfrak{M}_{g_0,1}$ whose reduced space is $\mathcal{SM}_{g_0,1}$. From this we build $\widetilde{\mathfrak{M}}_{g_0,1}$: since $\widetilde{\mathcal{M}}_{g_0,1} \to \mathcal{M}_{g_0,1}$ is a covering map, so is $\widetilde{\mathcal{SM}}_{g_0,1} \to \mathcal{SM}_{g_0,1}$, and we get a covering supermanifold $\widetilde{\mathfrak{M}}_{g_0,1} \to \mathfrak{M}_{g_0,1}$ as in section 2.1.2. Similarly, we can start with $\overline{\mathcal{U}} := \mathfrak{M}_{g_0,1} \times_{\mathfrak{M}_{g_0}} \mathfrak{M}_{g_0,1}$, whose reduced space is $S\overline{U}$. Since $U \to \overline{U}$, hence also $SU \to \overline{SU}$, are covering maps, we again get a covering supermanifold $\mathcal{U} \to \overline{\mathcal{U}}$, as in section 2.1.2. In fact, we can identify this explicitly as: $\mathcal{U} = \widetilde{\mathfrak{M}}_{g_0,1} \times_{\mathfrak{M}_{g_0}} \mathfrak{M}_{g_0,1}$. This has reduced space SU and comes with a map $\mathcal{F} : \mathcal{U} \to \widetilde{\mathfrak{M}}_{g_0,1}$ which is the projection onto the first factor.

(5) To be able to apply the results of 2.1.3 and of Proposition 3.3, we need a divisor $\mathcal{B} \subset \mathcal{U}$ whose reduction is the branch divisor $SB \subset SU$, which as we saw above is the pullback via $SU \to U \to \overline{U}$ of the diagonal $\Delta \subset \overline{U}$. So we need an appropriate divisor $\overline{\mathcal{B}} \subset \overline{\mathcal{U}} = \mathfrak{M}_{g_0,1} \times_{\mathfrak{M}_{g_0}} \mathfrak{M}_{g_0,1}$. The first guess might be to consider the diagonal; but this is not a divisor, it is a submanifold of codimension $(1|1)$. Instead, we need to invoke the duality of section 3.1 for SRS's, which converts points to divisors on a SRS and the diagonal to a divisor $\overline{\mathcal{B}} \subset \overline{\mathcal{U}}$. We then define $\mathcal{B} \subset \mathcal{U}$ as the inverse image of $\overline{\mathcal{B}}$.

(6) and (7): We have now constructed all the input needed for Proposition 3.3: a family of super Riemann surfaces, a branched covering of the reduced space with odd local degrees, and a thickening of the branch divisor. So we get a family of super Riemann surfaces that are branched covers of the original family, as claimed. By the universal property of moduli spaces, this gives the desired morphism of supermanifolds, from $\widetilde{\mathfrak{M}}_{g_0,1}$ to $\mathfrak{M}_g$.

5.5. Components. The spin moduli space $\mathcal{SM}_{g_0}$ has two components $\mathcal{SM}_{g_0}^{\pm}$, distinguished by the parity of the spin structure. Therefore, the same holds for the supermoduli space $\mathfrak{M}_{g_0}$. Related spaces such as $\widetilde{\mathfrak{M}}_{g_0,1}$ inherit at least two components. We will see here that there are actually more components than these obvious two:

PROPOSITION 5.1. *$\widetilde{\mathfrak{M}}_{g_0,1}^{+}$ has at least two components. Under the restriction $\widetilde{\mathfrak{M}}_{g_0,1}^{+} \to \mathfrak{M}_g$ of the map constructed in the previous section, these map to the two components $\mathfrak{M}_g^{\pm}$ of $\mathfrak{M}_g$.*

PROOF. We can see this very explicitly in case $g_0 = 1, d = 3, g = 2$. Here the base curve $C = E$ is elliptic. The branch point $p \in E$ determines a degree 2 map $E \to \mathbb{P}^1$. Let $p_0, p_1, p_2, p_3 = p$ be its four ramification points. We choose a coordinate z on $\mathbb{P}^1$ which takes the values $0, z_1, z_2, \infty$ at p_0, p_1, p_2, p_3, and write the equation of E as $y^2 = z(z - z_1)(z - z_2)$. The cover $\pi : \widetilde{C} \to E$ is essentially unique. It is given topologically as in section 5.2 by mapping $\pi_1(E \setminus p)$ to the symmetric group S_3 by sending $\alpha_1 \to (23)$, $\beta_1 \to (123)$. We can also describe it algebraically. Let $\pi_0 : \widetilde{\mathbb{P}^1} \to \mathbb{P}^1$ be a triple cover which is totally branched over ∞ and has simple branching over z_1, z_2. This is accomplished by setting z to be a cubic polynomial in the coordinate w on $\widetilde{\mathbb{P}^1}$: $z = (w-w_3)(w-w_4)(w-w_5)$, whose critical values are z_1, z_2 (and ∞, with multiplicity 2). For $i = 1, 2$, let w_i, w_i', respectively, be the values of w at the unramified and ramified points above z_i, so $z - z_i = (w - w_i)(w - w_i')^2$.

We then take $\widetilde{C}$ to be the normalization of the fiber product of E and $\widetilde{\mathbb{P}^1}$ over $\mathbb{P}^1$. Explicitly, the fiber product has the equation $y^2 = z(z-z_1)(z-z_2)$, and its normalization has the equation $y^2 = (w-w_1)(w-w_2)(w-w_3)(w-w_4)(w-w_5)$. In particular, this $\widetilde{C}$ is a genus 2 hyperelliptic curve, a double cover of $\widetilde{\mathbb{P}^1}$ ramified at points q_l, $\quad l = 1\ldots 5$ and q over the branch points w_l and ∞. Note that for $i = 1,2$, the inverse image of w_i is q_i (with multiplicity 2), but the inverse image of w_i' consists of two distinct points q_i', q_i'' forming a hyperelliptic pair. The map to E is totally ramified at q, whose image is the unique branch point p.

The three even spin structures on E are $L_j = p_0 - p_j$ for $j = 1,2,3$. Our recipe for lifting to a spin structure on $\widetilde{C}$ is $\widetilde{L} := \pi^*(L)(q)$. If we start with $L_1 = p_0 - p_1$, we get $\widetilde{L_1} = \mathcal{O}(q_3+q_4+q_5-q_1-q_1'-q_1''+q) = \mathcal{O}(q_2)$, which is an odd spin structure on $\widetilde{C}$: $h^0(\widetilde{L_3}) = 1$. Similarly, $\widetilde{L_2} = \mathcal{O}(q_1)$. But $\widetilde{L_3} = \mathcal{O}(q_3+q_4+q_5-3q+q) = \mathcal{O}(q_3+q_4-q_5)$ is an even spin structure on $\widetilde{C}$: $h^0(\widetilde{L_3}) = 0$.

To extend to higher genera g_0 and degrees d, we need some basic facts about spin structures on *singular* algebraic curves. These facts are important in string theory. A mathematical version was obtained by Cornalba [**31**], who has constructed a compactified moduli space $\overline{\mathcal{SM}_g}$ of stable spin curves. This has been studied in [**32**], and a useful review is in [**33**]. The compactification has two components $\overline{\mathcal{SM}_g}^{\pm}$ which are compactifications of $\mathcal{SM}_g{}^{\pm}$ respectively. Each component maps onto the Deligne-Mumford compactified moduli space $\overline{\mathcal{M}_g}$ of stable curves. A spin structure on a stable curve C consists of data C', L', β. Here C' is a 'blow up' of C along some subset Δ of the nodes of C: take the partial resolution N of C at Δ, and for each $p_i \in \Delta$ attach to N a smooth rational curve R_i meeting N transversally at the two branches above p_i. (Such curves C' are called decent or quasi stable.) The remaining data consist of a line bundle L' on C' whose total degree is $g-1$ and whose degree on each R_i is 1, and a homomorphism $\beta : L'^{\otimes 2} \to \omega_{C'}$ that vanishes on the R_i but on no other component of C'. Equivalently and perhaps more intuitively, a spin structure on C is specified by the torsion free sheaf (not necessarily a line bundle) $L := \nu_* L'$ on C, where $\nu : C' \to C$ is the map that collapses each R_i to p_i. This has rank 2 at the points of Δ, and is locally free elsewhere.

The behavior is very simple over points of compact type in $\overline{\mathcal{M}_g}$, i.e. curves whose dual graph is simply-connected (a tree). In this case Δ must consist of all the singular points of C. If such a curve C is the union of irreducible components C_i whose intersection pattern is determined by the dual graph, a spin structure L on C is uniquely specified by a collection of spin structures L_i on the C_i. The corresponding spin structure L is the direct sum of the direct images of the L_i: it has rank 2 (i.e. fails to be a line bundle) at every node of C. The parity of L is the sum of the parities of the L_i. (The extra complications for curves not of compact type arise from the possibility of the spin structure having rank 1 at some of the nodes. For example, for an irreducible C with a single node, let C' be its normalization and p, q the points above the node. Then any square root of $K_{C'}(p+q)$ can be glued – in two ways – to give a line bundle spin structure on C. The two cases that the spin structure has rank 2 or rank 1 at a node correspond in string theory to a degeneration of NS or Ramond type. Ramond degenerations are not possible if the dual graph is a tree because an irreducible curve always has an even number of Ramond punctures. The fact that Ramond degenerations occur in the natural compactification of the spin moduli space $\mathcal{SM}_g$ – and therefore also of the super Riemann surface moduli space $\mathfrak{M}_g$ – can be regarded as the reason that it is necessary to study Ramond punctures in string theory.)

We can now prove the proposition for triple covers of curves of genus $g_0 > 1$. Let C be a reducible curve consisting of an elliptic component E meeting a genus $g_0 - 1$ curve C' in a single point a. For the triple cover $\widetilde{C} \to C$ we start with the triple cover of E constructed above (but let us rename it $\widetilde{E}$), and glue it to three copies of C' at the three inverse images of a. We choose even spin structures $L_E, L_{C'}$ on E, C' and lift them to the four components $\widetilde{E}, C_j$ of $\widetilde{C}$. On the C_j we get even spin structures, but on $\widetilde{E}$ we can get either even or odd spins. By going to smooth deformations of $C, \widetilde{C}$ we conclude that $\widetilde{\mathcal{SM}}^+_{g_0}$ is reducible, with components mapping to both components $\mathcal{SM}^\pm_g$ of $\mathcal{SM}_g$. So $\widetilde{\mathfrak{M}}^+_{g_0,1}$ has at least two components. Under the restriction $\widetilde{\mathfrak{M}}^+_{g_0,1} \to \mathfrak{M}_g$ of the map constructed in the previous section, these map to the two components $\mathfrak{M}^\pm_g$ of $\mathfrak{M}_g$ as claimed.

We need one more modification to allow arbitrary degrees $d \geq 3$. We again start with a reducible curve C consisting of an elliptic component E meeting a genus $g_0 - 1$ curve C' in a single point a. We build the d sheeted covering $\widetilde{C}$ by gluing the following pieces. Over E we take the previous triple cover with single ramification point, $\widetilde{E}$, plus $d-3$ disjoint copies E_i of E. Over C' we take a $d-2$-sheeted unramified covering $\widetilde{C'}$ plus 2 disjoint copies C'_j of C'. We glue the $d-2$ points of a fiber of $\widetilde{C'} \to C'$ to one point on each of the E_i and to just one point on $\widetilde{E}$. The two remaining points of $\widetilde{E}$ are glued to the two C'_j. This produces a simply connected dual graph, so we again specify spin structures on C and $\widetilde{C}$ by specifying even spin structures $L_E, L_{C'}$ on E and C'. Now the effect of switching our choice of L_E from what we were calling L_1 to L_3 is to switch the parity on $\widetilde{E}$ without changing anything else. We conclude as before that in this general case too, $\widetilde{\mathfrak{M}}^+_{g_0,1}$ has at least two components, which map to the two components $\mathfrak{M}^\pm_g$ of $\mathfrak{M}_g$ as claimed. $\square$

5.6. The normal bundle sequence.

Fix a d-sheeted branched cover of curves $\pi : \widetilde{C} \to C$, with branch divisor $B = \sum_{i=1}^n p_i$ in C whose points are distinct and labelled from 1 to n, and ramification divisor $R = \sum_{i,j} a_{i,j}\widetilde{p}_{i,j}$ in $\widetilde{C}$, with $\sum_j a_{i,j} = d$ and $a_{i,j} \geq 1$. The $a_{i,j}$ specify the *ramification pattern* of π: $a_{i,j}$ is the number of sheets that come together at a ramification point $\widetilde{p}_{i,j}$. Let $g, \widetilde{g}$ be the genera of $C, \widetilde{C}$ respectively.

Now we allow the continuous parameters, i.e. the curves $\widetilde{C}, C$ and hence the map π (and in particular also the branch divisor B) to vary, holding fixed the discrete data of the ramification pattern, i.e. the $a_{i,j}$ and the genera $g, \widetilde{g}$. There is a moduli space $\widetilde{\mathcal{M}}_{g,n}$ parametrizing these covers. (Sometimes this is called a Hurwitz scheme.) It comes with a forgetful map

$$\widetilde{\mathcal{M}}_{g,n} \to \mathcal{M}_{g,n}$$
$$(\pi : \widetilde{C} \to C) \mapsto (C, B).$$

This map is a local isomorphism, in fact an unramified finite cover: given $\pi : \widetilde{C} \to C$ and a small deformation of (C, B), there is a unique lift to a deformation of

$\pi : \widetilde{C} \to C$. (Note that the points of B are not allowed to collide.) Let $\mathcal{C} \to \widetilde{\mathcal{M}}_{g,n}$ be the universal curve.

There is also a map

$$\iota : \widetilde{\mathcal{M}}_{g,n} \to \mathcal{M}_{\widetilde{g}}$$
$$(\pi : \widetilde{C} \to C) \mapsto \widetilde{C}.$$

When $\widetilde{g} > 1$, a curve $\widetilde{C}$ can have at most finitely many maps to curves C, and ι is a local embedding away from curves with extra automorphisms. The main result of this section is:

PROPOSITION 5.2. *The normal bundle sequence of* ι*:*

$$0 \to T\widetilde{\mathcal{M}}_{g,n} \xrightarrow{\iota_*} T\mathcal{M}_{\widetilde{g}|\widetilde{\mathcal{M}}_{g,n}} \longrightarrow N \to 0$$

splits.

PROOF. First consider the special case when the map $\pi : \widetilde{C} \to C$ is a Galois cover at one point of $\widetilde{\mathcal{M}}_{g,n}$, hence at all such points. In other words, assume there is a finite group G which acts faithfully on $\widetilde{C}$ with quotient $C = \widetilde{C}/G$. For now, assume also that G acts on the universal curve $\mathcal{C} \to \widetilde{\mathcal{M}}_{g,n}$. The action of G on $\widetilde{C}$ induces actions on $\pi_* \mathcal{O}_{\widetilde{C}}$ and on other natural objects such as $\pi_* K^{\ell}_{\widetilde{C}}$ (for various integer ℓ) and $T_{\widetilde{C}}\mathcal{M}_{\widetilde{g}}$. Similarly, the action of G on $\mathcal{C} \to \widetilde{\mathcal{M}}_{g,n}$ turns $T\mathcal{M}_{\widetilde{g}|\widetilde{\mathcal{M}}_{g,n}}$ into an equivariant G-bundle whose typical fiber is $T_{\widetilde{C}}\mathcal{M}_{\widetilde{g}}$. Therefore $T\mathcal{M}_{\widetilde{g}|\widetilde{\mathcal{M}}_{g,n}}$ decomposes as a direct sum:

$$T\mathcal{M}_{\widetilde{g}|\widetilde{\mathcal{M}}_{g,n}} = \bigoplus_{\rho \in G^\vee} V_\rho \otimes \rho.$$

Here ρ runs over the irreducible representations of G, and $V_\rho := \mathrm{Hom}^G(\rho, T\mathcal{M}_{\widetilde{g}|\widetilde{\mathcal{M}}_{g,n}})$ is the multiplicity bundle of ρ. Since the summand with $\rho = \mathbf{1}$ is $V_{\mathbf{1}} = T\widetilde{\mathcal{M}}_{g,n}$, we have a decomposition

$$T\mathcal{M}_{\widetilde{g}|\widetilde{\mathcal{M}}_{g,n}} = T\widetilde{\mathcal{M}}_{g,n} \oplus N$$

where N is the sum of the remaining summands.

In the above we assumed that the action of G on $\widetilde{C}$ extended to an action of G on the universal curve $\mathcal{C} \to \widetilde{\mathcal{M}}_{g,n}$. This may not be the case: monodromy around $\widetilde{\mathcal{M}}_{g,n}$ may take the action of G on $\widetilde{C}$ to another, conjugate action. The effect is that several of the V_ρ may have to be combined. Nevertheless, our argument goes through since the trivial representation $\rho = \mathbf{1}$ is not conjugate to any other.

In the general case, we can replace $\pi : \widetilde{C} \to C$ by its Galois closure $\widehat{\pi} : \widehat{C} \to C$: Away from the branch locus, i.e. over the open subset $C_0 := C \setminus B$, let $\overline{C}_0$ be the $n!$-sheeted unramified cover whose fiber over $p \in C_0$ consists of the $n!$ ways of ordering the n points of $\pi^{-1}(p)$. This $\overline{C}_0$ may be disconnected, so let $\widehat{C}_0$ be a connected component. It is an unramified Galois cover of C_0, with Galois group a subgroup of the symmetric group S_n. The complete curve $\widehat{C}$ is the unique (non-singular) compactification of $\widehat{C}_0$. It is still Galois over C with the same G, but of course it is ramified. Denote its genus by $\widehat{g}$.

In this situation, G does not act on $\widetilde{C}$, nor on $T\mathcal{M}_{\widetilde{g}|\widetilde{\mathcal{M}}_{g,n}}$. Nevertheless we can still make sense of the decomposition. We have an action of G on $\widehat{C}$ with quotient

C. Let $H \subset G$ be the stabilizer of an unramified point of $\widetilde{C}$. Then $\widehat{C}$ is also a Galois cover of $\widetilde{C} = \widehat{C}/H$, with group H. For each irreducible representation ρ of G there is an invariant subspace $\rho^H \subset \rho$, and we have compatible decompositions:

$$T_{\widehat{C}}\mathcal{M}_{\widehat{g}} = \bigoplus_{\rho \in G^\vee} V_\rho \otimes \rho$$

$$T_{\widetilde{C}}\mathcal{M}_{\widetilde{g}} = \bigoplus_{\rho \in G^\vee} V_\rho \otimes \rho^H.$$

Under these decompositions, the corresponding tangent space of $\widetilde{\mathcal{M}}_{g,n}$ is just $V_{\mathbf{1}} \otimes \mathbf{1}^H = V_{\mathbf{1}} \otimes \mathbf{1}$, so we have again exhibited $T\widetilde{\mathcal{M}}_{g,n}$ as a direct summand of the restriction of $T\mathcal{M}_{\widetilde{g}}$. □

5.7. Non-projectedness of $\mathfrak{M}_g$ and $\mathfrak{M}_{g,n}$.

We can now prove our main result, Theorem 1.1: the non-projectedness of super moduli space $\mathfrak{M}_g$ for $g \geq 5$.

PROOF. We do this by reducing from $\mathfrak{M}_g$ to its submanifold $\widetilde{\mathfrak{M}}_{2,1}$ constructed above. In fact, we now have in place all the ingredients for this reduction:

- the non-vanishing of the first obstruction class $\omega_2(\widetilde{\mathfrak{M}}_{2,1})$. (The non-vanishing of $\omega_2(\mathfrak{M}_{2,1})$ is Theorem 1.2, proved in section 4. The lifting to the covering space $\widetilde{\mathfrak{M}}_{2,1}$ was seen in Corollary 2.8.)
- the inclusion of supermanifolds $\widetilde{\mathfrak{M}}_{2,1} \to \mathfrak{M}_g$, proved in section 5.4; and
- the decomposition of the restriction to $\widetilde{\mathcal{SM}}_{2,1}$ of $T_+\mathfrak{M}_g$ into its tangential and normal pieces, proved in proposition 5.2 in section 5.6.

These three ingredients are precisely the inputs of the Proposition 2.12. We conclude the non-vanishing of the first obstruction class:

$$\omega := \omega_2(\mathfrak{M}_g) \neq 0 \in H^1(\mathcal{SM}_g, \mathrm{Hom}(\wedge^2 T_-\mathfrak{M}_g, T_+\mathfrak{M}_g)),$$

and hence the non-projectedness of $\mathfrak{M}_g$. □

We conclude with a proof of Theorem 1.3:

PROOF. We start with the extreme case $n = g-1$. Consider the space $\widetilde{\mathfrak{M}}_{2,1}$ which parametrizes pairs $(C,p) \in \mathfrak{M}_{2,1}$ plus an unramified cyclic n-sheeted cover $\pi : \widetilde{C} \to C$. Note that the genus of $\widetilde{C}$ is $g = n+1$. There is a natural embedding $i : \widetilde{\mathcal{M}}_{2,1} \to \mathcal{M}_{g,n}$ sending the above data to the curve $\widetilde{C}$ with the n marked points $\pi^{-1}(p)$. A priori, these n points are only cyclically ordered, so there are n distinct ways to order them. However, the cyclic automorphism group $\mathbb{Z}/n$ of $\widetilde{C}$ over C permutes these n orderings transitively, so we get a well defined image point in $\mathcal{M}_{g,n}$. In fact, the entire $\mathcal{M}_{g,n}$ admits an action of the cyclic group $\mathbb{Z}/n$ which cyclically permutes the n marked points. Our locus $\widetilde{\mathcal{M}}_{2,1}$ is a component of the fixed locus of this action. It follows that the normal bundle sequence for the embedding of $\widetilde{\mathcal{M}}_{2,1}$ in $\mathcal{M}_{g,n}$ is split: the tangent bundle is the +1-eigenbundle, while the normal bundle is the sum of the remaining eigenbundles. Finally, the embedding of $\widetilde{\mathcal{M}}_{2,1}$ in $\mathcal{M}_{g,n}$ lifts, as in section 5.4, to an embedding of $\widetilde{\mathfrak{M}}_{2,1}$ in $\mathfrak{M}_{g,n}$. As in the proof of Theorem 1.1 above, we now have the three ingredients needed for Proposition 2.12 to apply. We conclude the non-vanishing of the first obstruction class:

$$\omega := \omega_2(\mathfrak{M}_{g,g-1}) \neq 0,$$

and hence the non-projectedness of $\mathfrak{M}_{g,g-1}$.

For lower values of n, we map the image of $\widetilde{\mathcal{M}}_{2,1}$ to $\mathcal{M}_{g,g-1}$ as above, and then project to $\mathcal{M}_{g,n}$ by the map that preserves the first n of the $g-1$ marked points and omits the rest. Again, there is no problem in lifting to a map of supermoduli spaces. We can no longer interpret $\widetilde{\mathcal{M}}_{2,1}$ as the fixed locus of a group of automorphisms, but nevertheless we can conclude the splitting of the normal bundle sequence:

Very generally, let $f : Y \to Z$ be a fibration, and $i : X \to Y$ an immersion, such that $f \circ i : X \to Z$ is also an immersion.

We map the normal bundle sequence of bundles on X:

$$0 \to T_X \to i^*T_Y \to N_{X,Y} \to 0 \tag{5.2}$$

onto the normal bundle sequence:

$$0 \to T_X \to (f \circ i)^*T_Z \to N_{X,Z} \to 0, \tag{5.3}$$

and note that the kernel sequence is trivial:

$$0 \to 0 \to i^*T_{Y/Z} \to i^*T_{Y/Z} \to 0.$$

Now if sequence (5.2) is split by a map

$$N_{X,Y} \to i^*T_Y$$

we get an induced splitting

$$N_{X,Z} = N_{X,Y}/i^*T_{Y/Z} \to i^*T_Y/i^*T_{Y/Z} = (f \circ i)^*T_Z$$

of sequence (5.3).

We want to apply this to:

* $X = \widetilde{\mathcal{M}}_{2,1}$ which parametrizes pairs $(C, p) \in \mathcal{M}_{2,1}$ plus an unramified $(g-1)$-sheeted cover $\pi : \widetilde{C} \to C$.

* $Y = \mathcal{M}_{g,g-1}$

* $Z = \mathcal{M}_{g,n}$, for $g - 1 \geq n \geq 1$, with $f : Y \to Z$ preserving the first n of the $g - 1$ marked points and omitting the rest.

For this, we need to check that $i : X \to Y$ and the induced $f \circ i : X \to Z$ are immersions. Write $\pi^{-1}(p) = p_1 + \cdots + p_{g-1}$. Then we need injectivity of the maps on tangent spaces:

$$H^1(C, T_C(-p)) \overset{d(i)}{\to} H^1(\widetilde{C}, T_{\widetilde{C}}(-\pi^{-1}p)) \overset{d(f)}{\to} H^1(\widetilde{C}, T_{\widetilde{C}}(-(p_1 + \cdots + p_n))).$$

This commutes with the map on tangents of the moduli spaces of curves without marked points:

$$H^1(C, T_C) \to H^1(\widetilde{C}, T_{\widetilde{C}}),$$

and the latter is injective. These maps fit together into a commutative diagram with exact columns:

$$\begin{array}{ccccc} H^0(T_{C|p}) & \to & \oplus_{j=1}^{g-1} H^0(T_{\widetilde{C}|p_j}) & \to & \oplus_{j=1}^{n} H^0(T_{\widetilde{C}|p_j}) \\ \downarrow & & \downarrow & & \downarrow \\ H^1(C, T_C(-p)) & \overset{d(i)}{\to} & H^1(\widetilde{C}, T_{\widetilde{C}}(-\pi^{-1}p)) & \overset{d(f)}{\to} & H^1(\widetilde{C}, T_{\widetilde{C}}(-(p_1 + \cdots + p_n))) \\ \downarrow & & \downarrow & & \downarrow \\ H^1(C, T_C) & \to & H^1(\widetilde{C}, T_{\widetilde{C}}) & = & H^1(\widetilde{C}, T_{\widetilde{C}}). \end{array}$$

injectivity of the bottom map implies that $\mathrm{Ker}(d(f \circ i))$ must come from the vertical direction $H^0(T_{C|p}) = \mathrm{Ker}(H^1(C, T_C(-p)) \to H^1(C, T_C))$. But for each j, the map $H^0(T_{C|p}) \to H^0(T_{\widetilde{C}|p_j})$ is an isomorphism, so $\mathrm{Ker}(d(f \circ i))$ vanishes, showing that $f \circ i$ is indeed an immersion.

This shows that the normal sequence for $\widetilde{\mathcal{M}}_{2,1}$ in $\mathcal{M}_{g,n}$ splits for $g - 1 \geq n \geq 1$, so the non-vanishing of the obstruction for $\widetilde{\mathcal{M}}_{2,1}$ implies the same for $\mathcal{M}_{g,n}$. The theorem now follows as before from Proposition 2.12. (Note that the argument fails for $n = 0$, because the natural map $f \circ i : \widetilde{\mathcal{M}}_{2,1} \to \mathcal{M}_g$ factors through $\widetilde{\mathcal{M}}_2$ and is therefore not an immersion.) □

In stating this argument, we have ignored the fact that particular genus 2 curves with a marked point have exceptional automorphisms. To justify what we have asserted, one may either develop the theory for orbifolds, or restrict from X to an open subset of X that contains a fiber of $\widetilde{\mathcal{M}}_{2,1} \to \mathcal{M}_2$.

6. Acknowledgments

We thank Pierre Deligne for a careful reading of an earlier version of the manuscript, and for correcting several statements and proofs, especially in section 2. We are grateful to Gavril Farkas, Dick Hain, Brendan Hassett, Sheldon Katz, Igor Krichever, Dimitry Leites, Yuri Manin, Tony Pantev, Albert Schwarz, Liza Vishnyakova and Katrin Wendland for helpful discussions. RD acknowledges partial support by NSF grants DMS 0908487, DMS 1304962 and RTG 0636606. EW acknowledges partial support by NSF Grant PHY-0969448.

Appendix A. A detailed example in genus 5

In this appendix, we give an elementary construction of a family of triple covers $\widetilde{C} \to C$ with a single branch point, where $g(\widetilde{C}) = 5$, $g(C) = 2$. We analyze the parameters on which this construction depends and find, somewhat surprisingly, that the parameter space of such covers with fixed C is a *rational* curve in $\mathcal{M}_5$. This curve has some orbifold points, so it maps to the moduli space but not to the moduli stack. It has a cover of genus 19 over which a family of genus 5 triple covers exists. Finally, we examine the effect of adding spin structures to our curves, and verify that even spin structures on the genus 2 curve C can lead to both even and odd spin structures on the genus 5 curve $\widetilde{C}$.

A.1. The Galois closure. A triple cover $\rho : \widetilde{C} \to C$ with a single branch point $p \in C$ cannot be cyclic, so its Galois group is the symmetric group S_3 of permutations of $\{1, 2, 3\}$. Its Galois closure is therefore a smooth curve $\widetilde{\widetilde{C}}$ on which S_3 acts. (One way to obtain $\widetilde{\widetilde{C}}$ explicitly is by taking the self product $\widetilde{C} \times_C \widetilde{C}$, removing the diagonal $\widetilde{C}$, and taking the unique smooth compactification.)

The quotient by S_3 is the original C, and the quotient by the subgroup S_2 of permutations of $\{1,2\}$ is the original $\widetilde{C}$. This subgroup is not normal: there are three conjugate subgroups $(S_2)_i$, $i = 1,2,3$, and corresponding quotient curves $\widetilde{C}_i$. These are isomorphic to each other: if $\{i,j,k\}$ is a permutation of $\{1,2,3\}$, the involution $\overline{\tau}_k$ on $\overline{\widetilde{C}}$ induced by the transposition (ij) exchanges $\widetilde{C}_i$ and $\widetilde{C}_j$. But if we divide by the alternating subgroup $A_3 \subset S_3$, we get a new intermediate curve $\overline{C}$ which is a double cover of C and a triple quotient $\overline{\rho} : \overline{\widetilde{C}} \to \overline{C}$ of $\overline{\widetilde{C}}$. It comes with an involution τ, induced by any of the $\overline{\tau}_k$. A point of $\overline{\widetilde{C}}$ over some $q \neq p$ in C can be thought of as a labeling or ordering of the 3 points in $\rho^{-1}(q)$. The action of S_3 permutes the labels, and a point of the quotient $\overline{C}$ above q can be thought of as an orientation, or cyclic ordering, of that fiber. An easy monodromy argument shows that above p there is a single point $\widetilde{p}_i$ in each $\widetilde{C}_i$, two points $\overline{\overline{p}}$, $\overline{\overline{p}}' = \tau\overline{\overline{p}} \in \overline{C}$, and two points $\overline{\widetilde{p}}, \overline{\widetilde{p}}' \in \overline{\widetilde{C}}$. (Each of the transpositions $\tau_k = (ij)$ exchanges $\overline{\widetilde{p}}, \overline{\widetilde{p}}'$, while the 3-cycles preserve them.) It follows that $\overline{\widetilde{C}}$ is an unramified double cover of each $\widetilde{C}_i$, as is $\overline{C}$ over C. We display these curves, their genera and the maps between them in the following snapshot:

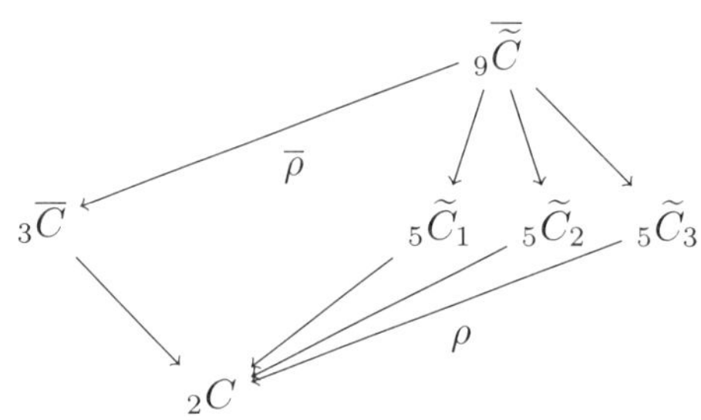

A.2. The construction. We now reverse the above analysis, obtaining a direct construction of the triple covers $\rho : \widetilde{C} \to C$ with a single branch point $p \in C$. This will allow us to describe the parameter spaces on which the construction depends.

Start with a genus 2 curve $C = {}_2C$ and an unramified double cover ${}_3\overline{C} \to C$, with fixed point free involution $\tau : \overline{C} \to \overline{C}$. (Since C has genus 2, the genus of $\overline{C}$ is $3 = 2 \times 2 - 1$, which we indicate with the left subscript.) Given a point $\overline{p} \in \overline{C}$ and some additional data, we construct a cyclic triple cover $\overline{\rho} : \overline{\widetilde{C}} \to \overline{C}$ which is totally ramified over $\overline{p}$ and (with the opposite orientation) over $\tau\overline{p}$, with a deck transformation $\sigma : \overline{\widetilde{C}} \to \overline{\widetilde{C}}$. The extra data consists of a line bundle $L \in \overline{J} := \mathrm{Pic}^0(\overline{C})$ with an isomorphism

$$L^{\otimes 3} \cong \mathcal{O}_{\overline{C}}(\overline{p} - \tau\overline{p}), \tag{A.1}$$

which we interpret as a 3 to 1 map from the total space of L to the total space of $\mathcal{O}_{\overline{C}}(\overline{p} - \tau\overline{p})$. Now $\mathcal{O}_{\overline{C}}(\overline{p} - \tau\overline{p})$ has a meromorphic section corresponding to the section 1 of $\mathcal{O}_{\overline{C}}$, and we let $\overline{\widetilde{C}}$ be its inverse image in the total space of L, a cyclic triple cover of $\overline{C}$. The automorphism σ is induced from multiplication by a cubic root of unity on L.

The involution $\tau : \overline{C} \to \overline{C}$ lifts to an involution $\widetilde{\tau} : \overline{\widetilde{C}} \to \overline{\widetilde{C}}$ if and only if τ^*L is isomorphic either to L or to L^{-1}. In the former case, $\widetilde{\tau}$ commutes with

σ, so the resulting $\overline{\widetilde{C}}$ is a Galois cover of C with Galois group $\mathbb{Z}/6$. In the latter case, $\widetilde{\tau}$ commutes σ to σ^{-1}, so the resulting $\overline{\widetilde{C}}$ is a Galois cover of C with Galois group the symmetric group S_3. (In general, the line bundle τ^*L^{-1} also satisfies condition (A.1) and therefore defines another cyclic triple cover $\overline{\widetilde{C}'} \to \overline{C}$ with the same ramification pattern as $\overline{\widetilde{C}}$. The involution $\tau : \overline{C} \to \overline{C}$ always lifts to an isomorphism $\overline{\widetilde{C}} \to \overline{\widetilde{C}'}$, in fact to three of them, and when

$$\tau^*L^{-1} \cong L \tag{A.2}$$

these give three involutions of $\overline{\widetilde{C}}$, each conjugating σ to σ^{-1}.)

A.3. A rational curve in $\mathcal{M}_5$. We will now analyze the parameters on which the construction of the previous section depends. Perhaps surprisingly, we find that the compact curve in moduli space parametrizing our triple covers (of a fixed curve C, and corresponding to a specified double cover $\overline{C} \to C$) is actually *rational.*

We need to choose a point $\overline{p} \in \overline{C}$ and a cubic root L of $\mathcal{O}_{\overline{C}}(\overline{p} - \tau\overline{p})$ as in (A.1). The set of those roots is a coset of the subgroup of points of order 3 in the Jacobian $\overline{J} = \mathrm{Pic}^0(\overline{C})$, which is isomorphic to $(\mathbb{Z}/3)^6$. So our data seems to live in a cover of $\overline{C}$ of degree 3^6. But this cover turns out to be reducible, and our additional condition (A.2) picks out a subcover of degree 9. In order to see this, we need to review some general results on Prym varieties of unramified double covers.

A.3.1. *Pryms.* Condition (A.2) is equivalent to $L \in \mathrm{Ker}(1 + \tau^*)$, where $1 + \tau^*$ is the endomorphism of $\overline{J}$ sending L to $L \otimes \tau^*L$. For a general unramified double cover $\pi : \overline{C} \to C$ with involution τ, Mumford [**34**] described $\mathrm{Ker}(1+\tau^*)$. It consists of four cosets of the *Prym* variety

$$P := (1 - \tau^*)\overline{J}. \tag{A.3}$$

It is convenient to use the *Norm* map $\mathrm{Nm}(\pi) : \overline{J} \to J$, which is the homomorphism that sends a line bundle $\mathcal{O}_{\overline{C}}(\overline{D})$ to $\mathcal{O}_C(D)$ where $D := \pi(\overline{D})$ is the image of the divisor D. Note that $\pi^* \circ \mathrm{Nm}(\pi) = 1 + \tau^*$, so $\mathrm{Ker}(1 + \tau^*)$ contains $\mathrm{Ker}(\mathrm{Nm}(\pi))$. In fact, $\mathrm{Ker}(1 + \tau^*) = (\mathrm{Nm}(\pi))^{-1}(K)$, where

$$K := \mathrm{Ker}(\pi^* : J \to \overline{J}) \cong \mathbb{Z}/2.$$

Finally, $\mathrm{Ker}(\mathrm{Nm}(\pi)) = P \cup P'$ where P' is the coset:

$$P' := (1 - \tau^*)\mathrm{Pic}^1(\overline{C}) \tag{A.4}$$

of P. (More generally, $1 - \tau^*$ maps divisors of even degree on $\overline{C}$ to P, and divisors of odd degree to P'.) So all in all we have:

$$\mathrm{Ker}(\mathrm{Nm}(\pi)) = P \cup P' = (1 - \tau^*)\mathrm{Pic}(\overline{C})$$

and

$$0 \to \mathrm{Ker}(\mathrm{Nm}(\pi)) \to \mathrm{Ker}(1 + \tau^*) \to K \to 0.$$

We now return to the conditions on our line bundle L. Condition (A.2) says that L is in $\mathrm{Ker}(1 + \tau^*)$ which, as we have just seen, consists of four components. The map $L \to L^{\otimes 3}$ sends each of these components to itself. Since the right hand side of (A.1) is in P' by (A.4), we see that L must be in P' too.

A.3.2. *Hyperelliptic Pryms.* In our case, we can make everything more explicit. Our genus 2 base curve C is hyperelliptic, a double cover of $\mathbb{P}^1$ branched at the 6 points $0, \infty, e_0 = 1, e_1, e_2, e_3$. The double cover $\overline{C}$ is determined by a $4+2$ partition of these 6 branch points: say $0, \infty$ vs. e_0, e_1, e_2, e_3. The double cover of $\mathbb{P}^1$ branched at $0, \infty$ is a rational curve $R = {}_0R$, and the double cover of $\mathbb{P}^1$ branched at e_0, e_1, e_2, e_3 is a genus 1 curve $E = {}_1E$. The cover $\overline{C}$ has three involutions $\tau_0, \tau_1, \tau_2 = \tau$, sitting in a symmetry group $\mathbb{Z}/2 \times \mathbb{Z}/2$, with quotient $\mathbb{P}^1$ and intermediate quotients R, E, C. We can include these in our snapshot:

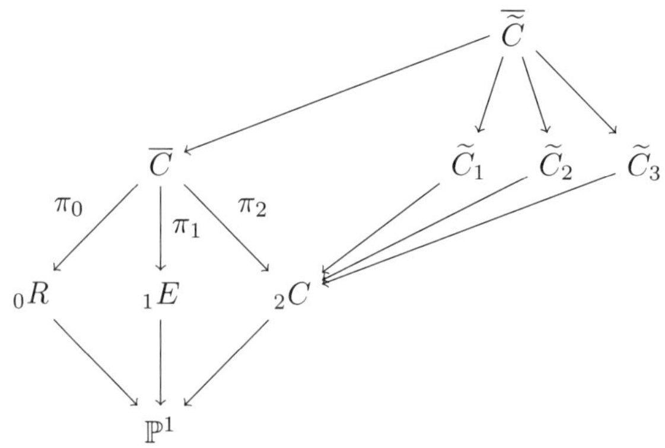

In order to avoid clutter, we will show only one of the 3 quotients $\widetilde{C}_i$, which we rename $\widetilde{C}$:

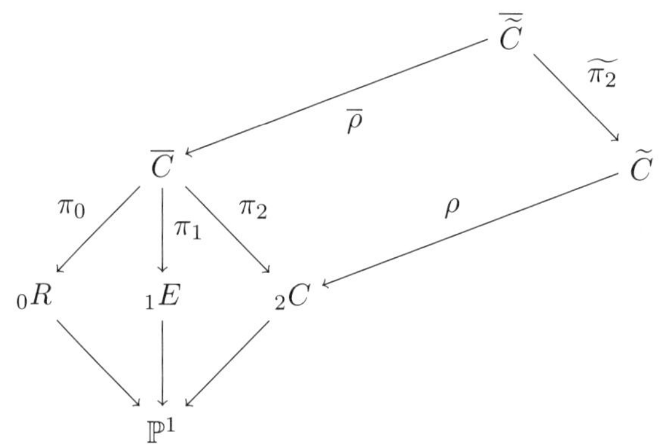

Since any degree 0 line bundle on $\overline{C}$ can be written as the sum of pullbacks from the quotients:

$$\mathrm{Pic}^0(\overline{C}) = (\pi_1)^*(\mathrm{Pic}^0(E)) + (\pi_2)^*(\mathrm{Pic}^0(C))$$

and $1 - \tau^*$ kills $\mathrm{Pic}^0(C)$, we get from(A.3) an isomorphism:

$$P = (1 - \tau^*)\overline{J} \cong (1 - \tau^*)(\pi_1)^*\mathrm{Pic}^0(E).$$

But since $(1 - \tau^*)(\pi_1)^* = 2(\pi_1)^*$ on $\mathrm{Pic}^0(E)$ and $(\pi_1)^*$ is injective, we see that

$$(\pi_1)^* : \mathrm{Pic}^0(E) \to P$$

is an isomorphism. Similarly we find that $E = \mathrm{Pic}^1(E)$ can be naturally identified with P' via a translate of $(\pi_1)^*$. The map is:

$$e \mapsto L_e := (\pi_1)^*\mathcal{O}_E(e) \otimes (H_{\overline{C}})^{-1}, \tag{A.5}$$

where $H_{\overline{C}} := (\pi_0)^*\mathcal{O}_R(1)$ is the hyperelliptic line bundle on $\overline{C}$, and it needs to be inserted in the above formula in order to yield a divisor of degree 0 on $\overline{C}$.

A.3.3. *The parameter space.* We can now describe the parameter space for our covers $\widetilde{C}$. Originally, we wanted pairs $(L, \overline{p})$ satisfying conditions (A.1), (A.2). The space of eligible line bundles L was identified in section A.3.1 with the shifted Prym P'. In our hyperelliptic setting this was translated in section A.3.2 to $E = P'$, the isomorphism being given by (A.5). Putting these together, we need to parametrize pairs $(e, \overline{p})$ satisfying the condition:

$$(\pi_1)^*\mathcal{O}_E(3e) \cong \mathcal{O}_{\overline{C}}(\overline{p} - \tau\overline{p}) \otimes (H_{\overline{C}})^3. \tag{A.6}$$

This is an equation in the Picard of $\overline{C}$. Keeping in mind that $(\pi_1)^*$ is injective, this is also equivalent to the equation in E:

$$3e \sim \pi_1\overline{p} + H_E, \tag{A.7}$$

where $\sim$ means linear equivalence on E, and we have used that for any $\overline{p} \in \overline{C}$,

$$H_{\overline{C}} \cong \mathcal{O}(\tau_2\overline{p} + \tau_1\overline{p}). \tag{A.8}$$

In order to parametrize solutions of (A.7), consider another copy of E, say E_r. We think of it as parametrizing cubic *roots* of points of E. The curves E_r, E are isomorphic, but we find it convenient to keep the distinction. Let $m_1 : E_r \overset{\cong}{\to} E$ be the isomorphism, and m_3 the multiplication by 3 map:

$$m_3 : E_r \to \mathrm{Pic}^1(E) = E, \quad e \mapsto \mathcal{O}_E(3e) \otimes (H_E)^{-1},$$

where H_E is the hyperelliptic line bundle on E, pullback of $\mathcal{O}_{\mathbb{P}^1}(1)$. (Note that we have not chosen a base point in E, only a degree 2 line bundle H_E, or equivalently the map to $\mathbb{P}^1$. The cubing map m_3 is nevertheless well defined.) We see that the natural parameter space for our triple covers $\widetilde{C} \to C$ is the fiber product:

$$\overline{C}_r := E_r \times_E \overline{C}.$$

This is a 9-sheeted unramified cover of $\overline{C}$, so its genus is $19 = 1 + 9 \times (3-1)$. Locally over $\overline{C}_r$ we can construct the family of triple covers $\widetilde{\overline{C}} \to \overline{C}$ and their quotients $\widetilde{C}$. Since the generic curve $\widetilde{C}$ has no non-trivial automorphisms, these local families automatically glue to a family of triple covers $\widetilde{C} \to C$ parametrized by $\overline{C}_r$.

The resulting map $\sigma : \overline{C}_r \to \mathcal{M}_5$ is clearly not an embedding. For one thing, the pairs $(L, \overline{p})$ and $(L^{-1} = \tau^*L, \tau\overline{p})$ give isomorphic covers. To understand the quotient, we note that the involution on E with quotient $\mathbb{P}^1$ lifts to an involution on E_r, so let $\mathbb{P}^1_r$ be the quotient. It is a 9 sheeted branched cover of the original $\mathbb{P}^1$, with ramification pattern $(2^4, 1)$ over each of the 4 branch points e_i of E over $\mathbb{P}^1$, and E_r is recovered as the normalization of the fiber product $\mathbb{P}^1_r \times_{\mathbb{P}^1} E$. We can complete this into a commutative box:

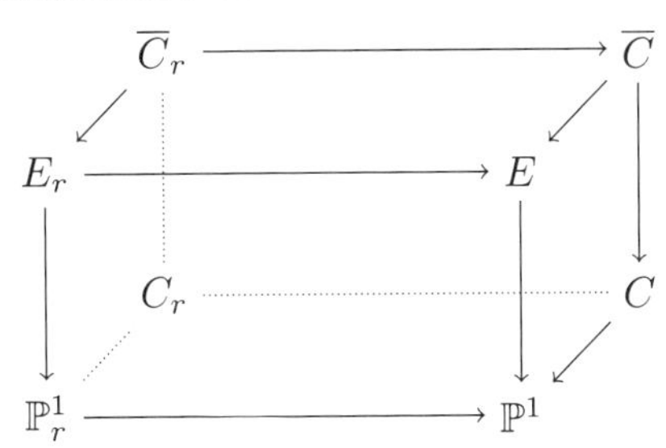

where the horizontal maps have degree 9, the others have degree 2. We see that the quotient of $\overline{C}_r$ by the above involution is what we have now labeled $C_r := \mathbb{P}^1_r \times_{\mathbb{P}^1} C$, a 9-sheeted unramified cover of C, hence of genus 10. However, there is a further symmetry: the hyperelliptic involution of C. Dividing by that, we see that the map of our family $\overline{C}_r$ to $\mathcal{M}_g$ factors through the rational curve $\mathbb{P}^1_r$, as claimed. Pictorially, the hyperelliptic involutions of C and E generate a group $\mathbb{Z}/2 \times \mathbb{Z}/2$ which acts on the entire box. In particular it acts on $\overline{C}_r$ which parametrizes the triple covers $\widetilde{C}$, and the map $\sigma : \overline{C}_r \to \mathcal{M}_5$ is invariant under this action, so the curves $\widetilde{C}$ over points in an orbit of $\mathbb{Z}/2 \times \mathbb{Z}/2$ are isomorphic.

A.4. The family. We are going to construct universal curves over $\overline{C}_r$, i.e. surfaces

$$\overline{\widetilde{\mathcal{C}}},\ \widetilde{\mathcal{C}}$$

and fibrations

$$\overline{\widetilde{\mathcal{C}}} \to \widetilde{\mathcal{C}} \to \overline{C}_r,$$

such that the fibers over each point of $\overline{C}_r$ are isomorphic to the corresponding curves $\overline{\widetilde{C}} \to \widetilde{C}$ constructed above. More precisely:

PROPOSITION A.1. *There is a commutative diagram:*

$$\begin{array}{ccc}
\overline{\widetilde{\mathcal{C}}} & \xrightarrow{\ \widetilde{\pi_2}\ } & \widetilde{\mathcal{C}} \\
\Big\downarrow{\scriptstyle \overline{\rho}} & & \Big\downarrow{\scriptstyle \rho} \\
\overline{C}_r \times \overline{C} & \xrightarrow[\ 1 \times \pi_2\ ]{} & \overline{C}_r \times C,
\end{array}$$

where $\overline{\widetilde{\mathcal{C}}}$, $\widetilde{\mathcal{C}}$ are smooth surfaces, the vertical maps are branched triple covers, and the fibers of $\overline{\widetilde{\mathcal{C}}}$, $\widetilde{\mathcal{C}}$ over points of $\overline{C}_r$ are the triple covers $\overline{\widetilde{C}} \to \overline{C}$, $\widetilde{C} \to C$, with two (respectively one) total ramification points, constructed in section A.2.

PROOF. To construct $\overline{\widetilde{\mathcal{C}}}$ we need a line bundle $\mathcal{L}$ on $\overline{C}_r \times \overline{C}$ satisfying the analog of (A.1):

$$\mathcal{L}^{\otimes 3} \cong p_{23}^* \mathcal{O}(\Delta - \Delta') =: \mathcal{D}, \tag{A.9}$$

where p_{23} is the projection:

$$p_{23} : \overline{C}_r \times \overline{C} = E_r \times_E \overline{C} \times \overline{C} \to \overline{C} \times \overline{C},$$

$\Delta \subset \overline{C} \times \overline{C}$ is the diagonal, and Δ' is the graph of $\tau_2 : \overline{C} \to \overline{C}$.

We claim that (A.9) is satisfied by the choice:

$$\mathcal{L} := \mathcal{O}(\Gamma) \otimes pr_2^* \pi_0^* \mathcal{O}_R(-1) \otimes pr_1^* \pi_{0,r}^* \mathcal{O}_{R_r}(-1),$$

where the maps are:

$$\overline{C}_r \times \overline{C} \xrightarrow{pr_2} \overline{C} \xrightarrow{\pi_0} R$$

and

$$\overline{C}_r \times \overline{C} \xrightarrow{pr_1} \overline{C}_r \xrightarrow{\pi_{0,r}} R_r,$$

while $\Gamma \subset \overline{C}_r \times \overline{C} = E_r \times_E \overline{C} \times \overline{C}$ is the effective divisor:

$$\Gamma := \quad \{(e, \overline{p}, \overline{q}) \quad | \quad \pi_1 \overline{p} = m_3 e,\ \pi_1 \overline{q} = e\}.$$

To prove (A.9), it suffices to verify that it holds when restricted to each horizontal curve $\overline{C}_r \times \{\overline{q}\}$ and each vertical curve $\{(e,\overline{p})\} \times \overline{C}$. Indeed, on the vertical curve $\mathcal{L}$ becomes

$$\pi_1^*\mathcal{O}_E(e) \otimes \pi_0^*\mathcal{O}_R(-1)$$

while $\mathcal{D}$ becomes

$$\mathcal{O}_{\overline{C}}(\overline{p} - \tau_2\overline{p}),$$

so the equality is just the condition (A.6). For the horizontal curves we need to work a little harder. It is convenient to focus on the diagram:

$$\text{(A.10)} \qquad \begin{array}{ccc} \overline{C}_r = E_r \times_E \overline{C} & \xrightarrow{\ p_2\ } & \overline{C} \\ {\scriptstyle p_1}\big\downarrow & & \big\downarrow{\scriptstyle \pi_1} \\ E_r & \xrightarrow[\ m_3\ ]{} & E, \end{array}$$

and to recall that for any $e \in E$,

$$\text{(A.11)} \qquad m_3^*\mathcal{O}_E(e) \cong m_1^*\mathcal{O}_E(3e) \otimes H^3_{E_r}.$$

Now on the horizontal curve $\overline{C}_r \times \{\overline{q}\}$, the line bundle $\mathcal{L}$ becomes

$$\text{(A.12)} \qquad p_1^*m_1^*\mathcal{O}_E(\pi_1\overline{q}) \otimes H^{-1}_{\overline{C}_r},$$

while $\mathcal{D}$ becomes

$$\begin{aligned} p_2^*\mathcal{O}_{\overline{C}}(\overline{q} - \tau_2\overline{q}) &= && \text{by identity (A.8)} \\ p_2^*(\mathcal{O}_{\overline{C}}(\overline{q} + \tau_1\overline{q}) \otimes H^{-1}_{\overline{C}}) &= && \\ p_2^*\mathcal{O}_{\overline{C}}(\pi_1^{-1}\pi_1\overline{q}) \otimes H^{-9}_{\overline{C}_r} &= && \\ p_2^*\pi_1^*\mathcal{O}_E(\pi_1\overline{q}) \otimes H^{-9}_{\overline{C}_r} &= && \text{by commutativity of (A.10)} \\ p_1^*m_3^*\mathcal{O}_E(\pi_1\overline{q}) \otimes H^{-9}_{\overline{C}_r} &= && \text{by (A.11)} \\ p_1^*m_1^*\mathcal{O}_E(3\pi_1\overline{q}) \otimes H^{-3}_{\overline{C}_r}, & && \end{aligned}$$

showing the needed equality of the vertical restrictions of $\mathcal{L}^{\otimes 3}$ and $\mathcal{D}$. We therefore have the right line bundle $\mathcal{L}$, so we get the desired triple cover $\overline{\rho} : \overline{\widetilde{\mathcal{C}}} \to \overline{C}_r \times \overline{C}$.

Next, we want to lift the involution $\tau = \tau_2 : \overline{C} \to \overline{C}$ to an involution $\widetilde{\tau} : \overline{\widetilde{\mathcal{C}}} \to \overline{\widetilde{\mathcal{C}}}$ which would allow us to construct the quotient $\widetilde{\pi_2} : \overline{\widetilde{\mathcal{C}}} \to \widetilde{\mathcal{C}}$. For this we need to know that $\mathcal{L}$ satisfies the global analog of condition (A.2) as well. Again, it suffices to check this on horizontal and vertical curves. On horizontal curves, this follows immediately from (A.12). On vertical curves, this is the original condition (A.2). This completes the construction of the universal curves $\overline{\widetilde{\mathcal{C}}}$, $\widetilde{\mathcal{C}}$ over $\overline{C}_r$.

□

Recall that we have an action of the group $\mathbb{Z}/2 \times \mathbb{Z}/2$ on $\overline{C}_r$ with quotient $\mathbb{P}^1_r$. This action lifts to $\overline{\widetilde{\mathcal{C}}}, \widetilde{\mathcal{C}}$, but it has fixed points there, so the smooth family does not descend to one over $\mathbb{P}^1_r$. (In fact, the base curve B of any non-locally trivial family of smooth curves must have genus $g(B) \geq 2$, since the period map lifts to a non-constant map from the universal cover of B to a bounded domain.)

A.5. Adding spin. We have constructed families $\widetilde{\overline{\mathcal{C}}}$ and $\widetilde{\mathcal{C}}$ of curves $\widetilde{\overline{C}}$ and $\widetilde{C}$, parametrized by $\overline{C}_r$. We want to promote these to families of spin curves $(\widetilde{\overline{C}}, \widetilde{\overline{N}})$ and $(\widetilde{C}, \widetilde{(N)})$, making the construction of section 5.3 explicit. The further promotion to super Riemann surfaces then follows Section 5.4. We will in particular recover the result of Section 5.5 which asures us that both spin components do arise.

Recall that our genus 2 base curve C is hyperelliptic, a double cover of $\mathbb{P}^1$ branched at the 6 points $B = \{0, \infty, e_0 = 1, e_1, e_2, e_3\}$. Let $p_i \in C$ denote the corresponding Weierstrass points, for $i \in B$. The double cover $\overline{C}$ is determined by a $4+2$ partition of these 6 branch points: say $0, \infty$ vs. e_0, e_1, e_2, e_3. The kernel of the pullback $\pi_2^* : J(C) \to J(\overline{C})$ is isomorphic to $\mathbb{Z}_2$, and we let μ be the non-zero element. It is a line bundle on C, given explicitly by $\mathcal{O}_C(p_0 - p_\infty)$. Its pullback $\widetilde{\mu} := \rho^*\mu$ is a line bundle of order 2 on $\widetilde{C}$, and is the non-trivial element in the kernel of $\widetilde{\pi}_2^*$.

Fix a point $\overline{p}$ of $\overline{C}$. The involution $\tau = \tau_2$ on $\overline{C}$ takes it to $\tau\overline{p}$, and both map by π_2 to the same point $p \in C$. Due to the total ramification, these three points have unique lifts to points $\widetilde{\overline{p}}, \widetilde{\tau\overline{p}} \in \widetilde{\overline{C}}$ and $\widetilde{p} \in \widetilde{C}$.

PROPOSITION A.2. *Each of the 16 spin structures N on C determines a relative spin structure (i.e. a square root of the relative canonical bundle) for the family $\widetilde{\mathcal{C}} \to \overline{C}_r$, so in particular a lift $\sigma_N : \overline{C}_r \to \mathcal{SM}_5$ of our $\sigma : \overline{C}_r \to \mathcal{M}_5$. Of the 10 even spin structures N, 6 give odd spin structures on the $\widetilde{C}$ and 4 give even spin structures on the $\widetilde{C}$.*

PROOF. Let N be a spin structure on C, i.e. a line bundle satisfying $N^2 \cong K_C$. It induces spin structures

$$\overline{N} := \pi_2^* N, \quad \widetilde{N} := \rho^* N(\widetilde{p}), \quad \widetilde{\overline{N}} := \overline{\rho}^* \pi_2^* N(\widetilde{\overline{p}} + \widetilde{\tau\overline{p}}) = \overline{\rho}^* \overline{N}(\widetilde{\overline{p}} + \widetilde{\tau\overline{p}}) = \widetilde{\pi}_2^* \widetilde{N}$$

on $\overline{C}$, $\widetilde{\overline{C}}$, and $\widetilde{C}$ respectively. We have two other natural spin structures: $N \otimes \mu$ on C and $\widetilde{N} \otimes \widetilde{\mu}$ on $\widetilde{C}$, which pull back to the same $\overline{N}, \widetilde{\overline{N}}$. The choice of $\widetilde{N}$ clearly gives a lift $\sigma_N : \overline{C}_r \to \mathcal{SM}_5$. Moreover, we get a global line bundle $\widetilde{\mathcal{N}}$ on $\widetilde{\mathcal{C}}$ whose square is the relative canonical bundle. We still need to compare the parities.

LEMMA A.3. *The spin structures $\overline{N}, \widetilde{\overline{N}}$ have the same parity.*

PROOF. Since $\overline{\rho}$ is cyclic, the direct image $\overline{\rho}_*(\widetilde{\overline{N}})$ decomposes under $\mathbb{Z}_3$:

$$\overline{\rho}_*(\widetilde{\overline{N}}) = \overline{N} \oplus (\overline{N} \otimes L) \oplus (\overline{N} \otimes L^{-1}),$$

where $L \in Pic(\overline{C})$ is the defining line bundle of the cyclic cover, satisfying (A.1), (A.2). But the two bundles $\overline{N} \otimes L$, $\overline{N} \otimes L^{-1}$ have the same parity (they are each other's Serre duals, and of zero Euler characteristic). So

$$h^0(\widetilde{\overline{C}}, \widetilde{\overline{N}}) = h^0(\overline{C}, \overline{\rho}_* \widetilde{\overline{N}}) = h^0(\overline{C}, \overline{N}) \qquad (\text{mod } 2).$$

□

LEMMA A.4. *The spin structures $\widetilde{N}$, $N \otimes \mu$ have opposite parity. (As do $\widetilde{N} \otimes \widetilde{\mu}$, N.)*

PROOF. These parities remain constant over connected families, so we may as well specialize to a convenient cover $\widetilde{C}$. We take it to be one of the orbifold points in the image $\sigma(\overline{C}_r) \subset \mathcal{M}_5$. Namely, we take the point $(e, \overline{p}) \in \overline{C}_r = E_r \times_E \overline{C}$ where e is one of the four ramification points of E_r over $\mathbb{P}^1_r$, so $m_3 e$ is one of the four ramification points of E over $\mathbb{P}^1$, say the one over $1 \in \mathbb{P}^1$, and $\overline{p}$ is one of the two points in $\pi_1^{-1}(e)$, with image $p_1 \in C$. The corresponding $\widetilde{C}$ is Galois over $\mathbb{P}^1$, with group S_3: it is the fiber product

$$\widetilde{C} \cong C \times_{\mathbb{P}^1} \widetilde{\mathbb{P}^1},$$

where $\widetilde{\mathbb{P}^1}$ is the triple cover of $\mathbb{P}^1$ with total ramification over $1 \in \mathbb{P}^1$ and simple ramification over $0, \infty \in \mathbb{P}^1$. (Such $\widetilde{\mathbb{P}^1}$ is uniquely specified by the above.) We see that in this special case where the cover is parametrized by one of the orbifold points, our previous snapshot can be extended:

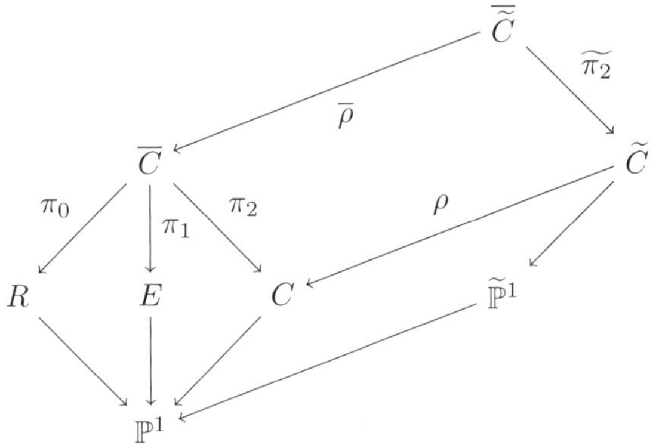

The advantage of this choice is that now $\widetilde{C}$ is hyperelliptic as is C, so we know everything about spin structures on them and we can check the claim directly. On a hyperelliptic curve of genus g with hyperelliptic line bundle H, the spin structures are of the form $\mathcal{O}(D) \otimes H^{(g-1-\ell)/2}$ where D is a subset of cardinality ℓ of the set of Weierstrass points, and $\ell \equiv g-1 \pmod 2$. The parity of this spin structure is then $(g+1-\ell)/2$.

Thus the 6 odd spin structures on our ${}_2C$ are the

$$N = \mathcal{O}_C(p_i), \;\; i \in B = \{0, \infty, e_0 = 1, e_1, e_2, e_3\},$$

and the 10 even ones are of the form $N = \mathcal{O}_C(p_i + p_j - p_k), \quad i, j, k \in B$. The line bundle μ is $\mu = \mathcal{O}_C(p_0 - p_\infty)$.

Our ${}_5\widetilde{C}$ has one Weierstrass point $\widetilde{p}_i$, $i = 0, \infty, 1$ above the corresponding p_i, and 3 Weierstrass point $\widetilde{p}^a_j$, $j = e_1, e_2, e_3$, $a = 1, 2, 3$ above the corresponding p_j, for a total of 12 Weierstrass points. These satisfy:

$$\rho^* \mathcal{O}_C(p_i) \cong \mathcal{O}_{\widetilde{C}}(\widetilde{p}_i) \otimes H, \quad i = 0, \infty, 1$$

and

$$\rho^* \mathcal{O}_C(p_j) \cong \mathcal{O}_{\widetilde{C}}(\widetilde{p}^1_j + \widetilde{p}^2_j + \widetilde{p}^3_j), \quad j = e_1, e_2, e_3.$$

It is therefore natural to write

$$\ell = \ell_0 + \ell_1 + \ell_2 \equiv 1 \pmod 2$$

where ℓ_0, ℓ_1, ℓ_2 are the numbers of points of D from the subsets $\{0, \infty\}, \{1\}, \{e_1, e_2, e_3\}$ respectively. The corresponding partition for $\widetilde{N} = \rho^* N(\widetilde{p}_1)$ is therefore

$$\widetilde{\ell} = \widetilde{\ell}_0 + \widetilde{\ell}_1 + \widetilde{\ell}_2$$

with

$$\widetilde{\ell}_0 = \ell_0,\ \widetilde{\ell}_1 = 1 - \ell_1,\ \widetilde{\ell}_2 = 3\ell_2$$

so

$$\text{parity}(\widetilde{N}) = \frac{5 - \ell_0 + \ell_1 - 3\ell_2}{2}.$$

On the other hand, for $N \otimes \mu$ we have

$$\begin{aligned} \ell_\mu &= \ell_{\mu,0} + \ell_{\mu,1} + \ell_{\mu,2} \\ &= (2 - \ell_0) + \ell_1 + \ell_2 \end{aligned}$$

so

$$\begin{aligned} \text{parity}(N \otimes \mu) &= \frac{1 + \ell_0 - \ell_1 - \ell_2}{2} \\ &= \text{parity}(\widetilde{L}) - (2 + 2\ell_1 - \ell) \\ &\equiv \text{parity}(\widetilde{L}) - 1 \pmod 2). \end{aligned}$$

□

To complete the proof of the Proposition, we therefore have to count the even spin structures N for which $N \otimes \mu$ is odd. In the notation of the previous proof, the condition is that ℓ_0 should be even. There are indeed four of these: $\mathcal{O}_C(p_k) \otimes \mu$, where k is one of e_0, e_1, e_2, e_3.

□

References

[1] P. Green, *On holomorphic graded manifolds*, Proc. Amer. Math. Soc. **85** (1982), no. 4, 587–590, DOI 10.2307/2044071. MR660609 (83m:32013)

[2] F. A. Berezin, *Introduction to superanalysis*, Mathematical Physics and Applied Mathematics, vol. 9, D. Reidel Publishing Co., Dordrecht, 1987. Edited and with a foreword by A. A. Kirillov; With an appendix by V. I. Ogievetsky; Translated from the Russian by J. Niederle and R. Kotecký; Translation edited by Dimitri Leĭtes. MR914369 (89b:58006)

[3] Y. I. Manin, *Gauge field theory and complex geometry*, Grundlehren der Mathematischen Wissenschaften [Fundamental Principles of Mathematical Sciences], vol. 289, Springer-Verlag, Berlin, 1988. Translated from the Russian by N. Koblitz and J. R. King. MR954833 (89d:32001)

[4] A. Yu. Vaĭntrob, *Deformations of complex structures on supermanifolds* (Russian), Funktsional. Anal. i Prilozhen. **18** (1984), no. 2, 59–60. MR745703 (85k:32041)

[5] M. J. Rothstein, *Deformations of complex supermanifolds*, Proc. Amer. Math. Soc. **95** (1985), no. 2, 255–260, DOI 10.2307/2044523. MR801334 (87d:32041)

[6] A. L. Onishchik, *On supermanifolds associated with the cotangent bundle*, Sūrikaisekikenkyūsho Kōkyūroku **1150** (2000), 45–53. Lie groups, geometric structures and differential equations—one hundred years after Sophus Lie (Japanese) (Kyoto/Nara, 1999). MR1793784

[7] M. A. Baranov and A. S. Shvarts, *Multiloop contribution to string theory* (Russian), Pis′ma Zh. Èksper. Teoret. Fiz. **42** (1985), no. 8, 340–342; English transl., JETP Lett. **42** (1985), no. 8, 419–421. MR875755 (88a:81152)

[8] M. A. Baranov and A. S. Schwarz, *On the multiloop contribution to the string theory*, Internat. J. Modern Phys. A **2** (1987), no. 6, 1773–1796, DOI 10.1142/S0217751X87000922. MR913613 (89e:81134)

[9] D. Friedan, *Notes on string theory and two-dimensional conformal field theory*, Workshop on unified string theories (Santa Barbara, Calif., 1985), World Sci. Publishing, Singapore, 1986, pp. 162–213. MR849106

[10] Yu. I. Manin, *Critical dimensions of string theories and a dualizing sheaf on the moduli space of (super) curves* (Russian), Funktsional. Anal. i Prilozhen. **20** (1986), no. 3, 88–89. MR868568 (88h:17021)

[11] A. A. Rosly, A. S. Schwarz, and A. A. Voronov, *Geometry of superconformal manifolds*, Comm. Math. Phys. **119** (1988), no. 1, 129–152. MR968484 (89m:58015)
[12] E. Witten, "Notes On Super Riemann Surfaces And Their Moduli," arXiv:1209.2459.
[13] S. N. Dolgikh, A. A. Rosly, and A. S. Schwarz, *Supermoduli spaces*, Comm. Math. Phys. **135** (1990), no. 1, 91–100. MR1086753 (92i:32005)
[14] E. D'Hoker and D. H. Phong, *The geometry of string perturbation theory*, Rev. Modern Phys. **60** (1988), no. 4, 917–1065, DOI 10.1103/RevModPhys.60.917. MR969998 (89m:81132)
[15] E. Witten, "Superstring Perturbation Theory Revisited," arXiv:1209.5461.
[16] E. D'Hoker and D. H. Phong, *Lectures on two-loop superstrings*, Superstring theory, Adv. Lect. Math. (ALM), vol. 1, Int. Press, Somerville, MA, 2008, pp. 85–123. MR2464928 (2010f:81219)
[17] M. Nagata, *Imbedding of an abstract variety in a complete variety*, J. Math. Kyoto Univ. **2** (1962), 1–10. MR0142549 (26 #118)
[18] B. Conrad, *Deligne's notes on Nagata compactifications*, J. Ramanujan Math. Soc. **22** (2007), no. 3, 205–257. MR2356346 (2009d:14002)
[19] J. Ebert and O. Randal-Williams, *Stable cohomology of the universal Picard varieties and the extended mapping class group*, Doc. Math. **17** (2012), 417–450. MR2946828
[20] O. Randal-Williams, *The Picard group of the moduli space of r-spin Riemann surfaces*, Adv. Math. **231** (2012), no. 1, 482–515, DOI 10.1016/j.aim.2012.04.027. MR2935396
[21] S. Gorchinskiy and F. Viviani, *A note on families of hyperelliptic curves*, Arch. Math. (Basel) **92** (2009), no. 2, 119–128, DOI 10.1007/s00013-008-2746-1. MR2481507 (2010b:14055)
[22] S. M. Natanzon, *Moduli of Riemann surfaces, real algebraic curves, and their superanalogs*, Translations of Mathematical Monographs, vol. 225, American Mathematical Society, Providence, RI, 2004. Translated from the 2003 Russian edition by Sergei Lando. MR2075914 (2005d:32020)
[23] R. Donagi and E. Witten, *Super Atiyah classes and obstructions to splitting of supermoduli space*, Pure Appl. Math. Q. **9** (2013), no. 4, 739–788. MR3263973
[24] E. Witten, "Notes On Supermanifolds and Integration," arXiv:1209.2199
[25] K. Kodaira, *A certain type of irregular algebraic surfaces*, J. Analyse Math. **19** (1967), 207–215. MR0216521 (35 #7354)
[26] M. F. Atiyah, *The signature of fibre-bundles*, Global Analysis (Papers in Honor of K. Kodaira), Univ. Tokyo Press, Tokyo, 1969, pp. 73–84. MR0254864 (40 #8071)
[27] F. Hirzebruch, *The signature of ramified coverings*, Global Analysis (Papers in Honor of K. Kodaira), Univ. Tokyo Press, Tokyo, 1969, pp. 253–265. MR0258060 (41 #2707)
[28] J. Bryan and R. Donagi, *Surface bundles over surfaces of small genus*, Geom. Topol. **6** (2002), 59–67 (electronic), DOI 10.2140/gt.2002.6.59. MR1885589 (2002m:14037)
[29] F. Oort, *Subvarieties of moduli spaces*, Invent. Math. **24** (1974), 95–119. MR0424813 (54 #12771)
[30] J. Harris and I. Morrison, *Moduli of curves*, Graduate Texts in Mathematics, vol. 187, Springer-Verlag, New York, 1998. MR1631825 (99g:14031)
[31] M. Cornalba, *Moduli of curves and theta-characteristics*, Lectures on Riemann surfaces (Trieste, 1987), World Sci. Publ., Teaneck, NJ, 1989, pp. 560–589. MR1082361 (91m:14037)
[32] L. Caporaso, C. Casagrande, and M. Cornalba, *Moduli of roots of line bundles on curves*, Trans. Amer. Math. Soc. **359** (2007), no. 8, 3733–3768 (electronic), DOI 10.1090/S0002-9947-07-04087-1. MR2302513 (2008i:14041)
[33] G. Farkas, *Theta characteristics and their moduli*, Milan J. Math. **80** (2012), no. 1, 1–24, DOI 10.1007/s00032-012-0178-7. MR2984107
[34] D. Mumford, *Prym varieties. I*, Contributions to analysis (a collection of papers dedicated to Lipman Bers), Academic Press, New York, 1974, pp. 325–350. MR0379510 (52 #415)

Departments of Mathematics and Physics, University of Pennsylvania, Philadelphia, Pennsylvania 19104
E-mail address: donagi@math.upenn.edu

Institute for Avanced Study, Princeton, New Jersey 08540
E-mail address: witten@ias.edu

Proceedings of Symposia in Pure Mathematics
Volume **90**, 2015
http://dx.doi.org/10.1090/pspum/090/01520

Generalised Moonshine and Holomorphic Orbifolds

Matthias R. Gaberdiel, Daniel Persson, and Roberto Volpato

Dedicated to the memory of Friedrich Hirzebruch

ABSTRACT. Generalised moonshine is reviewed from the point of view of holomorphic orbifolds, putting special emphasis on the role of the third cohomology group $H^3(G, U(1))$ in characterising consistent constructions. These ideas are then applied to the case of Mathieu moonshine, i.e. the recently discovered connection between the largest Mathieu group M_{24} and the elliptic genus of K3. In particular, we find a complete list of twisted twining genera whose modular properties are controlled by a class in $H^3(M_{24}, U(1))$, as expected from general orbifold considerations.

1. Introduction

Monstrous moonshine refers to a deep connection between modular forms, the Monster group $\mathbb{M}$, generalised Kac-Moody algebras and string theory. It unfolded over the course of 15 years, starting with the Conway-Norton conjecture in 1979 [**1**] and the subsequent construction of the Frenkel-Lepowsky-Meurman Monster module $\mathcal{V}^\natural$ [**2**], finally culminating in Borcherds complete proof of the moonshine conjecture [**3**]. In a nutshell, monstrous moonshine asserts that for each element $g \in \mathbb{M}$ of the Monster group, there exists a class function T_g (the McKay-Thompson series), which is a holomorphic modular function (more precisely, the hauptmodul for a genus zero subgroup of $SL(2,\mathbb{R})$, see [**1**]) on the upper-half-plane $\mathbb{H}$, and for which the Fourier coefficients are characters of representations of $\mathbb{M}$. For example, when $g = e$ (the identity element), the McKay-Thompson series T_e coincides with the modular-invariant J-function whose coefficients are dimensions of Monster group representations.

A few years after the original moonshine conjectures, Norton proposed [**4**] an extension that he dubbed generalised monstrous moonshine. Norton argued that to each commuting pair (g, h) of elements in $\mathbb{M}$ there should exist a holomorphic modular function $f(g, h; \tau)$ on $\mathbb{H}$, whose Fourier coefficients also carry representation-theoretic information about the Monster. The generalised moonshine conjecture was subsequently interpreted physically by Dixon, Ginsparg, Harvey [**5**] in terms of orbifolds of the Monster CFT $\mathcal{V}^\natural$. Although the conjecture has been proven for many special cases [**6–8**], the general case remains open (see however [**9–12**] for recent progress).

2010 *Mathematics Subject Classification.* Primary 81R10, 81R05.

A new moonshine phenomenon was conjectured in 2010 by Eguchi, Ooguri, Tachikawa (EOT) [**13**], subsequently dubbed Mathieu moonshine. In this case the Monster group $\mathbb{M}$ is replaced by the largest Mathieu group M_{24}, and the role of the modular J-function is played by the unique weak Jacobi form $\phi_{0,1}(\tau, z)$ of weight 0 and index 1 corresponding to the elliptic genus of K3. The analogue of the McKay-Thompson series, the so called twining genera $\phi_g(\tau, z)$, $g \in M_{24}$, were constructed in a series of papers [**14**–**17**], and it was verified that they have precisely the properties required for Mathieu moonshine to hold. Indeed, Gannon has recently shown [**18**] that all multiplicity spaces can be consistently decomposed into sums of irreducible representations of M_{24}, thereby proving the EOT conjecture.

Although this establishes Mathieu moonshine, there is a major outstanding question: what is the M_{24}-analogue of the Monster module $\mathcal{V}^\natural$? In [**19**] we gave evidence that some kind of holomorphic vertex operator algebra (VOA) should be underlying Mathieu moonshine. The main point was to extend the previous results on twining genera to the complete set of twisted twining genera $\phi_{g,h}(\tau, z)$, corresponding to the M_{24}-analogues of Norton's generalised moonshine functions $f(g, h; \tau)$ for the Monster. One of the key insights was that many of the properties of these functions, such as modularity, are controlled by a class in the third cohomology group $H^3(M_{24}, U(1))$, just as for orbifolds of holomorphic VOAs [**20**–**22**].

Our aim in this note is to give a short review of the generalised Mathieu moonshine phenomenon uncovered in [**19**], focussing on the main ideas rather than technical details. For completeness we include a discussion of holomorphic orbifolds and group cohomology which are the key ingredients in our work, as well as some background on Norton's generalised moonshine conjecture, which served as strong motivation for [**19**].

This short note is organised as follows. We begin in section 2 by discussing some features of orbifolds of holomorphic VOAs, explaining in particular the crucial role played by the cohomology group $H^3(G, U(1))$. In section 3 we then proceed to discuss generalised Mathieu moonshine. We define the twisted twining genera and list the properties they should satisfy. We show that there is a unique class in $H^3(M_{24}, U(1))$ that is compatible with the modular properties of the twining genera, and use this input to construct all twisted twining genera explicitly. Finally, we end in section 4 with a brief summary.

2. Holomorphic Orbifolds Generalised Moonshine

In this section we will review some pertinent properties of orbifolds of holomorphic VOAs, with particular focus on the role of group cohomology.

2.1. Preliminaries. Let $\mathcal{V}$ be a rational vertex operator algebra (VOA), and let $\mathcal{H}$ be a $\mathbb{Z}$-graded $\mathcal{V}$-module.[1] Rationality implies that $\mathcal{V}$ has only finitely many inequivalent simple modules $\mathcal{H}$, and that each graded component of $\mathcal{H}$ is finite-dimensional. By a *holomorphic* (or 'self-dual') VOA we shall mean the case that $\mathcal{V}$ has a *unique* such module, namely the adjoint module of $\mathcal{V}$ itself; in this case we shall also write $\mathcal{V}$ for this module. The partition function of a holomorphic VOA is a holomorphic section of a line bundle over the moduli space of Riemann surfaces.

[1]See for instance [**23**] for a nice introduction to VOAs.

The most prominent example of a holomorphic VOA is the moonshine module $\mathcal{V}^\natural$ [**2**], to which we shall return below.

Suppose we have a holomorphic VOA $\mathcal{V}$ with finite automorphism group G. We want to analyse the orbifold of $\mathcal{V}$ by G, denoted $\widehat{\mathcal{V}} = \mathcal{V}/G$. The first step consists in projecting onto the G-invariant sub-VOA

$$\mathcal{V}^G = \{\psi \in \mathcal{V} \mid g\psi = \psi, \forall g \in G\} .$$

The character of the VOA $\mathcal{V}^G$ is however not modular invariant, and to remedy this we must include twisted sectors. Since $\mathcal{V}$ is holomorphic the twisted sectors are just labelled by conjugacy classes in G, i.e. for each $g \in G$ there is a g-twisted simple $\mathcal{V}$-module (or g-twisted sector) $\mathcal{H}_g$ [**7**], which is an ordinary module for the G-invariant sub-VOA $\mathcal{V}^G$. The twisted sectors associated to group elements in the same conjugacy class are isomorphic.

Each automorphism $h \in G$ of the VOA $\mathcal{V}$ induces a linear map $\mathcal{H}_g \to \mathcal{H}_{hgh^{-1}}$ between twisted sectors. In particular, each twisted sector $\mathcal{H}_g$ carries a representation of the centraliser

$$C_G(g) := \{h \in G | hgh^{-1} = g\} \subseteq G$$

of g in G, though in general this will not be an honest representation. We will discuss this important subtlety below.

2.2. Twisted Twining Characters. Given a holomorphic VOA $\mathcal{V}$ of central charge c its partition function is defined by the usual formula

$$\mathscr{Z}_{\mathcal{V}}(\tau) = \mathrm{Tr}_{\mathcal{V}}(q^{L_0 - c/24}) ,$$

where $q = e^{2\pi i \tau}$ and L_0 is the Virasoro (Cartan) generator. Similarly, for each twisted sector $\mathcal{H}_g$ in the orbifold theory one may construct the associated twisted character (sometimes called 'characteristic function')

$$Z_{g,e}(\tau) = \mathrm{Tr}_{\mathcal{H}_g}(q^{L_0 - c/24}) ,$$

where e denotes the identity element in G. Moreover, since $\mathcal{H}_g$ is invariant under the centraliser subgroup $C_G(g)$ it makes sense to define, for all $h \in C_G(g)$, the twisted twining character

$$Z_{g,h}(\tau) = \mathrm{Tr}_{\mathcal{H}_g}\big(\rho(h)\, q^{L_0 - c/24}\big) ,$$

where $\rho : C_G(g) \to \mathrm{End}(\mathcal{H}_g)$ denotes the representation with which h acts on the twisted vector space $\mathcal{H}_g$.

Physically, the twisted twining character $Z_{g,h}$ corresponds to the path integral on a torus with modular parameter τ and boundary conditions twisted by (g, h) along the (a, b)-cycles of $\mathbb{T}^2$. Choosing periodic boundary conditions corresponds to setting $(g, h) = (e, e)$ and hence gives back the original partition function

$$Z_{e,e}(\tau) = \mathscr{Z}(\tau) .$$

Given the definition of the twisted twining characters $Z_{g,h}$ one should expect that they only depend on the conjugacy class of (g, h) in G, i.e. they should correspond to class functions

$$Z_{g,h}(\tau) = Z_{k^{-1}gk,k^{-1}hk}(\tau) , \qquad k \in G . \tag{2.1}$$

As we shall see (see section 2.3 below), in general this property will only be true up to a (multiplier) phase.

In contrast to $\mathscr{X}(\tau)$ the twisted twining characters $Z_{g,h}(\tau)$ are not invariant under the full modular group $SL(2,\mathbb{Z})$. Under a modular transformation

$$\tau \longmapsto \frac{a\tau+b}{c\tau+d}\,, \qquad \begin{pmatrix} a & b \\ c & d \end{pmatrix} \in SL(2,\mathbb{Z})\,,$$

the spin structures of the torus change such that the twists by g and h along the a- and b-cycles transform according to

$$(g,h) \longmapsto (g,h)\begin{pmatrix} a & b \\ c & d \end{pmatrix}^{-1} = (g^d h^{-c}, g^{-b} h^a)\,.$$

The twisted twining characters then transform among themselves as

$$Z_{g,h}\left(\frac{a\tau+b}{c\tau+d}\right) = \chi_{g,h}\left(\begin{smallmatrix} a & b \\ c & d \end{smallmatrix}\right) Z_{g^a h^c, g^b h^d}(\tau)\,, \tag{2.2}$$

where we have included the possibility of having a non-trivial multiplier system

$$\chi_{g,h}\,:\, SL(2,\mathbb{Z}) \longrightarrow U(1)\,.$$

The set of functions $\{Z_{g,h}\}$ thus forms a representation of $SL(2,\mathbb{Z})$.

2.3. Twisted Sectors and Projective Representations. As we have mentioned above, the states in the twisted sector $\mathcal{H}_g$ transform in a representation ρ of $C_G(g)$. However, this representation need not be an honest representation, but may only be projective. Recall that a projective representation ρ of a finite group H respects the group multiplication only up to a phase,

$$\rho(h_1)\,\rho(h_2) = c(h_1,h_2)\,\rho(h_1 h_2)\,,$$

where $c(h_1,h_2)$ is a $U(1)$-valued 2-cocycle, representing a class in $H^2(H,U(1))$. Thus we have, for each twisted sector $\mathcal{H}_g$, a class $c_g \in H^2(C_G(g),U(1))$, characterising the projectivity of the action of $C_G(g)$ in the g-twisted sector. One consequence of these phases is that the formula (2.1) must be modified; the correct generalisation is

$$Z_{g,h}(\tau) = \frac{c_g(h,k)}{c_g(k,k^{-1}hk)}\, Z_{k^{-1}gk,k^{-1}hk}(\tau)\,. \tag{2.3}$$

Furthermore, these phases modify the modular S and T-transformations as [**22**]

$$\begin{aligned} Z_{g,h}(\tau+1) &= c_g(g,h)\, Z_{g,gh}(\tau)\,, \\ Z_{g,h}(-1/\tau) &= \overline{c_h(g,g^{-1})}\, Z_{h,g^{-1}}(\tau)\,. \end{aligned} \tag{2.4}$$

Although the twisted twining genera $Z_{g,h}$ are not invariant under the full $SL(2,\mathbb{Z})$, they will be modular functions with respect to some arithmetic subgroup $\Gamma_{g,h} \subset SL(2,\mathbb{Z})$ which fixes the pair (g,h). The group $\Gamma_{g,h}$ will, in particular, contain a congruence subgroup $\Gamma(N) \subset SL(2,\mathbb{Z})$, for a suitable positive integer N. Each $Z_{g,h}$ therefore has a Fourier expansion of the form

$$Z_{g,h}(\tau) = \sum_n \mathrm{Tr}_{\mathcal{H}_{g,n}}\left(\rho(h)\right) q^{n/N},$$

where $\mathcal{H}_{g,n}$ is the grade n subspace of the twisted module $\mathcal{H}_g$ and $\mathrm{Tr}_{\mathcal{H}_{g,n}}(\rho(h))$ is a projective character of $C_G(g)$, i.e. a character of a graded representation of a central extension of $C_G(g)$. We can therefore decompose the different graded components $\mathcal{H}_{g,n}$ of the twisted module $\mathcal{H}_g$ into (finite) sums of irreducible projective representations R_j, each corresponding to the same 2-cocycle class c_g

$$\mathcal{H}_{g,n} = \bigoplus_j h_{g,n}^{(j)} R_j \ . \tag{2.5}$$

Here $h_{g,n}^{(j)}$ describes the multiplicity with which R_j occurs.

2.4. Group Cohomology of Holomorphic Orbifolds. In the previous section we have seen that the action of $h \in G$ on the twisted sector $\mathcal{H}_g$ is generically projective, and inequivalent choices are classified by $H^2(C_G(g), U(1))$. The appearance of this cohomology group can in fact be traced back to an even finer and more sophisticated underlying structure, namely the third cohomology group $H^3(G, U(1))$.

For every commuting pair $g, h \in G$ the fusion product between the associated twisted sectors induces an isomorphism $\mathcal{H}_g \boxtimes \mathcal{H}_h \to \mathcal{H}_{gh}$. For every triple $g, h, k \in G$ there exists a 3-cocycle $\alpha(g, h, k) \in H^3(G, U(1))$ which measures the failure of associativity in the choice of isomorphism for the triple fusion product [**21**]. The third cohomology group therefore classifies consistent holomorphic orbifolds. The class $[\alpha] \in H^3(G, U(1))$ determines many properties of the orbifold theory $\widehat{\mathcal{V}}$. In particular, it determines the particular central extension of $C_G(g)$ which controls the projective representations ρ in the g-twisted sector $\mathcal{H}_g$. Indeed, for every $h \in G$, the 3-cocycle α gives rise to a distinguished element $c_h \in H^2(C_G(h), U(1))$ through the formula [**20, 22**]

$$c_h(g_1, g_2) = \frac{\alpha(h, g_1, g_2)\ \alpha(g_1, g_2, (g_1 g_2)^{-1} h (g_1 g_2))}{\alpha(g_1, h, h^{-1} g_2 h)} \ . \tag{2.6}$$

Since the projective phases also control the modular properties of $Z_{g,h}$ (2.4), these are then also determined in terms of the class $[\alpha]$.

These phases actually lead to a number of interesting consequences. For example, for the special case when k, g and h are pairwise commuting elements in G, we get

$$Z_{g,h}(\tau) = \frac{c_g(h,k)}{c_g(k,h)}\, Z_{g,h}(\tau) \ ,$$

and thus $Z_{g,h}(\tau) = 0$ unless the 2-cocycle satisfies the regularity condition

$$c_g(h, k) = c_g(k, h) \tag{2.7}$$

for all $k \in G$ that commute both with g and h.

2.5. Application to Generalised Monstrous Moonshine. We will now discuss a specific example of the framework introduced above, that arises for the case when $\mathcal{V}$ is the Frenkel-Lepowsky-Meurman (FLM) Monster VOA $\mathcal{V}^\natural$ [**2**] with $c = 24$, whose automorphism group G is the Monster group $\mathbb{M}$.

In 1987 Norton proposed [**4**] a generalisation of monstrous moonshine in which he suggested that it was natural to associate a holomorphic function $f(g, h; \tau)$ to each commuting pair (g, h) of elements in $\mathbb{M}$. Norton argued that these functions should satisfy the following conditions:

(1) $f(g, h; \tau) = f(k^{-1} g k, k^{-1} h k; \tau)\ , \qquad k \in \mathbb{M}$

(2) $f(g,h;\tau) = \gamma f\big(g^a h^c, g^b h^d; \frac{a\tau+b}{c\tau+d}\big)$, $\left(\begin{smallmatrix} a & b \\ c & d \end{smallmatrix}\right) \in SL(2,\mathbb{Z})$
(Here γ is a 24'th root of unity.)
(3) *the coefficients in the q-expansion of $f(g,h;\tau)$ are characters of a graded projective representation of $C_{\mathbb{M}}(g)$*
(4) *$f(g,h;\tau)$ is either constant or a hauptmodul for some genus zero $\Gamma_{g,h} \subset SL(2,\mathbb{R})$*
(5) *$f(e,h;\tau) = T_h(\tau)$, where T_h is the McKay-Thompson series associated to h*

All of these conditions, with the exception of the genus zero property (4), can be understood within the framework of holomorphic orbifolds. The FLM Monster module $\mathcal{V}^\natural$ is a holomorphic VOA and so for each $g \in \mathbb{M}$ we have a unique g-twisted module $\mathcal{H}_g^\natural$ with an inherited grading

$$\mathcal{H}_g^\natural = \bigoplus_{n=-N}^{\infty} \mathcal{H}_{g,n}^\natural ,$$

where each $\mathcal{H}_{g,n}^\natural$ is a projective representation of $C_{\mathbb{M}}(g)$. For each twisted module $\mathcal{H}_g^\natural$ we can define the associated twisted twining character

$$Z_{g,h}^{(\mathcal{V}^\natural)}(\tau) = \mathrm{Tr}_{\mathcal{H}_g^\natural}(\rho(h)\, q^{L_0-1}) = \sum_{n=-N}^{\infty} \mathrm{Tr}_{\mathcal{H}_{g,n}^\natural}\big(\rho(h)\big)\, q^{n/N},$$

for a suitable positive integer N. By the properties of holomorphic orbifolds discussed above, this twisted twining character satisfies properties (1) – (3) and (5) of Norton, and it is therefore natural to suspect that $Z_{g,h}^{(\mathcal{V}^\natural)}(\tau) = f(g,h;\tau)$. This connection was first made by Dixon, Ginsparg and Harvey [**5**], and has subsequently been proven in many special cases [**6**–**8**], though the general conjecture remains open.

Since generalised moonshine can be understood within the framework of holomorphic orbifolds, one should expect that the third cohomology group $H^3(\mathbb{M}, U(1))$ plays an important role (see [**24**] for a related discussion). In particular, one might guess that Norton's condition (1) should be generalised to include the cohomological prefactor (2.3), involving a 2-cocycle $c_g \in H^2(C_{\mathbb{M}}(g), U(1))$. Moreover, the roots of unity γ appearing in the modular transformation (2) should be computable from some $\alpha \in H^3(\mathbb{M}, U(1))$, via the general formulae (2.4). This would then also suggest that the cases where $f(g,h;\tau)$ are constant (see condition (4)) are manifestations of an obstruction, for example of the type described above in (2.7).[2] Unfortunately, little is known about $H^3(\mathbb{M}, U(1))$, and thus it is difficult to confirm this directly.

3. Generalised Mathieu Moonshine

3.1. A Lightning Review of Mathieu Moonshine. In 2010, Eguchi, Ooguri and Tachikawa [**13**] conjectured a supersymmetric version of the moonshine phenomenon for a certain sporadic finite simple group, the Mathieu group M_{24}, where the role of the J-function is played by the elliptic genus of K3. The latter is most naturally defined as a refined partition function of a certain class of two-dimensional superconformal field theories with $\mathcal{N} = (4,4)$ superconformal

[2] We thank Terry Gannon for suggesting this idea to us.

symmetry, which have central charge $c = 6$, and can be realised as non-linear sigma models with target space K3. More precisely, the elliptic genus is a complex function on $\mathbb{H} \times \mathbb{C}$ defined as

$$\phi_{\mathrm{K3}}(\tau, z) = \mathrm{Tr}_{RR}\big((-1)^{F+\tilde{F}} q^{L_0 - \frac{c}{24}} \bar{q}^{\tilde{L}_0 - \frac{\tilde{c}}{24}} y^{2J_0^3}\big) , \qquad q = e^{2\pi i \tau}, \quad y = e^{2\pi i z} ,$$

where $L_0, \tilde{L}_0$ are the left- and right-moving Virasoro generators, $(-1)^{F+\tilde{F}}$ is the total worldsheet fermion number, J_0^3 is the Cartan generator of the affine $su(2)_1$ subalgebra of the left $\mathcal{N} = 4$ superconformal algebra, and the trace is taken over the Ramond-Ramond sector $\mathcal{H}_{\mathrm{RR}}$ of the theory. In general, the elliptic genus can be defined in any theory with (at least) $\mathcal{N} = 2$ superconformal symmetry and does not change under superconformal deformations of the theory. In a non-linear sigma model, this means that ϕ is independent of the choice of a metric and the Kalb-Ramond field of the target space, but it encodes information on the topology. For example, $\phi(\tau, z = 0)$ is the Euler number of the target space, so that in particular $\phi_{\mathrm{K3}}(\tau, 0) = 24$.

The only states that give rise to a non-vanishing contribution to ϕ_{K3} are the right-moving ground states, i.e. the eigenstates with zero eigenvalue for $\tilde{L}_0 - \frac{1}{4}$; this implies that K3 is holomorphic both in τ and z. The elliptic genus has good modular properties

$$\phi_{g,h}\Big(\frac{a\tau + b}{c\tau + d}, \frac{z}{c\tau + d}\Big) = e^{2\pi i \frac{cz^2}{c\tau + d}} \phi_{\mathrm{K3}}(\tau, z) , \qquad \left(\begin{smallmatrix} a & b \\ c & d \end{smallmatrix}\right) \in SL(2, \mathbb{Z}) ,$$

and because of the spectral flow automorphism of the $\mathcal{N} = 4$ superconformal algebra, possesses the elliptic transformation rules [**25**]

$$\phi_{\mathrm{K3}}(\tau, z + \ell\tau + \ell') = e^{-2\pi i(\ell^2\tau + 2\ell z)} \phi_{\mathrm{K3}}(\tau, z) \qquad \ell, \ell' \in \mathbb{Z} .$$

These are the defining properties of a weak Jacobi form of weight 0 and index 1 [**26**], and are sufficient to determine ϕ_{K3} up to normalisation, which in turn is fixed by the condition $\phi_{\mathrm{K3}}(\tau, 0) = 24$. Explicitly,

$$\phi_{\mathrm{K3}}(\tau, z) = 8\Big(\frac{\vartheta_2(\tau, z)^2}{\vartheta_2(\tau, 0)^2} + \frac{\vartheta_3(\tau, z)^2}{\vartheta_3(\tau, 0)^2} + \frac{\vartheta_4(\tau, z)^2}{\vartheta_4(\tau, 0)^2}\Big) ,$$

in terms of Jacobi theta functions [**26**]. The states contributing to the elliptic genus form a representation of the left $\mathcal{N} = 4$ superconformal algebra, so that ϕ_{K3} admits a decomposition into irreducible $\mathcal{N} = 4$ characters

$$\phi_{\mathrm{K3}}(\tau, z) = 20\, \mathrm{ch}_{\frac{1}{4}, 0}(\tau, z) - 2\, \mathrm{ch}_{\frac{1}{4}, \frac{1}{2}}(\tau, z) + \sum_{n=1}^{\infty} A_n\, \mathrm{ch}_{\frac{1}{4}+n, \frac{1}{2}}(\tau, z) .$$

Here, $\mathrm{ch}_{h,\ell}(\tau, z) = \mathrm{Tr}_{h,\ell}((-1)^F q^{L_0 - \frac{c}{24}} y^{2J_0^3})$ is the character of the Ramond $\mathcal{N} = 4$ representation whose highest weight vector is an eigenstate with eigenvalues h, ℓ under L_0 and J_0^3, respectively. By unitarity, the only possible values for (h, ℓ) are $(\frac{1}{4}, 0)$, $(\frac{1}{4}, \frac{1}{2})$ (short or BPS representations), and $(\frac{1}{4} + n, \frac{1}{2})$, $n = 1, 2, 3, \ldots$ [**27, 28**]. Finally, if $N(h, \ell; \bar{h}, \bar{\ell})$ is the multiplicity of the corresponding $\mathcal{N} = (4, 4)$ representation in the spectrum of the theory, then

$$A_n := \sum_{(h=\frac{1}{4}+n, \ell=\frac{1}{2}; \bar{h}, \bar{\ell})} \mathrm{ch}_{\bar{h},\bar{\ell}}(\bar{\tau}, 0) = N(\tfrac{1}{4} + n, \tfrac{1}{2}; \tfrac{1}{4}, 0) - 2N(\tfrac{1}{4} + n, \tfrac{1}{2}; \tfrac{1}{4}, \tfrac{1}{2})$$

is the $\mathbb{Z}_2$-graded multiplicity of the (h,ℓ) representations of the left $\mathcal{N}=4$ algebra. As it turns out, the A_n with $n \geq 1$ are all even positive integers. The most surprising property, however, is that the first few of them

$$\tfrac{1}{2}A_n = 45,\ 231,\ 770,\ 2277,\ 5796,\ \ldots$$

exactly match the dimensions of some irreducible representations of M_{24} [**13**]. There is a similar construction in the monstrous moonshine case: the J-function can be decomposed into Virasoro characters and the multiplicities of the first few Virasoro representations are dimensions of *irreducible* representations of the Monster group.[3] In analogy with the monstrous moonshine observation, it is then natural to conjecture that the space of states contributing to the elliptic genus carries an action of M_{24}, commuting with the $\mathcal{N}=4$ algebra, so that

$$\phi_{\mathrm{K3}}(\tau,z) = \sum_{(h,\ell)} \dim R_{h,\ell}\, \mathrm{ch}_{h,\ell}(\tau,z)\ ,$$

for some (possibly virtual) M_{24} representations $R_{h,\ell}$. Soon after the EOT observation, the analogues of the McKay-Thompson series, the twining genera

$$\phi_g(\tau,z) = \sum_{(h,\ell)} \mathrm{Tr}_{R_{h,\ell}}(g)\, \mathrm{ch}_{h,\ell}(\tau,z)\ , \qquad g \in M_{24} \tag{3.1}$$

have been considered [**14, 15**]. Each ϕ_g is expected to be a Jacobi form of weight 0 and index 1 (possibly up to a multiplier [**15, 16**]) for a group

$$\Gamma_0(N) := \left\{ \begin{pmatrix} a & b \\ c & d \end{pmatrix} \in SL(2,\mathbb{Z}) \mid c \equiv 0 \mod N \right\}\ ,$$

where $N = o(g)$ is the order of g. Explicitly [**16**],

$$\phi_g(\tau, z+\ell\tau+\ell') = e^{-2\pi i(\ell^2\tau+2\ell z)}\, \phi_g(\tau,z)\ , \qquad \ell,\ell' \in \mathbb{Z}\ , \tag{3.2}$$

$$\phi_g\Big(\frac{a\tau+b}{c\tau+d}, \frac{z}{c\tau+d}\Big) = e^{2\pi i \frac{cd}{N\ell(g)}}\, e^{2\pi i \frac{cz^2}{c\tau+d}}\, \phi_g(\tau,z)\ , \qquad \left(\begin{smallmatrix} a & b \\ c & d \end{smallmatrix}\right) \in \Gamma_0(N)\ , \tag{3.3}$$

where $\ell(g)$ is the length of the shortest cycle of g in the 24-dimensional permutation representation of M_{24} [**29**]. A complete list of twining genera satisfying (3.2) and (3.3) has been proposed in [**16, 17**], where the first few hundred M_{24}-representation $R_{h,\ell}$ have been computed explicitly. Finally, it was shown in [**18**] that all representations $R_{h,\ell}$ matching (3.1) for all $g \in M_{24}$ exist, and that the the only virtual representations correspond to the BPS characters

$$R_{\frac{1}{4},0} = \mathbf{23} - 3\cdot\mathbf{1}\ , \qquad R_{\frac{1}{4},\frac{1}{2}} = -2\cdot\mathbf{1}\ . \tag{3.4}$$

These results, in a sense, prove the EOT conjecture. The interpretation of this Mathieu moonshine, however, is still an open problem. The most obvious explanation would be the existence of a non-linear sigma model on K3 with symmetry group M_{24}. If such a theory existed, the twining genera could be identified with the traces

$$\phi_g(\tau,z) = \mathrm{Tr}_{RR}\big(g(-1)^{F+\tilde{F}} q^{L_0-\frac{c}{24}} \bar{q}^{\tilde{L}_0-\frac{\tilde{c}}{24}} y^{2J_0^3}\big)\ , \tag{3.5}$$

[3]In analogy with the original McKay observation, one could also decompose directly the Fourier coefficients of the elliptic genus in terms of dimensions of M_{24} representations, without any reference to $\mathcal{N}=4$ characters. However, in contrast with the J-function, in the decomposition of the elliptic genus one has to allow for M_{24} representations with negative multiplicity, corresponding to contributions from states with odd fermion number. Due to this complication, the connection with M_{24} becomes manifest only after decomposing into $\mathcal{N}=4$ characters.

and (3.2) and (3.3) would follow by standard CFT arguments. This possibility, however, has been excluded in [**30**], where the actual groups of symmetries of non-linear sigma models on K3 have been classified, and it was shown that none of them contains the Mathieu group M_{24}. More generally, one might conjecture the existence of some unknown CFT with $\mathcal{N}=(4,4)$ superconformal symmetry and carrying an action of M_{24} such that the twining genera ϕ_g are reproduced by (3.5) for all $g\in M_{24}$. However, because of the $-3\cdot\mathbf{1}$ in (3.4), this theory should contain fields in the R-R representation $(h,\ell;\bar h,\bar\ell)=(\frac14,0;\frac14,\frac12)$ of the $\mathcal{N}=(4,4)$ algebra, which, by spectral flow, correspond to fields in the NS-NS representation $(h,\ell;\bar h,\bar\ell)=(\frac12,\frac12;0,0)$. It has been argued in [**31**] that every theory with $\mathcal{N}=(4,4)$ superconformal symmetry at $c=6$ containing such fields is necessarily a non-linear sigma model on a torus, for which the elliptic genus vanishes. Thus it seems that a satisfactory explanation of Mathieu moonshine will need some more radically new idea.

3.2. Twisted Twining Genera: Definitions and Properties. As explained in the previous subsection, the twining genera ϕ_g, $g\in M_{24}$ satisfy all the properties expected for traces of the form (3.5) in a $\mathcal{N}=(4,4)$ theory with symmetry M_{24}. In such a (conjectural) theory, the twisted sector $\mathcal{H}_g$, for each $g\in M_{24}$, would form a representation ρ_g of the centraliser $C_{M_{24}}(g)$, whose action commutes with the $\mathcal{N}=(4,4)$ superconformal algebra. The characters

(3.6)
$$\phi_{g,h}(\tau,z)=\mathrm{Tr}_{\mathcal{H}_g}\big(\rho_g(h)\,(-1)^{F+\tilde F}q^{L_0-\frac{c}{24}}\bar q^{\tilde L_0-\frac{\tilde c}{24}}y^{2J_0^3}\big),\qquad g,h\in M_{24},\ gh=hg,$$

would be the $\mathcal{N}=4$ counterpart of the twisted twining partition functions $Z_{g,h}$ considered in section 2, and should obey analogous properties.

As we have stressed above, a superconformal field theory with the properties above is not known. However, following the philosophy of the previous subsection, we will show that functions $\phi_{g,h}$ exist, satisfying all the properties expected for characters of the form (3.6). This is very convincing evidence in favour of a generalised Mathieu moonshine, analogous to Norton's conjecture in the Monster case.

The definition of the twisted twining genera $\phi_{g,h}$ in terms of (3.6) suggests that they should satisfy the following properties:

(1) Elliptic and modular properties:

$$\phi_{g,h}(\tau,z+\ell\tau+\ell')=e^{-2\pi i(\ell^2\tau+2\ell z)}\,\phi_{g,h}(\tau,z)\qquad \ell,\ell'\in\mathbb{Z}\tag{3.7}$$

$$\phi_{g,h}\Big(\frac{a\tau+b}{c\tau+d},\frac{z}{c\tau+d}\Big)=\chi_{g,h}\big(\begin{smallmatrix}a&b\\c&d\end{smallmatrix}\big)\,e^{2\pi i\frac{cz^2}{c\tau+d}}\,\phi_{g^ah^c,g^bh^d}(\tau,z)\,,\tag{3.8}$$

where $\big(\begin{smallmatrix}a&b\\c&d\end{smallmatrix}\big)\in SL(2,\mathbb{Z})$ and $\chi_{g,h}:SL(2,\mathbb{Z})\to U(1)$ is a multiplier. In particular, each $\phi_{g,h}$ is a weak Jacobi form of weight 0 and index 1 with multiplier $\chi_{g,h}$ under a subgroup $\Gamma_{g,h}$ of $SL(2,\mathbb{Z})$.

(2) Invariance under conjugation of the pair g,h in M_{24},

$$\phi_{g,h}(\tau,z)=\xi_{g,h}(k)\,\phi_{k^{-1}gk,k^{-1}hk}(\tau,z)\,,\qquad k\in M_{24}\,,\tag{3.9}$$

where $\xi_{g,h}(k)$ is a phase.

(3) If $g \in M_{24}$ has order N, the twisted twining genera $\phi_{g,h}$ have an expansion of the form

$$\phi_{g,h}(\tau,z) = \sum_{\substack{r\in\lambda_g+\mathbb{Z}/N \\ r\geq 0}} \mathrm{Tr}_{\mathcal{H}_{g,r}}\big(\rho_{g,r}(h)\big)\, \mathrm{ch}_{h=\frac{1}{4}+r,\ell}(\tau,z)\ , \tag{3.10}$$

where $\lambda_g \in \mathbb{Q}$, and $\mathrm{ch}_{h,\ell}(\tau,z)$ are elliptic genera of Ramond representations of the $\mathcal{N}=4$ superconformal algebra at central charge $c=6$. (Here $\ell=\frac{1}{2}$, except possibly for $h=\frac{1}{4}$, where $\ell=0$ is also possible — if both $\ell=0,\frac{1}{2}$ appear for $r=0$, it is understood that there are two such terms in the above sum.) Furthermore, each vector space $\mathcal{H}_{g,r}$ is finite dimensional, and it carries a projective representation $\rho_{g,r}$ of the centraliser $C_{M_{24}}(g)$ of g in M_{24}, such that

$$\rho_{g,r}(g) = e^{2\pi i r}\ , \qquad \rho_{g,r}(h_1)\,\rho_{g,r}(h_2) = c_g(h_1,h_2)\,\rho_{g,r}(h_1h_2)\ ,$$

for all $h_1,h_2 \in C_{M_{24}}(g)$. Here $c_g : C_{M_{24}}(g)\times C_{M_{24}}(g) \to U(1)$ is independent of r, and satisfies the cocycle condition

$$c_g(h_1,h_2)\,c_g(h_1h_2,h_3) = c_g(h_1,h_2h_3)\,c_g(h_2,h_3)$$

for all $h_1,h_2,h_3 \in C_{M_{24}}(g)$.

(4) For $g=e$, where e is the identity element of M_{24}, the functions $\phi_{e,h}$ correspond to the twining genera (3.1). In particular, $\phi_{e,e}$ is the K3 elliptic genus.

(5) The multipliers $\chi_{g,h}$, the phases $\xi_{g,h}$, and the 2-cocycles c_g associated with the projective representations $\rho_{g,r}$ are completely determined (by the same formulas as for holomorphic orbifolds) in terms of a 3-cocycle α representing a class in $H^3(M_{24},U(1))$.

3.3. The role of $H^3(M_{24},U(1))$, obstructions and computation of $\phi_{g,h}$. The third cohomology group of M_{24} was only recently computed with the result [**32**][4]

$$H^3(M_{24},U(1)) \cong \mathbb{Z}_{12}\ .$$

The fact that this group is known explicitly plays a crucial role in our analysis. The specific cohomology class $[\alpha]\in H^3(M_{24},U(1))$ that is relevant in our context is uniquely determined by the condition that it reproduces the multiplier system for the twining genera $\phi_{e,h}$ as described in [**16**], namely

$$\chi_{e,h}\big(\begin{smallmatrix} a & b \\ c & d\end{smallmatrix}\big) = e^{\frac{2\pi i c d}{o(h)\ell(h)}}\ , \qquad \big(\begin{smallmatrix} a & b \\ c & d\end{smallmatrix}\big) \in \Gamma_0(o(h))\ . \tag{3.11}$$

Here, $o(h)$ is the order of h and $\ell(h)$ is the length of the smallest cycle, when $h\in M_{24}$ is regarded as a permutation of 24 symbols [**29**]. Indeed, since $\ell(12B)=12$, it follows that α must correspond to a generator of $H^3(M_{24},U(1))$. With the help of the software GAP [**33**], we have verified that a generator reproducing the mutliplier phases (3.11) exists and is unique [**19**].

Once the 3-cocycle α is known, one can use (3.8) and (3.9) to deduce the precise modular properties of each twisted twining genus $\phi_{g,h}$. It turns out that, in many cases, these properties can only be satisfied if $\phi_{g,h}$ vanishes identically [**19**]. In

[4]Note that for a finite group G one has the isomorphisms

$$H_{n-1}(G,\mathbb{Z}) \cong H^n(G,\mathbb{Z}), \qquad H^n(G,\mathbb{Z}) \cong H^{n-1}(G,U(1))\ ,$$

which in particular imply that $H_3(M_{24},\mathbb{Z}) \cong H^3(M_{24},U(1))$.

particular, there are two kinds of potential obstructions that can force a certain twisted twining genus to vanish

(i) Consider three pairwise commuting elements $g, h, k \in M_{24}$. By (3.9),

$$\phi_{g,h}(\tau, z) = \xi_{g,h}(k)\phi_{g,h}(\tau, z) .$$

Therefore, if $\xi_{g,h}(k) \neq 1$, we conclude that $\phi_{g,h}(\tau, z) = 0$.

(ii) Consider a commuting pair of elements $g, h \in M_{24}$, and suppose that $k \in M_{24}$ exists such that $k^{-1}g^{-1}k = g$ and $k^{-1}h^{-1}k = h$. Then, by (3.9) and (3.8)

$$\phi_{g,h}(\tau, -z) = \chi_{g,h}\left(\begin{smallmatrix} -1 & 0 \\ 0 & -1 \end{smallmatrix}\right) \phi_{g^{-1},h^{-1}}(\tau, z) = \chi_{g,h}\left(\begin{smallmatrix} -1 & 0 \\ 0 & -1 \end{smallmatrix}\right) \xi_{g^{-1},h^{-1}}(k)\, \phi_{g,h}(\tau, z) .$$

By (3.10), and using the fact that the $\mathcal{N} = 4$ characters are even functions of z, i.e. $\mathrm{ch}_{h,\ell}(\tau, -z) = \mathrm{ch}_{h,\ell}(\tau, z)$, we obtain

$$\phi_{g,h}(\tau, -z) = \phi_{g,h}(\tau, z) . \tag{3.12}$$

Therefore, if

$$\chi_{g,h}\left(\begin{smallmatrix} -1 & 0 \\ 0 & -1 \end{smallmatrix}\right) \xi_{g^{-1},h^{-1}}(k) \neq 1 ,$$

$\phi_{g,h}$ must vanish.

In all cases where $\phi_{g,h}$ is not obstructed, we define $\Gamma_{g,h} \subseteq SL(2, \mathbb{Z})$ to be the subgroup of $SL(2, \mathbb{Z})$ that leaves (g, h) fixed or maps it to (g^{-1}, h^{-1}), up to conjugation in M_{24}, i.e.

$$\Gamma_{g,h} = \left\{ \begin{pmatrix} a & b \\ c & d \end{pmatrix} \in SL(2, \mathbb{Z}) \mid \exists k \in M_{24}, \right.$$
$$\left. (g^a h^c, g^b h^d) = (k^{-1}gk, k^{-1}hk) \text{ or } (k^{-1}g^{-1}k, k^{-1}h^{-1}k) \right\} .$$

Then, by (3.8), (3.9) and (3.12), $\phi_{g,h}$ must be a weak Jacobi form of weight 0 and index 1 under $\Gamma_{g,h}$, possibly with a multiplier. It turns out that, whenever $g \neq e$, the spaces of such Jacobi forms are either zero- or one-dimensional, and the normalisation can be easily fixed by requiring that a decomposition of the form (3.10) exists (note that, since the representations $\rho_{g,r}$ are projective, the phase of the normalisation is ambiguous). This allows us to determine $\phi_{g,h}$ for all commuting pairs $g, h \in M_{24}$. The results are summarised in the next subsection.

3.4. Generalised Mathieu Moonshine: Statement of Results. In order to describe the twisted twining genera $\phi_{g,h}$ for all commuting pairs of elements $g, h \in M_{24}$, we first note that the functions associated to different such pairs are not necessarily independent. In particular, because of (3.8) and (3.9), we have relations between pairs that are conjugated by some element $k \in M_{24}$, $(g, h) \sim (k^{-1}gk, k^{-1}hk)$, or related by a modular transformation

$$(g, h) \sim (g^a h^c, g^b h^d) , \qquad \begin{pmatrix} a & b \\ c & d \end{pmatrix} \in SL(2, \mathbb{Z}) .$$

It follows that it is sufficient to determine just 55 twisted twining genera. Of these, 21 can be chosen to be of the form $\phi_{e,h}$ and therefore correspond to the twining genera computed in [**14**–**17**]. As for the remaining 34 'genuinely twisted' genera, 28 of them must vanish due to one of the obstructions described in section 3.3. The remaining six twisted twining genera can be computed as discussed in the previous subsection and have the form

$$\phi_{2B,8A} = 2\frac{\eta(2\tau)^2}{\eta(\tau)^4}\vartheta_1(\tau,z)^2\,, \qquad \phi_{2B,4A} = 4\frac{\eta(2\tau)^2}{\eta(\tau)^4}\vartheta_1(\tau,z)^2\,,$$

$$\phi_{4B,4A_1} = 2\sqrt{2}\frac{\eta(2\tau)^2}{\eta(\tau)^4}\vartheta_1(\tau,z)^2\,, \qquad \phi_{4B,4A_2} = 2\sqrt{2}\frac{\eta(2\tau)^2}{\eta(\tau)^4}\vartheta_1(\tau,z)^2\,,$$

$$\phi_{3A,3B} = 0\,, \qquad \phi_{3A,3A} = 0\,,$$

where the subscripts denote the conjugacy classes of the elements g, h (see [**19**] for more details).

Once all the twisted twining genera satisfying (3.8) and (3.9) are known, one has to verify that they admit a decomposition of the form (3.10). More precisely, one has to show that, for each $g \in M_{24}$, there exist projective representations $\rho_{g,r}$ of the centraliser $C_{M_{24}}(g)$ that match with (3.10). Furthermore, the projective equivalence class of these representations must be the one determined by the 3-cocycle α.

In [**19**], the first 500 such representations were computed for each twisted sector (see the ancillary files of the arXiv version of the paper), and were shown to satisfy these properties. The only virtual representations that were found correspond to the BPS states in the untwisted ($g = e$) sector that appeared already in the original Mathieu moonshine (see (3.4)). Using the methods of [**18**], it should be possible to prove the existence of the representations $\rho_{g,r}$ for all r, and to confirm that there are indeed no virtual representations beyond the ones in (3.4). In any case, these results already provide very convincing evidence in favour of generalised Mathieu moonshine.

4. Conclusions

In this short note we have reviewed the construction of the twisted twining genera for Mathieu moonshine. As we have explained, the twisted twining genera we have constructed behave very analogously to the twisted twining characters of holomorphic orbifolds; in particular, the various transformation properties of both are controlled by an element in $H^3(G, U(1))$. We regard this as convincing evidence for the idea that some (superconformal) VOA should underlie and explain Mathieu moonshine. However, as we have also mentioned, this VOA cannot just be a sigma-model on K3, and it must have some unusual features in order to evade the arguments at the end of section 3.1. Understanding the structure of this VOA is, in our opinion, the central open problem in elucidating Mathieu moonshine.

Acknowledgments

We thank Miranda Cheng, Mathieu Dutour, Graham Ellis, Terry Gannon, Jeff Harvey, Stefan Hohenegger, Axel Kleinschmidt, Ashoke Sen and Don Zagier for useful conversations and correspondences. We also thank Henrik Ronnellenfitsch for the collaboration [**19**] on which this review is largely based.

References

[1] J. H. Conway and S. P. Norton, *Monstrous moonshine*, Bull. London Math. Soc. **11** (1979), no. 3, 308–339, DOI 10.1112/blms/11.3.308. MR554399 (81j:20028)

[2] I. Frenkel, J. Lepowsky, and A. Meurman, *Vertex operator algebras and the Monster*, Pure and Applied Mathematics, vol. 134, Academic Press, Inc., Boston, MA, 1988. MR996026 (90h:17026)

[3] R. E. Borcherds, *Monstrous moonshine and monstrous Lie superalgebras*, Invent. Math. **109** (1992), no. 2, 405–444, DOI 10.1007/BF01232032. MR1172696 (94f:11030)

[4] S. Norton, "Generalized moonshine," Proc. Sympos. Pure Math. **47**, 209 The Arcata Conference on Representations of Finite Groups (Arcata, Calif., 1986), Amer. Math. Soc., Providence, RI (1987).

[5] L. Dixon, P. Ginsparg, and J. Harvey, *Beauty and the beast: superconformal symmetry in a Monster module*, Comm. Math. Phys. **119** (1988), no. 2, 221–241. MR968697 (90b:81119)

[6] M. P. Tuite, *Generalised Moonshine and abelian orbifold constructions*, Moonshine, the Monster, and related topics (South Hadley, MA, 1994), Contemp. Math., vol. 193, Amer. Math. Soc., Providence, RI, 1996, pp. 353–368, DOI 10.1090/conm/193/02380. MR1372731 (97f:11029)

[7] C. Dong, H. Li, and G. Mason, *Modular-invariance of trace functions in orbifold theory and generalized Moonshine*, Comm. Math. Phys. **214** (2000), no. 1, 1–56, DOI 10.1007/s002200000242. MR1794264 (2001k:17043)

[8] G. Höhn, "Generalized moonshine for the baby Monster," Habilitation thesis, Mathematisches Institut, Universität Freiburg (2003), `http://www.math.ksu.edu/ gerald/papers/baby8.ps`.

[9] S. Carnahan, *Generalized moonshine I: genus-zero functions*, Algebra Number Theory **4** (2010), no. 6, 649–679, DOI 10.2140/ant.2010.4.649. MR2728485 (2012b:11065)

[10] S. Carnahan, *Generalized moonshine, II: Borcherds products*, Duke Math. J. **161** (2012), no. 5, 893–950, DOI 10.1215/00127094-1548416. MR2904095

[11] S. Carnahan, "Generalized moonshine III: equivariant intertwining operators," to appear.

[12] S. Carnahan, "Generalized moonshine IV: monstrous Lie algebras," `arXiv:1208.6254 [math.RT]`.

[13] T. Eguchi, H. Ooguri, and Y. Tachikawa, *Notes on the $K3$ surface and the Mathieu group M_{24}*, Exp. Math. **20** (2011), no. 1, 91–96, DOI 10.1080/10586458.2011.544585. MR2802725 (2012e:58039)

[14] M. C. N. Cheng, *$K3$ surfaces, $\mathcal{N} = 4$ dyons and the Mathieu group M_{24}*, Commun. Number Theory Phys. **4** (2010), no. 4, 623–657, DOI 10.4310/CNTP.2010.v4.n4.a2. MR2793423 (2012e:11076)

[15] M. R. Gaberdiel, S. Hohenegger, and R. Volpato, *Mathieu twining characters for $K3$*, J. High Energy Phys. **9** (2010), 058, 20, DOI 10.1007/JHEP09(2010)058. MR2776956 (2012h:11066)

[16] M. R. Gaberdiel, S. Hohenegger, and R. Volpato, *Mathieu Moonshine in the elliptic genus of $K3$*, J. High Energy Phys. **10** (2010), 062, 24, DOI 10.1007/JHEP10(2010)062. MR2780524 (2012h:58027)

[17] T. Eguchi and K. Hikami, *Note on twisted elliptic genus of $K3$ surface*, Phys. Lett. B **694** (2011), no. 4-5, 446–455, DOI 10.1016/j.physletb.2010.10.017. MR2748168 (2012g:58044)

[18] T. Gannon, "Much ado about Mathieu," `arXiv:1211.5531 [math.RT]`.

[19] M. R. Gaberdiel, D. Persson, H. Ronellenfitsch, and R. Volpato, *Generalized Mathieu Moonshine*, Commun. Number Theory Phys. **7** (2013), no. 1, 145–223, DOI 10.4310/CNTP.2013.v7.n1.a5. MR3108775

[20] R. Dijkgraaf and E. Witten, *Topological gauge theories and group cohomology*, Comm. Math. Phys. **129** (1990), no. 2, 393–429. MR1048699 (91g:81133)

[21] R. Dijkgraaf, V. Pasquier, and P. Roche, *Quasi Hopf algebras, group cohomology and orbifold models*, Nuclear Phys. B Proc. Suppl. **18B** (1990), 60–72 (1991), DOI 10.1016/0920-5632(91)90123-V. Recent advances in field theory (Annecy-le-Vieux, 1990). MR1128130 (92m:81238)

[22] P. Bántay, *Orbifolds and Hopf algebras*, Phys. Lett. B **245** (1990), no. 3-4, 477–479, DOI 10.1016/0370-2693(90)90676-W. MR1070067 (91k:81147)

[23] T. Gannon, *Moonshine beyond the Monster: The bridge connecting algebra, modular forms and physics*, Cambridge Monographs on Mathematical Physics, Cambridge University Press, Cambridge, 2006. MR2257727 (2008a:17032)

[24] G. Mason, "Orbifold conformal field theory and the cohomology of the Monster," unpublished.

[25] T. Eguchi, H. Ooguri, A. Taormina, and S.-K. Yang, *Superconformal algebras and string compactification on manifolds with* SU(n) *holonomy*, Nuclear Phys. B **315** (1989), no. 1, 193–221, DOI 10.1016/0550-3213(89)90454-9. MR985505 (90i:81114)

[26] M. Eichler and D. Zagier, *The theory of Jacobi forms*, Progress in Mathematics, vol. 55, Birkhäuser Boston, Inc., Boston, MA, 1985. MR781735 (86j:11043)

[27] T. Eguchi and A. Taormina, *Unitary representations of the $N = 4$ superconformal algebra*, Phys. Lett. B **196** (1987), no. 1, 75–81, DOI 10.1016/0370-2693(87)91679-0. MR910253 (88j:17022)

[28] T. Eguchi and A. Taormina, *Character formulas for the $N = 4$ superconformal algebra*, Phys. Lett. B **200** (1988), no. 3, 315–322, DOI 10.1016/0370-2693(88)90778-2. MR926866 (89h:81181)

[29] M. C. N. Cheng and J. F. R. Duncan, *On Rademacher sums, the largest Mathieu group and the holographic modularity of moonshine*, Commun. Number Theory Phys. **6** (2012), no. 3, 697–758, DOI 10.4310/CNTP.2012.v6.n3.a4. MR3021323

[30] M. R. Gaberdiel, S. Hohenegger, and R. Volpato, *Symmetries of K3 sigma models*, Commun. Number Theory Phys. **6** (2012), no. 1, 1–50, DOI 10.4310/CNTP.2012.v6.n1.a1. MR2955931

[31] W. Nahm and K. Wendland, *A hiker's guide to $K3$. Aspects of $N = (4,4)$ superconformal field theory with central charge $c = 6$*, Comm. Math. Phys. **216** (2001), no. 1, 85–138, DOI 10.1007/PL00005548. MR1810775 (2002h:81235)

[32] M. Dutour Sikirić and G. Ellis, *Wythoff polytopes and low-dimensional homology of Mathieu groups*, J. Algebra **322** (2009), no. 11, 4143–4150, DOI 10.1016/j.jalgebra.2009.09.031. MR2556144 (2010j:20082)

[33] The GAP Group, "GAP – Groups, Algorithms, and Programming, version 4.5.6," (2012) `http://www.gap-system.org`.

Institut für Theoretische Physik, ETH Zürich, CH-8093 Zürich, Switzerland
E-mail address: `gaberdiel@itp.phys.ethz.ch`

Fundamental Physics, Chalmers University of Technology, 412 96, Gothenburg, Sweden
E-mail address: `daniel.persson@chalmers.se`
URL: `http://www.danper.se`

Max-Planck-Institut für Gravitationsphysik, Am Mühlenberg 1, 14476 Golm, Germany
E-mail address: `volpato@slac.stanford.edu`

Proceedings of Symposia in Pure Mathematics
Volume **90**, 2015
http://dx.doi.org/10.1090/pspum/090/01521

The First Chern Class of the Verlinde Bundles

Alina Marian, Dragos Oprea, and Rahul Pandharipande

To the memory of Friedrich Hirzebruch

Abstract. A formula for the first Chern class of the Verlinde bundle over the moduli space of smooth genus g curves is given. A finite-dimensional argument is presented in rank 2 using geometric symmetries obtained from strange duality, relative Serre duality, and Wirtinger duality together with the projective flatness of the Hitchin connection. A derivation using conformal-block methods is presented in higher rank. An expression for the first Chern class over the compact moduli space of curves is obtained.

Contents

1. Introduction

1.1. The slopes of the Verlinde complexes. Let $\mathcal{M}_g$ be the moduli space of nonsingular curves of genus $g \geq 2$. Over $\mathcal{M}_g$, we consider the relative moduli space of rank r slope-semistable bundles of degree $r(g-1)$,

$$\nu : \mathcal{U}_g(r, r(g-1)) \to \mathcal{M}_g \ .$$

The moduli space comes equipped with a canonical universal theta bundle corresponding to the divisorial locus

$$\Theta_r = \{(C, E) : h^0(E) \neq 0\}.$$

2010 *Mathematics Subject Classification.* Primary 14H10, 14H60, 14D20; Secondary 14N35.
Key words and phrases. Moduli of vector bundles, moduli of curves, conformal blocks.
Supported by NSF grant DMS 1001604 and a Sloan Foundation Fellowship.
Supported by NSF grants DMS 1001486, DMS 1150675 and a Sloan Foundation Fellowship.
Supported by grant ERC-2012-AdG-320368-MCSK.

Pushing forward the pluritheta series, we obtain a canonical Verlinde complex[1]

$$\mathbb{V}_{r,k} = \mathbf{R}\nu_\star\left(\Theta_r^k\right)$$

over $\mathcal{M}_g$. For $k \geq 1$, $\mathbb{V}_{r,k}$ is a vector bundle.

The Verlinde bundles are known to be projectively flat [**Hi**]. Therefore, their Chern characters satisfy the identity

$$\mathrm{ch}(\mathbb{V}_{r,k}) = \text{ rank } \mathbb{V}_{r,k} \cdot \exp\left(\frac{c_1(\mathbb{V}_{r,k})}{\text{rank } \mathbb{V}_{r,k}}\right). \tag{1.1}$$

The rank of $\mathbb{V}_{r,k}$ is given by the well-known Verlinde formula, see [**B**]. We are interested here in calculating the slope

$$\mu(\mathbb{V}_{r,k}) = \frac{c_1(\mathbb{V}_{r,k})}{\text{rank } \mathbb{V}_{r,k}} \in H^2(\mathcal{M}_g, \mathbb{Q}).$$

Since the Picard rank of $\mathcal{M}_g$ is 1, we can express the slope in the form

$$\mu(\mathbb{V}_{r,k}) = s_{r,k}\, \lambda$$

where $\lambda \in H^2(\mathcal{M}_g, \mathbb{Q})$ is the first Chern class of the Hodge bundle. We seek to determine the rational numbers $s_{r,k} \in \mathbb{Q}$. By Grothendieck-Riemann-Roch for the push-forward defining the Verlinde bundle, $s_{r,k}$ is in fact a rational function in k.

MAIN FORMULA. *The Verlinde slope is*

$$\mu(\mathbb{V}_{r,k}) = \frac{r(k^2-1)}{2(k+r)}\lambda\,. \tag{1.2}$$

The volume of the moduli space $\mathcal{U}_C(r, r(g-1))$ of bundles over a fixed curve with respect to the symplectic form induced by the canonical theta divisor is known to be given in terms of the irreducible representations χ of the group SU_r :

$$\mathrm{vol}_r = \int_{\mathcal{U}_C(r,r(g-1))} \exp(\Theta) = c_r \cdot \sum_\chi \left(\frac{1}{\dim \chi}\right)^{2g-2}$$

for the constant

$$c_r = (2\pi)^{-r(r-1)(g-1)}(1!\,2! \cdots (r-1)!)^{-(g-1)}.$$

Taking the $k \to \infty$ asymptotics in formula (1.2) and using (1.1), we obtain as a consequence an expression for the cohomological push-forward:

$$\nu_\star\left(\exp(\Theta)\right) = \mathrm{vol}_r \cdot \exp\left(\frac{r}{2}\lambda\right)\,.$$

This is a higher rank generalization of an equality over the relative Jacobian observed in [**vdG**].

In **Part I** of this paper, we are concerned with a finite-dimensional geometric proof of the Main Formula. In **Part II**, we give a derivation via conformal blocks. We also extend the formula over the boundary of the moduli space. Let us now detail the discussion.

[1]To avoid technical difficulties, it will be convenient to use the coarse moduli schemes of semistable vector bundles throughout most of the paper. Nonetheless, working over the moduli stack yields an equivalent definition of the Verlinde complexes, see Proposition 8.4 of [**BL**].

For the finite dimensional argument, we note four basic symmetries of the geometry:

(i) Relative level-rank duality for the moduli space of bundles over $\mathcal{M}_g$ will be shown to give the identity

$$s_{r,k} + s_{k,r} = \frac{kr-1}{2} \ .$$

(ii) Relative duality along the the fibers of $\mathcal{SU}_g(r, \mathcal{O}) \to \mathcal{M}_g$ leads to

$$s_{r,k} + s_{r,-k-2r} = -2r^2.$$

(iii) The initial conditions in rank 1, and in level 0 are

$$\mu(\mathbb{V}_{1,k}) = \frac{k-1}{2} \ , \quad \mu(\mathbb{V}_{r,0}) = -\frac{1}{2} \ .$$

(iv) The projective flatness of the Verlinde bundle.

The four features of the geometry will be shown to determine the Verlinde slopes completely in the rank 2 case, proving:

THEOREM 1.1. *The Verlinde bundle* $\mathbb{V}_{2,k}$ *has slope*

$$\mu(\mathbb{V}_{2,k}) = \frac{k^2-1}{k+2} \lambda \ .$$

In arbitrary rank, the symmetries entirely determine the slopes in the Main Formula (1.2) under one additional assumption. This assumption concerns the roots of the Verlinde polynomial

$$v_g(k) = \chi(\mathcal{SU}_C(r, \mathcal{O}), \Theta^k)$$

giving the SU_r Verlinde numbers at level k. Specifically, with the exception of the root $k = -r$ which should have multiplicity exactly $(r-1)(g-1)$, all the other roots of $v_g(k)$ should have multiplicity less than $g-2$. Numerical evidence suggests this is true.

Over a fixed curve C, the moduli spaces of bundles with fixed determinant $\mathcal{SU}_C(2r, \mathcal{O}_C)$ and $\mathcal{SU}_C(2r, \omega_C^r)$ are isomorphic. Relatively over $\mathcal{M}_g$ such an isomorphism does not hold. Letting Θ denote the canonical theta divisor in

$$\nu : \mathcal{SU}_g(2r, \omega^r) \to \mathcal{M}_g \ ,$$

we may investigate the slope of

$$\mathbb{W}_{2r,k} = \mathbf{R}\nu_\star(\Theta^k) \ .$$

The following statement is equivalent to Main Formula (1.2) via Proposition 3.5 of Section 3.5. As will be clear in the proof, the equivalence of the two statements corresponds geometrically to the relative version of Wirtinger's duality for level 2 theta functions.

THEOREM 1.2. *The Verlinde bundle* $\mathbb{W}_{2r,k}$ *has slope*

$$\mu(\mathbb{W}_{2r,k}) = \frac{k(2rk+1)}{2(k+2r)} \lambda \ .$$

In **Part II**, we deduce the Main Formula from a representation-theoretic perspective by connecting results in the conformal-block literature. In particular, essential to the derivation are the main statements in [**T**]. There, an action of a suitable Atiyah algebra, an analogue of a sheaf of differential operators, is used to describe the projectively flat WZW connection. Next, results of Laszlo [**L**] identify conformal blocks and the bundles of theta functions aside from a normalization ambiguity. An integrality argument fixes the variation over moduli of the results of [**L**], yielding the main slope formula. This is explained in Section 5.

Finally, in the last section, we consider the extension of the Verlinde bundle over the compact moduli space $\overline{\mathcal{M}}_g$ via conformal blocks. The Hitchin connection is known to acquire regular singularities along the boundary [**TUY**]. The formulas for the first Chern classes of the bundles of conformal blocks are given in Theorem 6.1 of Section 6. They specialize to the genus 0 expressions of [**F**] in the simplified form of [**Mu**].

Related work. In genus 0, the conformal block bundles have been studied in recent years in connection to the nef cone of the moduli space $\overline{\mathcal{M}}_{0,n}$, see [**AGS**], [**AGSS**], [**F**], [**Fe**], [**GG**], [**Sw**]. In higher genus, the conformal block bundles have been considered in [**S**] in order to study certain representations arising from Lefschetz pencils. The method of [**S**] is to use Segal's loop-group results. In rank 2, our formulas correct Proposition 4.2 of [**S**].

There are at least two perspectives on the study of the *higher* Chern classes of the Verlinde bundle. A first approach is pursued in [**FMP**] by carrying out the Thaddeus wall crossings relatively over the moduli space of pointed curves $\mathcal{M}_{g,1}$. Projective flatness then yields nontrivial relations in the tautological ring $R^\star(\mathcal{M}_{g,1})$. Whether these relations always lie in the Faber-Zagier set [**PP**] is an open question.

A different point of view is taken in [**MOPPZ**]. By the fusion rules, the Chern character of the Verlinde bundle defines a semisimple CohFT. The Givental-Teleman theory provides a classification of the CohFT up to the action of the Givental group. The CohFT is uniquely determined by the projective flatness condition and the first Chern class calculation. The outcome is an explicit formula for the higher Chern classes extending the result of Theorem 6.1 below. However, since the projective flatness is used as input, no nontrivial relations in $R^\star(\mathcal{M}_{g,1})$ are immediately obtained.

1.2. Acknowlegements. We thank Carel Faber for the related computations in [**FMP**] and Ivan Smith for correspondence concerning [**S**]. Our research was furthered during the *Conference on Algebraic Geometry* in July 2013 at the University of Amsterdam. We thank the organizers for the very pleasant environment.

Part I: Finite-dimensional methods

2. Jacobian geometry

In this section, we record useful aspects of the geometry of relative Jacobians over the moduli space of curves. The results will be used to derive the slope identities of Section 3.

Let $\mathcal{M}_{g,1}$ be the moduli space of nonsingular 1-pointed genus $g \geq 2$ curves, and let

$$\pi : \mathcal{C} \to \mathcal{M}_{g,1}, \quad \sigma : \mathcal{M}_{g,1} \to \mathcal{C}$$

be the universal curve and the tautological section respectively. We set $\bar{g} = g-1$ for convenience. The following line bundle will play an important role in subsequent calculations:

$$\mathcal{L} \to \mathcal{M}_{g,1}, \qquad \mathcal{L} = \left(\det \mathbf{R}\pi_\star \mathcal{O}_{\mathcal{C}}(\bar{g}\sigma)\right)^{-1}.$$

An elementary Grothendieck-Riemann-Roch computation applied to the morphism π yields

$$c_1(\mathcal{L}) = -\lambda + \binom{g}{2}\Psi,$$

where

$$\Psi \in H^2(\mathcal{M}_{g,1}, \mathbb{Q})$$

is the cotangent class.

Consider $p : \mathcal{J} \to \mathcal{M}_{g,1}$ the relative Jacobian of degree 0 line bundles. We let

$$\widehat{\Theta} \to \mathcal{J}$$

be the line bundle associated to the divisor

$$\{(C, p, L) \text{ with } H^0(C, L(\bar{g}\,p)) \neq 0\}, \tag{2.1}$$

and let

$$\theta = c_1(\widehat{\Theta})$$

be the corresponding divisor class. We show

LEMMA 2.1. $p_\star\left(e^{n\theta}\right) = n^g e^{\frac{n\lambda}{2}}$.

PROOF. Since the pushforward sheaf $p_\star(\widehat{\Theta})$ has rank 1 and a nowhere-vanishing section obtained from the divisor (2.1), we see that

$$p_\star\left(\widehat{\Theta}\right) = \mathcal{O}_{\mathcal{M}_{g,1}}.$$

The relative tangent bundle of

$$p : \mathcal{J} \to \mathcal{M}_{g,1}$$

is the pullback of the dual Hodge bundle $\mathbb{E}^\vee \to \mathcal{M}_{g,1}$, with Todd genus

$$\operatorname{Todd} \mathbb{E}^\vee = e^{-\frac{\lambda}{2}},$$

see [**vdG**]. Hence, Grothendieck-Riemann-Roch yields

$$p_\star(e^\theta) = e^{\frac{\lambda}{2}}.$$

The Lemma follows immediately. □

Via Grothendieck-Riemann-Roch for $p_\star\left(\widehat{\Theta}^k\right)$, we obtain the following corollary of Lemma 2.1.

COROLLARY 2.2. *We have*

$$s_{1,k} = \frac{k-1}{2}.$$

We will later require the following result obtained as a consequence of Wirtinger duality. Let $(-1)^\star\theta$ denote the pull-back of θ by the involution -1 in the fibers of p.

LEMMA 2.3. $p_\star\left(e^{n(\theta + (-1)^\star\theta)}\right) = (2n)^g e^{2n\, c_1(\mathcal{L})}$.

PROOF. We begin by recalling the classical Wirtinger duality for level 2 theta functions. For a principally polarized abelian variety $(A, \widehat{\Theta})$, we consider the map

$$\mu : A \times A \to A \times A$$

given by

$$\mu(a, b) = (a + b, a - b).$$

We calculate the pullback line bundle

$$\mu^\star(\widehat{\Theta} \boxtimes \widehat{\Theta}) = \widehat{\Theta}^2 \boxtimes (\widehat{\Theta} \otimes (-1)^\star\widehat{\Theta}). \tag{2.2}$$

The unique section of $\widehat{\Theta} \boxtimes \widehat{\Theta}$ gives a natural section of the bundle (2.2), inducing by Künneth decomposition an isomorphism

$$H^0(A, \widehat{\Theta}^2)^\vee \to H^0(A, \widehat{\Theta} \otimes (-1)^\star\widehat{\Theta}),$$

see [**M**].

We carry out the same construction for the relative Jacobian

$$\mathcal{J} \to \mathcal{M}_{g,1}.$$

Concretely, we let

$$\mu : \mathcal{J} \times_{\mathcal{M}_{g,1}} \mathcal{J} \to \mathcal{J} \times_{\mathcal{M}_{g,1}} \mathcal{J}$$

be relative version of the map above. The fiberwise identity (2.2) needs to be corrected by a line bundle twist from $\mathcal{M}_{g,1}$:

$$\mu^\star(\widehat{\Theta} \boxtimes \widehat{\Theta}) = \widehat{\Theta}^2 \boxtimes \left(\widehat{\Theta} \otimes (-1)^\star\widehat{\Theta}\right) \otimes \mathcal{T}. \tag{2.3}$$

We determine

$$\mathcal{T} = \mathcal{L}^{-2}$$

by constructing a section

$$s : \mathcal{M}_{g,1} \to \mathcal{J} \times_{\mathcal{M}_{g,1}} \mathcal{J},$$

for instance

$$s(C, p) = (\mathcal{O}_C, \mathcal{O}_C).$$

Pullback of (2.3) by s then gives the identity

$$\mathcal{L}^2 = \mathcal{L}^2 \otimes \mathcal{L}^2 \otimes \mathcal{T}$$

yielding the expression for $\mathcal{T}$ claimed above. Pushing forward (2.3) to $\mathcal{M}_{g,1}$ we obtain the relative Wirtinger isomorphism

$$\left(p_\star(\widehat{\Theta}^2)\right)^\vee \cong p_\star\left(\widehat{\Theta} \otimes (-1)^\star\widehat{\Theta}\right) \otimes \mathcal{L}^{-2}.$$

We calculate the Chern characters of both bundles via Grothendieck-Riemann-Roch. We find

$$\left(p_\star(e^{2\theta})e^{-\frac{\lambda}{2}}\right)^\vee = p_\star(e^{\theta+(-1)^\star\theta}) \cdot e^{-\frac{\lambda}{2}} \cdot e^{-2c_1(\mathcal{L})}.$$

We have already seen that

$$p_\star(e^{2\theta}) = 2^g e^\lambda,$$

hence the above identity becomes

$$p_\star(e^{\theta+(-1)^\star\theta}) = 2^g e^{2c_1(\mathcal{L})}.$$

The formula in the Lemma follows immediately. $\square$

3. Slope identities

3.1. Notation. In the course of the argument, we will consider the following spaces of bundles over the moduli space $\mathcal{M}_{g,1}$ of pointed genus g curves:

$$\mathcal{SU}_{g,1}(r,\mathcal{O}) = \mathcal{SU}_g(r,\mathcal{O}) \times_{\mathcal{M}_g} \mathcal{M}_{g,1}, \quad \mathcal{U}_{g,1}(r,r\bar{g}) = \mathcal{U}_g(r,r\bar{g}) \times_{\mathcal{M}_g} \mathcal{M}_{g,1}.$$

To keep the notation simple, we will use ν to denote all bundle-forgetting maps from the relative moduli spaces of bundles to the space of (possibly pointed) nonsingular curves.

Over the relative moduli space $\mathcal{U}_{g,1}(r,r\bar{g})$ there is a natural determinant line bundle

$$\Theta_r \to \mathcal{U}_{g,1}(r,r\bar{g}),$$

endowed with a canonical section vanishing on the locus

$$\theta_r = \{E \to C \text{ with } H^0(C,\, E) \neq 0\}.$$

We construct analogous theta bundles for the moduli space of bundles with trivial determinant, and decorate them with the superscript "+" for clarity. Specifically, we consider the determinant line bundle and corresponding divisor

$$\Theta_r^+ \to \mathcal{SU}_{g,1}(r,\mathcal{O}), \quad \theta_r^+ = \{(C,\, p,\, E \to C) \text{ with } H^0(C,\, E(\bar{g}p)) \neq 0\}.$$

Pushforward yields an associated Verlinde bundle

$$\mathbb{V}_{r,k}^+ = \mathbf{R}\nu_\star\left(\left(\Theta_r^+\right)^k\right) \to \mathcal{M}_{g,1}.$$

This bundle is however not defined over the unpointed moduli space $\mathcal{M}_g$.

While the first Chern class of $\mathbb{V}_{r,k}$ is necessarily a multiple of λ, the first Chern class of $\mathbb{V}_{r,k}^+$ is a combination of λ and the cotangent class

$$\Psi \in H^2(\mathcal{M}_{g,1},\mathbb{Q}).$$

3.2. Strange duality. Using a relative version of the level-rank duality over moduli spaces of bundles on a smooth curve, we first prove the following slope symmetry.

PROPOSITION 3.1. *For any positive integers k and r, we have*

$$s_{k,r} + s_{r,k} = \frac{kr-1}{2}.$$

Proof. Let

$$\tau : \mathcal{SU}_{g,1}(r,\mathcal{O}) \times_{\mathcal{M}_{g,1}} \mathcal{U}_{g,1}(k,k\bar{g}) \longrightarrow \mathcal{U}_{g,1}(kr,kr\bar{g})$$

be the tensor product map,

$$\tau(E,F) = E \otimes F.$$

Over each fixed pointed curve $(C,p) \in \mathcal{M}_{g,1}$ we have, as explained for instance in [**B**],

$$\tau^\star\Theta_{kr} \simeq \left(\Theta_r^+\right)^k \boxtimes \Theta_k^r \quad \text{on } \mathcal{SU}_C(r,\mathcal{O}) \times \mathcal{U}_C(k,k\bar{g}). \tag{3.1}$$

The natural divisor

$$\tau^\star\theta_{kr} = \{(E,\, F) \text{ with } H^0(E \otimes F) \neq 0\}$$

induces the strange duality map, defined up to multiplication by scalars,

$$H^0\left(\mathcal{SU}_C(r,\mathcal{O}), (\Theta_r^+)^k\right)^\vee \longrightarrow H^0\left(\mathcal{U}_C(k,k\bar{g}), \Theta_k^r\right). \tag{3.2}$$

This map is known to be an isomorphism [**Bel**], [**MO**], [**P**].

Relatively over $\mathcal{M}_{g,1}$ we write, using the fixed-curve pullback identity (3.1),

$$\tau^\star \Theta_{kr} \simeq \left(\Theta_r^+\right)^k \boxtimes \Theta_k^r \otimes \nu^\star \mathcal{T} \quad \text{on } \mathcal{SU}_{g,1}(r, \mathcal{O}) \times_{\mathcal{M}_{g,1}} \mathcal{U}_{g,1}(k, k\bar{g}), \tag{3.3}$$

for a line bundle twist

$$\mathcal{T} \to \mathcal{M}_{g,1}.$$

We will determine

$$\mathcal{T} = \mathcal{L}^{kr}, \text{ so that } c_1(\mathcal{T}) = kr\left(\lambda - \binom{g}{2}\Psi\right).$$

To show this, we pull back (3.3) via the section

$$s : \mathcal{M}_{g,1} \to \mathcal{SU}_{g,1}(r, \mathcal{O}) \times_{\mathcal{M}_{g,1}} \mathcal{U}_{g,1}(k, k\bar{g}), \ \ s(C, p) = (\mathcal{O}_C^{\oplus r}, \ \mathcal{O}_C(\bar{g}p)^{\oplus k}),$$

obtaining

$$\mathcal{L}^{kr} \simeq \mathcal{L}^{kr} \otimes \mathcal{L}^{kr} \otimes \mathcal{T},$$

hence the claimed expression for $\mathcal{T}$.

Pushing forward (3.3) now, we note, as a consequence of (3.2), the isomorphism of Verlinde vector bundles over $\mathcal{M}_{g,1}$,

$$\left(\mathbb{V}_{r,k}^+\right)^\vee \simeq \mathbb{V}_{k,r} \otimes \mathcal{T}.$$

We conclude

$$-\mu\left(\mathbb{V}_{r,k}^+\right) = \mu\left(\mathbb{V}_{k,r}\right) + c_1(\mathcal{T}),$$

hence

$$-\mu\left(\mathbb{V}_{r,k}^+\right) = \mu\left(\mathbb{V}_{k,r}\right) + kr\left(\lambda - \binom{g}{2}\Psi\right).$$

The equation, alongside the following Lemma, allows us to conclude Proposition 3.1. $\square$

LEMMA 3.2. *We have*

$$\begin{aligned} \mu\left(\mathbb{V}_{r,k}\right) &= \mu\left(\mathbb{V}_{r,k}^+\right) + \frac{kr-1}{2}\lambda - krc_1(\mathcal{L}) \\ &= \mu\left(\mathbb{V}_{r,k}^+\right) + \frac{3kr-1}{2}\lambda - kr\binom{g}{2}\Psi. \end{aligned}$$

Proof. To relate $\mu\left(\mathbb{V}_{r,k}^+\right)$ and $\mu\left(\mathbb{V}_{r,k}\right)$ we use a slightly twisted version of the tensor product map τ in the case $k = 1$. More precisely we have the following diagram, where the top part is a fiber square

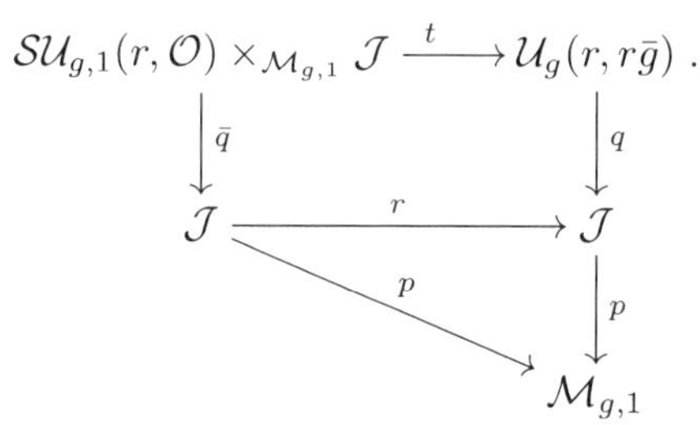

Here, as in the previous section, we write

$$p : \mathcal{J} \to \mathcal{M}_{g,1}$$

for the relative Jacobian of degree 0 line bundles, while r denotes multiplication by r on $\mathcal{J}$. Furthermore, for a pointed curve (C, p),

$$t(E, L) = E \otimes L(\bar{g}p), \quad q(E) = \det(E(-\bar{g}p)).$$

Finally, $\bar{q}$ is the projection onto $\mathcal{J}$. The pullback equation (3.3) now reads

$$t^\star \Theta_r \simeq \Theta_r^+ \boxtimes \widehat{\Theta}^r \otimes \mathcal{L}^{-r},$$

where, keeping with the previous notation, $\widehat{\Theta} \to \mathcal{J}$ is the theta line bundle associated with the divisor

$$\theta := \{(C, p, L \to C) \text{ with } H^0(C, L(\bar{g}p)) \neq 0\}.$$

Using the pullback identity and the Cartesian diagram, we conclude

$$r^\star q_\star(\Theta_r^k) = \bar{q}_\star\left(\left(\Theta_r^+\right)^k \boxtimes \widehat{\Theta}^{kr} \otimes \mathcal{L}^{-kr}\right) = p^\star \mathbb{V}_{r,k}^+ \otimes \widehat{\Theta}^{kr} \otimes \mathcal{L}^{-kr} \text{ on } \mathcal{J}. \tag{3.4}$$

We are however interested in calculating

$$\operatorname{ch} \mathbb{V}_{r,k} = \operatorname{ch} \nu_\star \Theta_r^k = \operatorname{ch} p_\star(q_\star \Theta_r^k).$$

We have recorded in Lemma 2.1 the Todd genus of the the relative tangent bundle of

$$p : \mathcal{J} \to \mathcal{M}_{g,1}$$

to be

$$\operatorname{Todd} \mathbb{E}^\vee = e^{-\frac{\lambda}{2}}.$$

Grothendieck-Riemann-Roch then gives

$$\operatorname{ch} \mathbb{V}_{r,k} = e^{-\frac{\lambda}{2}}\, p_\star(\operatorname{ch}(q_\star \Theta_r^k)).$$

We further write, on $\mathcal{J}$,

$$\operatorname{ch}(q_\star \Theta_r^k) = \frac{1}{r^{2g}}\, r^\star \operatorname{ch}(q_\star \Theta_r^k) = \frac{1}{r^{2g}} \operatorname{ch}(r^\star q_\star \Theta_r^k) = \frac{1}{r^{2g}}\, e^{kr\theta - krc_1(\mathcal{L})}\, p^\star \operatorname{ch} \mathbb{V}_{r,k}^+,$$

where (3.4) was used. We obtain

$$\operatorname{ch} \mathbb{V}_{r,k} = \frac{1}{r^{2g}}\, e^{-\frac{\lambda}{2}}\, e^{-krc_1(\mathcal{L})} \left(p_\star e^{kr\theta}\right) \operatorname{ch} \mathbb{V}_{r,k}^+ \text{ on } \mathcal{M}_{g,1}.$$

The final p-pushforward in the identity above was calculated in Lemma 2.1. Substituting, we obtain

$$\operatorname{ch} \mathbb{V}_{r,k} = \frac{k^g}{r^g}\, e^{\frac{(kr-1)\lambda}{2}}\, e^{-krc_1(\mathcal{L})} \operatorname{ch} \mathbb{V}_{r,k}^+ \text{ on } \mathcal{M}_{g,1}.$$

Therefore,

$$\mu(\mathbb{V}_{r,k}) = \mu\left(\mathbb{V}_{r,k}^+\right) + \frac{kr-1}{2}\lambda - krc_1(\mathcal{L}),$$

which is the assertion of Lemma 3.2. $\square$

3.3. Relative Serre duality. We will presently deduce another identity satisfied by the numbers $s_{r,k}$ using relative Serre duality for the forgetful morphism

$$\nu : \mathcal{SU}_{g,1}(r, \mathcal{O}) \to \mathcal{M}_{g,1} \ .$$

PROPOSITION 3.3. *We have*

$$s_{r,k} + s_{r,-k-2r} = -2r^2 \ .$$

PROOF. By relative duality, we have

$$\mathbb{V}^+_{r,k} = \mathbf{R}\nu_\star\left(\left(\Theta^+_r\right)^k\right) \cong \mathbf{R}\nu_\star\left(\left(\Theta^+_r\right)^{-k} \otimes \omega_\nu\right)^\vee [(r^2-1)(g-1)] \ .$$

We determine the relative dualizing sheaf of the morphism ν. As explained in Theorem E of [**DN**], the fibers of the morphism

$$\nu : \mathcal{SU}_{g,1}(r, \mathcal{O}) \to \mathcal{M}_{g,1}$$

are Gorenstein, hence the relative dualizing sheaf is a line bundle. Furthermore, along the fibers of ν, the canonical bundle equals $\left(\Theta^+_r\right)^{-2r}$. Thus, up to a line bundle twist $\mathcal{T} \to \mathcal{M}_{g,1}$, we have

$$\omega_\nu = \left(\Theta^+_r\right)^{-2r} \otimes \nu^\star \mathcal{T}. \tag{3.5}$$

The twist $\mathcal{T}$ will be found via a Chern class calculation to be

$$c_1(\mathcal{T}) = -(r^2+1)\lambda + 2r^2\binom{g}{2}\Psi.$$

Since

$$\left(\mathbb{V}^+_{r,k}\right)^\vee \cong \mathbf{R}\nu_\star((\Theta^+_r)^{-k} \otimes \omega_\nu)[(r^2-1)(g-1)] = \mathbb{V}^+_{r,-k-2r} \otimes \mathcal{T}[(r^2-1)(g-1)],$$

we obtain taking slopes that

$$-\mu(\mathbb{V}^+_{r,k}) = \mu(\mathbb{V}^+_{r,-k-2r}) + \left(-(r^2+1)\lambda + 2r^2\binom{g}{2}\Psi\right).$$

The proof is concluded using Lemma 3.2.

To determine the twist $\mathcal{T}$, we begin by restricting (3.5) to the smooth stable locus of the moduli space of bundles

$$\nu : \mathcal{SU}^s_{g,1}(r, \mathcal{O}) \to \mathcal{M}_{g,1}.$$

There, the relative dualizing sheaf is the dual determinant of the relative tangent bundle. By Corollary 4.3 of [**DN**], adapted to the relative situation, the Picard group of the coarse moduli space and the Picard group of the moduli stack are naturally isomorphic. We therefore consider (3.5) over the moduli stack of stable bundles. (We do not introduce separate notation for the stack, for simplicity.) Let

$$\mathcal{E} \to \mathcal{SU}^s_{g,1}(r, \mathcal{O}) \times_{\mathcal{M}_{g,1}} \mathcal{C}$$

denote the universal vector bundle of rank r over the stable part of the moduli stack. We write

$$\pi : \mathcal{SU}^s_{g,1}(r, \mathcal{O}) \times_{\mathcal{M}_{g,1}} \mathcal{C} \to \mathcal{SU}^s_{g,1}(r, \mathcal{O})$$

for the natural projection. Clearly,

$$\Theta^+_r = (\det \mathbf{R}\pi_\star(\mathcal{E} \otimes \mathcal{O}_{\mathcal{C}}(\bar{g}\sigma)))^{-1}.$$

The relative dualizing sheaf of the morphism ν is expressed as

$$\omega_\nu = \mathbf{R}\pi_\star Hom(\mathcal{E}, \mathcal{E})_{(0)} = \mathbf{R}\pi_\star Hom(\mathcal{E}, \mathcal{E}) - \mathbf{R}\pi_\star \mathcal{O}.$$

We therefore have

$$c_1(\omega_\nu) = c_1(\mathbf{R}\pi_\star Hom(\mathcal{E},\mathcal{E})) - \lambda.$$

Using ω for the relative dualizing sheaf along the fibers of π, we calculate

$$\begin{aligned}
c_1(\omega_\nu) \quad + \quad & 2rc_1(\Theta_r^+) = c_1\left(\mathbf{R}\pi_\star End\mathcal{E}\right) - \lambda - 2rc_1\left(\mathbf{R}\pi_\star(\mathcal{E}\otimes\mathcal{O}_C(\bar{g}\sigma))\right) \\
= \quad & \pi_\star\left[\left(1-\frac{\omega}{2}+\frac{\omega^2}{12}\right)\left(r^2+((r-1)c_1(\mathcal{E})^2-2rc_2(\mathcal{E})\right)\right]_{(2)} - \lambda \\
- \quad & 2r\pi_\star\left[\left(1-\frac{\omega}{2}+\frac{\omega^2}{12}\right)\left(r+c_1(\mathcal{E})+\frac{1}{2}c_1(\mathcal{E})^2-c_2(\mathcal{E})\right)\left(1+\bar{g}\sigma-\frac{\bar{g}^2}{2}\sigma\,\Psi\right)\right]_{(2)} \\
= \quad & -(r^2+1)\lambda + 2r^2\binom{g}{2}\Psi + \pi_\star(r\omega\cdot c_1(\mathcal{E}) - 2r\bar{g}\,\sigma\cdot c_1(\mathcal{E}) - c_1(\mathcal{E})^2).
\end{aligned}$$

Since the determinant of $\mathcal{E}$ is trivial on the fibers of π, we may write

$$\det\mathcal{E} = \pi^\star\mathcal{A}$$

for a line bundle $\mathcal{A}\to\mathcal{SU}^s_{g,1}(r,\mathcal{O})$ with first Chern class

$$\alpha = c_1(\mathcal{A}).$$

We calculate

$$\pi_\star(r\omega\cdot c_1(\mathcal{E}) - 2r\bar{g}\,\sigma\cdot c_1(\mathcal{E}) - c_1(\mathcal{E})^2) = 2r\bar{g}\alpha - 2r\bar{g}\alpha - \pi_\star(\alpha^2) = 0,$$

and conclude

$$\nu^\star c_1(\mathcal{T}) = c_1(\omega_\nu) + 2rc_1(\Theta_r^+) = -(r^2+1)\lambda + 2r^2\binom{g}{2}\Psi.$$

This equality holds in the Picard group of the stable locus of the moduli stack and of the coarse moduli space. Since the strictly semistables have codimension at least 2, the equality extends to the entire coarse space $\mathcal{SU}_{g,1}(r,\mathcal{O})$. Finally, pushing forward to $\mathcal{M}_{g,1}$, we find the expression for the twist $\mathcal{T}$ claimed above. □

3.4. Initial conditions. The next calculation plays a basic role in our argument.

LEMMA 3.4. *We have*

$$s_{r,0} = -\frac{1}{2}.$$

PROOF. Since the Verlinde number for $k=0$ over the moduli space $\mathcal{U}_C(r, r\bar{g})$ is zero, the slope appears to have poles if computed directly. Instead, we carry out the calculation via the fixed determinant moduli space. The trivial bundle has no higher cohomology along the fibers of

$$\nu : \mathcal{SU}_{g,1}(r,\mathcal{O}) \to \mathcal{M}_{g,1}$$

by Kodaira vanishing. To apply the vanishing theorem, we use that the fibers of ν have rational singularities, and the expression of the dualizing sheaf of Proposition 3.3. Hence,

$$\nu_\star(\mathcal{O}) = \mathcal{O}_{\mathcal{M}_{g,1}}.$$

Therefore

$$\mu(\mathbb{V}^+_{r,0}) = 0$$

which then immediately implies $s_{r,0} = -\frac{1}{2}$ by Lemma 3.2. □

3.5. Pluricanonical determinant. We have already investigated moduli spaces of bundles with trivial determinant. Here, we assume that the determinant is of degree equal to rank times $\bar{g}$ and is a multiple of the canonical bundle. The conditions require the rank to be even. Thus, we are concerned with the slopes of the complexes

$$\mathbb{W}_{2r,k} = \mathbf{R}\nu_\star\left(\Theta_{2r}^k\right),$$

where

$$\nu : \mathcal{SU}_g(2r, \omega^r) \to \mathcal{M}_g.$$

The following slope identity is similar to that of Lemma 3.2:

PROPOSITION 3.5. *We have*

$$\mu(\mathbb{W}_{2r,k}) = \mu(\mathbb{V}_{2r,k}) + \frac{\lambda}{2}\,.$$

In particular, via Theorem 1.1, *we have*

$$\mu(\mathbb{W}_{2,k}) = \frac{k(2k+1)}{2(k+2)}\lambda\,.$$

PROOF. Just as in the proof of Lemma 3.2, we relate $\mu\left(\mathbb{W}_{2r,k}\right)$ and $\mu\left(\mathbb{V}_{2r,k}\right)$ via the tensor product map t:

$$\begin{array}{ccc}
\mathcal{SU}_{g,1}(2r,\omega^r)\times_{\mathcal{M}_{g,1}}\mathcal{J} & \xrightarrow{\ t\ } & \mathcal{U}_g(2r,2r\bar{g})\,. \\
\downarrow{\scriptstyle \bar{q}} & & \downarrow{\scriptstyle q} \\
\mathcal{J} & \xrightarrow{\ 2r\ } & \mathcal{J} \\
& \searrow{\scriptstyle p} & \downarrow{\scriptstyle p} \\
& & \mathcal{M}_{g,1}
\end{array}$$

We keep the same notation as in Lemma 3.2, letting

$$p : \mathcal{J} \to \mathcal{M}_{g,1}$$

denote the relative Jacobian of degree 0 line bundles, and writing $2r$ for the multiplication by $2r$ on $\mathcal{J}$. Furthermore, for a pointed curve (C,p),

$$t(E,L) = E\otimes L,\quad q(E) = \det E\otimes\omega_C^{-r}$$

Finally, $\bar{q}$ is the projection onto $\mathcal{J}$. Recall that $\widehat{\Theta}$ denotes the theta line bundle on the relative Jacobian associated with the divisor

$$\theta := \{(C,p,L) \text{ with } H^0(C, L(\bar{g}p)) \neq 0\}.$$

It is clear that $(-1)^\star\widehat{\Theta}$ has the associated divisor

$$(-1)^\star\theta = \{(C,p,L) \text{ with } H^0(C, L\otimes\omega_C(-\bar{g}p)) \neq 0\}.$$

For a fixed pointed curve (C,p), we have the fiberwise identity

$$t^\star\Theta_{2r} = \Theta_{2r}\boxtimes\left(\widehat{\Theta}\otimes(-1)^\star\widehat{\Theta}\right)^r$$

on $\mathcal{SU}_C(2r,\omega^r)\times\mathcal{J}_C$. Relatively over $\mathcal{M}_{g,1}$, the same equation holds true up to a twist $\mathcal{T}\to\mathcal{M}_{g,1}$:

$$t^\star\Theta_{2r} \simeq \Theta_{2r}\boxtimes\left(\widehat{\Theta}\otimes(-1)^\star\widehat{\Theta}\right)^r\otimes\mathcal{T}.$$

We claim that

$$\mathcal{T} = \mathcal{L}^{-2r}.$$

Indeed, the twist can be found in the usual way, using a suitable section

$$s : \mathcal{M}_{g,1} \to \mathcal{SU}_{g,1}(2r, \omega^r) \times_{\mathcal{M}_{g,1}} \mathcal{J},$$

for instance

$$s(C,p) = (\omega_C^r(-\bar{g}p)^{\oplus r} \oplus \mathcal{O}_C(\bar{g}p)^{\oplus r}, \mathcal{O}_C).$$

Pulling back by s, we obtain the identity

$$(\mathcal{L} \otimes \mathcal{M})^r = (\mathcal{L} \otimes \mathcal{M})^r \otimes (\mathcal{L} \otimes \mathcal{M})^r \otimes \mathcal{T}$$

where

$$\mathcal{L} = \det\left(\mathbf{R}\pi_\star(\mathcal{O}_\mathcal{C}(\bar{g}\sigma))\right)^{-1}, \quad \mathcal{M} = \det\left(\mathbf{R}\pi_\star\left(\omega_\mathcal{C}(-\bar{g}\sigma)\right)\right)^{-1}.$$

In fact, by relative duality, $\mathcal{M} \cong \mathcal{L}$, so we conclude

$$\mathcal{T} = \mathcal{L}^{-2r}.$$

Using the pullback identity and the Cartesian diagram, we find that over $\mathcal{J}$ we have

$$\begin{aligned} (3.6) \qquad (2r)^\star q_\star \Theta_{2r}^k &= \bar{q}_\star \left(\Theta_{2r}^k \boxtimes \left(\widehat{\Theta} \otimes (-1)^\star \widehat{\Theta}\right)^{kr} \otimes \mathcal{L}^{-2kr}\right) \\ &= p^\star \mathbb{W}_{2r,k} \otimes \left(\widehat{\Theta} \otimes (-1)^\star \widehat{\Theta}\right)^{kr} \otimes \mathcal{L}^{-2kr} \end{aligned}$$

Next, we calculate

$$\operatorname{ch} \mathbb{V}_{2r,k} = \operatorname{ch} \nu_\star \Theta_{2r}^k = \operatorname{ch} p_\star(q_\star \Theta_{2r}^k)$$

via Grothendieck-Riemann-Roch:

$$\operatorname{ch} \mathbb{V}_{2r,k} = e^{-\frac{\lambda}{2}} p_\star(\operatorname{ch}(q_\star \Theta_{2r}^k)).$$

We further evaluate, on $\mathcal{J}$,

$$\begin{aligned} \operatorname{ch}(q_\star \Theta_{2r}^k) &= \frac{1}{(2r)^{2g}} (2r)^\star \operatorname{ch}(q_\star \Theta_{2r}^k) = \frac{1}{(2r)^{2g}} \operatorname{ch}((2r)^\star q_\star \Theta_r^k) \\ &= \frac{1}{(2r)^{2g}} e^{kr(\theta + (-1)^\star \theta) - 2krc_1(\mathcal{L})} p^\star \operatorname{ch} \mathbb{W}_{2r,k}, \end{aligned}$$

where (3.6) was used. We obtain

$$\operatorname{ch} \mathbb{V}_{2r,k} = \frac{1}{(2r)^{2g}} e^{-\frac{\lambda}{2}} e^{-2krc_1(\mathcal{L})} \left(p_\star e^{kr(\theta + (-1)^\star \theta)}\right) \operatorname{ch} \mathbb{W}_{2r,k} \text{ on } \mathcal{M}_{g,1}.$$

The p-pushforward in the identity above is given by Lemma 2.3. Substituting, we find

$$\operatorname{ch} \mathbb{V}_{2r,k} = \left(\frac{k}{2r}\right)^g e^{-\frac{\lambda}{2}} \operatorname{ch} \mathbb{W}_{2r,k},$$

and taking slopes it follows that

$$\mu(\mathbb{V}_{2r,k}) = \mu(\mathbb{W}_{2r,k}) - \frac{\lambda}{2}.$$

□

4. Projective flatness and the rank two case

4.1. Projective flatness. By the Grothendieck-Riemann-Roch theorem for singular varieties due to Baum-Fulton-MacPherson [**BFM**], the Chern character of $\mathbb{V}_{r,k}$ is a polynomial in k with entries in the cohomology classes of $\mathcal{M}_g$. (Alternatively, we may transfer the calculation to a smooth moduli space of degree 1 bundles using a Hecke modification at a point as in [**BS**], and then invoke the usual Grothendieck-Riemann-Roch theorem.) Taking account of the projective flatness identity (1.1),

$$\mathrm{ch}(\mathbb{V}_{r,k}) = \mathrm{rank}\ \mathbb{V}_{r,k} \cdot \exp\left(s_{r,k}\lambda\right),$$

we therefore write

$$\mathrm{ch}_i(\mathbb{V}_{r,k}) = \sum_{j=0}^{r^2\bar{g}+i+1} k^j \alpha_{i,j} = (\mathrm{rank}\ \mathbb{V}_{r,k}) \frac{s_{r,k}^i}{i!} \lambda^i \text{ for } i \geq 0, \quad \alpha_{i,j} \in H^{2i}(\mathcal{M}_g).$$

As the Vandermonde determinant is nonzero, for each i we can express $\alpha_{i,j}$ in terms of λ^i. Since $\lambda^{g-2} \neq 0$, we deduce that

$$(\mathrm{rank}\ \mathbb{V}_{r,k})\, s_{r,k}^i\,, \quad 0 \leq i \leq g-2,$$

is a polynomial in k of degree $r^2\bar{g} + i + 1$, with coefficients that may depend on r and g. The following is now immediate:

(i) For each r we can write

$$s_{r,k} = \frac{a_r(k)}{b_r(k)}$$

as quotient of polynomials of minimal degree, with

$$\deg a_r(k) - \deg b_r(k) \leq 1.$$

Setting $v_{g,r}(k) = \mathrm{rank}\ \mathbb{V}_{r,k}$, we also have

$$b_r(k)^{g-2} \text{ divides } v_{g,r}(k)$$

as polynomials in $\mathbb{Q}[k]$.

In addition, the following properties of the function $s_{r,k}$ have been established in the previous sections:

(ii) $s_{1,k} = \frac{k-1}{2}$, $s_{r,0} = -\frac{1}{2}$,

(iii) $s_{r,k} + s_{k,r} = \frac{kr-1}{2}$ for all $k, r \geq 1$,

(iv) $s_{r,k} + s_{r,-k-2r} = -2r^2$ for all $r \geq 1$ and all k.

Clearly, the function

$$s_{r,k} = \frac{r(k^2-1)}{2(k+r)}$$

of formula (1.2) satisfies symmetries (ii)-(iv). Therefore, the shift

$$s'_{r,k} = s_{r,k} - \frac{r(k^2-1)}{2(k+r)}$$

satisfies properties similar to (i)-(iv):

(i)$'$ $s'_{r,k}$ is a rational function of k,

(ii)$'$ $s'_{1,k} = 0$ for all k, and $s'_{r,0} = 0$ for all $r \geq 1$,

(iii)$'$ $s'_{r,k} + s'_{k,r} = 0$ for $r, k \geq 1$,

(iv)$'$ $s'_{r,k} + s'_{r,-k-2r} = 0$ for all $r \geq 1$ and all k.

4.2. The rank two analysis. To prove Theorem 1.1, we now show that $s'_{2,k} = 0$ for all k. Of course, $s'_{2,0} = 0$ by (ii)$'$. Also by (ii)$'$, we know that $s'_{1,2} = 0$, hence by (iii)$'$ we find

$$s'_{2,1} = 0.$$

Similarly,

$$s'_{2,2} = 0$$

also by (iii)$'$. Using (iv)$'$, we obtain that

$$s'_{2,0} = s'_{2,1} = s'_{2,2} = s'_{2,-4} = s'_{2,-5} = s'_{2,-6} = 0.$$

Finally, we make use of the projective flatness of $\mathbb{V}_{2,k}$. The Verlinde formula reads [**B**]

$$v_{g,2}(k) = k^g \left(\frac{k+2}{2}\right)^{g-1} \left(\sum_{j=1}^{k+1} \frac{1}{\sin^{2g-2} \frac{j\pi}{k+2}}\right).$$

The polynomial $v_{g,2}(k)$ admits $k = 0$ as a root of order g and $k = -2$ as a root of order $(g-1)$. Indeed, it was shown by Zagier that

$$\widehat{v}_g(k+2) = \sum_{j=1}^{k+1} \left(\frac{1}{\sin \frac{j\pi}{k+2}}\right)^{2g-2}$$

is a polynomial in $k+2$ such that

$$\widehat{v}_g(0) < 0,$$

see Remark 1 on page 4 of [**Z**].

Let us write

$$b_2(k) = (k+2)^m k^n B(k)$$

for a polynomial B which does not have 0 and -2 as roots. By property (i) above, we obtain

$$m \leq \frac{g-1}{g-2} \implies m \leq 1.$$

Similarly

$$n \leq \frac{g}{g-2} \implies n \leq 1$$

unless $g = 3, 4$. Also, $B(k)^{g-2}$ divides the Verlinde polynomial $\widehat{v}_g(k+2)$ which has degree $4g - 3 - (g-1) - g = 2g - 2$. Thus

$$(g-2) \deg B \leq 2g - 2 \implies \deg B \leq 2$$

except possibly when $g = 3, 4$. In conclusion

$$s'_{2,k} = \frac{a_2(k)}{B(k)(k+2)^m k^n} - \frac{k^2-1}{k+2} = \frac{A(k)}{B(k)(k+2)k}$$

for a polynomial

$$A(k) = a_2(k)(k+2)^{1-m} k^{1-n} - (k^2-1)B(k).$$

Since

$$\lim_{k\to\infty} \frac{s_{2,k}}{k} < \infty \implies \lim_{k\to\infty} \frac{s'_{2,k}}{k} < \infty,$$

we must have

$$\deg A - \deg B \leq 3.$$

Since

$$\deg B \leq 2 \implies \deg A \leq 5.$$

Furthermore, we have already observed that

$$A(-6) = A(-5) = A(-4) = A(0) = A(1) = A(2) = 0.$$

This implies $A = 0$ hence $s'_{2,k} = 0$ as claimed.

The cases $g = 3$ and $g = 4$ have to be considered separately. First, when $g = 4$ we obtain

$$m \leq 1, \ n \leq 2$$

and $B(k)^2$ divides the polynomial $\widehat{v}_4(k+2)$. By direct calculation via the Verlinde formula we find

$$\widehat{v}_4(x) = \frac{2x^6 + 21x^4 + 168x^2 - 191}{945}.$$

This implies $B = 1$, and thus

$$s'_{2,k} = \frac{A(k)}{k^2(k+2)}$$

with

$$\deg A \leq 4.$$

Since $A = 0$ for 6 different values, it follows as before that $A = 0$ hence $s'_{2,k} = 0$.

When $g = 3$, the Verlinde flatness does not give us useful information. In this case, one possible argument is via relative Thaddeus flips, for which we refer the reader to the preprint [**FMP**]. Along these lines, although we do not explicitly show the details here, the genus 3 slope formula was in fact checked by direct calculation. □

Part II: Representation-theoretic methods

5. The slope of the Verlinde bundles via conformal blocks

We derive here the Main Formula (1.2) using results in the extensive literature on conformal blocks. In particular, the central statement of [**T**] is used in an essential way. The derivation is by direct comparison of the bundle $\mathbb{V}^+_{r,k}$ of generalized theta functions with the bundle of covacua

$$\mathcal{B}_{r,k} \to \mathcal{M}_{g,1}$$

defined using the representation theory of the affine Lie algebra $\widehat{\mathfrak{sl}}_r$. Over pointed curves (C, p), the fibers of the dual bundle $\mathcal{B}^\vee_{r,k}$ give the spaces of generalized theta functions

$$H^0(\mathcal{SU}_C(r, \mathcal{O}), (\Theta^+_r)^k).$$

Globally, the identification $\mathcal{B}^\vee_{r,k} \simeq \mathbb{V}^+_{r,k}$ will be shown below to hold only up to a twist. The explicit identification of the twist and formula (1.2) will be deduced together.

5.1. The bundles of covacua. For a self-contained presentation, we start by reviewing briefly the definition of $\mathcal{B}_{r,k}$. Fix a smooth pointed curve (C, p), and write K for the field of fractions of the completed local ring $\mathcal{O} = \widehat{\mathcal{O}}_{C,p}$. For notational simplicity, we set

$$\mathfrak{g} = \mathfrak{sl}_r,$$

and write $(|)$ for the suitably normalized Killing form. The loop algebra is the central extension

$$\widehat{L\mathfrak{g}} = \mathfrak{g} \otimes K \oplus \mathbb{C} \cdot c$$

of $\mathfrak{g} \otimes K$, endowed with the bracket

$$[X \otimes f, Y \otimes g] = [X, Y] \otimes fg + (X|Y) \cdot \operatorname{Res}(g\, df) \cdot c.$$

Two natural subalgebras of the loop algebra $\widehat{L\mathfrak{g}}$ play a role:

$$\widehat{L^+\mathfrak{g}} = \mathfrak{g} \otimes \mathcal{O} \oplus \mathbb{C} \cdot c \hookrightarrow \widehat{L\mathfrak{g}}$$

and

$$L_C\,\mathfrak{g} = \mathfrak{g} \otimes \mathcal{O}_C(C - p) \hookrightarrow \widehat{L\mathfrak{g}}.$$

For each positive integer k, we consider the basic representation H_k of $\widehat{L\mathfrak{g}}$ at level k, defined as follows. The one-dimensional vector space $\mathbb{C}$ is viewed as a module over the universal enveloping algebra $U(\widehat{L^+\mathfrak{g}})$ where the center c acts as multiplication by k, and $\mathfrak{g}$ acts trivially. We set

$$V_k = U(\widehat{L\mathfrak{g}}) \otimes_{U(\widehat{L^+\mathfrak{g}})} \mathbb{C}.$$

There is a unique maximal $\widehat{L\mathfrak{g}}$-invariant submodule

$$V_k' \hookrightarrow V_k.$$

The basic representation is the quotient

$$H_k = V_k / V_k'.$$

The finite-dimensional space of covacua for (C, p), dual to the space of conformal blocks, is given in turn as a quotient

$$B_{r,k} = H_k / L_C\mathfrak{g}\, H_k.$$

When the pointed curve varies, the loop algebra as well as its two natural subalgebras relativize over $\mathcal{M}_{g,1}$. The above constructions then give rise to the finite-rank vector bundle

$$\mathcal{B}_{r,k} \to \mathcal{M}_{g,1},$$

endowed with the projectively flat WZW connection.

5.2. Atiyah algebras. The key theorem in [**T**] uses the language of Atiyah algebras to describe the WZW connection on the bundles $\mathcal{B}_{r,k}$. We review this now, and refer the reader to [**Lo**] for a different account.

An Atiyah algebra over a smooth base S is a Lie algebra which sits in an extension

$$0 \to \mathcal{O}_S \xrightarrow{i} \mathcal{A} \xrightarrow{\pi} \mathcal{T}_S \to 0.$$

If $L \to S$ is a line bundle, then the sheaf of first order differential operators acting on L is an Atiyah algebra

$$\mathcal{A}_L = \operatorname{Diff}^1(L),$$

via the symbol exact sequence.

We also need an analogue of the sheaf of differential operators acting on tensor powers L^c for all rational numbers c, even though these line bundles don't actually make sense. To this end, if $\mathcal{A}$ is an Atiyah algebra and $c \in \mathbb{Q}$, then $c\mathcal{A}$ is by definition the Atiyah algebra

$$c\mathcal{A} = (\mathcal{O}_S \oplus \mathcal{A})/(c, 1)\mathcal{O}_S$$

sitting canonically in an exact sequence

$$0 \to \mathcal{O}_S \to c\mathcal{A} \to \mathcal{T}_S \to 0.$$

The sum of two Atiyah algebras $\mathcal{A}$ and $\mathcal{B}$ is given by

$$\mathcal{A} + \mathcal{B} = \mathcal{A} \times_{\mathcal{T}_S} \mathcal{B}/(i_{\mathcal{A}}(f), -i_{\mathcal{B}}(f)) \text{ for } f \in \mathcal{O}_S.$$

When c is a positive integer, $c\mathcal{A}$ coincides with the sum $\mathcal{A}+\ldots+\mathcal{A}$, but $c\mathcal{A}$ is more generally defined for all $c \in \mathbb{Q}$. In particular, $c\mathcal{A}_L$ makes sense for any $c \in \mathbb{Q}$ and any line bundle $L \to S$.

An action of an Atiyah algebra $\mathcal{A}$ on a vector bundle $\mathcal{V}$ is understood to enjoy the following properties

(i) each section a of $\mathcal{A}$ acts as a first order differential operator on $\mathcal{V}$ with symbol given by $\pi(a) \otimes \mathbf{1}_{\mathcal{V}}$;
(ii) the image of $1 \in \mathcal{O}_S$ i.e. $i(1)$ acts on $\mathcal{V}$ via the identity.

It is immediate that the action of an Atiyah algebra on $\mathcal{V}$ is tantamount to a projectively flat connection in $\mathcal{V}$. Furthermore, if two Atiyah algebras $\mathcal{A}$ and $\mathcal{B}$ act on vector bundles $\mathcal{V}$ and $\mathcal{W}$ respectively, then the sum $\mathcal{A} + \mathcal{B}$ acts on $\mathcal{V} \otimes \mathcal{W}$ via

$$(a, b) \cdot v \otimes w = av \otimes w + v \otimes bw.$$

We will make use of the following:

LEMMA 5.1. *Let $c \in \mathbb{Q}$ be a rational number and $L \to S$ be a line bundle. If the Atiyah algebra $c\mathcal{A}_L$ acts on a vector bundle $\mathcal{V}$, then the slope $\mu(\mathcal{V}) = \det \mathcal{V}/\operatorname{rank} \mathcal{V}$ is determined by*

$$\mu(\mathcal{V}) = c\, L\ .$$

PROOF. Replacing the pair $(\mathcal{V}, L)$ by a suitable tensor power we reduce to the case $c \in \mathbb{Z}$ via the observation preceding the Lemma. Then, we induct on c, adding one copy of the Atiyah algebra of L at a time. The base case $c = 0$ corresponds to a flat connection in $\mathcal{V}$. Indeed, the Atiyah algebra of $\mathcal{O}_S$ splits as $\mathcal{O}_S \oplus \mathcal{T}_S$ and an action of this algebra of $\mathcal{V}$ is equivalent to differential operators ∇_X for $X \in \mathcal{T}_S$, such that

$$[\nabla_X, \nabla_Y] = \nabla_{[X,Y]},$$

hence to a flat connection. □

Consider the rational number

$$c = \frac{k(r^2-1)}{r+k},$$

which is the charge of the Virasoro algebra acting on the basic level k representation H_k of $\widehat{L\mathfrak{g}}$. The representation H_k entered the construction of the bundles of covacua $\mathcal{B}_{r,k}$. The main result of [**T**] is the fact that the Atiyah algebra

$$\frac{c}{2}\mathcal{A}_L$$

acts on the bundle of covacua $\mathcal{B}_{r,k}$ where $\mathcal{A}_L$ is the Atiyah algebra associated to the determinant of the Hodge bundle

$$L = \det \mathbb{E}.$$

By Lemma 5.1, we deduce the slope

$$\mu(\mathcal{B}_{r,k}) = \frac{k(r^2-1)}{2(r+k)}\lambda.$$

In fact, by the proof of Lemma 5.1, the bundle

$$\mathcal{B}_{r,k}^{2(r+k)} \otimes L^{-k(r^2-1)}$$

is flat.

5.3. Identifications and the slope calculation. We now explain how the above calculation implies the Main Formula (1.2) via the results of Section 5.7 of [**L**].

Crucially, Laszlo proves that the projectivization of $\mathcal{B}_{r,k}^{\vee}$ coincides with the projectivization of the bundle $\mathbb{V}_{r,k}^{+}$ coming from geometry. In fact, Laszlo shows that for a suitable line bundle $\mathcal{L}_r$ over

$$\mathcal{SU}_{g,1}(r,\mathcal{O}) \to \mathcal{M}_{g,1}$$

we have

$$\mathcal{B}_{r,k}^{\vee} = \pi_\star(\mathcal{L}_r^k),$$

where fiberwise, over a fixed pointed curve, $\mathcal{L}_r^k$ coincides with the usual theta bundle $\left(\Theta_r^+\right)^k$. Hence,

$$\mathcal{L}_r^k = \left(\Theta_r^+\right)^k \otimes \mathcal{T}_{r,k}$$

for some line for some line bundle twist $\mathcal{T}_{r,k} \to \mathcal{M}_{g,1}$ over the moduli stack. At the heart of this identification is the double quotient construction of the moduli space of bundles over a curve

$$\mathcal{SU}_C(r,\mathcal{O}) = L_C G \backslash \widehat{LG} / \widehat{L^+G}$$

with the theta bundle Θ_r^+ being obtained by descent of a natural line bundle $\mathcal{Q}_r$ from the affine Grassmannian

$$\mathcal{Q}_r \to \widehat{LG} / \widehat{L^+G}.$$

Here $\widehat{LG}$ and $\widehat{L^+G}$ are the central extensions of the corresponding loop groups. The construction is then carried out relatively over $\mathcal{M}_{g,1}$, such that $\mathcal{Q}_r^k$ descends to the line bundle

$$\mathcal{L}_r^k \to \mathcal{SU}_{g,1}(r,\mathcal{O}).$$

It follows from here that fiberwise $\mathcal{L}_r^k$ coincides with the usual theta bundle $\left(\Theta_r^+\right)^k$.

Collecting the above facts, we find that

$$\mathcal{B}_{r,k}^{\vee} = \mathbb{V}_{r,k}^{+} \otimes \mathcal{T}_{r,k}.$$

Therefore

$$-\mu(\mathcal{B}_{r,k}) = \mu(\mathbb{V}_{r,k}^{+}) + c_1(\mathcal{T}_{r,k}).$$

Using Lemma 3.2 we conclude that

$$-\frac{k(r^2-1)}{2(r+k)}\lambda = \mu(\mathbb{V}_{r,k}) - \frac{kr-1}{2}\lambda + krc_1(\mathcal{L}) + c_1(\mathcal{T}_{r,k}).$$

Simplifying, this yields

$$\mu(\mathbb{V}_{r,k}) = \frac{r(k^2-1)}{2(r+k)}\lambda - c_1(\mathcal{T}_{r,k}) - krc_1(\mathcal{L}).$$

Now, the left hand side is a multiple of λ, namely $s_{r,k}\lambda$. The right hand side must be a multiple of λ as well. With

$$s'_{r,k} = s_{r,k} - \frac{r(k^2-1)}{2(r+k)}.$$

we find that

$$s'_{r,k}\lambda = -c_1(\mathcal{T}_{r,k}) - krc_1(\mathcal{L}).$$

This implies that $s'_{r,k}$ must be an integer by comparison with the right hand side, because the Picard group of $\mathcal{M}_g$ is generated over $\mathbb{Z}$ by λ for $g \geq 2$, see [**AC2**]. The fact that $s'_{r,k} \in \mathbb{Z}$ is enough to prove

$$s'_{r,k} = 0,$$

which is what we need.

Indeed, as explained in Section 4.1, Grothendieck-Riemann-Roch for the push-forwards giving the Verlinde numbers shows that

$$\lim_{k\to\infty} \frac{s'_{r,k}}{k} < \infty.$$

Writing

$$s'_{r,k} = a_r(k)/b_r(k)$$

with $\deg a_r(k) \leq \deg b_r(k) + 1$, we see by direct calculation that

$$\lim_{k\to\infty} s'_{r,k+1} - 2s'_{r,k} + s'_{r,k-1} = 0.$$

Since the expression in the limit is an integer, it must equal zero. By induction, it follows that

$$s'_{r,k} = A_r k + B_r$$

for constants A_r, B_r that may depend on the rank and the genus. Since

$$s'_{r,0} = s'_{r,-2r} = 0$$

by the initial condition in Lemma 3.4 and by Proposition 3.3, it follows that $A_r = B_r = 0$ hence $s'_{r,k} = 0$.

As a consequence, we have now also determined the twist $\mathcal{T}_{r,k} = \mathcal{L}^{-kr}$. Therefore, the bundle of conformal blocks is expressed geometrically as

$$\mathcal{B}^\vee_{r,k} = \mathbb{V}^+_{r,k} \otimes \mathcal{L}^{-kr}.$$

We remark furthermore that the latter bundle descends to $\mathcal{M}_g$. To see this, one checks that

$$(\Theta^+_r)^k \otimes \mathcal{L}^{-kr}$$

restricts trivially over the fibers of $\mathcal{SU}_{g,1}(r, \mathcal{O}) \to \mathcal{SU}_g(r, \mathcal{O})$. This is a straightforward verification.

6. Extensions over the boundary

The methods of [**T**] can be used to find the first Chern class of the bundle of conformal blocks over the compactification $\overline{\mathcal{M}}_g$. The resulting formula is stated in Theorem 6.1 below. In particular the first Chern class contains nonzero boundary contributions, contrary to a claim of [**S**].

In genus 0, formulas for the Chern classes of the bundle of conformal blocks were given in [**F**], and have been recently brought to simpler form in [**Mu**]. In higher genus, the expressions we obtain using [**T**] specialize to the simpler formulas of [**Mu**].

As it is necessary to consider parabolics, we begin with some terminology on partitions. We denote by $\mathcal{P}_{r,k}$ the set of Young diagrams with at most r rows and at most k columns. Enumerating the lengths of the rows, we write a diagram μ as

$$\mu = (\mu^1, \ldots, \mu^r), \ k \geq \mu^1 \geq \cdots \geq \mu^r \geq 0.$$

The partition μ is viewed as labeling the irreducible representation of the group $SU(r)$ with highest weight μ, which we denote by V_μ. Two partitions which differ by the augmentation of the rows by a common number of boxes yield isomorphic representations. We will identify such partitions in $\mathcal{P}_{r,k}$, writing $\sim$ for the equivalence relation. There is a natural involution

$$\mathcal{P}_{r,k} \ni \mu \mapsto \mu^\star \in \mathcal{P}_{r,k}$$

where $\mu^\star$ is the diagram whose row lengths are

$$k \geq k - \mu^r \geq \ldots \geq k - \mu^1 \geq 0.$$

Further, to allow for an arbitrary number of markings, we consider multipartitions

$$\underline{\mu} = (\mu_1, \ldots, \mu_n)$$

whose members belong to $\mathcal{P}_{r,k}/\sim$. Finally, for a single partition μ, we write

$$\mathsf{w}_\mu = -\frac{1}{2(r+k)}\left(\sum_{i=1}^r \mu_i^2 - \frac{1}{r}\left(\sum_{i=1}^r \mu_i\right)^2 + \sum_{i=1}^r (r-2i+1)\mu_i\right)$$

for the suitably normalized action of the Casimir element on the representation V_μ.

In this setup, we let

$$\mathcal{B}_{g,\underline{\mu}} \to \overline{\mathcal{M}}_{g,n}$$

be the bundle of covacua, obtained analogously to the construction of Section 5.1 using representations of highest weight $\underline{\mu}$, see [**T**]. To simplify notation, we do not indicate dependence on r, k and n explicitly: these can be read off from the multipartition $\underline{\mu}$. We set

$$v_g(\underline{\mu}) = \text{rank } \mathcal{B}_{g,\underline{\mu}}$$

to be the parabolic Verlinde number.

We determine the first Chern class $c_1(\mathcal{B}_{g,\underline{\mu}})$ over $\overline{\mathcal{M}}_{g,n}$ in terms of the natural generators:

$$\lambda, \Psi_1, \ldots, \Psi_n$$

and the boundary divisors. To fix notation, we write as usual:

- δ_{irr} for the class of the divisor corresponding to irreducible nodal curves;

- $\delta_{h,A}$ for the boundary divisor corresponding to reducible nodal curves, with one component having genus h and containing the markings of the set A.

Note that each subset $A \subset \{1, 2, \ldots, n\}$ determines a splitting

$$\underline{\mu}_A \cup \underline{\mu}_{A^c}$$

of the multipartition $\underline{\mu}$ corresponding to the markings in A and in its complement A^c. Finally, we define the coefficients

$$c_{\mathrm{irr}} = \sum_{\nu \in \mathcal{P}_{r,k}/\sim} \mathsf{w}_\nu \cdot \frac{v_{g-1}(\underline{\mu}, \nu, \nu^\star)}{v_g(\underline{\mu})}$$

and

$$c_{h,A} = \sum_{\nu \in \mathcal{P}_{r,k}/\sim} \mathsf{w}_\nu \cdot \frac{v_h(\underline{\mu}_A, \nu) \cdot v_{g-h}(\underline{\mu}_{A^c}, \nu^\star)}{v_g(\underline{\mu})}.$$

THEOREM 6.1. *Over $\overline{\mathcal{M}}_{g,n}$ the slope of the bundle of covacua is*

$$\mathrm{slope}(\mathcal{B}_{g,\underline{\mu}}) = \frac{k(r^2-1)}{2(r+k)}\lambda + \sum_{i=1}^{n} \mathsf{w}_{\mu_i} \Psi_i - c_{irr}\delta_{irr} - \sum_{h,A} c_{h,A}\delta_{h,A}. \tag{6.1}$$

In the formula, the repetition $\delta_{h,A} = \delta_{g-h,A^c}$ is not allowed, so that each divisor appears only once.

PROOF. The formula written above is correct over the open stratum $\mathcal{M}_{g,n}$. Indeed, the main theorem of [**T**], used in the presence of parabolics, shows that the bundle of covacua

$$\mathcal{B}_{g,\underline{\mu}} \to \mathcal{M}_{g,n}$$

admits an action of the Atiyah algebra

$$\frac{k(r^2-1)}{2(r+k)}\mathcal{A}_L + \sum_{i=1}^{n} \mathsf{w}_{\mu_i} \mathcal{A}_{\mathcal{L}_i}.$$

As before

$$L = \det \mathbb{E}$$

is the determinant of the Hodge bundle and the $\mathcal{L}_i$ denote the cotangent lines over $\mathcal{M}_{g,n}$. Therefore, by Lemma 5.1, we have

$$\mathrm{slope}(\mathcal{B}_{g,\underline{\mu}}) = \frac{k(r^2-1)}{2(r+k)}\lambda + \sum_{i=1}^{n} \mathsf{w}_{\mu_i} \Psi_i$$

over $\mathcal{M}_{g,n}$.

It remains to confirm that the boundary corrections take the form stated above. Since the derivation is identical for all boundary divisors, let us only find the coefficient of δ_{irr}. To this end, observe the natural map

$$\xi : \mathcal{M}_{g-1,n+2} \to \overline{\mathcal{M}}_{g,n}$$

whose image is contained in the divisor δ_{irr}. The map is obtained by gluing together the last two markings which we denote $\bullet$ and $\star$. We pull back (6.1) under ξ. For the left hand side, we use the fusion rules of [**TUY**]:

$$\xi^\star \mathcal{B}_{g,\underline{\mu}} = \bigoplus_{\nu \in \mathcal{P}_{r,k}/\sim} \mathcal{B}_{g-1,\underline{\mu},\nu,\nu^\star}.$$

Thus, the left hand side becomes

$$\begin{aligned}
&\sum_{\nu \in \mathcal{P}_{r,k}/\sim} \frac{v_{g-1}(\underline{\mu}, \nu, \nu^\star)}{v_g(\underline{\mu})} \cdot \operatorname{slope}(\mathcal{B}_{g-1,\underline{\mu},\nu,\nu^\star}) \\
&= \sum_{\nu \in \mathcal{P}_{r,k}/\sim} \frac{v_{g-1}(\underline{\mu}, \nu, \nu^\star)}{v_g(\underline{\mu})} \cdot \left(\frac{k(r^2-1)}{2(r+k)} \lambda + \sum_{i=1}^n \mathsf{w}_{\mu_i} \Psi_i + \mathsf{w}_\nu \Psi_\bullet + \mathsf{w}_{\nu^\star} \Psi_\star \right) \\
&= \frac{k(r^2-1)}{2(r+k)} \lambda + \sum_{i=1}^n \mathsf{w}_{\mu_i} \Psi_i + \sum_{\nu \in \mathcal{P}_{r,k}/\sim} \frac{v_{g-1}(\underline{\mu}, \nu, \nu^\star)}{v_g(\underline{\mu})} \cdot (\mathsf{w}_\nu \Psi_\bullet + \mathsf{w}_{\nu^\star} \Psi_\star).
\end{aligned}$$

The fusion rules have been used in the third line to compare the ranks of the Verlinde bundles. For the right hand side, we record the following well-known formulas [**AC1**]:

(i) $\xi^\star \lambda = \lambda$;
(ii) $\xi^\star \Psi_i = \Psi_i$ for $1 \leq i \leq n$;
(iii) $\xi^\star \delta_{\mathrm{irr}} = -\Psi_\bullet - \Psi_\star$;
(iv) $\xi^\star \delta_{h,A} = 0$.

These yield the following expression for the right hand side of (6.1):

$$\frac{k(r^2-1)}{2(r+k)} \lambda + \sum_{i=1}^n \mathsf{w}_{\mu_i} \Psi_i - c_{\mathrm{irr}}(-\Psi_\bullet - \Psi_\star).$$

For $g-1 \geq 2$, $\Psi_\star$ and $\Psi_\bullet$ are independent in the Picard group of $\mathcal{M}_{g-1,n+2}$, see [**AC2**], hence we can identify their coefficient c_{irr} uniquely to the formula claimed above. The case of the other boundary corrections is entirely similar. □

REMARK 6.2. The low genus case $g \leq 2$ not covered by the above argument can be established by the following approach. Once a correct formula for the Chern class has been proposed, a proof can be obtained by induction on the genus and number of markings. Indeed, with some diligent bookkeeping, it can be seen that the expression of the Theorem restricts to the boundary divisors compatibly with the fusion rules in [**TUY**]. To finish the argument, we invoke the Hodge theoretic result of Arbarello-Cornalba [**AC1**] stating the boundary restriction map

$$H^2(\overline{\mathcal{M}}_{g,n}) \to H^2(\overline{\mathcal{M}}_{g-1,n+2}) \bigoplus_{h,A} H^2(\overline{\mathcal{M}}_{h,A\cup\{\bullet\}} \times \overline{\mathcal{M}}_{g-h,A^c\cup\{\star\}})$$

is injective, with the exception of the particular values $(g,n) = (0,4)$, $(0,5)$, $(1,1)$, $(1,2)$, which may be checked by hand.

In fact, the slope expression of the Theorem is certainly correct in the first three cases by [**F**], [**Mu**]. When $(g,n) = (1,2)$, we already know from [**T**] that the slope takes the form

$$\operatorname{slope}(\mathcal{B}_{\mu_1,\mu_2}) = \frac{k(r^2-1)}{2(r+k)} \lambda + \mathsf{w}_{\mu_1} \Psi_1 + \mathsf{w}_{\mu_2} \Psi_2 - c_{\mathrm{irr}} \delta_{\mathrm{irr}} - c\Delta,$$

where δ_{irr} and Δ are the two boundary divisors in $\overline{\mathcal{M}}_{1,2}$. The coefficients c_{irr} and c are determined uniquely in the form stated in the Theorem by restricting $\mathcal{B}_{\mu_1,\mu_2}$ to the two boundary divisors δ_{irr} and Δ (and not only to their interiors as was done above) via the fusion rules. The verification is not difficult for the particular case $(1,2)$.

References

[AC1] E. Arbarello and M. Cornalba, *Calculating cohomology groups of moduli spaces of curves via algebraic geometry*, Inst. Hautes Études Sci. Publ. Math. **88** (1998), 97–127 (1999). MR1733327 (2001h:14030)

[AC2] E. Arbarello and M. Cornalba, *The Picard groups of the moduli spaces of curves*, Topology **26** (1987), no. 2, 153–171, DOI 10.1016/0040-9383(87)90056-5. MR895568 (88e:14032)

[AGS] V. Alexeev, A. Gibney, and D. Swinarski, *Higher-level $\mathfrak{sl}_2$ conformal blocks divisors on $\overline{M}_{0,n}$*, Proc. Edinb. Math. Soc. (2) **57** (2014), no. 1, 7–30, DOI 10.1017/S0013091513000941. MR3165010

[AGSS] M. Arap, A. Gibney, J. Stankewicz, and D. Swinarski, *sl_n level 1 conformal blocks divisors on $\overline{M}_{0,n}$*, Int. Math. Res. Not. IMRN **7** (2012), 1634–1680. MR2913186

[BFM] P. Baum, W. Fulton, and R. MacPherson, *Riemann-Roch for singular varieties*, Inst. Hautes Études Sci. Publ. Math. **45** (1975), 101–145. MR0412190 (54 #317)

[B] A. Beauville, *Vector bundles on curves and generalized theta functions: recent results and open problems*, Current topics in complex algebraic geometry (Berkeley, CA, 1992/93), Math. Sci. Res. Inst. Publ., vol. 28, Cambridge Univ. Press, Cambridge, 1995, pp. 17–33. MR1397056 (97h:14015)

[BL] A. Beauville and Y. Laszlo, *Conformal blocks and generalized theta functions*, Comm. Math. Phys. **164** (1994), no. 2, 385–419. MR1289330 (95k:14011)

[BS] A. Bertram and A. Szenes, *Hilbert polynomials of moduli spaces of rank* 2. *Vector bundles. II*, Topology **32** (1993), no. 3, 599–609, DOI 10.1016/0040-9383(93)90011-J. MR1231966 (94m:14024)

[Bel] P. Belkale, *Strange duality and the Hitchin/WZW connection*, J. Differential Geom. **82** (2009), no. 2, 445–465. MR2520799 (2010j:14065)

[DN] J.-M. Drezet and M. S. Narasimhan, *Groupe de Picard des variétés de modules de fibrés semi-stables sur les courbes algébriques* (French), Invent. Math. **97** (1989), no. 1, 53–94, DOI 10.1007/BF01850655. MR999313 (90d:14008)

[F] N. Fakhruddin, *Chern classes of conformal blocks*, Compact moduli spaces and vector bundles, Contemp. Math., vol. 564, Amer. Math. Soc., Providence, RI, 2012, pp. 145–176, DOI 10.1090/conm/564/11148. MR2894632

[Fe] M. Fedorchuk, *Cyclic covering morphisms on $\overline{M}_{0,n}$*, preprint, arXiv:1105.0655.

[FMP] C. Faber, A. Marian, R. Pandharipande, *Verlinde flatness and relations in $H^\star(\mathcal{M}_g)$*, available at http://www.math.ethz.ch/~rahul/vertaut.pdf.

[GG] N. Giansiracusa and A. Gibney, *The cone of type A, level 1, conformal blocks divisors*, Adv. Math. **231** (2012), no. 2, 798–814, DOI 10.1016/j.aim.2012.05.017. MR2955192

[Hi] N. J. Hitchin, *Flat connections and geometric quantization*, Comm. Math. Phys. **131** (1990), no. 2, 347–380. MR1065677 (91g:32022)

[L] Y. Laszlo, *Hitchin's and WZW connections are the same*, J. Differential Geom. **49** (1998), no. 3, 547–576. MR1669720 (2000e:14012)

[Lo] E. Looijenga, *From WZW models to modular functors*, Handbook of Moduli. Vol. II, Adv. Lect. Math. (ALM), vol. 25, Int. Press, Somerville, MA, 2013, pp. 427–466. MR3184182

[M] D. Mumford, *Prym varieties. I*, Contributions to analysis (a collection of papers dedicated to Lipman Bers), Academic Press, New York, 1974, pp. 325–350. MR0379510 (52 #415)

[Mu] S. Mukhopadhyay, *Rank-Level Duality of Conformal Blocks*, ProQuest LLC, Ann Arbor, MI, 2013. Thesis (Ph.D.)–The University of North Carolina at Chapel Hill. MR3153462

[MO] A. Marian and D. Oprea, *The level-rank duality for non-abelian theta functions*, Invent. Math. **168** (2007), no. 2, 225–247, DOI 10.1007/s00222-006-0032-z. MR2289865 (2007k:14070)

[MOPPZ] A. Marian, D. Oprea, R. Pandharipande, A. Pixton, D. Zvonkine, *The Chern character of the Verlinde bundle*, J. Reine Angew. Math., to appear.

[PP] R. Pandharipande, A. Pixton, *Relations in the tautological ring of the moduli space of curves*, preprint, arXiv:1301.4561.

[P] C. Pauly, *Strange duality revisited*, preprint, arXiv:1204.1186.

[S] I. Smith, *Symplectic four-manifolds and conformal blocks*, J. London Math. Soc. (2) **71** (2005), no. 2, 503–515, DOI 10.1112/S0024610705006307. MR2122441 (2005i:57034)

[Sw] D. Swinarski, *sl_2 conformal block divisors and the nef cone of $\overline{M}_{0,n}$*, preprint, arXiv:1107.5331.

[TUY] A. Tsuchiya, K. Ueno, and Y. Yamada, *Conformal field theory on universal family of stable curves with gauge symmetries*, Integrable systems in quantum field theory and statistical mechanics, Adv. Stud. Pure Math., vol. 19, Academic Press, Boston, MA, 1989, pp. 459–566. MR1048605 (92a:81191)

[T] Y. Tsuchimoto, *On the coordinate-free description of the conformal blocks*, J. Math. Kyoto Univ. **33** (1993), no. 1, 29–49. MR1203889 (95c:14023)

[vdG] G. van der Geer, *Cycles on the moduli space of abelian varieties*, Moduli of curves and abelian varieties, Aspects Math., E33, Vieweg, Braunschweig, 1999, pp. 65–89, DOI 10.1007/978-3-322-90172-9_4. MR1722539 (2001b:14009)

[Z] D. Zagier, *Elementary aspects of the Verlinde formula and of the Harder-Narasimhan-Atiyah-Bott formula*, Proceedings of the Hirzebruch 65 Conference on Algebraic Geometry (Ramat Gan, 1993), Israel Math. Conf. Proc., vol. 9, Bar-Ilan Univ., Ramat Gan, 1996, pp. 445–462. MR1360519 (96k:14005)

Department of Mathematics, Northeastern University
E-mail address: a.marian@neu.edu

Department of Mathematics, University of California, San Diego
E-mail address: doprea@math.ucsd.edu

Department of Mathematics, ETH Zürich
E-mail address: rahul@math.ethz.ch

Proceedings of Symposia in Pure Mathematics
Volume **90**, 2015
http://dx.doi.org/10.1090/pspum/090/01532

Framing the Di-logarithm (Over $\mathbb{Z}$)

Albert Schwarz, Vadim Vologodsky, and Johannes Walcher

Abstract. Motivated by their role for integrality and integrability in topological string theory, we introduce the general mathematical notion of "s-functions" as integral linear combinations of poly-logarithms. 2-functions arise as disk amplitudes in Calabi-Yau D-brane backgrounds and form the simplest and most important special class. We describe s-functions in terms of the action of the Frobenius endomorphism on formal power series and use this description to characterize 2-functions in terms of algebraic K-theory of the completed power series ring. This characterization leads to a general proof of integrality of the framing transformation, via a certain orthogonality relation in K-theory. We comment on a variety of possible applications. We here consider only power series with rational coefficients; the general situation when the coefficients belong to an arbitrary algebraic number field is treated in a companion paper.

...as if this function alone among all others possessed a sense of humor (D. Zagier [**1**])

1. Introduction

A remarkable aspect of exact calculations in supersymmetric and topological quantum theories is the blending of discrete and analytic information. What we mean is that while, on the one hand, the microscopic Lagrangian formulation of some given supersymmetric observable makes manifest the holomorphic dependence on the parameters, and can, in some cases, be used to derive the behavior under certain duality transformations (or under analytic continuation), the typical answers localize to finite sums (or at the most, finite-dimensional integrals) over configurations of classical supersymmetric solutions (BPS states).

In other words, the expansion coefficients of the supersymmetric amplitude around the appropriate limit often admit an *a priori* perhaps unexpected interpretation as counting dimensions of certain vector spaces of BPS states (or, more generally, an index of some operator on such spaces). That, conversely, the generating functions for these dimensions have interesting analytic and modular properties, is a remarkable fact that can be understood, at least in part, as a consequence of the underlying duality symmetries of the microscopic formulation.

In these respects, supersymmetric partition functions are reminiscent of functions of interest in analytic number theory, and in fact there are many cases in which the two are very closely related. This has led to a number of results on,

---

2010 *Mathematics Subject Classification.* Primary 81T45; Secondary 33B30, 13F35.

most notably, modular forms finding applications in diverse areas of mathematical physics related to supersymmetric field and string theories. This interest has also led to a number of new mathematical results. The ways in which the functions of interest are related to some geometric situation in both physics and number theory are also often similar. The rampant speculation about the deeper meaning of such coincidences is best restrained by pointing out that there remain large classes of very deep number theoretic functions (such as L-functions, ζ-functions) whose relevance for supersymmetric quantum theory are much less clear.

In this paper, we study elementary algebraic properties of a certain class of functions that we call "s-functions" (where, at least for now, s is a positive integer). We extract this notion from the appearance of s-function in perturbative computations in topological string theory, where they are building blocks of supersymmetric generating functions. We define an s-function as an integral linear combination of s-logarithms. We give later an equivalent definition in terms of Frobenius map on (formal) power series with rational coefficients. That second definition can easily be generalized to the case when we allow coefficients to lie in arbitrary local or global number fields, see [**5**].

One of our main results concerns the most important special case, $s = 2$, and is an integrality statement of a certain algebraic transformation of 2-functions (viewed as formal power series) that we call "framing". We point out that while special cases of this framing transformation are known in the context of open topological string theory (where we borrowed the name, see [**2**]), the generality in which it applies has not been pointed out in the literature to our knowledge (Although it might be known to experts. We made our initial observations after reading [**3**].) Secondly, we will give a mathematical proof of this framing property, using an interpretation of the notion of 2-function in algebraic K-theory. (M. Kontsevich informed us that he also obtained this interpretation and used it to prove some integrality theorems.)

Finally, we point out a few further generalizations of our setup and constructions. One of these generalizations involves extending the field of definition of the coefficients from $\mathbb{Q}$ to a more general number field. The relevance of such extensions was first observed in [**4**], and we will elaborate on them in a companion paper [**5**].

At the moment the only immediate applications of our results that we are aware of come from open topological string theory and mirror symmetry. We suspect however that the concepts we introduce might play a role in other contexts as well. As a particular example, refs. [**3, 6**] lead us to expect certain connections with the theory of Mahler measures. In a different way, the characterization of 2-functions in algebraic K-theory is reminiscent of a certain integrability condition recently explored by Gukov and Sulkowski [**7**].

In mathematical terms, integrality of framing is the following statement. Say

$$W(z) = \sum_{d=1}^{\infty} n_d \mathrm{Li}_2(z^d) \tag{1.1}$$

is an integral linear combination of standard di-logarithms,

$$\mathrm{Li}_2(z) = \sum_{k=1}^{\infty} \frac{z^k}{k^2} \tag{1.2}$$

Namely, the coefficients n_d in (1.1) are integers, and we may as well assume that the series (1.1) is convergent. We introduce the power series

$$Y(z) = \exp\big(-z\frac{d}{dz}W(z)\big)\,, \tag{1.3}$$

with constant term 1 around $z=0$. Then the relation

$$\tilde{z} = -zY(z) \tag{1.4}$$

may be inverted (formally, and as a convergent power series),

$$z = -\tilde{z}\tilde{Y}(\tilde{z}) \tag{1.5}$$

to yield another power series $\tilde{Y}(\tilde{z})$ around $\tilde{z}=0$ with constant term 1. Expanding

$$\tilde{W}(\tilde{z}) = \int \frac{d\tilde{z}}{\tilde{z}}\big(-\log\tilde{Y}(\tilde{z})\big) = \sum_{d=1}^{\infty}\tilde{n}_d\mathrm{Li}_2(\tilde{z}^d) \tag{1.6}$$

defines coefficients $\tilde{n}_d$ which are *a priori* rational numbers. It is elementary to see that $\tilde{Y}(\tilde{z})$ has integer coefficients, and as a consequence that $d\tilde{n}_d\in\mathbb{Z}$. We claim that, in fact,

$$\tilde{n}_d\in\mathbb{Z} \tag{1.7}$$

(whenever $n_d\in\mathbb{Z}$). See section 3 for the sketch of a proof of this statement.

In physical terms, we think of $W(z)$ as a contribution to the space-time superpotential from D-branes wrapping supersymmetric cycles in some Calabi-Yau compactification of string theory preserving $\mathcal{N}=1$ supersymmetry in 4 dimensions. In that context, z is a chiral superfield whose vacuum expectation value parametrizes the moduli space of open/closed string vacua, and corresponds to a geometric modulus of the configuration. It was shown long time ago by Ooguri and Vafa [**8**], generalizing work by Gopakumar and Vafa [**9**], that, for an appropriate choice of parametrization, one expects the coefficients n_d in the expansion (1.1) to count dimensions of spaces of appropriate BPS states, hence the integrality.

One of the features of the setup of Ooguri and Vafa is the dependence of the superpotential (and the BPS invariants) on an integer parameter, f, known as "the framing". Algebraically, the framing results from an ambiguitiy in the identification of the open string modulus [**2**]. Namely, framing by f amounts to replacing

$$\log z \to \log z_f = \log z + fz\partial_z W(z)\,,\qquad f\in\mathbb{Z} \tag{1.8}$$

Although it might not be immediately obvious, the transformations (1.8) are very closely related to (namely, generated by) (1.4), (1.5). We will explain this connection, and also review the general setup in somewhat more detail, in section 2.

The dependence on f of open topological string amplitudes in the Ooguri-Vafa setup is explained through the (large-N) duality with Chern-Simons theory and knot invariants. (The framing of a knot is the choice of non-vanishing section of the normal bundle of a knot in a three-manifold, and gets identified with f under this duality.) For this duality to operate, it is important that the underlying Calabi-Yau manifold be *non-compact*. Mathematically, the framing dependence of the enumerative (open Gromov-Witten) invariants is only well understood when the manifold is *toric* [**11**].

In a series of works, from [**10**] to [**4**], it was shown that the expansion (1.1) continues to be valid in principle when the underlying Calabi-Yau manifold is *compact*, and *non-toric*, but requires a number of modifications in practice. Most prominently, the parameter z should be a closed string modulus that remains massless at tree-level. Secondly, the standard di-logarithm (1.2) has to be "twisted" in general to take into account the symmetries of the open string vacuum structure. On the other hand, neither the geometric setup (in the A-model) nor the actual calculation (in the B-model) seem to involve any ambiguity that could be identified as "framing". Most recently, it was pointed out in [**4**] that the generic B-model setup will predict invariants n_d that are *irrational* numbers valued in some algebraic number field (a finite extension of $\mathbb{Q}$, fixed for each geometric situation). The symmetry of the space of vacua and the associated twist of the di-logarithm was related to the Galois group of that finite field extension. It is possible to prove some integrality statements also in this case (see [**5**] for detail).

The statement that D-brane superpotentials given by geometric formulas of [**2**, **10**, **12**] indeed admit a decomposition of the form (1.1) into integral pieces was proven mathematically in ref. [**15**], using and extending earlier work by the same authors [**13**, **14**]. The central aspect of that series of works was to relate the BPS numbers n_d to the action of the Frobenius automorphism on p-adic cohomology of the Calabi-Yau (together with an algebraic cycle).

The proofs of [**15**] show integrality of open topological string amplitudes, separately for any value of the framing, in situations in which this concept is well-defined. One of the main messages of the present paper is that the integrality of the framing transformation is more general, and in fact not tied to a particular geometric siutation. But the methods for proving the integrality statements developed in [**15**] continue to apply. This means in particular that we can define a "framed" superpotential and enumerative invariants even when we do not know a geometric interpretation for the integer ambiguity f. We interpret this fact, together with the observation that instanton numbers are related to Mahler measures by a framing transformation [**3**, **6**], as a hint that framing is an important intrinsic property of the di-logarithm.

As further support, we mention that the integrality of the framing transformation is naturally expressed as a certain torsion/orthogonality condition in algebraic K-theory, see section 3, and can also be given a Hodge theoretic interpretation. Framing can also be generalized to the multi-variable situation, in which it depends in an interesting way on the additional data of a symmetric bilinear form. Finally, while in this paper we are concerned mainly with the situation in which the coefficients are actually rational numbers, the generalization to arbitrary number fields is rather straightforward. We will consider it in a separate paper [**5**], where further mathematical details may also be found.

2. Dilogarithm, s-functions, and topological strings

(The mathematically inclined reader may gain from skipping the odd (numbered) subsections which provide some physics motivation for our definitions.)

2.1. A-model. From the point of view of the A-model, the origin of the formula (1.1) is, intuitively, easy to understand. Consider a Calabi-Yau threefold X and a Lagrangian submanifold $L \subset X$. We view L as the support of a topological D-brane in the A-model, which we may want to use as an ingredient in a superstring theory construction. As is well-known, the classical deformation space of L modulo Hamiltonian isotopy (or, preserving the "special Lagrangian" condition, if one exists) is unobstructed and of dimension equal to $b_1(L)$. Worldsheet instanton corrections however induce a space-time superpotential that schematically takes the form

$$W = \ldots + \sum_{u:(D,\partial D)\to(X,L)} e^{\int_D u^*\omega}\,\mathrm{Tr}\big[Pe^{\oint_{\partial D} u^*A}\big] \tag{2.1}$$

and depends (via the symplectic form ω) on the Kähler moduli of X and (via the (unitary) connection A) on the choice of a flat bundle over L. Here, $\ldots$ denotes certain (subtle) classical terms that we will neglect in this paper, so the sum is over all (non-constant) holomorphic maps

$$u : (D, \partial D) \to (X, L) \tag{2.2}$$

from the disk D to X mapping the boundary ∂D to L.

It is well known that the expected (virtual) dimension of the space of such holomorphic maps is zero for any class $\beta = u_*([D, \partial D]) \in H_2(X, L)$, and hence one expects to write[1]

$$W(Q) = \sum_{\beta\in H_2(X,L)} m_\beta q^\beta \tag{2.3}$$

where $\log q$ is the appropriate combination of moduli of (X, L), and m_β is the "number" of holomorphic maps in a fixed class β (open Gromov-Witten invariants). While the general definition of m_β is plagued with difficulty, it is in any case clear that the moduli space $\mathcal{M}(\beta)$ of such maps will contain components of positive dimension, if β is not primitive. Namely, say $\beta = k\beta'$ with β' integral and $k > 1$. Then any $u' \in \mathcal{M}(\beta')$ may be composed with a degree k covering map $c : (D, \partial D) \to (D, \partial D)$ to give a map $u = u' \circ c \in \mathcal{M}(\beta)$. Since the maps c come in families (of dimension $2k - 2$), so $\mathcal{M}(\beta)$ will contain components of positive dimension.

The formula (1.1) is a reflection of these multi-covers. (Even though it might not be strictly true that all holomorphic maps can be factorized in this way, see [**16**], the success of the formula suggests that this is effectively the case.) The general statement is that a BPS state corresponding (intuitively) to an embedded disk with boundary on L in the class β, together with all its multi-covers, makes a contribution to W of the form

$$W_\beta(q) = \sum \frac{1}{k^2} q^{k\beta} = \mathrm{Li}_2(q^\beta)\,, \tag{2.4}$$

so that if n_β is the (integer!) degeneracy of BPS states of charge β, the total superpotential is

$$W(q) = \sum_\beta n_\beta W_\beta(q) = \sum_\beta n_\beta \mathrm{Li}_2(q^\beta) \tag{2.5}$$

Eq. (1.1) is recovered when $H_2(X, L)$ has rank one, and $q = z$.

The prototypical example of a multi-cover formula like (2.4) is known from the beginning of mirror symmetry [**17**] as the Aspinwall-Morrison formula [**18**]. It states that the large volume (A-model) expansion of the $\mathcal{N} = 2$ prepotential (i.e., the genus 0 Gromov-Witten potential) takes the form

$$F^{(0)} = \sum M_d q^d = \sum N_d \mathrm{Li}_3(q^d) \tag{2.6}$$

[1] Note that the Tr in (2.1) really depends on the homotopy class of u in $\pi_2(X, L)$. The formula (2.3) makes sense if the fundamental group of L is abelian, so that $\pi_2(X, L) = H_2(X, L)$.

with *integer* N_d. The M_d are rational numbers, which is obvious from both the definition in Gromov-Witten theory, as well as from the B-model formulas (involving differential equations with rational coefficients). The integrality of the N_d however is harder to see. It was proven mathematically in [**13**–**15**, **19**]. A physical explanation was given in [**9**] by relating the N_d to the degeneracy of BPS states. The generalization of the Aspinwall-Morrison formula to arbitrary genus g was shown to involve the poly-logarithm Li_{3-2g}.

2.2. s-functions. The central importance of these multi-cover formula in relating the perturbative topological string amplitudes to the degeneracy of BPS states motivates us to introduce the following notion: If s is a positive integer, we call a power series

$$V(z) = \sum_{d=1}^{\infty} m_d z^d \in \mathbb{Q}[[z]] \tag{2.7}$$

with rational coefficients m_d an s-function if it can be written as an integral linear combination of s-logarithms.

$$V(z) = \sum_{d=1}^{\infty} n_d \mathrm{Li}_s(z^d) , \qquad \mathrm{Li}_s(z) := \sum_{k=1}^{\infty} k^{-s} z^k \tag{2.8}$$

with $n_d \in \mathbb{Z}$. It is convenient to define the logarithmic derivative,

$$\delta_z = \frac{d}{d \ln z} \tag{2.9}$$

So $\delta_z \mathrm{Li}_s(z) = \mathrm{Li}_{s-1}(z)$, and if $V(z)$ is an s-function, $\delta_z V(z)$ is an $(s-1)$-function. For the topological string, the relevant values are $s = 3$ for genus 0 (closed string tree-level) invariants, $s = 2$ for disk invariants (open string tree-level), and $s = 1$ for all one-loop amplitudes (open or closed).

Sometimes it is convenient to consider s-functions with respect to prime number p requiring that the denominators of the coefficients m_d are not divisible by p (the coefficients are p-integral).

2.3. B-model. As mentioned in the introduction, framing originally entered topological string theory through the relation between local toric manifolds and Chern-Simons gauge theory and knot invariants. Framing has also been explained in (toric) A-model [**11**] as a choice of linearization of the torus action required to make the localization calculation of open Gromov-Witten invariants well-defined. The operation itself is however most straightforward to explain geometrically in the B-model.

The B-model mirror of a general toric Calabi-Yau threefold has the form

$$\{uv = H(x, y)\} \subset \mathbb{C} \times \mathbb{C} \times \mathbb{C}^* \times \mathbb{C}^* \ni (u, v, x, y) \tag{2.10}$$

where we do not need to write explicitly the dependence on complex structure parameters. Geometrically, (2.10) is a conic bundle over $\mathbb{C}^* \times \mathbb{C}^*$ with discriminant locus given by the (non-compact, *i.e.*, punctured) curve

$$\mathcal{C} = \{H(x, y) = 0\} \subset \mathbb{C}^* \times \mathbb{C}^* \tag{2.11}$$

Aganagic and Vafa [**12**] study B-type D-branes in this geometry wrapped on one component of a reducible fiber, say $u = 0$, varying over $\mathcal{C}$. They identify certain "semi-classical" regimes of these branes as punctures of the curve, and show that the superpotential expanded near such a point is given by an Abel-Jacobi computation on $\mathcal{C}$. For simplicity, let us say the interest is in a puncture at $y = 1$. Then the superpotential is given by the formula

$$\delta_z W(z) = -\log y(z) \tag{2.12}$$

where z (the open string modulus) is a local coordinate on the curve such that $z = 0$ corresponds to the puncture. This choice is made such that the superpotential is critical, *i.e.*, $y = 1$, at $z = 0$. In some simple cases, z coincides with x in (2.10).) As pointed out in [**2**], however, this prescription is ambiguous: If z is such a good coordinate, then so is any combination

$$z_f = z(-1)^f y^f \tag{2.13}$$

with integer f. Up to the sign, this is equation (1.8) from the introduction.[2]

Let us pause briefly here to explain the relevance of the mirror map: W as defined by (2.12) is a 2-function in the sense of the previous subsection when expanded, not only around $z = 0$, but also in the appropriate flat closed string coordinates around a degeneration of the curve. Assuming that the family of curves is defined over $\mathbb{Q}$, the 2-function property of W (after the mirror map) was shown in general in [**15**]. However, as pointed out in general in [**2**], the mirror map is in fact independent of the open string coordinate itself, and as a consequence framing commutes with the mirror map. This observation also explains why we are using the traditional B-model notation z interchangeably with the A-model q for the argument of our s-functions.

Our main point now is to abstract the ambiguity (2.13) to the following general "framing transformation", parameterized by an integer f.

2.4. The group of framing transformations. Say $V(z)$ is an s-function with $s \geq 1$. Define

$$Y = \exp\bigl(-(\delta_z)^{s-1} V\bigr) \in \mathbb{Z}[[z]] \tag{2.14}$$

(The integrality of the power series follows from $\delta_z^{s-1}\mathrm{Li}_s(z) = \mathrm{Li}_1(z) = -\ln(1-z)$.) When $s = 2$ and $V = W$ is the superpotential of (the mirror of) a toric D-brane configuration, then the corresponding $Y = \exp(-\delta_z W) = y(z)$ will by the above construction satisfy an algebraic equation

$$H(z, Y) = 0 \tag{2.15}$$

while the "framed superpotential" $\delta_{z_f} W_f = -\log Y_f$ can be identified with a solution of the equation

$$H_f(z_f, Y_f) = H(z_f(-Y_f)^{-f}, Y_f) = 0 \tag{2.16}$$

Even though the equations are algebraic, it is natural (for the purposes of mirror symmetry, for example) to think of solutions $Y(z)$ and $Y_f(z_f)$ as local power series around $z = 0$, $z_f = 0$, respectively. By the construction, both $Y(0) = 1$, and $Y_f(0) = 1$. The relation between the two is then simply $Y_f(z_f) = Y(z)$, $z_f = (-Y)^f z$. Eliminating z, and renaming z_f as z, this means that we can obtain $Y_f(z)$ as the solution to the equation in formal power series,

$$Y_f = Y\bigl(z(-Y_f)^f\bigr) \tag{2.17}$$

[2]The sign is thrown in to preserve integrality also at $p = 2$, see below.

The "framed 2-function" is then the power series

$$W_f(z) = \int \frac{dz}{z}(-\log Y_f(z)) \tag{2.18}$$

and, as we prove below, in fact is also a 2-function. Since the relations (2.17), (2.18) make sense independent of the existence of an algebraic equation of the type (2.15), we may take this as the general definition of framing, even for cases in which no such equation is known to exist.

We note a few elementary properties of this definition. First of all, framing defines a group action $\mathbb{Z} \ni f : W \to W_f$. Indeed, using the definitions, we find

$$\begin{aligned}(Y_f)_{f'} &= Y_f\big(z\big(-(Y_f)_{f'}\big)^{f'}\big)\\ &= Y\big(z\big(-(Y_f)_{f'}\big)^{f'}\big(-(Y_f)_{f'}\big)^{f}\big)\\ &= Y\big(z\big(-(Y_f)_{f'}\big)^{f+f'}\big)\end{aligned} \tag{2.19}$$

Thus, the equation for $(Y_f)_{f'}$ is exactly the defining equation for $Y_{f+f'}$.

Secondly, to make contact with the introduction, we write eqs. (1.4), (1.5) in the form

$$z = -\tilde{z}\tilde{Y} = zY\big(-\tilde{z}\tilde{Y}\big)\tilde{Y} \tag{2.20}$$

which is equivalent to

$$Y(-\tilde{z}\tilde{Y}) = \tilde{Y}^{-1} \tag{2.21}$$

Comparison with (2.17) shows that

$$\tilde{Y} = (Y_{-1})^{-1} \tag{2.22}$$

By substituting Y_1 for Y into this equation, we learn that $\widetilde{Y_1} = Y^{-1}$, and since $Y \mapsto \tilde{Y}$ is obviously involutive, this implies

$$Y_1 = \widetilde{Y^{-1}} \tag{2.23}$$

We conclude that framing transformations in the sense of (2.17) in fact are generated by the single transformation (1.4), (1.5), together with the operation $Y \mapsto Y^{-1}$ (which in terms of the 2-function, corresponds simply to $W \mapsto -W$). Hence, we will refer to (1.4), (1.5) also as "framing".

3. K-theoretical description of 2-functions. Integrality of framing

The purpose of this section is to give a description of 2-functions in terms of K-theory and to derive from this description the integrality of the framing transformation:

THEOREM 3.1. *If $W(z) \in \mathbb{Q}[[z]]$ is a 2-function , then its image under framing, $\tilde{W}(z)$ is also a 2-function .*

To prove this theorem it is sufficient to check for all prime numbers p that the fact that $W(z)$ is a 2-function with respect to p implies that $\tilde{W}(z)$ is also a 2-function with respect to p.

A complete proof of this statement will be given in [**5**]. The K-theoretic proof that we will sketch in this section works only for odd primes.

We begin with a lightning review of algebraic K-theory.

3.1. Algebraic K-theory. Orthogonality relation. For a ring A, the group $K_1(A)$ is defined as the abelianization of the infinite linear group $GL(A)$:

$$K_1(A) = GL(A)/[GL(A), GL(A)]. \tag{3.1}$$

If A is a Euclidean domain (in particular, it is a commutative ring), then the group $K_1(A)$ is isomorphic to the group $A^\times$ of invertible elements of A. (For arbitrary commutative rings we have an embedding $A^\times \to K_1(A)$ induced by the embedding $A^\times = GL_1(A) \hookrightarrow GL(A)$, and a map $K_1(A) \to A^\times$ induced by the determinant map $\det : GL(A) \to A^\times$.)

Notice that $K_1(A)$ is usually regarded as an additive group, but in our situation it is isomorphic to a multiplicative group $A^\times$, and so the multiplicative notation is more convenient.

The group $K_2(A)$, which we will write *additively*, is defined for an arbitrary ring via the universal central extension of the commutator subgroup $E(A) = [GL(A), GL(A)]$. Thus it fits into the sequence

$$K_2(A) \to \mathrm{St}(A) \to E(A) \to GL(A) \to K_1(A) \tag{3.2}$$

where $\mathrm{St}(A)$ is the Steinberg group (when A is Euclidean, we may think of the "universal cover" of $E(A) = SL(A) = \mathrm{Ker}(\det)$).

For an arbitrary ring A there exists a pairing $K_1(A) \otimes_{\mathbb{Z}} K_1(A) \to K_2(A)$. Via the embedding $A^\times \to K_1(A)$, this pairing induces a (skew) pairing of invertible elements of A in $K_2(A)$ (an antisymmetric bilinear map $\phi : A^\times \otimes A^\times \to K_2(A)$).

An important property of ϕ is that the pairing of two invertible elements f, g vanishes if $f + g = 1$. Let us denote by J the subgroup of $A^\times \otimes A^\times$ generated by elements of the form $f \otimes (1 - f)$; the above statement means that $J \subset \mathrm{Ker}\phi$. If A is a field, then $J = \mathrm{Ker}\phi$ (by Matsumoto's theorem).

Using the notation

$$K_2^0(A) = A^\times \otimes A^\times / J$$

we can consider the pairing ϕ as a composition of maps $A^\times \otimes A^\times \to K_2^0(A)$ and $K_2^0(A) \to K_2(A)$. We will work with the first map considered as a pairing on $A^\times$; we denote this pairing by $\{f, g\}$. (It is possible to work also with ϕ, but this makes the proof more complicated.)

We will characterize 2-functions in terms of this pairing.

Let us take two invertible elements $f \in A^\times, g \in A^\times$. By definition f and g are orthogonal if the element $2\{f, g\} = \{f^2, g\} = \{f, g^2\}$ vanishes. We have used the bilinearity of the pairing ϕ expressed by formulas $\{f_1 f_2, g\} = \{f_1, g\} + \{f_2, g\}, \{f, g_1 g_2\} = \{f, g_1\} + \{f, g_2\}$. These unusual formulas come from the fact that the operation in $A^\times$ is written as multiplication, while the operations in $K_2(A)$ and $K_2^0(A)$ are written additively.[3] Notice that it follows from bilinearity that an invertible element f that is orthogonal to elements $g \in A^\times$ and $h \in A^\times$ is orthogonal to their product gh. It is also obvious that automorphisms of the ring A preserve orthogonality.

The relevant ring for us is $A = \mathbb{Z}((q))$, the ring of formal Laurent series in one variable q, with integer coefficients. The ring A has a natural topology, that induces a topology in $A^\times$. This allows us to modify the notion of orthogonality: we will say that f, g are orthogonal in the new sense if there exists a sequence of pairs

[3]Some other useful properties include $\{f, -f\} = 0$, $\{f, 1\} = 0$, and anti-symmetry $\{f, g\} = -\{g, f\}$.

(f_n, g_n) such that f_n tends to f, g_n tends to g and f_n is orthogonal to g_n in the old sense .

Notice that starting with a 2-function $W(q) = \sum_{d=1} n_d \mathrm{Li}_2(q^d)$ represented as a sum of di-logarithms we construct an invertible element of A by the formula

$$Y(q) = \exp(-\delta_q W(q)) = \prod (1-q^d)^{dn_d}.$$

If the sum of di-logarithms is finite the element $Y(q)$ is orthogonal to q. It is sufficient to check this statement for every factor. The fact that q is orthogonal to $(1-q^d)^d$ can be derived from the following chain of identities:

$$\{q, (1-q^d)^d\} = \{q^d, 1-q^d\} = 0. \tag{3.3}$$

If the sum of di-logarithms is infinite then $Y(q)$ is orthogonal to q in the new sense (infinite product is a limit of finite products). *In what follows we will understand the orthogonality in the new sense.*

Let us notice that q^m is orthogonal to q. It is sufficient to check this for $m = 1$. This follows from the identiy

$$\{q, (-q)\} = \{q, (1-q)\} + \{q^{-1}, (1-q^{-1})\} = 0,$$

which in turn follows from $-q = \frac{1-q}{1-q^{-1}}$. We see that $2\{q, q\} = \{q, (-q)^2\} = 2\{q, (-q)\} = 0$.

We obtain that q is orthogonal to an expression of the form $q^m \prod (1-q^d)^{dn_d}$. One can prove (Sec 5 and **[5]**) that the inverse statement is also correct:

THEOREM 3.2. *If* $Y(q) = \exp(-\delta_q W(q))$ *is orthogonal to* q *then the series* $W(q) \in \mathbb{Q}[[q]]$ *is a 2-function for all odd primes .*

In other words q is orthogonal to Y where Y behaves like q^m as q tends to 0 if and only if $\int \log(Y/q^m) d\log q$ is a 2-function for all odd primes.

3.2. Integrality of framing (Proof of Theorem 3.1). Notice that any change of variables $\tilde{q} = q + a_2 q^2 + a_3 q^3 + \cdots$ with $a_i \in \mathbb{Z}$ induces an automorphism of the algebra $A = \mathbb{Z}((q))$. This automorphism preserves the orthogonality relation.

Using this fact we can describe all orthogonal pairs (f, g). It is sufficient to consider only the case when f behaves like q as q tends to zero. Then we can take f as a new variable; in terms of this variable g has an expression of the form $f^m \prod (1-f^d)^{dn_d}$ and integrates to a 2-function $\sum n_d \mathrm{Li}_2(f^d)$. This follows from Theorem 3.2 above, applied to the ring $\mathbb{Z}((f))$, which is isomorphic to A as remarked above.

Let us now consider all orthogonal pairs (f, g) where both f and g behave like q as $q \to 0$. The orthogonality relation is symmetric, hence f and g are on equal footing. Therefore, we can construct two different 2-functions (one from expression of g in terms of f, another from expression of f in terms of g.) These two 2-functions are related by framing transformation (1.4), (1.5). Thus we see that Theorem 3.1 is a simple consequence of the K-theoretical description of 2-functions and of the symmetry of the orthogonality relation.

Notice that in the proof we used orthogonal pairs where both f and g behave like q as $q \to 0$ (this is important for symmetry). In the orthogonal pair $(q, Y(q))$ the function $Y(q)$ does not satisfy this condition, therefore in the construction of the framing transformation it should replaced by $qY(q)$.

4. Frobenius automorphism and local s-functions

The purpose of this section is to reformulate the definition of an s-function in terms of the action of the Frobenius endomorphism acting on (formal) power series. Such a reformulation was crucial in the proofs of integrality theorems in [**13**–**15**]. We will apply it here to sketch a proof of the description of 2-functions in terms of algebraic K-theory (Theorem 3.2) that was used in the derivation of integrality of framing (see [**5**] for more detail).

First of all we will check that $V(z) \in \mathbb{Q}[[z]]$ is an s-function if and only if for any prime number p the formal series

$$\frac{1}{p^s}(V(z^p)) - V(z) \tag{4.1}$$

is p-integral (the denominators of the coefficients are not divisible by p). The proof is an easy consequence of Möbius inversion formula (and a trivial generalization of the special statements for $s = 2, 3$).

Recall that if (a_d) and (b_d) are two sequences such that

$$a_d = \sum_{k|d} b_k \tag{4.2}$$

(where the sum is over all divisors of d), then

$$b_d = \sum_{k|d} \mu(\tfrac{d}{k}) a_k = \sum_{k|d} \mu(k) a_{k/d} \tag{4.3}$$

where μ is the Möbius function: $\mu(k) = 0$ if k is not squarefree, $\mu(k) = (-1)^r$ if $k = p_1 \cdots p_r$ is the product of r distinct prime factors. The important property of μ is that

$$\sum_{k|d} \mu(k) = \begin{cases} 1 & \text{if } d = 1 \\ 0 & \text{if } d > 1 \end{cases} \tag{4.4}$$

which itself follows from the fact that if $r > 0$

$$\sum_{l|k} \mu(l) = \sum_{l|p_2\cdots p_r} \mu(l) + \sum_{l|p_2\cdots p_r} \mu(p_1 l) \tag{4.5}$$

Then (4.3) follows from the computation

$$\sum_{k|d} b_k = \sum_{l|k|d} \mu(\tfrac{k}{l}) a_l = \sum_{l|d} a_l \sum_{\substack{k|d \\ l|k}} \mu(\tfrac{k}{l}) = \sum_{l|d} a_l \sum_{k|\frac{d}{l}} \mu(k) = a_d \tag{4.6}$$

Returning to s-functions, we compare coefficients of z^d in

$$V(z) = \sum_{d=1}^{\infty} m_d z^d = \sum_{d,k=1}^{\infty} \frac{n_d}{k^s} z^{dk} \tag{4.7}$$

to conclude that

$$d^s m_d = \sum_{k|d} k^s n_k \tag{4.8}$$

where the sum is over all divisors of d. Applying Möbius inversion, we find

$$d^s n_d = \sum_{k|d} \mu(\tfrac{d}{k}) k^s m_k \tag{4.9}$$

On the other hand, the statement that (4.1) be p-integral for all primes p is equivalent to the condition that

$$m_d - \frac{1}{p^s} m_{d/p} \quad \text{be } p\text{-integral for all } p,\ d \tag{4.10}$$

it being understood that $m_{d/p} = 0$ if $p \nmid d$. To see that the integrality of the n_d implies this condition, we note that (4.8) implies

$$m_d - \frac{1}{p^s} m_{d/p} = \frac{1}{d^s} \sum_{\substack{k|d \\ k\nmid d/p}} k^s n_k \tag{4.11}$$

The sum is restricted to those k divisible by as many powers of p as d, and therefore the right hand side is p-integral if the $n_k \in \mathbb{Z}$.

Conversally, since $\mu(k) = 0$ if k is divisible by p^2, and $\mu(pk) = -\mu(k)$ if $p \nmid k$, we see that the formula (4.9) may be rewritten as

$$n_d = \sum_{k|d} \mu(k) \frac{m_{d/k}}{k^s} = \sum_{\substack{k|d \\ p\nmid k}} \frac{\mu(k)}{k^2} \big(m_{d/k} - \frac{1}{p^s} m_{d/(pk)}\big) \tag{4.12}$$

with the same understanding that $m_{d/(pk)} = 0$ if $p \nmid d$. We see that if (4.10) holds, the n_d are p-integral for any p, hence integral.

We can reformulate the above statement in p-adic terms. Let us denote by $V_p(z)$ the p-adic reduction of $V(z)$, *i.e.*, the series obtained from $V(z)$ by viewing all coefficients as p-adic numbers. (Really, this is the same series.) We also denote by $\mathrm{Fr}_p : \mathbb{Q}[[z]] \to \mathbb{Q}[[z]]$ the Frobenius endomorphism fixing $\mathbb{Q}$ and sending z to z^p. Then the characterization (4.1) of s-functions is equivalent to the statement that for any p,

$$M_p(z) = \frac{1}{p^s} \mathrm{Fr}_p V_p(z) - V_p(z) \in \mathbb{Z}_p[[z]] \tag{4.13}$$

is a series with p-adic integral coefficients. We may call a function that satisfies (4.13) (only) for some fixed prime p a local s-function.

One can express the coefficients n_d (or, better to say, their p-adic reductions) in terms coefficients of the series $m_p(z)$, see Lemma 3 of [**15**].) It follows immediately from this formula that p-adic integrality of coefficients of $M_p(z)$ for all p guarantees the integrality of n_p.

5. Regulators. K-theoretic characterization of 2-functions

Fix an odd prime number p.

Let us define a map

$$(f, g)_p : \mathbb{Z}_p((z))^\times \otimes \mathbb{Z}_p((z))^\times \to \mathbb{Q}_p((z))/\mathbb{Z}_p((z)) \tag{5.1}$$

to be a unique bilinear skew-symmetric pairing such that for every $g \in \mathbb{Z}_p[[z]]^\times$ and every $f \in \mathbb{Z}_p((z))^\times$, one has

(5.2)

$$(f, g)_p = \int (\frac{1}{p^2} \mathrm{Fr}^*(\log g d \log f) - \log g d \log f) - \frac{1}{p} \mathrm{Fr}^*(\log g)(\frac{1}{p} \mathrm{Fr}^*(\log f) - \log f).$$

Here Fr^* is the "Frobenius lifting":

$$\mathrm{Fr}^*(h(z)) = h(z^p).$$

The key property of the pairing $(f,g)_p$ is that it is invariant under any change of variables of the form $z \to h(z) = a_1 z + a_2 z^2 + \cdots$, $a_1 \in \mathbb{Z}_p^\times$. In particular, if $(f,g)_p = 0$ (*i.e.*, $(f(z),g(z))_p$ is p-adically integral) then the same is true for $(f(h(z)),g(h(z)))_p$. This can be derived from a more general fact: the right-hand side of formula (5.2) viewed as an element of the quotient group $\mathbb{Q}_p((z))/\mathbb{Z}_p((z))$ does not get changed if one replaces Fr^* by an arbitrary Frobenius lifting of the form

$$\mathrm{F}^*(h(z)) = h(z^p(1+pr(z)),$$

for some $r(z)] \in \mathbb{Z}_p[[z]]$.

One can derive from this fact (or prove directly) that $(1-f,f)_p = 0$ for every $f \in \mathbb{Z}_p((z))^\times$ such that $1-f$ is also in $\mathbb{Z}_p((z))^\times$. Thus, (5.1) factors through a homomorphism

$$K_2^0(\mathbb{Z}_p((z))) \to \mathbb{Q}_p((z))/\mathbb{Z}_p((z)). \tag{5.3}$$

Finally, one can easily check that $(f,g)_p$ is continuous in each variable with respect to "z-adic" topology on $\mathbb{Z}_p((z))$.

The above properties are sufficient to prove Theorem 2. Indeed, if $Y(z)$ is orthogonal to z then $(z,Y(z))_p = 0$ for every odd prime p. On the other hand, we have that

$$(z,Y(z))_p = \frac{1}{p^2}\mathrm{Fr}^* W(z) - W(z).$$

Therefore $W(z)$ is a 2-function.

REMARK 5.1. One can check that (5.3) factors uniquely through a homomorphism

$$K_2(\mathbb{Z}_p((z))) \to \mathbb{Q}_p((z))/\mathbb{Z}_p((z)).$$

6. Generalizations

One of our central contentions in writing this note is that s-functions are interesting algebraic objects in their own right, independent of relations to physics of topological strings. We also claim that the value $s = 2$ is special. We give further evidence in this section by pointing out some very natural generalizations of our discussion so far. One possible application that we will not (p)review in any detail here is to the theory of Mahler measures. Indeed, the reader will find it easy to verify that the relation between the so-called Modular Mahler Measures of [**6**] and the instanton expansion of certain "exceptional non-critical strings" [**20**], which was pointed out by Stienstra [**3**], is nothing but an elementary framing transformation (1.4), (1.5). Given our results, it is clear that the relation will be valid in much greater generality than the examples presented in [**3**]. It would be interesting to explore this further.

6.1. Arithmetic twists. The generalization that motivated our initial observations concerning 2-functions is related to the results of [**4**]. In that work, it was pointed out that the A-model instanton expansion of the superpotential associated with a general D-brane on a compact Calabi-Yau three-fold is not rational (let alone integral in the usual sense). Instead, the coefficients were found to be contained in the algebraic number field K over which the curve representing the D-brane in the B-model was defined. However, it was also observed that with an appropriate modification of the Ooguri-Vafa multi-cover formula, (2.5), at least integrality in the algebraic sense could be preserved. The expansion proposed in [**4**] was of the form

$$W(q) = \sum \tilde{n}_d q^d = \sum n_d \mathrm{Li}_2^{\mathrm{D}}(q^d) \tag{6.1}$$

where n_d are algebraic integers[4] and $\mathrm{Li}_2^{\mathrm{D}}$, dubbed the "D-logarithm" is a certain (formal) power series with coefficients in K that depend n_d, and its images under the Galois group of the extension $K/\mathbb{Q}$. More precisely, the definition of the D-logarithm given in [**4**] depended on the lifting $\bmod p^2$ of the Frobenius automorphism at each unramified prime p, so in that sense the coefficients n_d (though integer for any choice of lifting) depend on several infinities of choices, and would not appear as true geometric invariants. The noteworthy exception is provided by abelian extensions, where $\mathrm{Li}_2^{\mathrm{D}}$ could be taken to be the di-logarithm twisted by a Dirichlet character χ, of the form

$$\sum \chi(k) \frac{q^k}{k^2} \tag{6.2}$$

As will be clear, this can be rewritten as an integral linear combination of ordinary di-logarithms evaluated at appropriate roots of unity, which can therefore be viewed as a canonical basis in which to decompose the superpotential. It remains rather unclear at this point whether such a basis exists also for general extensions with non-abelian Galois group.

On the other hand, however, one may formulate the integrality statement of [**4**] without explicitly referring to any "D-logarithm". Moreover, the proofs of [**15**] can rather straightforwardly be adapted to prove that integrality statement as well. We will explain this in the forthcoming paper [**5**].

Finally, and this is most relevant with respect to the present note, it turns out that the framing transformation can also be defined, and preserves integrality (in the algebraic sense), for 2-functions with coefficients in an arbitrary number field [**5**].

6.2. Multi-variable case. As we have pointed out before, if V is an s-function (for $s > 2$), then $W_V = \delta^{s-2} V$ is a 2-function, and its framed version $\widetilde{W_V} = \widetilde{\delta^{s-2} V}$ is also a 2-function. It is a natural question to ask whether this 2-function also comes from an s-function, namely whether there exists an s-function $\tilde{V}$ such that $\widetilde{W_V} = \delta^{s-2}\tilde{V}$. It is not hard to see that really this is not the case (for instance, framing Li_s for $s > 3$ in this way returns at most a 3-function). Thus, among s-functions for other values of s, 2-functions are distinguished by the integrality of framing.

Perhaps the most direct way to see that framing naturally only makes sense for 2-functions is to consider the generalization to the multi-variable situation. With rational coefficients, we say, as before, that a formal power series $W \in \mathbb{Q}[[z_1, \ldots z_r]]$ is a 2-function if it can be written as an integral linear combination of di-logarithms,

$$W(z_1, \ldots, z_r) = \sum_{d_1, \ldots, d_r} n_{d_1, \ldots, d_r} \mathrm{Li}_2\big(z_1^{d_1} \cdots z_r^{d_r}\big) \tag{6.3}$$

[4] Recall that an algebraic number x can be identified with a root of a polynomial $P(x) \in \mathbb{Q}[x]$. If P has coefficients in $\mathbb{Z}$, leading coefficient 1, and is irreducible, then x is an algebraic integer.

Defining $\delta_i \equiv \frac{d}{d\ln z_i}$, and following eq. (1.3), we introduce

$$Y_i = \exp\big(\delta_i W\big) \tag{6.4}$$

Since the $\delta_i W$ are 1-functions, the Y_i naturally have integer coefficients. An interesting distinction from the one-variable case is that we find an additional degree of freedom when identifying the "framed" variables $\tilde{z}_i$ with Y_i, in analogy to (1.4). Namely, say $(\kappa^{ij})_{i,j=1,\dots r}$ is a symmetric matrix with integer coefficients. Define $\sigma_i = (-1)^{\kappa_{ii}}$, and

$$\tilde{z}_i = \sigma_i z_i \prod_{j=1}^{r} Y_i^{\kappa_{ij}} = \sigma_i z_i \exp\big(\kappa^{ij}\delta_j W\big) \tag{6.5}$$

We may invert this relation as before, and upon writing

$$z_i = \sigma_i \tilde{z}_i \exp\big(\kappa^{ij}\tilde{\delta}_j \tilde{W}\big) \tag{6.6}$$

we find that $\tilde{W} \in \mathbb{Q}[[\tilde{z}_1, \dots, \tilde{z}_r]]$ is also a 2-function. This assertion can be proved rather straightforwardly by realizing the multi-dimensional operation as a combination of elementary one-dimensional framing, leading to the identification of the group of framing transformations with the additive group of symmetric integral matrices.

Now it is clear that if we had started from a multi-variable s-function with $s > 2$, we would in (6.4) have obtained more Y's from multi-derivatives than variables, so the identification would not be one-to-one. Thus, again, $s = 2$ is special.

Acknowledgments

We thank the organizers of String-Math 2012 for the invitation to speak in Bonn, which led to the present collaboration, and Maxim Kontsevich for valuable discussions and communications. J.W. thanks Fernando Rodriguez-Villegas for drawing attention to "modular Mahler measures" in Summer of 2011, and Henri Darmon for several helpful conversations. The research of J.W. is supported in part by an NSERC discovery grant and a Tier II Canada Research Chair, the research of A. Sch. was supported by NSF grant.

References

[1] D. Zagier, *The dilogarithm function*, Frontiers in number theory, physics, and geometry. II, Springer, Berlin, 2007, pp. 3–65, DOI 10.1007/978-3-540-30308-4_1. MR2290758 (2008h:33005)

[2] M. Aganagic, A. Klemm, and C. Vafa, *Disk instantons, mirror symmetry and the duality web*, Z. Naturforsch. A **57** (2002), no. 1-2, 1–28, DOI 10.1515/zna-2002-9-1001. MR1906661 (2003f:81183)

[3] J. Stienstra, *Mahler measure variations, Eisenstein series and instanton expansions*, Mirror symmetry. V, AMS/IP Stud. Adv. Math., vol. 38, Amer. Math. Soc., Providence, RI, 2006, pp. 139–150. MR2282958 (2008d:11095)

[4] J. Walcher, *On the arithmetic of D-brane superpotentials. Lines and conics on the mirror quintic*, Commun. Number Theory Phys. **6** (2012), no. 2, 279–337, DOI 10.4310/CNTP.2012.v6.n2.a2. MR2993120

[5] A. Schwarz, V.Vologodsky, and J. Walcher, *Integrality of framing and geometric origin of 2-functions (with algebraic coefficients)*, to appear

[6] F. R. Villegas, *Modular Mahler measures. I*, Topics in number theory (University Park, PA, 1997), Math. Appl., vol. 467, Kluwer Acad. Publ., Dordrecht, 1999, pp. 17–48. MR1691309 (2000e:11085)

[7] S. Gukov and P. Sułkowski, *A-polynomial, B-model, and quantization*, J. High Energy Phys. **2** (2012), 070, front matter+56. MR2996110
[8] H. Ooguri and C. Vafa, *Knot invariants and topological strings*, Nuclear Phys. B **577** (2000), no. 3, 419–438, DOI 10.1016/S0550-3213(00)00118-8. MR1765411 (2001i:81254)
[9] R. Gopakumar and C. Vafa, "M theory and topological strings. 1, 2.," arXiv:hep-th/9809187, arXiv:hep-th/9812127
[10] J. Walcher, *Opening mirror symmetry on the quintic*, Comm. Math. Phys. **276** (2007), no. 3, 671–689, DOI 10.1007/s00220-007-0354-8. MR2350434 (2008m:14111)
[11] S. Katz and C.-C. M. Liu, *Enumerative geometry of stable maps with Lagrangian boundary conditions and multiple covers of the disc*, Adv. Theor. Math. Phys. **5** (2001), no. 1, 1–49. MR1894336 (2003e:14047)
[12] M. Aganagic and C. Vafa, "Mirror symmetry, D-branes and counting holomorphic discs," arXiv:hep-th/0012041
[13] M. Kontsevich, A. Schwarz, and V. Vologodsky, *Integrality of instanton numbers and p-adic B-model*, Phys. Lett. B **637** (2006), no. 1-2, 97–101, DOI 10.1016/j.physletb.2006.04.012. MR2230876 (2007k:14115)
[14] A. Schwarz and V. Vologodsky, *Frobenius transformation, mirror map and instanton numbers*, Phys. Lett. B **660** (2008), no. 4, 422–427, DOI 10.1016/j.physletb.2008.01.006. MR2396265 (2008m:14079)
[15] A. Schwarz and V. Vologodsky, *Integrality theorems in the theory of topological strings*, Nuclear Phys. B **821** (2009), no. 3, 506–534, DOI 10.1016/j.nuclphysb.2009.05.014. MR2547214 (2010i:81360)
[16] L. Lazzarini, *Existence of a somewhere injective pseudo-holomorphic disc*, Geom. Funct. Anal. **10** (2000), no. 4, 829–862, DOI 10.1007/PL00001640. MR1791142 (2003a:32044)
[17] P. Candelas, X. C. de la Ossa, P. S. Green, and L. Parkes, *A pair of Calabi-Yau manifolds as an exactly soluble superconformal theory*, Nuclear Phys. B **359** (1991), no. 1, 21–74, DOI 10.1016/0550-3213(91)90292-6. MR1115626 (93b:32029)
[18] P. S. Aspinwall and D. R. Morrison, *Topological field theory and rational curves*, Comm. Math. Phys. **151** (1993), no. 2, 245–262. MR1204770 (94h:32033)
[19] V. Vologodsky, "On the N-integrality of instanton numbers," preprint
[20] A. Klemm, P. Mayr and C. Vafa, "BPS states of exceptional noncritical strings," arXiv:hep-th/9607139; W. Lerche, P. Mayr and N. P. Warner, "Noncritical strings, Del Pezzo singularities and Seiberg-Witten curves," Nucl. Phys. B **499**, 125 (1997) arXiv:hep-th/9612085

Department of Mathematics, University of California, Davis, California
E-mail address: `schwarz@math.ucdavis.edu`

Department of Mathematics, University of Oregon, Eugene, Oregon
E-mail address: `vologod@gmail.com`

Departments of Physics, and Mathematics and Statistics, McGill University, Montreal, Quebec, Canada
E-mail address: `Johannes.Walcher@mcgill.ca`

Proceedings of Symposia in Pure Mathematics
Volume **90**, 2015
http://dx.doi.org/10.1090/pspum/090/01522

Symmetry-Surfing the Moduli Space of Kummer K3s

Anne Taormina and Katrin Wendland

Dedicated to Prof. Dr. Friedrich Hirzebruch, October 17, 1927 - May 27, 2012, in admiration and gratitude:
To an extraordinary scientist, an unforgettable teacher, and a model of altruism.

ABSTRACT. A maximal subgroup of the Mathieu group M_{24} arises as the combined holomorphic symplectic automorphism group of all Kummer surfaces whose Kähler class is induced from the underlying complex torus. As a subgroup of M_{24}, this group is the stabilizer group of an octad in the Golay code. To meaningfully combine the symmetry groups of distinct Kummer surfaces, we introduce the concepts of Niemeier markings and overarching maps between pairs of Kummer surfaces. The latter induce a prescription for symmetry-surfing the moduli space, while the former can be seen as a first step towards constructing a vertex algebra that governs the elliptic genus of K3 in an M_{24}-compatible fashion. We thus argue that a geometric approach from K3 to Mathieu Moonshine may bear fruit.

Introduction

This work is motivated by several mysteries related to the Mathieu Moonshine phenomenon. Central to this phenomenon is the elliptic genus of K3, which encodes topological data on K3 surfaces and at the same time is expected to organise a selection of states in $N = (4,4)$ superconformal field theories (SCFTs) on K3 into representations of the Mathieu group M_{24}. The existence of the relevant representations follows from Gannon's result [**Gan12**], which in turn builds on the work of Cheng, Gaberdiel-Hohenegger-Volpato and Eguchi-Hikami [**Che10, GHV10a, GHV10b, EH11**]. The precise construction of those representations in terms of conformal field theory data, however, has been completely elusive so far, since the detailed nature of the states governing the elliptic genus has not been pinned down. Indeed, the elliptic genus is a topological invariant generalizing the genera of multiplicative sequences that were introduced by F. Hirzebruch [**Hir66**]. It can be viewed as the regularized index of a $U(1)$-equivariant Dirac operator on the loop

---

2010 *Mathematics Subject Classification.* Primary 14J28; Secondary 81T40, 81T60.

A.T. thanks the University of Freiburg for their hospitality, and acknowledges a Leverhulme Research Fellowship RF/2012-335.

We thank Ron Donagi, Matthias Gaberdiel and Roberto Volpato for very helpful discussions. We also thank the Heilbronn Institute and the International Centre for Mathematical Sciences in Edinburgh as well as the (other) organisers of the Heilbronn Day and Workshop on 'Algebraic geometry, modular forms and applications to physics', where part of this work was done.

K.W. acknowledges an ERC Starting Independent Researcher Grant StG No. 204757-TQFT.

space of K3 [**AKMW87**, **Wit87**]. It also arises from the supertrace over the subsector of Ramond-Ramond states of every superconformal field theory on K3, and hence it counts states with signs [**EOTY89**, **Kap05**]. That the net contribution should yield a well-defined representation of any group, let alone of M_{24}, is mysterious. However, from the properties of twining and twisted-twining genera it has been argued that one should actually expect this representation to be realized in terms of a vertex algebra $\widehat{\mathcal{X}}$ [**GPRV12**]. We share that view, although not the recent claim by some experts exclusively expecting holomorphic vertex algebras in this context, and casting doubts on whether K3 surfaces bear any key to the Mathieu Moonshine Mysteries [**Gan12**, **GPV13**].

In fact, we argue that the resolution of certain aspects of Mathieu Moonshine might benefit from deepening our understanding of the implications of Mukai's work [**Muk88**], and from building on the insights offered by Kondo [**Kon98**]. Of course, Mukai has proved in [**Muk88**] that every holomorphic symplectic symmetry group of a K3 surface is a subgroup of the group M_{24}. But he also proved that all these symmetry groups are smaller than M_{24} by orders of magnitude. In fact, all of them are subgroups of M_{23}. In [**TW11**] we advertised the idea that presumably, M_{24} could be obtained by combining the holomorphic symplectic symmetry groups of distinct K3 surfaces at different points of the moduli space. As a test bed, we proved the existence of an overarching map Θ which allows to combine the holomorphic symplectic symmetry groups of two special, distinct Kummer surfaces in terms of their induced actions on the Niemeier lattice N of type A_1^{24}. We also proved that this combined action on N yields the largest possible group that can arise by means of such an overarching map. This group is $(\mathbb{Z}_2)^4 \rtimes A_7$, which we therefore called the OVERARCHING FINITE SYMMETRY GROUP of Kummer surfaces. It contains as proper subgroups all holomorphic symplectic symmetry groups of Kummer surfaces which are equipped with the dual Kähler class induced from the underlying torus.

In this note, in Section 1 we briefly recall the Kummer construction and gather the information appearing in [**TW11**] that is useful for the present work. In Section 2, we introduce the concept of Niemeier markings and generalize the ideas summarized above by showing that the technique introduced for two specific examples of Kummer surfaces in [**TW11**], namely the tetrahedral and the square Kummer K3, generalizes to other pairs of Kummer surfaces. As an application of this technique, Section 3 constructs three overarching maps for three pairs of Kummer surfaces with maximal symmetry. Section 4 shows that for any pair of Kummer K3s, one can find representatives in the smooth universal cover of the moduli space of hyperkähler structures such that there exists an overarching map analogous to the one constructed in [**TW11**]. Moreover, there always exists a continuous path between the two representatives of our Kummer surfaces, such that Θ is compatible with all holomorphic symplectic symmetries along the path. This is the idea of symmetry-surfing the moduli space, alluded to in the title of the present paper.

Our surfing procedure allows us to combine the action of all holomorphic symplectic symmetry groups of Kummer surfaces with induced dual Kähler class by means of their induced actions on the lattice N. In fact, this action is independent of all choices of overarching maps. We also prove in Section 4 that the combined action of all these groups is given by a faithful representation of $(\mathbb{Z}_2)^4 \rtimes A_8$ on N.

The subgroup $(\mathbb{Z}_2)^4 \rtimes A_7$, i.e. the overarching finite symmetry group of Kummer surfaces, is the stabilizer subgroup of $(\mathbb{Z}_2)^4 \rtimes A_8$ for one root in the Niemeier lattice N, just as the subgroup M_{23} of the Mathieu group M_{24} is the stabilizer subgroup of M_{24}, which naturally acts on N, for one root in N. We view this as evidence that the Mathieu Moonshine phenomenon is tied to the largest Mathieu group M_{24} rather than M_{23}, as also argued by Gannon [**Gan12**].

In Section 5, we highlight the relevance of our geometric approach, and in particular of the Niemeier markings, in the quest for a vertex algebra that governs the elliptic genus of K3 at lowest order. To this effect, we establish a link between our work on Kummer surfaces and a special class of $N = (4,4)$ SCFTs at central charge $c = \overline{c} = 6$, namely $\mathbb{Z}_2$-orbifolds of toroidal conformal field theories[1]. This necessitates a transition from geometry to superconformal theory language, which we describe in Appendix A. The upshot is that our surfing idea is natural: the symmetry groups act on the twisted ground states of the $\mathbb{Z}_2$-orbifold conformal field theories, and that action completely determines these symmetries. The twisted ground states can be viewed as a stable part of the Hilbert space when one surfs between $\mathbb{Z}_2$-orbifolds. As such the twisted ground states collect the various symmetry groups just like the Niemeier lattice does by means of our Niemeier markings. In passing we explain how the very idea of constructing a vertex algebra from the field content of SCFTs on K3, which simultaneously governs the elliptic genus and symmetries, motivates why we restrict our attention to symmetry groups that are induced from geometric symmetries in some geometric interpretation, that is, to subgroups of M_{24}.

1. Kummer surfaces and quaternions

An interesting class of K3 surfaces is obtained through the KUMMER CONSTRUCTION, which amounts to taking a $\mathbb{Z}_2$-orbifold of any complex torus T of dimension 2, and minimally resolving the singularities that arise from the orbifold procedure. More specifically, let $T = T(\Lambda) = \mathbb{C}^2/\Lambda$ with $\Lambda \subset \mathbb{C}^2$ denote a lattice of rank 4 over $\mathbb{Z}$, and with generators $\vec{\lambda}_i$, $i \in \{1, \ldots, 4\}$. The group $\mathbb{Z}_2$ acts naturally on $\mathbb{C}^2$ by $(z_1, z_2) \mapsto (-z_1, -z_2)$ and thereby on $T(\Lambda)$. Using Euclidean coordinates $\vec{x} = (x_1, x_2, x_3, x_4)$, where $z_1 = x_1 + ix_2$ and $z_2 = x_3 + ix_4$, points on the quotient $T(\Lambda)/\mathbb{Z}_2$ are identified according to

$$\vec{x} \sim \vec{x} + \sum_{i=1}^{4} n_i \vec{\lambda}_i, \quad n_i \in \mathbb{Z}, \qquad \vec{x} \sim -\vec{x}.$$

Hence $T(\Lambda)/\mathbb{Z}_2$ has 16 singularities of type A_1, located at the fixed points of the $\mathbb{Z}_2$-action. These fixed points are conveniently labelled by the HYPERCUBE $\mathbb{F}_2^4 \cong \frac{1}{2}\Lambda/\Lambda$, where $\mathbb{F}_2 = \{0, 1\}$ is the finite field with two elements, as

$$\vec{F}_{\vec{a}} := \left[\tfrac{1}{2} \sum_{i=1}^{4} a_i \vec{\lambda}_i \right] \in T(\Lambda)/\mathbb{Z}_2, \quad \vec{a} = (a_1, a_2, a_3, a_4) \in \mathbb{F}_2^4. \tag{1.1}$$

[1]To avoid clumsy terminology, we simply refer to those SCFTs $\mathcal{C}$ on K3 which are obtained by the standard $\mathbb{Z}_2$-orbifold procedure from a toroidal theory as "$\mathbb{Z}_2$-orbifolds".

DEFINITION 1.1. The complex surface X_Λ obtained by minimally resolving the 16 singularities of $T(\Lambda)/\mathbb{Z}_2$ is a K3 surface (see e.g. [**Nik75**]) called a KUMMER SURFACE[2].

According to the above definition, the Kummer surface X_Λ carries the complex structure induced from the universal cover $\mathbb{C}^2$ of T. It may also be equipped with a Kähler structure[3], and this is natural if one is interested in the description of *finite* groups of symplectic automorphisms of Kummer surfaces. We specify such a Kähler structure by choosing a so-called DUAL KÄHLER CLASS ω, that is, a homology class which is Poincaré dual to a Kähler class. Indeed, first recall the following:

DEFINITION 1.2. Consider a K3 surface X. A map $f\colon X \longrightarrow X$ of finite order is called a SYMPLECTIC AUTOMORPHISM if and only if f is biholomorphic and it induces the identity map on $H^{2,0}(X,\mathbb{C})$.

If ω is a dual Kähler class on X and the induced map $f_*\colon H_*(X,\mathbb{R}) \longrightarrow H_*(X,\mathbb{R})$ leaves ω invariant, then f is a HOLOMORPHIC SYMPLECTIC AUTOMORPHISM with respect to ω.

When a dual Kähler class ω on X has been specified, then the group of holomorphic symplectic automorphisms of X with respect to ω is called the SYMMETRY GROUP of X.

As an application of the Torelli theorem for K3 surfaces, the discussion of holomorphic symplectic automorphisms f of a K3 surface X can be entirely rephrased in terms of the induced lattice automorphisms f_* of the full integral homology lattice $H_*(X,\mathbb{Z})$ (these and other results on geometry and symmetries of Kummer K3s are standard; for a summary, see e.g. [**TW11**, Thm. 3.2.2]). Then (see [**TW11**, Prop. 3.2.4] for a proof),

PROPOSITION 1.3. *Consider a K3 surface X, and denote by G a group of symplectic automorphisms of X. Then G is finite if and only if X possesses a dual Kähler class which is invariant under G.*

Throughout this work, we focus on Kummer surfaces X_{Λ,ω_0}, by which we mean that as Kähler structure on X_Λ we choose the one induced from the standard Kähler structure of the torus $T(\Lambda)$ inherited from the Euclidean metric on its universal cover $\mathbb{C}^2$. Here, ω_0 denotes the corresponding dual Kähler class on X_Λ. This restricts the symmetry groups of Kummer surfaces that can be obtained, but is sufficient to argue for the existence of a combined symmetry group $(\mathbb{Z}_2)^4 \rtimes A_8$ in Section 4.

The generic structure of the symmetry group G of the Kummer surface X_{Λ,ω_0} is a semi-direct product $G = G_t \rtimes G_T$ (see, for example, [**TW11**, Prop. 3.3.4]). The normal subgroup $G_t \cong (\mathbb{Z}_2)^4$ of G is the so-called translational automorphism group which is induced from the shifts by half lattice vectors $\frac{1}{2}\vec{\lambda}$, $\vec{\lambda} \in \Lambda$, on the underlying torus $T = T(\Lambda)$. The group G_T is the normalizer of G_t in G. It is the group of symmetries of the Kummer surface induced by the holomorphic symplectic automorphisms of the torus T fixing $0 \in \mathbb{C}^2/\Lambda = T$. That is, $G_T \cong G'_T/\mathbb{Z}_2$, where

[2] We denote by $\pi\colon T \dashrightarrow X$ the corresponding rational map of degree 2, and by $\pi_*\colon H_*(T,\mathbb{Z}) \longrightarrow H_*(X,\mathbb{Z})$ the induced map on homology.

[3] For most parts of our work, the Kähler class is degenerate in the sense that it corresponds to an orbifold limit of Kähler metrics.

G'_T is the group of linear holomorphic symplectic automorphisms of T. These groups and their possible actions on a torus T have been classified by Fujiki [**Fuj88**], who proves that G'_T is isomorphic to a subgroup of one of the following groups: the cyclic groups $\mathbb{Z}_4, \mathbb{Z}_6$, the binary dihedral groups $\mathcal{O}$ and $\mathcal{D}$ of order 8 and 12, and the binary tetrahedral group $\mathcal{T}$. This actually implies that the symmetry group G is a subgroup of $(\mathbb{Z}_2)^4 \rtimes A_6$, where A_6 is the alternating group on six elements. Moreover[4], $\mathcal{T}$ acts only on the so-called tetrahedral torus, while $\mathcal{D}$ acts only on the so-called triangular torus. $\mathcal{O}$ can act on the square torus or on the tetrahedral torus, where it is realized as a subgroup of $\mathcal{T}$. Finally the action of the cyclic groups $\mathbb{Z}_4$ and $\mathbb{Z}_6$ agrees with that of a cyclic subgroup of $\mathcal{O}$, $\mathcal{D}$ or $\mathcal{T}$, possibly on a torus that does not enjoy the full dihedral or tetrahedral symmetry. In summary, the maximal groups that can occur are $\mathcal{O}$, $\mathcal{D}$ and $\mathcal{T}$.

By definition, any element of G must leave the complex structure and the dual Kähler class ω_0 of the Kummer surface X_{Λ,ω_0} invariant. Hence in terms of real local coordinates $\vec{x} = (x_1, x_2, x_3, x_4)$ as above and with respect to standard real coordinate vector fields $\vec{e}_1, \dots, \vec{e}_4$, using the notations of [**TW11**, Section 3], G must preserve each of the following 2-cycles in $H_2(X_{\Lambda,\omega_0}, \mathbb{R})$,

$$\text{(1.2)}\quad \Omega_1 = e_1 \vee e_3 - e_2 \vee e_4, \quad \Omega_2 = e_1 \vee e_4 + e_2 \vee e_3 \text{ and } \omega_0 = e_1 \vee e_2 + e_3 \vee e_4.$$

Equivalently, every symmetry group G must preserve the hyperkähler structure which is specified by the nowhere vanishing holomorphic 2-form and the Kähler class on X_{Λ,ω_0}. We can work with local holomorphic coordinates (z_1, z_2) that are induced from the underlying torus. The invariant classes hence are given by $dz_1 \wedge dz_2$, and $\frac{1}{2i}(dz_1 \wedge d\overline{z}_1 + dz_2 \wedge d\overline{z}_2)$. Moreover, $G_T \cong G'_T/\mathbb{Z}_2$ where G'_T acts linearly. In other words, G'_T is a finite subgroup of $SU(2)$. Once a group $G'_T \subset SU(2)$ preserving the lattice Λ has been identified such that $\mathbb{Z}_2 \subset G'_T$, then $G_T \cong G'_T/\mathbb{Z}_2$ acts faithfully on the Kummer surface X_{Λ,ω_0}.

It is not surprising that quaternions provide an elegant framework to describe the groups $G_T \cong G'_T/\mathbb{Z}_2$ we are interested in when symmetry-surfing [**Fuj88, Bri98**]. Indeed, we recall a formalism taken from [**Bri98**] which is tailored to recover the maximal groups G'_T classified by Fujiki, i.e. $G'_T \cong \mathcal{O}, \mathcal{D}, \mathcal{T}$. It moreover provides a unified description of the lattice Λ for each torus on which one of these groups can act as automorphism group. In fact, each lattice Λ is given in terms of unit quaternion generators, and the automorphisms act by quaternionic left multiplication.

The link between the skew field of quaternions $\mathbb{H}$ and lattices $\Lambda \subset \mathbb{R}^4$ is through the natural isomorphism

$$\text{(1.3)}\qquad \mathbb{R}^4 \longrightarrow \mathbb{H}, \quad q = (q_0, q_1, q_2, q_3) \longmapsto q_0 + q_1 i + q_2 j - q_3 k,$$

with $\mathbb{H} = \{q = q_0 + q_1 i + q_2 j + q_3 k \mid q_\mu \in \mathbb{R}, \mu \in \{0, \dots, 3\}\}$. The unit quaternions form a group which is isomorphic to $SU(2)$, and under the identification (1.3) its regular representation on $\mathbb{R}^4 \cong \mathbb{C}^2$ is realized by left multiplication on $\mathbb{H} \cong \mathbb{R}^4$. One immediately checks that with this faithful representation, every unit quaternion leaves the standard holomorphic two-form $dz_1 \wedge dz_2$ and Kähler class $\frac{1}{2i}(dz_1 \wedge d\overline{z}_1 + dz_2 \wedge d\overline{z}_2)$ on $\mathbb{R}^4 \cong \mathbb{C}^2$ invariant. Hence this identification allows us to realize each of our groups G'_T in terms of a finite group of unit quaternions.

[4]See the end of this section, items 1.-3., for the precise definitions of the relevant lattices and group actions.

Assume now that $\Lambda \subset \mathbb{R}^4 \cong \mathbb{H}$ is a lattice of rank 4 which carries the faithful action of an automorphism group $G'_T \subset SU(2)$, where G'_T is one of the maximal groups $\mathcal{O}$, $\mathcal{D}$, $\mathcal{T}$ from Fujiki's classification. By the properties of these maximal groups, we can assume without loss of generality that G'_T has generators a, b, c that are represented by unit quaternions of the form

$$\begin{aligned} \hat{a} &= \cos(\tfrac{\pi}{m}) - i\sin(\tfrac{\pi}{r}) + j\cos(\tfrac{\pi}{n}), \\ \hat{b} &= j, \\ \hat{c} &= \cos(\tfrac{\pi}{n}) + j\cos(\tfrac{\pi}{m}) + k\sin(\tfrac{\pi}{r}), \end{aligned} \tag{1.4}$$

with the constraint $\cos^2(\frac{\pi}{m}) + \cos^2(\frac{\pi}{n}) = \cos^2(\frac{\pi}{r})$, where the numbers $m,\, n,\, r \in \mathbb{Z}$ determine the group G'_T [**Cox74**]. Moreover, for the lattice $\Lambda \subset \mathbb{R}^4 \cong \mathbb{H}$ we can choose the unit quaternion generators 1, $\hat{a}$, $\hat{b}$, $\hat{c}$. Hence in terms of $\mathbb{R}^4$, we let

$$\begin{aligned} \vec{\lambda_1} &= (1,0,0,0), & \vec{\lambda_2} &= \left(\cos(\tfrac{\pi}{m}), -\sin(\tfrac{\pi}{r}), \cos(\tfrac{\pi}{n}), 0\right), \\ \vec{\lambda_3} &= (0,0,1,0), & \vec{\lambda_4} &= \left(\cos(\tfrac{\pi}{n}), 0, \cos(\tfrac{\pi}{m}), -\sin(\tfrac{\pi}{r})\right), \end{aligned}$$

be the generators of Λ.

We now summarise the data needed for symmetry-surfing the moduli space of Kummer surfaces. We describe the three maximal symmetry groups $G_T \cong G'_T/\mathbb{Z}_2$ of Kummer surfaces induced by the holomorphic symplectic automorphisms of some torus $T = T(\Lambda)$ fixing $0 \in \mathbb{C}^2/\Lambda = T$, along with the possible lattices Λ:

(1) **Dihedral group $\mathbf{D_2} \cong \mathcal{O}/\mathbb{Z}_2 \cong \mathbb{Z}_2 \times \mathbb{Z}_2$**

Take the lattice Λ to be $\Lambda_0 := \operatorname{span}_{\mathbb{Z}}\{1, \hat{a} = i, \hat{b} = j, \hat{c} = k\}$, with $\{\hat{a}, \hat{b}, \hat{c}\}$ generating the quaternionic group $G'_T \cong Q_8$ of order 8. It is immediate that Q_8 is the automorphism group of Λ_0, which is the lattice yielding the square Kummer surface X_0 in [**TW11**]. There, an equivalent description of the generators of the binary dihedral group $\mathcal{O}$ was given by

$$\alpha_1\colon \quad (z_1, z_2) \longmapsto (iz_1, -iz_2), \qquad \alpha_2\colon \quad (z_1, z_2) \longmapsto (-z_2, z_1), \tag{1.5}$$

both of which are of order 4.

(2) **Alternating group $\mathbf{A_4} \cong \mathcal{T}/\mathbb{Z}_2$**

The lattice Λ may be generated by $\{1,\, \hat{a} = \cos(\frac{\pi}{3}) - i\sin(\frac{5\pi}{4}) + j\cos(\frac{\pi}{3}),\, \hat{b} = j,\, \hat{c} = \cos(\frac{\pi}{3}) + j\cos(\frac{\pi}{3}) + k\sin(\frac{5\pi}{4})\}$, hence the four lattice vectors that generate Λ may be chosen as $\vec{\lambda_1} = (1,0,0,0)$, $\vec{\lambda_2} = (\frac{1}{2}, \frac{1}{\sqrt{2}}, \frac{1}{2}, 0)$, $\vec{\lambda_3} = (0,0,1,0)$ and $\vec{\lambda_4} = (\frac{1}{2}, 0, \frac{1}{2}, \frac{1}{\sqrt{2}})$.

One shows that the orbit of $\vec{\lambda_1}$ under the group $G'_T = \mathcal{T}$ yields 24 unit lattice vectors. This lattice is isometric to the lattice $\Lambda_1 := \Lambda_{D_4}$ used in [**TW11**] to construct the tetrahedral Kummer surface $X_1 = X_{D_4}$ from the torus $T(\Lambda_{D_4})$. We will use this Kummer surface in what follows, hence we recall the generators of Λ_{D_4}:

$$\vec{\lambda_1} = (1,0,0,0),\ \vec{\lambda_2} = (0,1,0,0),\ \vec{\lambda_3} = (0,0,1,0),\ \vec{\lambda_4} = \tfrac{1}{2}(1,1,1,1). \tag{1.6}$$

Generators of the binary tetrahedral group $\mathcal{T}$ may be taken to be

$$\begin{aligned} \gamma_1\colon &\quad (z_1, z_2) &\longmapsto&\quad (iz_1, -iz_2), \\ \gamma_2\colon &\quad (z_1, z_2) &\longmapsto&\quad (-z_2, z_1), \\ \gamma_3\colon &\quad (z_1, z_2) &\longmapsto&\quad \tfrac{i+1}{2}(i(z_1 - z_2), -(z_1 + z_2)). \end{aligned} \tag{1.7}$$

These generators satisfy the relations $\gamma_1^4 = \gamma_2^4 = \mathbb{1}$ and $\gamma_3^3 = \mathbb{1}$. Note that the minimum number of generators for the group $\mathcal{T}$ is 2, and indeed, one has $\gamma_2 = \gamma_1^2\gamma_3\gamma_1(\gamma_3)^{-1}$.

(3) **Permutation group $\mathbf{S_3} \cong \mathcal{D}/\mathbb{Z}_2$**

Take the lattice Λ_2 generated by $\{1, \hat{a} = -\cos(\frac{\pi}{3}) + i\sin(\frac{\pi}{3}), \hat{b} = j, \hat{c} = -j\cos(\frac{\pi}{3}) - k\sin(\frac{\pi}{3})\}$, hence the four lattice vectors that generate Λ_2 may be chosen as

(1.8)
$$\vec{\lambda_1} = (1,0,0,0), \quad \vec{\lambda_2} = (-\tfrac{1}{2}, \tfrac{\sqrt{3}}{2}, 0, 0), \quad \vec{\lambda_3} = (0,0,1,0), \quad \vec{\lambda_4} = (0,0,-\tfrac{1}{2}, \tfrac{\sqrt{3}}{2}).$$

The orbit of $\vec{\lambda}_1$ under the binary dihedral group $G'_T \cong \mathcal{D}$ yields 12 unit vectors in Λ_2. The Kummer surface obtained from $T(\Lambda_2)$ is the triangular Kummer surface X_2. The generators of $\mathcal{D}$ have order 3 and 4, respectively, and they are given by

$$\begin{aligned} \beta_1\colon (z_1, z_2) &\longmapsto (\zeta z_1, \zeta^{-1} z_2), \\ \beta_2\colon (z_1, z_2) &\longmapsto (-z_2, z_1), \end{aligned} \tag{1.9}$$

where $\zeta := e^{2\pi i/3}$.

2. Overarching maps and Niemeier markings

The description of symmetries of K3 surfaces is most efficient in terms of lattices. To this end, recall that the geometric action of a symmetry group G of a K3 surface X is fully captured by its action on the lattice $L_G = (L^G)^\perp \cap H_*(X,\mathbb{Z})$, where $L^G := H_*(X,\mathbb{Z})^G$. This follows from the Torelli theorem (see the discussion of Def. 1.2) and the very definition of L^G as the sublattice of $H_*(X,\mathbb{Z})$ on which G acts trivially.

On the other hand, if X_{Λ,ω_0} is a Kummer surface with its induced dual Kähler class, then the induced action of G on the Kummer lattice $\Pi \subset H_*(X,\mathbb{Z})$ bears all information about the action of G (see [**TW11**, Prop. 3.3.3]):

Proposition 2.1. Consider a Kummer surface X_{Λ,ω_0} with its induced dual Kähler class. Let $\Pi \subset H_*(X,\mathbb{Z})$ denote the Kummer lattice, that is, the smallest primitive sublattice of the integral K3 homology which contains the 16 classes $E_{\vec{a}}$, $\vec{a} \in \mathbb{F}_2^4$, that are obtained from blowing up the fixed points $\vec{F}_{\vec{a}}$ of the $\mathbb{Z}_2$-action on the underlying torus (1.1).
Then every symmetry of X induces a permutation of the $E_{\vec{a}}$. This permutation is given by an affine linear transformation of the labels $\vec{a} \in \mathbb{F}_2^4$, which in turn uniquely determines the symmetry.

In the case of Kummer surfaces we thus have two competing lattices Π and L_G which conveniently encode the action of the symmetry group G of X_{Λ,ω_0}. In [**TW11**] we argue that neither does L_G contain the rank 16 Kummer lattice, nor does, in general, the Kummer lattice contain L_G. Instead, combining the two, in [**TW11**, Prop. 3.3.6] we introduce the lattice M_G, which is generated by L_G and Π along with the vector $\upsilon_0 - \upsilon$, where υ_0, υ are generators of $H_0(X,\mathbb{Z})$ and $H_4(X,\mathbb{Z})$ with[5] $\langle \upsilon_0, \upsilon \rangle = 1$. We argue that in the Kummer case we can generalize and improve some extremely useful techniques introduced by Kondo [**Kon98**] to

[5]On $H_*(X,\mathbb{Z})$, we use the standard quadratic form which is induced by the intersection form.

this enlarged lattice M_G. Indeed, we prove that this lattice allows a primitive embedding into the Niemeier lattice $N(-1)$ with root lattice A_1^{24} [**TW11**, Thm. 3.3.7], where the decoration (-1) indicates that the roots of $N(-1)$ have length square -2. This embedding allows us to view the symmetry group G as a group of lattice automorphisms of $N(-1)$: the action of G on $N(-1)$ is defined such that the embedding $\iota_G\colon M_G \hookrightarrow N(-1)$ is G-equivariant, and G acts trivially on the orthogonal complement of $\iota_G(M_G)$ in $N(-1)$. Since the automorphism group of $N(-1)$, up to reflections in the roots of $N(-1)$, is the Mathieu group M_{24}, this conveniently realizes every symmetry group G of a Kummer K3 as a subgroup of M_{24}.

In what follows, we use the notations and conventions of [**TW11**] throughout. In particular, we fix the Kummer lattice Π within the abstract lattice $H_*(X,\mathbb{Z})$ as well as its image under ι_G in $N(-1)$ for every Kummer surface, independently of the parameters of the underlying torus. More precisely, we fix a unique MARKING for all our Kummer surfaces, that is, an explicit isometry of the lattice $H_*(X,\mathbb{Z})$ with a standard even, unimodular lattice of signature $(4,20)$. As is explained in [**TW11**, Sect. 2.2], the Kummer construction induces a natural such marking, which in particular fixes the position of Π within the lattice $H_*(X,\mathbb{Z})$. In this setting, among the data specifying each Kummer surface we have to include the choice of generators $\vec{\lambda}_1,\dots,\vec{\lambda}_4 \in \mathbb{R}^4$ for the lattice Λ of the underlying torus $T = T(\Lambda)$. Note that the choice of such a fixed marking amounts to the transition to a smooth universal cover of the moduli space of hyperkähler structures on K3. Similarly to $\Pi \subset H_*(X,\mathbb{Z})$, we also fix the position of $\widetilde{\Pi}(-1) := \iota_G(\Pi)$ in $N(-1)$ such that $\widetilde{\Pi}$ is common to all Kummer surfaces. To do so, in [**TW11**, (2.14)] we construct a bijection $I:\mathcal{I}\backslash\mathcal{O}_9 \longrightarrow \mathbb{F}_2^4$ between the 16 elements of the set $\mathcal{I} := \{1,2,\dots,24\}$ that do not belong to our choice of reference octad $\mathcal{O}_9 := \{3,5,6,9,15,19,23,24\}$ from the Golay code and the vertices of the hypercube $\mathbb{F}_2^4$. In [**TW11**, Prop. 2.3.4] we prove that the $\mathbb{Q}$-linear extension of $\iota_G(E_{\vec{a}}) := f_{I^{-1}(\vec{a})}$ yields an isometry between Π and $\widetilde{\Pi}(-1)$, where $\{f_n,\, n\in\mathcal{I}\}$, denotes a root basis of the root lattice A_1^{24} in $N(-1)$. Thus we have fixed the position of $\widetilde{\Pi}$ within N for all Kummer surfaces, similarly to fixing the position of Π within the abstract lattice $H_*(X,\mathbb{Z})$. This motivates the

DEFINITION 2.2. With notations as above, for a Kummer surface X_{Λ,ω_0} with symmetry group G, an isometric embedding $\iota_G\colon M_G \hookrightarrow N(-1)$ such that $\iota_G(E_{\vec{a}}) = f_{I^{-1}(\vec{a})}$ for all $\vec{a}\in\mathbb{F}_2^4$ is called a NIEMEIER MARKING.

By the above, every Kummer surface X allows a Niemeier marking [**TW11**, Prop. 4.1.1]. In general, the embedding ι_G is not uniquely determined. However, the action of G on N, which is induced by the requirement that ι_G is G-equivariant, is independent of all choices: indeed, $\iota_G(E_{\vec{a}}) = f_{I^{-1}(\vec{a})}\ \forall\vec{a}\in\mathbb{F}_2^4$ fixes the action of G on the lattice $\widetilde{\Pi}\subset N$, and by the arguments presented in the discussion of [**TW11**, Cor. 3.3.8] this already uniquely determines the action of G on all of N.

In particular, consider the translational symmetry group $G_t \cong (\mathbb{Z}_2)^4$ discussed in Section 1. Its action on the roots f_n, $n\in\mathcal{I}$, of N, which is common to all Kummer surfaces, is generated by the following permutations [**TW11**, Prop. 4.1.1]:

(2.1)
$$G_t := (\mathbb{Z}_2)^4 : \left\{\begin{array}{rcl} \iota_1 &=& (1,11)(2,22)(4,20)(7,12)(8,17)(10,18)(13,21)(14,16),\\ \iota_2 &=& (1,13)(2,12)(4,14)(7,22)(8,10)(11,21)(16,20)(17,18),\\ \iota_3 &=& (1,14)(2,17)(4,13)(7,10)(8,22)(11,16)(12,18)(20,21),\\ \iota_4 &=& (1,17)(2,14)(4,12)(7,20)(8,11)(10,21)(13,18)(16,22). \end{array}\right.$$

Now recall Mukai's seminal result [**Muk88**] that the symmetry group of every K3 surface is isomorphic to a subgroup of one of eleven subgroups of the Mathieu group M_{24}, the largest one of which has 960 elements. Hence symmetry groups of K3 surfaces are by orders of magnitude smaller than the group M_{24}, whose appearance one expects from Mathieu Moonshine. Therefore, in [**TW11**] we propose to use Niemeier markings to combine the symmetry groups of distinct Kummer surfaces by means of their actions on the Niemeier lattice N. To underpin this idea by lattice identifications, we propose to extend a given Niemeier marking ι_G to a linear bijection $\Theta : H_*(X,\mathbb{Z}) \longrightarrow N(-1)$, which restricts to an isometry on the largest possible sublattice of $H_*(X,\mathbb{Z})$. More precisely, we propose to construct a map Θ which induces Niemeier markings of all K3 surfaces along a smooth path in the smooth universal cover of the moduli space of hyperkähler structures on K3. If this path connects two distinct Kummer K3s X_A and X_B, then we call Θ_{AB} an OVERARCHING MAP for X_A and X_B. This is the key to exhibit an overarching symmetry in the moduli space of Kummer K3s. We say that an overaching map Θ_{AB} for Kummer surfaces X_A and X_B allows us to SURF from one of the corresponding Kummer surfaces to the other IN MODULI SPACE.

For two Kummer surfaces X_A, X_B with complex and Kähler structures induced from the underlying torus and with symmetry groups G_A, G_B, respectively, we will argue below that the following holds: under appropriate additional assumptions, one can construct an overarching map Θ_{AB} which restricts to a Niemeier marking, that is to an isometric G_k-equivariant embedding $\iota_{G_k}\colon M_{G_k} \hookrightarrow N(-1)$, for both $k=A$ and $k=B$, just like the map Θ constructed in [**TW11**] for the tetrahedral Kummer surface $X_1 = X_{D_4}$ and the square Kummer surface X_0. That Θ restricts to the desired Niemeier markings is sufficient to ensure that Θ_{AB} is an overarching map according to the above definition. Indeed, we can always find a path in the smooth universal cover of the moduli space which connects X_A and X_B, such that all intermediate points of the path are Kummer surfaces with the minimal symmetry group $G = G_t \cong (\mathbb{Z}_2)^4$. The group G_t is compatible with Θ_{AB} by construction. See [**TW11**, Thm. 4.4.2] for an example – one solely needs to ensure that $\mathrm{span}_{\mathbb{C}}\{\Omega_1, \Omega_2, \omega_0\}^\perp \cap \pi_* H_2(T,\mathbb{Z}) = \{0\}$ along the path.

To determine sufficient conditions on the existence of Θ_{AB}, first note that by the above, see also [**TW11**, Thm. 3.3.7], the lattices M_{G_k} share the Kummer lattice Π and the vector $v_0 - v$. By the Definition 2.2 of Niemeier markings $\iota_G\colon M_G \hookrightarrow N(-1)$, we require $\Theta_{AB}(E_{\vec{a}}) = f_{I^{-1}(\vec{a})}$ for all $\vec{a} \in \mathbb{F}_2^4$. As mentioned above, G_k-equivariance of ι_{G_k} then already fixes the action of G_k on N. An overarching map Θ_{AB} hence only exists if there is an index $n_0 \in \mathcal{O}_9$, such that f_{n_0} is invariant under the action of both groups G_A, G_B, such that $\Theta_{AB}(v_0 - v) = f_{n_0}$ is consistent with G_k-equivariance.

For the complementary lattices $\widehat{K}_{G_k} := ((\pi_*(H_2(T,\mathbb{Z}))^{(G_k)_T})^\perp \cap \pi_* H_2(T,\mathbb{Z})$ for $k \in \{A,B\}$ introduced in [**TW11**, Thm. 3.3.7], choose bases $I^\perp_{i_k,k}$, $i_k \in \{1,\dots,N_k\}$, where $N_k \leq 3$ by construction. If all the vectors $I^\perp_{1,A},\dots,I^\perp_{N_A,A}, I^\perp_{1,B},\dots,I^\perp_{N_B,B}$ are linearly independent, then we claim that under one final assumption we can find an overarching map Θ_{AB} for X_A and X_B as desired[6]. Indeed, as in [**TW11**, §4.1],

[6]We will see below that the assumption of linear independence can be relaxed, but for simplicity of exposition we first consider this case.

for each of the six two-cycles[7] λ_{ij}, we first choose a set $Q_{ij} \subset \mathcal{I}$ of four labels, such that

$$\Theta_{AB}(\pi_*\lambda_{ij}) = \sum_{n\in Q_{ij}} f_n \mod 2N(-1)$$

is compatible with the required G_k-equivariance. In fact, for each λ_{ij}, this constraint only leaves a choice between two complementary sets $Q_{ij} \subset \mathcal{O}_9$ which are explicitly listed in [**TW11**, (4.3)]. Choose these quadruplets of labels such that for each Q_{ij}, $n_0 \notin Q_{ij}$. Analogously to [**TW11**, Prop. 4.2.5] this defines a map $\overline{I}$ through $\overline{I}(\pi_*\lambda_{ij}) := Q_{ij}$ and $\overline{I}(\lambda+\lambda') := \overline{I}(\lambda) + \overline{I}(\lambda')$ by symmetric differences of sets. Since isometric embeddings $\iota_{G_k}\colon M_{G_k} \hookrightarrow N(-1)$ exist by [**TW11**, Prop. 4.1.1], we can now find appropriate candidates $\Theta_{AB}(I^\perp_{i_k,k}) \in N(-1)$ such that Θ_{AB} restricts to an isometry on both lattices $\widehat{K}_{G_k}$. Indeed, up to contributions of the form 2Δ with $\Delta \in N(-1)$, each $\Theta_{AB}(I^\perp_{i_k,k})$ is a linear combination of roots f_j with $j \in \overline{I}(I^\perp_{i_k,k})$. Under the final assumption that all the $\Theta_{AB}(I^\perp_{i_k,k})$ constructed in this manner are linearly independent, clearly Θ_{AB} can be extended to an overarching map as desired.

All our assumptions hold true in two of the three cases for which we shall construct overarching maps and exhibit overarching symmetries in Section 3 below. In one case, the vectors $I^\perp_{1,A}, \dots, I^\perp_{N_A,A}, I^\perp_{1,B}, \dots, I^\perp_{N_B,B}$ fail to be linearly independent. However, the linear dependence results from a repetition of vectors, $I^\perp_{a,A} = I^\perp_{b,B}$, so by listing every vector only once, linear independence is achieved, and the argument goes through as above.

This technique allows us to find overarching maps between any two Kummer surfaces, as we shall see in the next two sections. More precisely, for any pair of Kummer surfaces we can find representatives X_A and X_B in the smooth universal cover of the moduli space of hyperkähler structures, such that an overarching map for X_A and X_B exists. Hence we can surf between any two points in moduli space.

3. Construction of overarching maps

In Section 1, we have identified three distinct Kummer surfaces X_k, $k \in \{0,1,2\}$, whose associated tori $T = \mathbb{C}^2/\Lambda$ have maximal symmetry. In order to explore the overarching symmetry for the moduli space of Kummer surfaces by surfing from X_0 to X_1 and X_2, and from X_1 to X_2, we apply the recipe given in Section 2 to construct three overarching maps $\Theta_{k\ell}$, $0 \le k < \ell \le 2$, that yield overarching symmetry groups for the three pairs of Kummer surfaces (X_k, X_ℓ). As was explained in Section 2, the construction of an overarching map requires the existence of a root $f_{n_0} \in N(-1)$, $n_0 \in \mathcal{O}_9$, that is invariant under the action of G_k and G_ℓ. In the cases of interest to us here, the value of n_0 varies from map to map, but we carefully note down all possible choices, since this will be crucial in the subsequent section. We first summarize the construction of the overarching map Θ_{01} valid for the square and tetrahedral Kummer surfaces, which appeared with some additional details in [**TW11**]. Then we proceed to the construction of the other two maps, Θ_{02} and Θ_{12}, which are new. This exercise paves the way to Section 4, where we argue that one can combine various overarching groups and obtain an action of a

[7]Recall that for $T = T(\Lambda)$, $\lambda_{ij} := \lambda_i \vee \lambda_j \in H_2(T,\mathbb{Z})$ denotes the integral two-cycle specified by the lattice vectors $\vec{\lambda}_i, \vec{\lambda}_j \in \Lambda$.

maximal subgroup $(\mathbb{Z}_2)^4 \rtimes A_8$ of M_{24} on the Niemeier lattice $N(-1)$, overarching the entire Kummer moduli space.

3.1. Overarching the square and tetrahedral Kummer K3s. The full symmetry group of the square Kummer surface X_0 is the group $G_0 := (\mathbb{Z}_2)^4 \rtimes (\mathbb{Z}_2 \times \mathbb{Z}_2)$ of order 64, while that of the tetrahedral Kummer surface $X_1 := X_{D_4}$ is the group $G_1 := (\mathbb{Z}_2)^4 \rtimes A_4$ of order 192. By the discussion in the previous section, there exist Niemeier markings ι_{G_k}, $k \in \{0, 1\}$, which allow the definition of induced actions of the groups G_k on the Niemeier lattice $N(-1)$, independently of all choices. Indeed, for the respective generators listed at the end of Section 1, according to [**TW11**, Sects. 4.2, 4.3] we obtain

(3.1)
$$(G_0)_T := \mathbb{Z}_2{\times}\mathbb{Z}_2 : \begin{cases} \alpha_1 = (4,8)(6,19)(10,20)(11,13)(12,22)(14,17)(16,18)(23,24), \\ \alpha_2 = (2,21)(3,9)(4,8)(10,12)(11,14)(13,17)(20,22)(23,24), \end{cases}$$

(3.2)
$$(G_1)_T := A_4 : \begin{cases} \gamma_1 = (2,8)(7,18)(9,24)(10,22)(11,13)(12,17)(14,20)(15,19), \\ \gamma_2 = (2,18)(7,8)(9,19)(10,17)(11,14)(12,22)(13,20)(15,24), \\ \gamma_3 = (2,12,13)(4,16,21)(7,17,20)(8,22,14)(9,19,24)(10,11,18). \end{cases}$$

The construction of the map Θ_{01} requires that one root f_{n_0} with $n_0 \in \mathcal{O}_9$ is invariant under G_0 and G_1. One checks that indeed $n_0 := 5$ is the only label in $\mathcal{O}_9$ which is fixed by both groups.

According to [**TW11**, (4.9),(4.21)], the generators of the rank 3 lattices $\widehat{K}_{G_1}$ and $\widehat{K}_{G_0}$ are

$$\begin{array}{lllllll} & I^{\perp}_{1,1} &=& \pi_*\lambda_{14} + \pi_*\lambda_{24} - \pi_*\lambda_{23}, & I^{\perp}_{1,0} &=& \pi_*\lambda_{14} - \pi_*\lambda_{23}, \\ (3.3) & I^{\perp}_{2,1} &=& \pi_*\lambda_{13} + \pi_*\lambda_{24} + \pi_*\lambda_{34}, & I^{\perp}_{2,0} &=& \pi_*\lambda_{13} + \pi_*\lambda_{24}, \\ & I^{\perp}_{3,1} &=& -\pi_*\lambda_{12} + \pi_*\lambda_{14} + \pi_*\lambda_{34}, & I^{\perp}_{3,0} &=& \pi_*\lambda_{34} - \pi_*\lambda_{12}. \end{array}$$

From [**TW11**, (4.3)] we read that $n_0 = 5 \notin Q_{ij}$ implies

(3.4)
$$\begin{array}{lllllllll} Q_{12} &=& \{3,6,15,19\}, & Q_{13} &=& \{6,15,23,24\}, & Q_{14} &=& \{3,9,15,24\}, \\ Q_{34} &=& \{6,9,15,19\}, & Q_{24} &=& \{15,19,23,24\}, & Q_{23} &=& \{3,9,15,23\}. \end{array}$$

Hence the map $\overline{I}$ described in Section 2 is

(3.5)
$$\begin{array}{lllllllll} \overline{I}(I^{\perp}_{1,1}) &=& \{15,19\}, & \overline{I}(I^{\perp}_{2,1}) &=& \{9,15\}, & \overline{I}(I^{\perp}_{3,1}) &=& \{15,24\}, \\ \overline{I}(I^{\perp}_{1,0}) &=& \{23,24\}, & \overline{I}(I^{\perp}_{2,0}) &=& \{6,19\}, & \overline{I}(I^{\perp}_{3,0}) &=& \{3,9\}. \end{array}$$

Our choice of images of the generators (3.3) under Θ_{01} must ensure that Θ_{01} restricts to an isometry on both lattices $\widehat{K}_{G_k}$. Therefore, note that the quadratic form on $\widehat{K}_{G_1}$ with respect to the basis $I^{\perp}_{i,1}$, $i \in \{1,2,3\}$, and that on $\widehat{K}_{G_0}$ with respect to the basis $I^{\perp}_{i,0}$, $i \in \{1,2,3\}$, are

$$(3.6) \qquad \widehat{K}_{G_1} : \begin{pmatrix} -4 & -2 & -2 \\ -2 & -4 & -2 \\ -2 & -2 & -4 \end{pmatrix}, \qquad \widehat{K}_{G_0} : \begin{pmatrix} -4 & 0 & 0 \\ 0 & -4 & 0 \\ 0 & 0 & -4 \end{pmatrix}$$

according to [**TW11**, (4.20)] and [**TW11**, (4.27)]. Then the following gives linearly independent candidates for the $\Theta_{01}(I^\perp_{i_k,k}) \in N(-1)$ as desired:

$$\Theta_{01}:\quad \left\{ \begin{array}{llllllll} I^\perp_{1,1} & \longmapsto & f_{19}-f_{15}, & I^\perp_{2,1} & \longmapsto & f_9 - f_{15}, & I^\perp_{3,1} & \longmapsto & f_{24}-f_{15}, \\ I^\perp_{1,0} & \longmapsto & f_{24}-f_{23}, & I^\perp_{2,0} & \longmapsto & f_{19}-f_6, & I^\perp_{3,0} & \longmapsto & f_9 - f_3. \end{array}\right. \tag{3.7}$$

Equivalently,

$$\Theta_{01}:\quad \left\{ \begin{array}{lllll} \pi_*\lambda_{12} & \longmapsto & 2q_{12} & = & f_3+f_6-f_{15}-f_{19}, \\ \pi_*\lambda_{34} & \longmapsto & 2q_{34} & = & f_6+f_9-f_{15}-f_{19}, \\ \pi_*\lambda_{13} & \longmapsto & 2q_{13} & = & -f_6+f_{15}-f_{23}+f_{24}, \\ \pi_*\lambda_{24} & \longmapsto & 2q_{24} & = & -f_{15}+f_{19}+f_{23}-f_{24}, \\ \pi_*\lambda_{14} & \longmapsto & 2q_{14} & = & f_3-f_9-f_{15}+f_{24}, \\ \pi_*\lambda_{23} & \longmapsto & 2q_{23} & = & f_3-f_9-f_{15}+f_{23}. \end{array}\right. \tag{3.8}$$

On the Kummer lattice Π, we set $\Theta_{01}(E_{\vec{a}}) = f_{I^{-1}(\vec{a})}$, as always. Finally, a consistent choice for the images of υ, υ_0 is

$$\Theta_{01}:\quad \left\{ \begin{array}{lll} \upsilon_0 & \longmapsto & \frac{1}{2}(f_3+f_5+f_6+f_9-f_{15}-f_{19}-f_{23}-f_{24}), \\ \upsilon & \longmapsto & \frac{1}{2}(f_3-f_5+f_6+f_9-f_{15}-f_{19}-f_{23}-f_{24}). \end{array}\right.$$

This completes the construction of the map Θ_{01} which is compatible with the symmetry groups of the square (G_0) and tetrahedral (G_1) Kummer surfaces. Viewed as a linear bijection $\Theta_{01}: H_*(X_0,\mathbb{Z}) \longrightarrow N(-1)$, its restriction $\Theta_{01}|_{M_{G_0}}$ yields a G_0-equivariant and isometric embedding of M_{G_0} in $N(-1)$. Viewed instead as a linear bijection $\Theta_{01}: H_*(X_1,\mathbb{Z}) \longrightarrow N(-1)$, its restriction $\Theta_{01}|_{M_{G_1}}$ yields a G_1-equivariant and isometric embedding of M_{G_1} in $N(-1)$. This property of the overarching map Θ_{01} gives us ground to argue that there is an overarching symmetry group for the square and tetrahedral Kummer surfaces, whose action is encoded in the same Niemeier lattice $N(-1)$ through the generators (2.1) of the translational symmetry group G_t common to all Kummer surfaces, in addition to the generators (3.1) *and* (3.2). The group generated this way is a copy of $(\mathbb{Z}_2)^4 \rtimes A_7 \subset M_{24}$.

3.2. Overarching the square and the triangular Kummer K3s. The full symmetry group of the triangular Kummer surface X_2 is the group $G_2 := (\mathbb{Z}_2)^4 \rtimes S_3$ of order 96, see (1.8) and (1.9). Independently of the choice of Niemeier marking, the induced action of G_2 on the Niemeier lattice is generated by

$$(G_2)_T := S_3:\quad \left\{ \begin{array}{lll} \beta_1 & = & (2,17,14)(4,7,8)(10,16,12)(11,13,21)(18,20,22)(5,24,23), \\ \beta_2 & = & (2,21)(3,9)(4,8)(10,12)(11,14)(13,17)(20,22)(23,24). \end{array}\right. \tag{3.9}$$

The construction of an overarching map Θ_{02} for X_0 and X_2 requires a root f_{n_0} with $n_0 \in \mathcal{O}_9$ which is invariant under G_0 and G_2. From (3.1) and (3.9) we observe that $\alpha_2 = \beta_2$ and that $n_0 = 15$ is the only label in $\mathcal{O}_9$ which is invariant under both groups.

To calculate the generators of the lattice $\widehat{K}_{G_2}$ following the techniques explained in [**TW11**], we first need to determine generators of the lattice $(\pi_* H_2(T,\mathbb{Z}))^{(G_2)_T}$. With the basis $\vec{\lambda}_1,\ldots,\vec{\lambda}_4$ for the triangular lattice given in (1.8), we obtain primitive generators of that lattice as

$$\pi_*\lambda_{13} - \pi_*\lambda_{24}, \quad \pi_*\lambda_{13} + \pi_*\lambda_{23} + \pi_*\lambda_{14}, \quad \pi_*\lambda_{12} + \pi_*\lambda_{34},$$

and hence the orthogonal complement $\widehat{K}_{G_2}$ of $(\pi_*H_2(T,\mathbb{Z}))^{(G_2)_T}$ in $\pi_*H_2(T,\mathbb{Z})$ is generated by the lattice vectors

(3.10)

$$I_{1,2}^\perp := \pi_*\lambda_{12} - \pi_*\lambda_{34}, \quad I_{2,2}^\perp := \pi_*\lambda_{13} + \pi_*\lambda_{23} + \pi_*\lambda_{24}, \quad I_{3,2}^\perp := \pi_*\lambda_{14} - \pi_*\lambda_{23}.$$

From [**TW11**, (4.3)] we read that $n_0 = 15 \notin Q_{ij}$ implies

(3.11)

$$\begin{array}{llllll} Q_{12} = \{5,9,23,24\}, & Q_{13} = \{3,5,9,19\}, & Q_{14} = \{5,6,19,23\}, \\ Q_{34} = \{3,5,23,24\}, & Q_{24} = \{3,5,6,9\}, & Q_{23} = \{5,6,19,24\}. \end{array}$$

Hence the map $\overline{I}$ described in Section 2 is as in (3.5) for $I_{i,0}^\perp$, $i \in \{1,2,3\}$, and furthermore,

$$\overline{I}(I_{1,2}^\perp) = \{3,9\}, \quad \overline{I}(I_{2,2}^\perp) = \{5,24\}, \quad \overline{I}(I_{3,2}^\perp) = \{23,24\}.$$

We now need to choose the images in $N(-1)$ of the generators (3.10) under Θ_{02} such that Θ_{02} restricts to an isometry on the lattices $\widehat{K}_{G_0}$ and $\widehat{K}_{G_2}$. To do so, note that the quadratic form on $\widehat{K}_{G_0}$ with respect to the basis $I_{i,0}^\perp$, $i \in \{1,2,3\}$, and that of $\widehat{K}_{G_2}$ with respect to the basis $I_{i,1}^\perp$, $i \in \{1,2,3\}$, by (3.6) and (3.10) are

$$(3.12) \qquad \widehat{K}_{G_0}: \begin{pmatrix} -4 & 0 & 0 \\ 0 & -4 & 0 \\ 0 & 0 & -4 \end{pmatrix}, \qquad \widehat{K}_{G_2}: \begin{pmatrix} -4 & 0 & 0 \\ 0 & -4 & 2 \\ 0 & 2 & -4 \end{pmatrix}.$$

Moreover, we have $I_{1,0}^\perp = I_{3,2}^\perp$ and $I_{3,0}^\perp = I_{1,2}^\perp$, such that we can find candidates for $\Theta_{02}(I_{i_k,k}^\perp) \in N$ as desired:

(3.13)

$$\Theta_{02}: \left\{ \begin{array}{llllll} I_{1,0}^\perp \longmapsto f_{24} - f_{23}, & I_{2,0}^\perp \longmapsto f_6 - f_{19}, & I_{3,0}^\perp \longmapsto f_3 - f_9, \\ I_{1,2}^\perp \longmapsto f_3 - f_9, & I_{2,2}^\perp \longmapsto f_5 - f_{24}, & I_{3,2}^\perp \longmapsto f_{24} - f_{23}. \end{array} \right.$$

For example, we can choose the following map in order to induce (3.13):

$$(3.14) \qquad \Theta_{02}: \left\{ \begin{array}{lcll} \pi_*\lambda_{12} & \longmapsto & 2q_{12} & = f_5 + f_9 - f_{23} - f_{24}, \\ \pi_*\lambda_{34} & \longmapsto & 2q_{34} & = f_3 + f_5 - f_{23} - f_{24}, \\ \pi_*\lambda_{13} & \longmapsto & 2q_{13} & = f_3 + f_5 - f_9 - f_{19}, \\ \pi_*\lambda_{24} & \longmapsto & 2q_{24} & = -f_3 - f_5 + f_6 + f_9, \\ \pi_*\lambda_{14} & \longmapsto & 2q_{14} & = f_5 - f_6 + f_{19} - f_{23}, \\ \pi_*\lambda_{23} & \longmapsto & 2q_{23} & = f_5 - f_6 + f_{19} - f_{24}. \end{array} \right.$$

On the Kummer lattice Π, we set $\Theta_{02}(E_{\vec{a}}) = f_{I^{-1}(\vec{a})}$, as before. Finally, a consistent choice for the images of v, v_0 is

$$(3.15) \qquad \Theta_{02}: \left\{ \begin{array}{lcl} v_0 & \longmapsto & \frac{1}{2}\left(f_3 + f_5 + f_6 - f_9 + f_{15} - f_{19} - f_{23} - f_{24}\right), \\ v & \longmapsto & \frac{1}{2}\left(f_3 + f_5 + f_6 - f_9 - f_{15} - f_{19} - f_{23} - f_{24}\right). \end{array} \right.$$

This completes the construction of the overarching map Θ_{02} for the square and the triangular Kummer surfaces. Again, the overarching map Θ_{02} leads to an overarching symmetry group, whose action is encoded in the same Niemeier lattice $N(-1)$

through the generators (2.1) of the translational symmetry group G_t common to all Kummer surfaces, in addition to the generators (3.1) *and* (3.9). The resulting group is a copy of $(\mathbb{Z}_2)^4 \rtimes \mathcal{D} \subset M_{24}$, where $\mathcal{D}$ denotes the binary dihedral group of order 12, as before.

3.3. Overarching the tetrahedral and triangular Kummer K3s. The construction of an overarching map Θ_{12} for X_1 and X_2 requires a root f_{n_0} with $n_0 \in \mathcal{O}_9$ which is invariant under G_1 and G_2, whose generators are given in (3.9) and (3.2). The only label in $\mathcal{O}_9$ which is invariant under both these groups is $n_0 = 6$.

The generators of the rank 3 lattice $\widehat{K}_{G_1}$ are given in (3.3), and those of the lattice $\widehat{K}_{G_2}$ by (3.10). From [**TW11**, (4.3)] we read that $n_0 = 6 \notin Q_{ij}$ implies

(3.16)
$$\begin{array}{llllll} Q_{12} = \{5,9,23,24\}, & Q_{13} = \{3,5,9,19\}, & Q_{14} = \{3,9,15,24\}, \\ Q_{34} = \{3,5,23,24\}, & Q_{24} = \{15,19,23,24\}, & Q_{23} = \{3,9,15,23\}. \end{array}$$

Hence the map $\overline{I}$ described in Section 2 is

$$\begin{array}{lll} \overline{I}(I_{1,1}^\perp) = \{15,19\}, & \overline{I}(I_{2,1}^\perp) = \{9,15\}, & \overline{I}(I_{3,1}^\perp) = \{15,24\}, \\ \overline{I}(I_{1,2}^\perp) = \{3,9\}, & \overline{I}(I_{2,2}^\perp) = \{5,24\}, & \overline{I}(I_{3,2}^\perp) = \{23,24\}. \end{array}$$

We now need to choose the images in N of the generators $I_{i,1}^\perp$, $i \in \{1,2,3\}$ and $I_{i,2}^\perp$, $i \in \{1,2,3\}$ under Θ_{12} such that Θ_{12} restricts to an isometry on the lattices $\widehat{K}_{G_1}$ and $\widehat{K}_{G_2}$. Given the quadratic form (3.6) for $\widehat{K}_{G_1}$ and (3.12) for $\widehat{K}_{G_2}$, the following gives linearly independent candidates for $\Theta_{12}(I_{i_k,k}^\perp) \in N$:

(3.17)
$$\Theta_{12}\colon \quad \left\{ \begin{array}{llllll} I_{1,1}^\perp \longmapsto f_{19} - f_{15}, & I_{2,1}^\perp \longmapsto f_9 - f_{15}, & I_{3,1}^\perp \longmapsto f_{24} - f_{15}, \\ I_{1,2}^\perp \longmapsto f_3 - f_9, & I_{2,2}^\perp \longmapsto f_5 - f_{24}, & I_{3,2}^\perp \longmapsto f_{24} - f_{23}. \end{array} \right.$$

Equivalently,

(3.18)
$$\Theta_{12}\colon \quad \left\{ \begin{array}{lcll} \pi_*\lambda_{12} \longmapsto & 2q_{12} + 2f_3 - 2f_{15} & = & -f_5 - f_9 + f_{23} + f_{24} + 2f_3 - 2f_{15}, \\ \pi_*\lambda_{34} \longmapsto & 2q_{34} - 2f_{15} & = & f_3 - f_5 + f_{23} + f_{24} - 2f_{15}, \\ \pi_*\lambda_{13} \longmapsto & 2q_{13} + 2f_{15} - 2f_{23} & = & -f_3 + f_5 + f_9 - f_{19} + 2f_{15} - 2f_{23}, \\ \pi_*\lambda_{24} \longmapsto & 2q_{24} & = & -f_{15} + f_{19} + f_{23} - f_{24}, \\ \pi_*\lambda_{14} \longmapsto & 2q_{14} & = & f_3 - f_9 - f_{15} + f_{24}, \\ \pi_*\lambda_{23} \longmapsto & 2q_{23} & = & f_3 - f_9 - f_{15} + f_{23}. \end{array} \right.$$

On the Kummer lattice Π, we set $\Theta_{12}(E_{\vec{a}}) = f_{I^{-1}(\vec{a})}$. Finally, a consistent choice for the images of v, v_0 is

$$\text{(3.19)} \qquad \Theta_{12}\colon \quad \left\{ \begin{array}{ll} v_0 \longmapsto \frac{1}{2}\left(f_3 + f_5 + f_6 - f_9 - f_{15} - f_{19} + f_{23} - f_{24}\right), \\ v \longmapsto \frac{1}{2}\left(f_3 + f_5 - f_6 - f_9 - f_{15} - f_{19} + f_{23} - f_{24}\right). \end{array} \right.$$

This completes the construction of the overarching map Θ_{12} which is compatible with the symmetry groups of the tetrahedral (G_1) and triangular (G_2) Kummer surfaces. Hence there is an overarching symmetry group for the tetrahedral and the triangular Kummer surfaces, whose action is encoded in the same Niemeier lattice $N(-1)$ through the generators (2.1) of the translational symmetry group G_t

common to all Kummer surfaces, in addition to the generators (3.2) *and* (3.9). The group thus generated is a copy of $(\mathbb{Z}_2)^4 \rtimes A_7 \subset M_{24}$.

4. Overarching the moduli space of Kummer K3s by $(\mathbb{Z}_2)^4 \rtimes \mathbf{A}_8$

In this section we argue that our surfing procedure allows us to surf between any two points of the moduli space of Kummer K3s. More precisely, for any two Kummer surfaces with induced dual Kähler class, we can find representatives in the smooth universal cover of the moduli space of hyperkähler structures, such that an overarching map between the two representatives exists. This allows us to combine all symmetry groups of such Kummer surfaces to a larger, overarching group.

To see this, let us first consider an arbitrary Kummer surface $X_{\widetilde{\Lambda},\omega_0}$ with induced dual Kähler class, and let $\widetilde{G}$ denote its symmetry group. According to our discussion in Section 1, $\widetilde{G} = (\mathbb{Z}_2)^4 \rtimes (\widetilde{G}'_T/\mathbb{Z}_2)$, where $\widetilde{G}'_T \subset SU(2)$ is the linear automorphism group of the lattice $\widetilde{\Lambda}$. Moreover, $\widetilde{G}'_T$ is a subgroup of one of the three maximal linear automorphism groups of complex tori, the binary tetrahedral group $\mathcal{T}$ or one of the dihedral groups $\mathcal{D}$, $\mathcal{O}$ of order 12 and 8.

Let $G'_T = \mathcal{O}, \mathcal{T}$ or $\mathcal{D}$, such that $\widetilde{G}'_T \subset G'_T$, and let $\Lambda = \Lambda_0$, Λ_1 or Λ_2 denote the corresponding choice of lattice from Section 1 which has G'_T as its linear automorphism group. Fujiki's classification [**Fuj88**] implies that we can choose G'_T and Λ in such a way that there is a smooth deformation of Λ into $\widetilde{\Lambda}$, call it Λ^t with $t \in [0,1]$ and $\Lambda^0 = \Lambda$, $\Lambda^1 = \widetilde{\Lambda}$, such that the linear automorphism group of each Λ^t with $t \neq 0$ is $\widetilde{G}'_T$. The quaternionic language introduced in Section 1 is particularly useful to check this. For example, if $\widetilde{G}'_T = \mathbb{Z}_4$, then by Fujiki's results we can choose coordinates such that the action of this group on $\mathbb{C}^2$ is generated by our symmetry α_1 of (1.5), and we can choose $G'_T = \mathcal{O}$ with $\Lambda = \Lambda_0$ the lattice of the square torus. One finds generators $\vec{\lambda}_1^t, \ldots, \vec{\lambda}_4^t$ for the lattices Λ^t as desired such that $\vec{\lambda}_2^t = \alpha_1(\vec{\lambda}_1^t)$ and $\vec{\lambda}_3^t = \alpha_1(\vec{\lambda}_4^t)$ for every $t \in [0,1]$.

This deformation argument implies that by use of our fixed marking, the invariant sublattices of the integral torus homology, $L^{\widetilde{G}'_T} = H_*(T,\mathbb{Z})^{\widetilde{G}'_T}$ and $L^{G'_T} = H_*(T,\mathbb{Z})^{G'_T}$, obey $L^{G'_T} \subset L^{\widetilde{G}'_T}$. Hence for the symmetry group G of X_{Λ,ω_0} and for the lattices that yield our Niemeier markings we have $M_{\widetilde{G}} \subset M_G$, see Def. 2.2 and the discussion preceding it. From this it follows that one can find a representative of $X_{\widetilde{\Lambda},\omega_0}$ in the smooth universal cover of the moduli space of hyperkähler structures such that every Niemeier marking $\iota_G \colon M_G \hookrightarrow N(-1)$ of the maximally symmetric Kummer surface X_{Λ,ω_0} restricts to a Niemeier marking $\iota_{\widetilde{G}} := \iota_{G|M_{\widetilde{G}}}$ of the Kummer surface $X_{\widetilde{\Lambda},\omega_0}$. Hence any overarching map Θ for the maximally symmetric Kummer surface X_{Λ,ω_0} and any other Kummer K3 X also allows to surf from $X_{\widetilde{\Lambda},\omega_0}$ to X.

Now consider two distinct Kummer surfaces $\widetilde{X}_A$ and $\widetilde{X}_B$ with their induced dual Kähler classes. By the above, we can choose maximally symmetric Kummer surfaces X_A and X_B from the square, the tetrahedral and the triangular Kummer surfaces, such that the following holds: there are representatives of $\widetilde{X}_A$ and $\widetilde{X}_B$ in the smooth universal cover of the moduli space of hyperkähler structures such that any Niemeier marking of X_A restricts to a Niemeier marking of $\widetilde{X}_A$, and analogously for X_B and $\widetilde{X}_B$. Then by the above, the overarching map Θ_{AB} for X_A

and X_B which was constructed[8] in Section 3 also overarches $\widetilde{X}_A$ and $\widetilde{X}_B$. In other words, we can surf from $\widetilde{X}_A$ to $\widetilde{X}_B$.

We conclude that by means of our overarching maps we can surf the entire moduli space of hyperkähler structures of Kummer surfaces. In particular, we can combine the actions of all symmetry groups of Kummer surfaces with induced dual Kähler class by means of their action on the Niemeier lattice N. Recall from Section 2 that by construction, every overarching map Θ_{AB} between Kummer surfaces X_A and X_B with symmetry groups G_A and G_B assigns a fixed root $\Theta_{AB}(v_0 - v) = f_{n_0} \in N(-1)$ to the root $v_0 - v \in H_*(X, \mathbb{Z})$, where $n_0 \in \mathcal{O}_9$ is a label in our reference octad from the Golay code[9]. This root f_{n_0} is fixed under the induced actions of both G_A and G_B. For the overarching group G_{AB} obtained from G_A and G_B, which by construction is a subgroup of the stabilizer group $(\mathbb{Z}_2)^4 \rtimes A_8$ of the octad $\mathcal{O}_9$, this implies that G_{AB} additionally fixes one label $n_0 \in \mathcal{O}_9$. Hence G_{AB} is a subgroup of $(\mathbb{Z}_2)^4 \rtimes A_7$, the group which we call the OVERARCHING SYMMETRY GROUP OF KUMMER K3S. In Section 3 we have seen that for two pairs of distinct Kummer surfaces with maximal symmetry, the overarching group yields $G_{AB} = (\mathbb{Z}_2)^4 \rtimes A_7$. The third pair has overarching group $(\mathbb{Z}_2)^4 \rtimes \mathcal{D}$. Moreover, in each case there exists precisely one label in $\mathcal{O}_9$ which is fixed by both G_A and G_B. This label, however, is different for each of the three pairs of Kummer K3s with maximal symmetry. It follows that the combined symmetry group for all Kummer K3s with induced dual Kähler class is $(\mathbb{Z}_2)^4 \rtimes A_8$.

5. Interpretation and outlook

Let us now explain how our construction fits into the quest for the expected representation of M_{24} on a vertex algebra which governs the elliptic genus of K3. As mentioned in the Introduction, the elliptic genus arises from the contribution to the partition function of any superconformal field theory on K3 which counts states in the Ramond-Ramond sector with signs according to fermion numbers. This part of the partition function is modular invariant on its own, inducing the well-known modularity properties of the elliptic genus. The very construction of the elliptic genus, in addition, amounts to a projection onto those states which are Ramond ground states on the antiholomorphic side. The usual rules for fermion numbers imply that the OPE between any two fields in the Ramond sector yields contributions from the Neveu-Schwarz sector only. Hence the expected vertex algebra can certainly not arise in the Ramond-Ramond sector. Of course we can spectral flow the relevant fields into the Neveu-Schwarz sector, where (prior to all projections) they indeed form a closed vertex algebra[10] $\widehat{\mathcal{X}}$. Note that the choice of a spectral flow requires the choice of a holomorphic and an antiholomorphic $U(1)$-current within the superconformal algebra of our SCFT. For definiteness, we use the spectral flow which maps Ramond-Ramond ground states to (chiral, chiral) states.

---

[8]If $X_A = X_B$, then there is nothing left to be shown.

[9]This fixed label n_0 is responsible for the fact that each G_k is a subgroup of M_{23}, as we emphasized in [**TW11**].

[10]Here and in the following, we loosely refer to the space of fields which create states in the Neveu-Schwarz sector, equipped with the OPE, as a "vertex algebra", which however is *not* a holomorphic VOA.

The resulting vertex algebra $\widehat{\mathcal{X}}$ certainly governs the elliptic genus. Its space of states contains the states underlying the well-known (chiral, chiral) algebra $\mathcal{X}$ of Lerche-Vafa-Warner [**LVW89**], which accounts for the contributions to the lowest order terms of the elliptic genus. In Appendix A we describe the (chiral, chiral) algebra $\mathcal{X}$ (see (A.1)) more concretely in the context relevant to this work, namely in $\mathbb{Z}_2$-orbifold conformal field theories $\mathcal{C} = \mathcal{T}/\mathbb{Z}_2$ on K3, where $\mathcal{T}$ denotes the underlying toroidal theory. As expanded upon in Appendix A, the very truncation to the (chiral, chiral) algebra $\mathcal{X}$ makes $\mathcal{X}$ completely independent of all moduli. In principle, this is a desired effect when aiming at constructing a vertex algebra which governs the elliptic genus, since the latter is independent of all moduli. However, from the action of a linear map on $\mathcal{X}$ (generated by the fields in (A.1), independently of all moduli), it is not clear whether or not it is a symmetry, while the Mathieu Moonshine phenomenon dictates that we consider symmetries of some underlying vertex algebra.

We shall come back to this 'bottom up' discussion further down, but we first take a closer look at the 'top-down' approach, and consider the action of symmetries of $\mathcal{C}$ on the (chiral, chiral) algebra $\mathcal{X}$ generated by (A.1). We impose a number of rather severe assumptions on such symmetries, in order to ensure that they descend to symmetries of a candidate vertex algebra that governs the elliptic genus. As mentioned in the Introduction, this graded vertex algebra at leading order is the (chiral,chiral) algebra $\mathcal{X}$. Following [**LVW89**] we identify $\mathcal{X}$ with the cohomology of a K3 surface X. Associated to every Calabi-Yau manifold Y, there is the chiral de Rham complex [**MSV99**] which furnishes a sheaf of vertex algebras governing the elliptic genus of Y and containing the usual de Rham complex of Y at leading order [**BL00**, **Bor01**]. We thus find it natural[11] to restrict our attention to symmetries of $\mathcal{C}$ that descend to the chiral de Rham complex of X. To this end, we assume that our SCFT $\mathcal{C}$ comes with a choice of generators of the $N = (4,4)$ superconformal algebra, which in particular fixes the $U(1)$-currents and a preferred $N = (2,2)$ subalgebra. As mentioned above, this is already necessary when we choose the spectral flow to $\mathcal{X}$. Recall that the choice of $U(1)$-currents amounts to the choice of a complex structure in any geometric interpretation of $\mathcal{C}$ [**AM94**]. We furthermore use the notion advertised by [**GPRV12**], which requires symmetries to fix the superconformal algebra of $\mathcal{C}$ pointwise.[12] To identify $\mathcal{X}$ with the cohomology of a K3 surface X, we need to perform a large volume limit [**Wit82**, **LVW89**]. More generally, according to [**Kap05**], the space of states singled out by the elliptic genus is mapped to the appropriate cohomology of the chiral de Rham complex of X only in the large volume limit. In order to perform such a large volume limit, we need to choose a geometric interpretation of $\mathcal{C}$.

Summarising, in view of constructing a vertex algebra from the fields in $\mathcal{C}$, such that $\mathcal{X}$ governs the leading order terms of the elliptic genus, we restrict our attention to symmetries that are induced from geometric symmetries. This justifies why so far, in our work, we have searched for explanations of Mathieu Moonshine phenomena within the context of geometric symmetries only.

[11]By [**BL00**, **FS07**], the CFT orbifold procedure descends to the chiral de Rham complex; this should be the source for the behavior of the twining genera in Mathieu Moonshine, at least for those symmetries that are induced from geometric ones.

[12]This, for instance, excludes equivalences of SCFTs induced by mirror symmetry, which acts as an outer automorphism on the superconformal algebra.

As a further potential justification for this restriction recall the notion of 'exceptional' symmetry groups of sigma models on K3, that is, symmetry groups of such SCFTs which are not realizable as subgroups of M_{24}, obtained from the classification in [**GHV10a**]. According to [**GV12**], in many cases 'exceptional' symmetry is linked to certain quantum symmetries which as we shall argue cannot be induced from any classical geometric symmetries. Indeed, these symmetries in [**GV12**] are characterized by the property that they generate a group G, such that orbifolding the K3 model by G yields a toroidal SCFT. We remark that there is no geometric counterpart of such an orbifold construction, which would have to yield a complex four-torus as an orbifold of a K3 surface. Indeed, the odd cohomology of a complex four-torus cannot be restored by blowing up quotient singularities in an orbifold by a symplectic automorphism group of a K3 surface. However, this is only a potential justification for our restriction to geometric symmetries since, according to [**GV12**], 'exceptional' symmetry groups also occur in a few cases where to date it is not known whether or not such purely non-geometric quantum symmetries are responsible for the 'exceptionality' of the symmetry group. Although the group M_{24} itself contains elements that can never act in terms of a geometric symmetry on K3, we are optimistic that every element of M_{24} can be obtained as a composition of 'geometric' symmetries.

We wish to emphasize that it is immediately clear that the (chiral, chiral) algebra $\mathcal{X}$ cannot carry a representation of M_{24}. Indeed, (A.2) is the basis of a four-dimensional subspace of the 24-dimensional space $\mathcal{X}$ which is invariant under all symmetries that are of interest here, but by the known properties of representations of M_{24}, this group can only act trivially on the remaining 20-dimensional space. Hence a vertex algebra which governs the massless leading order terms of the elliptic genus, and which at the same time carries the expected representation of M_{24}, must be related to $\mathcal{X}$ by some nontrivial map. The Niemeier markings and the overarching maps which were constructed in [**TW11**] should be viewed as a first approach towards constructing such a map. This claim is based on the observation that, from a geometric viewpoint, the introduction of Niemeier markings is necessary to combine symmetry groups of Kummer surfaces into larger groups. Indeed, it follows from Mukai's results that for any finite group $\widehat{G}$ of lattice automorphisms of $H_*(X,\mathbb{Z})$ that is not a subgroup of one of the eleven maximal groups listed in [**Muk88**], the lattice $L_{\widehat{G}} := (H_*(X,\mathbb{Z})^{\widehat{G}})^\perp \cap H_*(X,\mathbb{Z})$ is indefinite and thus violates the signature requirements for symmetries of K3 surfaces. Therefore, we never expected M_{24} to act on $H_*(X,\mathbb{Z})$ either. It would be interesting to see if the massive sector of the elliptic genus is also subject to a 'no-go theorem' when working in the framework of $\mathbb{Z}_2$-orbifold CFTs on K3. A priori, the situation could be different, as the original Mathieu Moonshine observation [**EOT11**] states that in the elliptic genus, the multiplicities of *massive* characters of the $N=4$ superconformal algebra yield dimensions of representations of M_{24}. In a forthcoming work [**TW13**] we present evidence in favour of our expectation that the *massive* fields which contribute to the elliptic genus are related to a representation of M_{24} in a much more immediate fashion.

We now return to the 'bottom-up' approach, and investigate more closely the action of symmetry groups on the vertex algebra $\mathcal{X}$, to explain in terms of CFT data how our Niemeier markings and overarching maps are relevant in the context of SCFTs on K3. To this end note that the entire group $\mathrm{SL}(2,\mathbb{C})$ acts naturally on the truncated vertex algebra $\mathbb{C}\otimes\mathcal{X}$ of (A.1), preserving $U(1)$-charges. However, a given element of $\mathrm{SL}(2,\mathbb{C})$ may not have an extension to a symmetry of the full SCFT $\mathcal{C}$. Whether or not this is the case cannot be determined from the action on the fields listed in (A.1). Indeed, this depends on the moduli of $\mathcal{C}$, but the vertex algebra $\mathcal{X}$ has lost its dependence on all moduli due to the truncation, as described earlier. However, as we explain in Appendix A, one may introduce the analog $\mathcal{X}^{\mathbb{Z}}$ of the lattice of integral homology in the vector space $\mathcal{X}$, and use its structure to determine whether or not an element of $\mathrm{SL}(2,\mathbb{C})$ acts as a symmetry of $\mathcal{C}$.

By the above, we are only interested in symmetry groups G that are induced by geometric symmetries, and in line with our work so far, we restrict our attention to those that are induced[13] from the underlying toroidal CFT $\mathcal{T}$. By definition, a symmetry of a SCFT must be compatible with all OPEs in that theory. In particular, the standardized OPE (A.4) must be preserved. Following the arguments presented in Appendix A, this implies that each of our symmetry groups G acts as a group of lattice automorphisms on $\mathcal{X}^{\mathbb{Z}}$, such that this lattice of fields in our SCFT contains a sublattice $\mathcal{X}_G^{\mathbb{Z}}$ which bears all relevant information about the G-action on our SCFT. This lattice can be identified with the lattice M_G which is central to our construction, in that our Niemeier markings isometrically replicate it as a sublattice of the Niemeier lattice $N(-1)$. This allows a more elegant description of G as a subgroup of M_{24}, and it enables us to combine the symmetry groups from distinct K3 theories to a larger, overarching group. In other words, our Niemeier marking describes precisely the action of geometric symmetry groups on the vertex algebra which governs the elliptic genus to leading order terms. This justifies the relevance of our construction in the context of our quest to unravel some of the mysteries of the Mathieu Moonshine phenomenon.

The picture that we offer here shows how the beautiful interplay between geometry and conformal field theory may yield some keys to the Mathieu Moonshine Mysteries. Such an interplay is expected. On the one hand, the elliptic genus is a purely geometric quantity. On the other hand, this quantity also appears in the context of SCFTs on K3, where its decomposition into $N=4$ characters is natural. Notably, it is only after decomposing the elliptic genus into $N=4$ characters that one observes the Mathieu Moonshine phenomenon [**EOT11**].

We expect that order by order, the elliptic genus dictates the construction of representations of M_{24} on appropriately truncated vertex algebras arising from SCFTs on K3. In other words, the very representations of M_{24} that are observed in the elliptic genus are intrinsic to these SCFTs. The reason why the emerging group is M_{24} is still unclear, but we expect it to be rooted in the structure of these SCFTs, where geometry dictates the symmetries which can act on these representations. By symmetry-surfing the moduli space of SCFTs on K3, we expect that the natural representations of geometric symmetry groups on these vertex algebras combine to the action of M_{24}.

[13]This includes the symmetries induced by half lattice shifts in the underlying toroidal theory $\mathcal{T}$.

Our construction of overarching maps in [**TW11**] should be viewed as a very first step towards finding such vertex algebras for the leading order terms of the elliptic genus. In the present work, we show that our overarching maps indeed allow us to combine all relevant symmetry groups, as long as we restrict to $\mathbb{Z}_2$-orbifold conformal field theories on K3 and their geometric interpretations on Kummer K3s, and to symmetries that are induced geometrically from the underlying toroidal theories. Indeed, since one can easily associate a vertex algebra to the Niemeier lattice N, one could claim that we have solved the problem of constructing a vertex algebra that furnishes the expected symmetries. However, of course we pay dearly since this vertex algebra does not govern the leading order terms of the elliptic genus in any obvious way. Still our approach paves the way to defining the desired vertex algebra. As we have explained above, we expect vertex algebras associated with all remaining orders of the elliptic genus to relate directly to the respective representations of M_{24}, and we present evidence in favour of this expectation in [**TW13**].

Appendix A. Transition to superconformal field theory

Throughout our work, we use homological data to describe geometric symmetries of K3 surfaces. This is natural, since the techniques are well-established in algebraic geometry, but also since the well-known properties of (chiral, chiral) algebras [**Wit82, LVW89**] recover (co)homological data from sigma model interpretations of SCFTs. This is particularly straightforward for the $\mathbb{Z}_2$-orbifold conformal field theories which are relevant to our investigations. Since our work is motivated by Mathieu Moonshine [**EOT11**], which is rooted in conformal field theory, and since the role of the *integral* (co)homology in (chiral, chiral) algebras seems not so well established, we gather in this appendix the tools needed to make a smooth transition to superconformal field theory.

We first need to fix some notations. Every toroidal conformal field theory $\mathcal{T}$ possesses two free Dirac fermions on the holomorphic side, which we denote by $\chi_+^1(z)$, $\chi_+^2(z)$. The complex conjugate fields are denoted $\chi_-^1(z)$, $\chi_-^2(z)$, such that

$$\chi_+^i(z)\chi_-^j(w) \sim \frac{\delta_{ij}}{z-w}, \qquad i,j \in \{1,2\},$$

while the antiholomorphic counterparts are denoted $\overline{\chi}_\pm^1(\overline{z})$, $\overline{\chi}_\pm^2(\overline{z})$. The corresponding holomorphic - antiholomorphic combinations are more appropriate for our purposes,

$$\xi_1 := \frac{1}{2}\left(\chi_+^1 + \overline{\chi}_+^1\right), \quad \xi_2 := \frac{1}{2i}\left(\chi_+^1 - \overline{\chi}_+^1\right), \quad \xi_3 := \frac{1}{2}\left(\chi_+^2 + \overline{\chi}_+^2\right), \quad \xi_4 := \frac{1}{2i}\left(\chi_+^2 - \overline{\chi}_+^2\right).$$

Moreover, in every $\mathbb{Z}_2$-orbifold conformal field theory $\mathcal{C} = \mathcal{T}/\mathbb{Z}_2$ on K3, there is a 16-dimensional space of twisted ground states, generated by fields $T_{\vec{a}}$ in the Ramond-Ramond sector, where the label $\vec{a} \in \mathbb{F}_2^4$ refers to the fixed point $\vec{F}_{\vec{a}}$ as in (1.1) at which the respective field is localized. For ease of notation we denote by $\widetilde{T}_{\vec{a}}$, $\vec{a} \in \mathbb{F}_2^4$, the (chiral, chiral) fields which the $T_{\vec{a}}$ flow to under our choice of spectral flow. Then the following is a list of 24 fields which generate the (chiral, chiral) algebra in every theory $\mathcal{C} = \mathcal{T}/\mathbb{Z}_2$ on K3:

$$\text{(A.1)} \qquad \xi_1\xi_2\xi_3\xi_4, \quad \xi_i\xi_j \ (1 \leq i < j \leq 4), \quad \mathbb{1}; \quad \widetilde{T}_{\vec{a}} \ (\vec{a} \in \mathbb{F}_2^4),$$

where $\mathbb{1}$ denotes the vacuum field, and where we may restrict our attention to the real vector space $\mathcal{X}$ generated by these 24 fields. After truncation of the OPE to chiral primaries [**LVW89**], the fields listed in (A.1) form a closed vertex algebra $\mathcal{X}$ over $\mathbb{R}$. Note that this very truncation makes the vertex algebra completely independent of all moduli.

We remark that the real and imaginary parts[14] of the four fields with $U(1)$-charges $(2,2)$, $(2,0)$, $(0,2)$, $(0,0)$ in (A.1),

$$\xi_1\xi_2\xi_3\xi_4,\ \xi_1\xi_3-\xi_2\xi_4,\ \xi_1\xi_4+\xi_2\xi_3,\ \mathbb{1}, \tag{A.2}$$

remain invariant under every symmetry of $\mathcal{C}$. These fields are naturally identified with the cycles $\pi_*\upsilon^T$, Ω_1, Ω_2, $\pi_*\upsilon_0^T\in\pi_*H_*(T,\mathbb{R})$ on K3, with Ω_1, Ω_2 as in (1.2) and υ^T, υ_0^T generators of $H_4(T,\mathbb{Z})$ and $H_0(T,\mathbb{Z})$ such that $\langle\upsilon^T,\upsilon_0^T\rangle=1$. The invariance of Ω_1, Ω_2 under symmetries means that in a given geometric interpretation, one restricts attention to symplectic automorphisms (see [**TW11**] for further details). In the description of the moduli space of SCFTs on a K3 surface X of [**AM94**], our SCFT $\mathcal{C}$ is specified by the relative position of a positive definite fourplane in $H^*(X,\mathbb{R})$ with respect to $H^*(X,\mathbb{Z})$. The two-forms Ω_1, Ω_2 generate a two-dimensional subspace of that fourplane, while the choice of υ^T and υ_0^T amounts to the choice of a geometric interpretation of the toroidal theory $\mathcal{T}$ which induces a natural geometric interpretation of its $\mathbb{Z}_2$-orbifold $\mathcal{C}$ (see [**NW01, Wen01**]). Here, the four fields in (A.2) are the real and imaginary parts of the images of the four charged Ramond-Ramond ground states under our choice of spectral flow. These four Ramond-Ramond ground states also furnish a fourplane that can be used to describe the moduli space of superconformal field theories on K3 [**NW01**]. Note however that the fourplane of [**AM94**] is not the one generated by the four vectors in (A.2).

The vector space $\mathcal{X}$ can be identified with the real K3 homology $H_*(X,\mathbb{R})$, where the $\xi_i\xi_j$ with $1\leq i<j\leq 4$ are mapped to our generators $e_i\vee e_j$ of $\pi_*H_2(T,\mathbb{R})$, and the $\widetilde{T}_{\vec{a}}$ are in 1: 1 correspondence with the cycles $E_{\vec{a}}$ that arise from the minimal resolution of $T/\mathbb{Z}_2$ (see [**NW03**] for the subtleties in this identification, due to the B-field that is induced by orbifolding).

One may, in addition, introduce the analog of the lattice of integral homology for the vector space $\mathcal{X}$, thereby recovering the dependence on the moduli. To appreciate this, note that before truncation the OPE between twist fields $T_{\vec{b}}$ and $T_{\vec{b}'}$ with $\vec{b}$, $\vec{b}'\in\mathbb{F}_2^4$ yields, to leading order, a primary field $W_{\vec{b}-\vec{b}'}(z,\overline{z})$ which does depend on the moduli. This is best measured by means of the OPE between the free bosonic superpartners of the Dirac fermions $\xi_1,\dots,\xi_4$ and $W_{\vec{b}-\vec{b}'}(z,\overline{z})$. For convenience of notation, we introduce real, holomorphic $U(1)$-currents $j_1(z),\dots,j_4(z)$, which arise as the superpartners of the real and the imaginary parts of $2\chi_+^1(z)$, $2\chi_+^2(z)$, respectively, and note that the relevant OPE then has the form

$$j_k(z)W_{\vec{a}}(w,\overline{w})\sim\frac{W_{\vec{a}}(w,\overline{w})}{z-w}\sum_{l=1}^{4}a_l\lambda_k^l\quad\text{ for }\vec{a}=(a_1,\dots,a_4)\in\mathbb{F}_2^4.$$

Here, $\lambda_1^l,\dots,\lambda_4^l$ are the Euclidean coordinates of generators $\vec{\lambda}_1,\dots,\vec{\lambda}_4$ of a rank 4 lattice $\Lambda\subset\mathbb{R}^4$, if the underlying toroidal SCFT $\mathcal{T}$ has a geometric interpretation on the torus $T=\mathbb{R}^4/\Lambda$, where we identify $\mathbb{R}^4$ with $\mathbb{C}^2$ as usual. We observe that in the

[14] Here and in the following, for a field $\eta\in\mathbb{C}\otimes\mathcal{X}$ with $\eta=\eta_1+i\eta_2$ and η_1, $\eta_2\in\mathcal{X}$, we call η_1, η_2 the real and the imaginary part of η.

truncation procedure yielding the (chiral, chiral) algebra $\mathcal{X}$ of (A.1), the moduli-dependent fields $W_{\vec{a}}(z,\overline{z})$ are projected to zero, and therefore the dependence on the moduli disappears from $\mathcal{X}$. However, one may introduce new fields

$$J_k(z) := \sum_{l=1}^{4} \mu_k^l j_l(z) \qquad \text{for } k \in \{1,\dots,4\}, \tag{A.3}$$

where $\mu_1^l,\dots,\mu_4^l$ are the Euclidean coordinates of the basis $\vec{\mu}_1,\dots,\vec{\mu}_4$ dual[15] to $\vec{\lambda}_1,\dots,\vec{\lambda}_4$, such that the OPEs with the fields $W_{\vec{a}}(w,\overline{w})$ take the standardized "integral" form

$$J_k(z)W_{\vec{a}}(w,\overline{w}) \sim \frac{a_k}{z-w}W_{\vec{a}}(w,\overline{w}), \qquad k \in \{1,\dots,4\}. \tag{A.4}$$

The fermionic superpartners $\widetilde{\Psi}_1(z),\dots,\widetilde{\Psi}_4(z)$ of the new fields $J_1(z),\dots,J_4(z)$ and their antiholomorphic counterparts $\overline{\Psi}_1(\overline{z}),\dots,\overline{\Psi}_4(\overline{z})$ yield a lattice with generators

$$\widetilde{\Psi}_1\widetilde{\Psi}_2\widetilde{\Psi}_3\widetilde{\Psi}_4,\ \widetilde{\Psi}_k\widetilde{\Psi}_l\overline{\Psi}_m\overline{\Psi}_n,\ \widetilde{\Psi}_k\overline{\Psi}_l,\dots$$

over $\mathbb{Z}$. However, to determine a lattice which plays the role of the integral homology of the Kummer surface X, one needs to recall that the identification[16] of $\mathcal{X}$ with $H^*(X,\mathbb{R})$ rests on the correspondence $\chi_+^k \leftrightarrow dz_k,\ ,\overline{\chi}_+^k \leftrightarrow d\overline{z}_k$ for $k\in\{1,\,2\}$, with local holomorphic coordinates z_1, z_2 on X. This correspondence holds exactly on flat manifolds and in a large radius limit [**Wit82, LVW89**]. Hence at large radii, the *real*[17] fermionic fields $\widetilde{\Psi}_k$ are identified with the $\overline{\Psi}_k$, and thus with

$$\Psi_k := \sum_{l=1}^{4} \mu_k^l \xi_l \qquad \text{for } k \in \{1,\dots,4\}.$$

This leaves us with the lattice $\mathcal{Y}^{\mathbb{Z}}$ generated over $\mathbb{Z}$ by

$$\Psi_1\Psi_2\Psi_3\Psi_4, \quad \Psi_i\Psi_j,\ (1\le i<j\le 4), \quad \mathbb{1},$$

which is the analog of the lattice $\pi_*H_*(T,\mathbb{Z}) \subset H_*(X,\mathbb{R})$. Using the twist fields $\widetilde{T}_{\vec{a}}, \vec{a}\in\mathbb{F}_2^4$, as additional generators that correspond to the vectors $E_{\vec{a}}$ in the Kummer lattice, and then performing the usual gluing procedure, one obtains a lattice $\mathcal{X}^{\mathbb{Z}}$ which can be identified with $H_*(X,\mathbb{Z}) \subset H_*(X,\mathbb{R})$. In particular, the relative position of $\mathcal{X}^{\mathbb{Z}}$ with respect to the basis (A.1) of $\mathcal{X}$ determines the respective point in the moduli space. For the SCFT associated with the square Kummer surface[18], we can choose the eight fields $\xi_1\xi_2\xi_3\xi_4$, $\xi_i\xi_j$ $(1\le i<j\le 4)$, $\mathbb{1}$ as generators of the lattice $\mathcal{Y}^{\mathbb{Z}}$.

Now note that each symmetry of a Kummer surface X_{Λ,ω_0} as studied in our work induces a symmetry of a SCFT $\mathcal{C}=\mathcal{T}/\mathbb{Z}_2$, with $\mathcal{T}$ a toroidal theory[19] associated with the torus $\mathbb{R}^4/\Lambda$. By construction, our geometric symmetry groups G

[15] Here, we identify $\mathbb{R}^4 \cong (\mathbb{R}^4)^*$ by means of the standard Euclidean scalar product.

[16] From [**LVW89**], we obtain an immediate identification with cohomology, which however is equivalent to homology by Poincaré duality.

[17] For open strings, one can view χ_+^k and its antiholomorphic partner $\overline{\chi}_+^k$ as complex conjugates, where the left and right modes combine into standing waves. In this language, we are simply reviewing the emergence of charge lattices for D-branes.

[18] with vanishing B-field on the underlying toroidal theory

[19] This leaves a choice of the B-field $B_{\mathcal{T}}$ in the toroidal theory $\mathcal{T}$, which must be invariant under our symmetry; of course $B_{\mathcal{T}}=0$ is always admissible.

enjoy an induced action as groups of lattice automorphisms on the lattice $\mathcal{X}^{\mathbb{Z}}$. By definition, the symmetries of a CFT are compatible with all OPEs, hence they must in particular leave the standardized OPEs (A.4) invariant. Since our symmetries are induced by geometric symmetries of the toroidal theory $\mathcal{T}$, they act linearly on the $J_k(z)$ and they permute the fields $\pm W_{\vec{a}}(z, \overline{z})$. It follows that such symmetries act as lattice automorphisms on the lattice generated by the $J_k(z)$. The same thus holds for the lattice generated by their superpartners $\Psi_k(z)$ and for the lattice $\mathcal{Y}^{\mathbb{Z}}$ mentioned above. Since our symmetries also permute the twist fields $\pm\widetilde{T}_{\vec{a}}$ amongst each other in a manner compatible with gluing, altogether it follows that they must act as automorphisms of the lattice $\mathcal{X}^{\mathbb{Z}}$. By the above, the vector space $\mathcal{X}$ can be identified with the K3 homology, and $\mathcal{X}^{\mathbb{Z}}$ can be identified with the integral homology. In particular, the lattice $\mathcal{X}^{\mathbb{Z}}$ possesses a sublattice $\mathcal{X}^{\mathbb{Z}}_G$ which can be identified with the lattice M_G that is so crucial to our construction, see Def. 2.2. The action of G on $\mathcal{X}^{\mathbb{Z}}_G$ bears all relevant information about the G-action on our SCFT. Our construction hence realizes the very representation of G on $\mathcal{X}$ in terms of the action of a subgroup G of M_{24} on the Niemeier lattice N. In other words, our Niemeier marking describes precisely the action of the relevant symmetry groups on the (chiral, chiral) algebra.

Certainly from the description of the moduli space of SCFTs in terms of cohomological data [**AM94**, **NW01**], we are lead to expect that the role of the (chiral, chiral) algebra $\mathcal{X}$ along with the lattice $\mathcal{X}^{\mathbb{Z}}$ in its underlying vector space generalizes to arbitrary K3 theories.

References

[AKMW87] O. Alvarez, T. P. Killingback, M. Mangano, and P. Windey, *String theory and loop space index theorems*, Comm. Math. Phys. **111** (1987), no. 1, 1–10. MR896755 (88j:58119)

[AM94] P. S. Aspinwall and D. R. Morrison, *String theory on K3 surfaces*, Mirror symmetry, II, AMS/IP Stud. Adv. Math., vol. 1, Amer. Math. Soc., Providence, RI, 1997, pp. 703–716. MR1416354 (97i:81128)

[BL00] L. A. Borisov and A. Libgober, *Elliptic genera of toric varieties and applications to mirror symmetry*, Invent. Math. **140** (2000), no. 2, 453–485, DOI 10.1007/s002220000058. MR1757003 (2001j:58037)

[Bor01] L. A. Borisov, *Vertex algebras and mirror symmetry*, Comm. Math. Phys. **215** (2001), no. 3, 517–557, DOI 10.1007/s002200000312. MR1810943 (2002f:17046)

[Bri98] M. Brightwell, *Lattices and Automorphisms of Compact Complex Manifolds*, PhD Thesis (1998); `http://eleanor.lib.gla.ac.uk/ record=b181046/`.

[Che10] M. C. N. Cheng, *K3 surfaces, $\mathcal{N} = 4$ dyons and the Mathieu group M_{24}*, Commun. Number Theory Phys. **4** (2010), no. 4, 623–657, DOI 10.4310/CNTP.2010.v4.n4.a2. MR2793423 (2012e:11076)

[Cox74] H. S. M. Coxeter, *Regular complex polytopes*, Cambridge University Press, London-New York, 1974. MR0370328 (51 #6555)

[EH11] T. Eguchi and K. Hikami, *Note on twisted elliptic genus of K3 surface*, Phys. Lett. B **694** (2011), no. 4-5, 446–455, DOI 10.1016/j.physletb.2010.10.017. MR2748168 (2012g:58044)

[EOT11] T. Eguchi, H. Ooguri, and Y. Tachikawa, *Notes on the K3 surface and the Mathieu group M_{24}*, Exp. Math. **20** (2011), no. 1, 91–96, DOI 10.1080/10586458.2011.544585. MR2802725 (2012e:58039)

[EOTY89] T. Eguchi, H. Ooguri, A. Taormina, and S.-K. Yang, *Superconformal algebras and string compactification on manifolds with* SU(n) *holonomy*, Nuclear Phys. B **315** (1989), no. 1, 193–221, DOI 10.1016/0550-3213(89)90454-9. MR985505 (90i:81114)

[FS07] E. Frenkel and M. Szczesny, *Chiral de Rham complex and orbifolds*, J. Algebraic Geom. **16** (2007), no. 4, 599–624, DOI 10.1090/S1056-3911-07-00466-3. MR2357685 (2009i:14019)

[Fuj88] A. Fujiki, *Finite automorphism groups of complex tori of dimension two*, Publ. Res. Inst. Math. Sci. **24** (1988), no. 1, 1–97, DOI 10.2977/prims/1195175326. MR944867 (90f:32031)

[Gan12] T. Gannon, *Much ado about Mathieu*; `arXiv:1211.5531 [math.RT]`.

[GHV10a] M. R. Gaberdiel, S. Hohenegger, and R. Volpato, *Mathieu Moonshine in the elliptic genus of $K3$*, J. High Energy Phys. **10** (2010), 062, 24, DOI 10.1007/JHEP10(2010)062. MR2780524 (2012h:58027)

[GHV10b] M. R. Gaberdiel, S. Hohenegger, and R. Volpato, *Mathieu twining characters for K3*, JHEP **1009** (2010), 058; `arXiv:1006.0221 [hep-th]`.

[GPRV12] M.R. Gaberdiel, D. Persson, H. Ronellenfitsch, and R. Volpato, *Generalised Mathieu Moonshine*; `arXiv:1211.7074 [hep-th]`.

[GPV13] M.R. Gaberdiel, D. Persson, and R. Volpato, *Generalised Moonshine and Holomorphic Orbifolds*; `arXiv:1302.5425 [hep-th]`.

[GV12] M.R. Gaberdiel and R. Volpato, *Mathieu moonshine and orbifold K3s*; `arXiv:1206.5143 [hep-th]`.

[Hir66] F. Hirzebruch, *Topological methods in algebraic geometry*, Third enlarged edition. New appendix and translation from the second German edition by R. L. E. Schwarzenberger, with an additional section by A. Borel. Die Grundlehren der Mathematischen Wissenschaften, Band 131, Springer-Verlag New York, Inc., New York, 1966. MR0202713 (34 #2573)

[Kap05] A. Kapustin, *Chiral de Rham complex and the half-twisted sigma-model*; `hep-th/0504074`.

[Kon98] S. Kondō, *Niemeier lattices, Mathieu groups, and finite groups of symplectic automorphisms of $K3$ surfaces*, Duke Math. J. **92** (1998), no. 3, 593–603, DOI 10.1215/S0012-7094-98-09217-1. With an appendix by Shigeru Mukai. MR1620514 (99i:14042)

[LVW89] W. Lerche, C. Vafa, and N. P. Warner, *Chiral rings in $N = 2$ superconformal theories*, Nuclear Phys. B **324** (1989), no. 2, 427–474, DOI 10.1016/0550-3213(89)90474-4. MR1025424 (91d:81132)

[MSV99] F. Malikov, V. Schechtman, and A. Vaintrob, *Chiral de Rham complex*, Comm. Math. Phys. **204** (1999), no. 2, 439–473, DOI 10.1007/s002200050653. MR1704283 (2000j:17035a)

[Muk88] S. Mukai, *Finite groups of automorphisms of $K3$ surfaces and the Mathieu group*, Invent. Math. **94** (1988), no. 1, 183–221, DOI 10.1007/BF01394352. MR958597 (90b:32053)

[Nik75] V.V. Nikulin, *On Kummer surfaces*, Math. USSR Isv. **9** (1975), 261–275.

[NW01] W. Nahm and K. Wendland, *A hiker's guide to $K3$. Aspects of $N = (4,4)$ superconformal field theory with central charge $c = 6$*, Comm. Math. Phys. **216** (2001), no. 1, 85–138, DOI 10.1007/PL00005548. MR1810775 (2002h:81235)

[NW03] W. Nahm and K. Wendland, *Mirror symmetry on Kummer type $K3$ surfaces*, Commun. Math. Phys. **243** (2003), 557–582; `hep-th/0106104`.

[TW11] A. Taormina and K. Wendland, *The overarching finite symmetry group of Kummer surfaces in the Mathieu group M_{24}*, J. High Energy Phys. **8** (2013), 125, front matter+62. MR3106313

[TW13] A. Taormina and K. Wendland, *A twist in the M_{24} Moonshine story*; `arXiv:1303.3221 [hep-th]`.

[Wen01] K. Wendland, *Consistency of orbifold conformal field theories on $K3$*, Adv. Theor. Math. Phys. **5** (2001), no. 3, 429–456. MR1898367 (2003h:81208)

[Wit82] E. Witten, *Constraints on supersymmetry breaking*, Nuclear Phys. B **202** (1982), no. 2, 253–316, DOI 10.1016/0550-3213(82)90071-2. MR668987 (84j:81131)

[Wit87] E. Witten, *Elliptic genera and quantum field theory*, Commun. Math. Phys. **109** (1987), 525–536.

Centre for Particle Theory, Department of Mathematical Sciences, Durham University, Durham, DH1 3LE, United Kingdom
E-mail address: anne.taormina@durham.ac.uk

Mathematics Institute, University of Freiburg, D-79104 Freiburg, Germany
E-mail address: katrin.wendland@math.uni-freiburg.de

Proceedings of Symposia in Pure Mathematics
Volume **90**, 2015
http://dx.doi.org/10.1090/pspum/090/01514

Secret Symmetries of AdS/CFT

Alessandro Torrielli

ABSTRACT. We review special quantum group symmetries underlying the integrability of the AdS/CFT spectral problem, with particular emphasis on the *secret* - or *bonus* - symmetry. This is a particular symmetry of the Yangian type, however not accounted for by the standard Yangian. This symmetry has been observed not only in the spectrum, but also in the presence of D-branes and integrable boundaries, in scattering amplitudes, in the pure spinor formalism and, recently, in the quantum affine deformation.

1. Introduction

Gauge theories dominate our current understanding of fundamental interactions. However, the derivation of exact results at large coupling constant is a challenging task. This obstructs our full understanding of certain nonperturbative phenomena like confinement.

The discovery of integrable structures in QCD [**50**] and, more recently, in planar $\mathcal{N} = 4$ Supersymmetric Yang-Mills theory and AdS/CFT [**56**] has changed this scenario. The $\mathcal{N} = 4$ theory is a quantum conformal field theory. Its spectral data are encoded in the singular behavior of 2-point functions of composite operators. A highly non-trivial operator-mixing renders the calculation of these functions extremely challenging. In [**56**] it was observed that, in the planar limit, the problem translates into the equivalent one of finding the spectrum of certain integrable spin-chain Hamiltonians [**63**]. This spectrum consists of spin-wave excitations whose S-matrix can be determined exactly [**15**]. It would be overwhelming to give here a comprehensive list of the relevant references. They can be found in many of the available reviews (just to mention some of the most recent ones, see [**8**,**17**]). The mapping to a two-dimensional model could provide the key to the long sought-for exact solution to an interacting four-dimensional quantum field theory, and a remarkable insight into the strongly-coupled regime of gauge theories.

Although an impressive progress has been made, several fundamental questions still remain open. Most importantly, no rigorous proof of integrability is available, and the quantum Hamiltonian of the system is not known in closed form. Moreover, the algebraic structure underlying the theory is still somewhat mysterious,

2010 *Mathematics Subject Classification.* Primary 81R50; Secondary 17B37, 16T25 .

Surrey Preprint nr. DMUS-MP-13/03. The author thanks the UK EPSRC for funding provided under grants EP/H000054/1 and EP/K014412/1 and the STFC (Science and Technology Facilities Councils) for support under the Consolidated Grant project nr. ST/L000490/1 “Fundamental Implications of Fields, Strings and Gravity”.

as further fascinating dualities are currently being observed in Wilson loops and n-point functions. These dualities often have Yangians as their main characters [**40**]. It is possible that all the Yangians observed so far (sigma model, spin-chain, magnons, spacetime n-point functions) may merge into a yet to be fully uncovered larger quantum group[1] all these different aspects of the integrability of the theory rely upon.

Hopf algebras and quantum groups provide a powerful mathematical framework where to reformulate integrable systems. The non-abelian Yangian algebra commuting with the Hamiltonian organizes the spectrum in terms of the corresponding irreducible representations. The S-matrix is largely fixed by symmetry and inherits very specific features [**33**, **36**]. The Yangian has been very useful to derive results which would have otherwise needed a perhaps unsurmountable amount of work, such as all bound state S-matrices [**6**], to open a fast route to the Bethe equations [**30**] without the need of an explicit diagonalization procedure [**5**] and to prove certain conjectures [**14**] crucial for the formulation of the so-called Thermodynamic Bethe Ansatz (TBA) and Y-system solution to the finize-size problem [**9**, **11**, **24**, **43**, **44**, **60**].

Ideally, one would like to apply this algebraic framework to quantise the two-dimensional string sigma model, a formidable problem whose solution is going to be essential for clarifying the duality between strings and nonperturbative gauge theories. Furthermore, the understanding of the novel quantum group underlying the problem is likely to represent a significant advance in the field of Algebra and its relationship with integrable systems. The situation in AdS/CFT is quite peculiar because of conformal invariance and the planar limit. However, one hopes that the understanding of even one single interacting four-dimensional gauge theory in this special limit will be important for progress in less symmetric cases as well.

Note. A few reviews concerning Yangians in AdS/CFT are already available in the literature, see for instance [**16**, **39**, **67**, **68**].

2. Hopf Algebras and Integrable Systems

In this section we connect ideas concerning integrable systems to the theory of Hopf algebras.

We will start with the Lie (super)algebra $\mathfrak{g}$ of symmetries of the system, and consider its universal enveloping algebra $A \equiv U(\mathfrak{g})$, containing a *unit element* $\mathbb{1}$ with respect to the multiplication $\mu : A \otimes A \to A$. One also defines a *unit map* $\eta : \mathbb{C} \to A$, which completes the algebraic definition of a single-particle system.

We then equip our algebra with two more maps consistent with the Lie (super)algebra structure, obtaining a *bialgebra*. One map is the *coproduct* $\Delta : A \to A \otimes A$, which describes how symmetries act on two-particle states. The other map is the *counit* $\epsilon : A \to \mathbb{C}$. The requirement of *coassociativity*

$$(\Delta \otimes \mathbb{1})\Delta = (\mathbb{1} \otimes \Delta)\Delta \tag{2.1}$$

allows to uniquely define the action of the coproduct on multi-particle states as $\Delta^n = \ldots(\Delta \otimes \mathbb{1} \otimes \mathbb{1})(\Delta \otimes \mathbb{1})\Delta$.

One more map is then needed to define a *Hopf algebra*: the *antipode* map $\Sigma : A \to A$, which allows to define antiparticles as conjugated representations of the symmetry algebra. The antipode has to satisfy $\Sigma(ab) = (-)^{ab}\Sigma(b)\Sigma(a)$ - where

[1] P. Etingof, private communication.

multiplication is defined by μ - and compatibility with the coproduct. If a bialgebra admits an antipode, it is unique.

In the scattering theory the coproduct acts on, say, *in* states. Consequently, the *opposite coproduct* $P\Delta \equiv \Delta^{op}$, with P the graded permutation map $P(a \otimes b) = (-)^{ab} b \otimes a$, will act on *out* states. For generic quantum groups these two actions are not the same. When they are, such as for the local (Leibniz) rule $\Delta(a) = a \otimes \mathbb{1} + \mathbb{1} \otimes a$ familiar from elementary Quantum Mechanics, one speaks of *cocommutative* Hopf algebras.

Even when different, Δ and Δ^{op} produce tensor product representations of the same dimension which may be related by conjugation *via* an invertible element, which is the S-matrix, or R-matrix in mathematical textbooks. The latter is defined via the condition

$$\Delta^{op}(J)\, R \;=\; R\, \Delta(J)$$

applied to any element J of A. The Hopf-algebra is in this case called *quasi-cocommutative* and, if the S-matrix satisfies the 'bootstrap' condition, *quasi-triangular*. The S-matrix must also be compatible with the antipode map via the *crossing* condition. A theorem of Drinfeld's states that the bootstrap condition implies that the S-matrix satisfies the Yang-Baxter equation and the crossing condition.

The language of Hopf algebras is therefore particularly suitable for dealing with integrable systems, where the scattering is reduced to an algebraic problem. The theory of Hopf algebras provides powerful theorems to simultaneously treat arbitrary representations (particle content) of the symmetry algebra. The notion of *universal R-matrix*, which solves the quasi-cocommutativity condition directly in terms of the generators of A, is very important in this respect. The study of the universal R-matrix can reveal some of the hidden symmetries the system. In the following figures 1, 2 and 3 we summarize the relationship between integrable scattering and Hopf algebras:

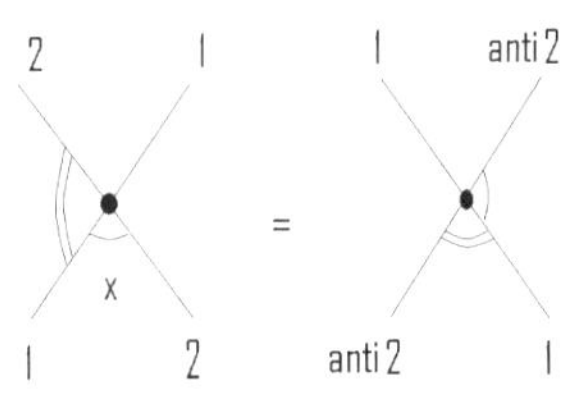

$$(\Sigma \otimes 1)R = R^{-1} = (1 \otimes \Sigma^{-1})R$$

FIGURE 1. Crossing relation. x is the difference in incoming rapidities, and it behaves like a hyperbolic angle.

2.1. The Hopf algebra of the AdS/CFT S-matrix. The algebra relevant to the AdS/CFT integrable system turns out to be the centrally-extended $\mathfrak{psl}(2|2)$ Lie superalgebra, which we call $\mathfrak{psl}(2|2)_c$. (Two copies of) this algebra are the residual symmetry upon the choice of a spin-chain vacuum [**15**], and upon fixing the gauge in the appropriate decompactification limit of the string sigma-model [**10**].

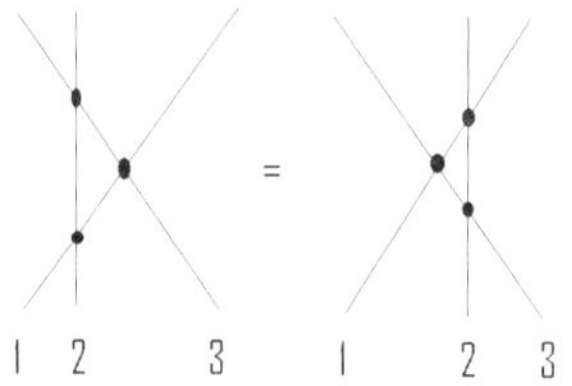

$$R_{12}R_{13}R_{23} = R_{23}R_{13}R_{12}$$

FIGURE 2. Yang-Baxter equation. It states the consistency of factorisation of multi-particle scattering.

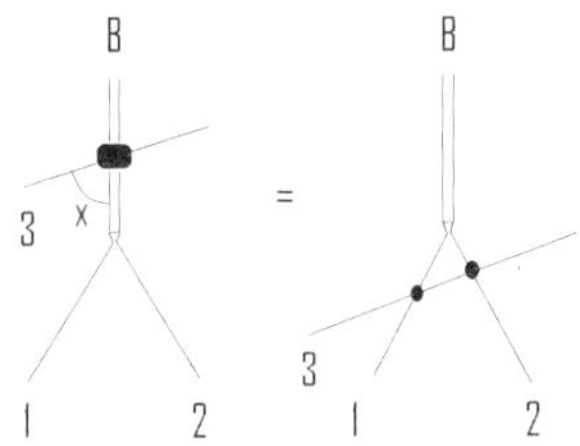

$$(\Delta \otimes 1)R = R_{13}\, R_{23}, \qquad (1 \otimes \Delta)R = R_{13}\, R_{12}.$$

FIGURE 3. Bootstrap condition. Scattering of a bound state of particles 1 and 2 from 3 factorises.

There are two sets of $\mathfrak{sl}(2)$ bosonic generators:

$$\begin{aligned} &[\mathbb{L}_1{}^1, \mathbb{L}_1{}^2] = 2\mathbb{L}_1{}^2, \qquad [\mathbb{L}_1{}^1, \mathbb{L}_2{}^1] = -2\mathbb{L}_2{}^1, \qquad [\mathbb{L}_1{}^2, \mathbb{L}_2{}^1] = \mathbb{L}_1{}^1, \\ &[\mathbb{R}_3{}^3, \mathbb{R}_3{}^4] = 2\mathbb{R}_3{}^4, \qquad [\mathbb{R}_3{}^3, \mathbb{R}_4{}^3] = -2\mathbb{R}_4{}^3, \qquad [\mathbb{R}_3{}^4, \mathbb{R}_4{}^3] = \mathbb{R}_3{}^3, \end{aligned} \tag{2.2}$$

while the commutation relations involving the supercharges $\mathbb{Q}$ and $\mathbb{G}$ are as follows:

$$\begin{aligned} &[\mathbb{L}_a{}^b, \mathbb{G}_c^\alpha] = \delta_c^b \mathbb{G}_a^\alpha - \tfrac{1}{2}\delta_a^b \mathbb{G}_c^\alpha, && [\mathbb{R}_\alpha{}^\beta, \mathbb{Q}_\gamma^a] = \delta_\gamma^\beta \mathbb{Q}_\alpha^a - \tfrac{1}{2}\delta_\alpha^\beta \mathbb{Q}_\gamma^A, \\ &[\mathbb{L}_a{}^b, \mathbb{Q}_\alpha^c] = -\delta_a^c \mathbb{Q}_\alpha^b + \tfrac{1}{2}\delta_a^b \mathbb{Q}_\alpha^c, && [\mathbb{R}_\alpha{}^\beta, \mathbb{G}_a^\gamma] = -\delta_\alpha^\gamma \mathbb{G}_a^\beta + \tfrac{1}{2}\delta_\alpha^\beta \mathbb{G}_a^\gamma, \\ &\{\mathbb{Q}_\alpha{}^a, \mathbb{Q}_\beta{}^b\} = \epsilon_{\alpha\beta}\epsilon^{ab}\mathbb{C}, && \{\mathbb{G}_a{}^\alpha, \mathbb{G}_b{}^\beta\} = \epsilon^{\alpha\beta}\epsilon_{ab}\mathbb{C}^\dagger, \\ &\{\mathbb{Q}_\alpha^a, \mathbb{G}_b^\beta\} = \delta_b^a \mathbb{R}_\alpha{}^\beta + \delta_\alpha^\beta \mathbb{L}_b{}^a + \tfrac{1}{2}\delta_b^a \delta_\alpha^\beta \mathbb{H}. \end{aligned} \tag{2.3}$$

The elements $\mathbb{H}$, $\mathbb{C}$ and $\mathbb{C}^\dagger$ are central. In unitary representations, $\mathbb{C}$ and $\mathbb{C}^\dagger$ are related by complex conjugation (as well as the supercharges $\mathbb{Q}$ and $\mathbb{G}$ in a suitable fashion).

The simple Lie superalgebra $A(1,1) \equiv \mathfrak{psl}(2|2)$ is the only basic classical simple Lie superalgebra which admits a three-dimensional central extension [**46**]. Leaving aside affine algebras, in fact, one has either no central extensions, or, for the series $A(n,n)$ with $n \neq 1$, a one-dimensional central extension to $\mathfrak{sl}(n+1|n+1)$. The latter coincides with the algebra of supertraceless $n+1|n+1 \times n+1|n+1$ matrices (in boson|fermion notation). In fact, the $n+1|n+1 \times n+1|n+1$-identity matrix is supertraceless.

One starts with an infinite spin-chain (asymptotic problem). The relevant representation in the spin-chain language is constructed out of the fundamental representation of $\mathfrak{psl}(2|2)_c$, but spin-chain sites can be created or destroyed by the action of the symmetry generators ("length-changing" or *dynamical* action). If we prepare the spin-chain in an eigenstate of momentum (magnon) $|p\rangle =$

$\sum_n e^{ipn} |\cdots Z Z \phi(n) Z \cdots\rangle$, Z being a complex combination of two out of the six real scalar fields in $\mathcal{N}=4$ SYM and ϕ (at position n) being one of the 4 possible states in the fundamental representation of $\mathfrak{psl}(2|2)_c$, the central charges act for instance schematically as

$$\text{(2.4)} \quad \mathbb{H}\,|p\rangle \;=\; \epsilon(p)\,|p\rangle, \qquad \mathbb{C}\,|p\rangle \;=\; c(p)\,|p\,Z^-\rangle, \qquad \mathbb{C}^\dagger\,|p\rangle \;=\; \bar{c}(p)\,|p\,Z^+\rangle,$$

where $Z^{+(-)}$ creates (destroys) one site to the right of the travelling excitation. More precisely, the length-changing is interpreted as a non-local modification of the coproduct [**41**, **58**]:

$$\mathbb{C}\otimes\mathbb{1}\,|p_1\rangle\otimes|p_2\rangle = \mathbb{C}\otimes\mathbb{1}\sum_{n_1<<n_2} e^{i\,p_1\,n_1+i\,p_2\,n_2}\,|\cdots Z\,Z\,\phi_1\,\underbrace{Z\cdots Z}_{n_2-n_1-1}\,\phi_2\,Z\cdots\rangle =$$

$$\text{(2.5)} \qquad (n_2\to n_2+1) \;=\; c(p_1)\,e^{ip_2}\,|p_1\rangle\otimes|p_2\rangle,$$

where the rescaling $n_2 \to n_2+1$ recasts the state into its original form with n_2-n_1 vacuum sites between the two excitations. In terms of the S-matrix, one has

$$\text{(2.6)} \qquad \mathrm{S}\,\Delta(\mathbb{C}) = \mathrm{S}\,[\mathbb{C}\otimes\mathbb{1}+\mathbb{1}\otimes\mathbb{C}] = \mathrm{S}\,[e^{ip_2}\mathbb{C}_{local}\otimes\mathbb{1}+\mathbb{1}\otimes\mathbb{C}_{local}],$$

where $\mathbb{C}_{local}$ is the *local* part of $\mathbb{C}$, acting as $\mathbb{C}_{local}|p\rangle = c(p)|p\rangle$, and analogously for $\Delta(\mathbb{C})\mathrm{S}$. As p_2 naturally pertains to particle 2, one reads off the non-trivial coproduct

$$\text{(2.7)} \qquad \Delta(\mathbb{C}_{local}) = \mathbb{C}_{local}\otimes e^{ip} + \mathbb{1}\otimes\mathbb{C}_{local}.$$

A consistent coproduct can be written for the whole $\mathfrak{psl}(2|2)_c$ algebra in terms of an additive quantum number $[[A]]$:

$$\text{(2.8)} \qquad \Delta(\mathfrak{J}^A) = \mathfrak{J}^A\otimes e^{i[[A]]p} + \mathbb{1}\otimes\mathfrak{J}^A, \qquad \Delta(e^{ip}) = e^{ip}\otimes e^{ip},$$

with the only non-zero values being $[[\mathbb{Q}]] = \frac{1}{2}$, $[[\mathbb{G}]] = -\frac{1}{2}$, $[[\mathbb{C}]] = 1$, $[[\mathbb{C}^\dagger]] = -1$. The antipode follows direclty from the Hopf algebra axioms as

$$\text{(2.9)} \qquad \Sigma(\mathfrak{J}^A) = -e^{-i[[A]]p}\mathfrak{J}^A, \qquad \Sigma(e^{ip}) = e^{-ip},$$

and is an idempotent map on $\mathfrak{psl}(2|2)_c$.

If we consider the invariance condition for the R-matrix $\Delta^{op}R = R\Delta$ under the coproduct we have just derived, we notice a peculiar feature connected to the presence of central elements. Since $\Delta(\mathbb{C})$ is central and R is invertible, one has

$$\text{(2.10)} \quad \Delta^{op}(\mathbb{C})\,R \;=\; R\,\Delta(\mathbb{C}) \;=\; \Delta(\mathbb{C})\,R \qquad \Longrightarrow \qquad \Delta^{op}(\mathbb{C}) \;=\; \Delta(\mathbb{C}).$$

One realises that (2.10) is equivalent to the physical requirement

$$\text{(2.11)} \qquad U^2 \equiv e^{ip}\,\mathbb{1} = \kappa\,\mathbb{C} \;+\; \mathbb{1}$$

for a constant κ depending on the Yang-Mills coupling g_{YM} [**15**]. Combining (2.11) with (2.8), one has (see also [**66**])

$$\text{(2.12)} \qquad \Delta(\mathbb{C}) = \mathbb{C}\otimes\mathbb{1}+\mathbb{1}\otimes\mathbb{C}+\kappa\,\mathbb{C}\otimes\mathbb{C} = \Delta^{op}(\mathbb{C}),$$

and similarly for $\mathbb{C}^\dagger$. This is equivalent to imposing that the total value of the central charges $\mathbb{C}$ and $\mathbb{C}^\dagger$ vanishes when the total momentum is set to zero: $\Delta(\mathbb{C}) = \Delta(\mathbb{C}^\dagger) = 0$ if $p_1+p_2=0$.

The S-matrix in the fundamental representation is completely fixed by invariance under (2.8), apart from an overall scalar factor, which is in turn constrained by crossing symmetry. From the Hopf-algebra antipode Σ one obtains the 'antiparticle' representation $\tilde{\mathfrak{J}}^A$, and the corresponding charge-conjugation matrix C:

$$\Sigma(\mathfrak{J}^A) = C^{-1}\,[\tilde{\mathfrak{J}}^A]^{st}\,C, \tag{2.13}$$

where M^{st} is the supertranspose of the matrix M. The original equation written by Janik [**47**] for the scalar factor then follows from the formula

$$(\Sigma\otimes\mathbb{1})\,R = (\mathbb{1}\otimes\Sigma^{-1})\,R = R^{-1} \tag{2.14}$$

applied to the antiparticle representation obtained from (2.13).

The coproduct (2.8) is consistent with the one derived from the semiclassical analysis of the dual string-theory sigma model [**10**, **49**]. The light-cone worldsheet supercharges contain a non-local contribution

$$\mathfrak{J}^A = \int_{-\infty}^{\infty} d\sigma\, J_0^A(\sigma)\, e^{i\,[[A]]\,\int_{-\infty}^{\sigma} d\sigma'\,\partial\chi^-(\sigma')}, \tag{2.15}$$

due to the fact that the light-cone field χ^- is not physical and one should rather use its derivative. If one considers two solitons well localized in $(-\infty,0)$ and $(0,\infty)$, respectively, the *semiclassical action* of the charges can roughly be written as (cf. [**51**, **52**])

$$\begin{aligned}
\mathfrak{J}^A{}_{|profile} &= \int_{-\infty}^{\infty} d\sigma\, J_0^A(\sigma)_{|profile}\, e^{i\,[[A]]\,\int_{-\infty}^{\sigma} d\sigma'\,\partial\chi^-(\sigma')_{|profile}}\\
&= \int_{-\infty}^{0} d\sigma\, J_0^A(\sigma) e^{i\,[[A]]\,\int_{-\infty}^{\sigma} d\sigma'\,\partial\chi^-(\sigma')} +\\
&\qquad \int_{0}^{\infty} d\sigma\, J_0^A(\sigma)\, e^{i\,[[A]]\,\int_{-\infty}^{0} d\sigma'\,\partial\chi^-(\sigma')}\, e^{i\,[[A]]\,\int_{0}^{\sigma} d\sigma'\,\partial\chi^-(\sigma')}\\
&\sim\ \mathfrak{J}_1^A + e^{i[[A]]\,p_1}\mathfrak{J}_2^A \quad\longrightarrow\quad \Delta(\mathfrak{J}^A) = \mathfrak{J}^A\otimes\mathbb{1} + e^{i[[A]]p}\otimes\mathfrak{J}^A,
\end{aligned}$$

where one has used the definition of the worldsheet momentum in terms of the field χ^- applied to the first excitation. This coproduct is then related to (2.8) via a non-local transformation.

2.2. Yangian symmetry. The AdS/CFT S-matrix possesses a further extended symmetry of the Yangian type. Let us begin by defining the Yangian $\mathcal{Y}(\mathfrak{g})$ of a simple Lie algebra $\mathfrak{g}$ in *Drinfeld's first* and *second* (isomorphic) realisations[2]. The first realisation was originally given by [**37**] and is very natural from the spin-chain perspective [**23**]. The second realisation [**38**] is particularly suitable for deriving the universal R-matrix [**48**]. We will not treat the *RTT* realisation.

2.3. Drinfeld's first realisation. The Yangian $\mathcal{Y}(g)$ is a deformation of the universal enveloping algebra of the loop algebra $\mathfrak{g}[u]$ associated to $\mathfrak{g}$. For the sake of illustration, we take $\mathfrak{g}$ to be a finite dimensional simple Lie algebra, with generators $\mathfrak{J}^A$ and commutation relations $[\mathfrak{J}^A,\mathfrak{J}^B] = f_C^{AB}\mathfrak{J}^C$, endowed with a non-degenerate invariant consistent supersymmetric bilinear form κ^{AB}, such as the Killing form $\kappa^{AB} = f_D^{AC} f_C^{BD}$. The Yangian is generated by level-zero $\mathfrak{J}^A$ (forming $\mathfrak{g}$) and level-one generators $\widehat{\mathfrak{J}}^A$ satisfying

$$[\mathfrak{J}^A,\mathfrak{J}^B] = f_C^{AB}\mathfrak{J}^C, \qquad [\mathfrak{J}^A,\widehat{\mathfrak{J}}^B] = f_C^{AB}\widehat{\mathfrak{J}}^C. \tag{2.16}$$

[2]The reader is referred to the standard literature (see for example [**53**, **57**]) for a treatment of this subject. For the generalisation to simple Lie superalgebras, see for instance [**25**, **26**, **42**, **65**, **69**].

The generators of higher levels are defined recursively by subsequent commutation, constrained by the Serre relations (for $\mathfrak{g} \neq \mathfrak{sl}(2)$):
(2.17)

$$[\widehat{\mathfrak{J}}^A, [\widehat{\mathfrak{J}}^B, \mathfrak{J}^C]] + [\widehat{\mathfrak{J}}^B, [\widehat{\mathfrak{J}}^C, \mathfrak{J}^A]] + [\widehat{\mathfrak{J}}^C, [\widehat{\mathfrak{J}}^A, \mathfrak{J}^B]] = \frac{1}{4} f_D^{AG} f_E^{BH} f_F^{CK} f_{GHK} \mathfrak{J}^{\{D} \mathfrak{J}^E \mathfrak{J}^{F\}}.$$

Curly brackets indicate complete symmetrization of indices, while index raising and lowering is done via κ^{AB} and its inverse, respectively. For $\mathfrak{sl}(2)$ the above Serre relations trivialise and are substituted by others (see [**53**]).

The Yangian is invariant under the *shift* automorphism

$$\mathfrak{J}^A \to \mathfrak{J}^A, \qquad \widehat{\mathfrak{J}}^A \to \widehat{\mathfrak{J}}^A + c\,\mathfrak{J}^A, \tag{2.18}$$

with c a constant. It is also a Hopf algebra, with the coproduct being uniquely determined by

$$\Delta(\mathfrak{J}^A) = \mathfrak{J}^A \otimes \mathbb{1} + \mathbb{1} \otimes \mathfrak{J}^A, \qquad \Delta(\widehat{\mathfrak{J}}^A) = \widehat{\mathfrak{J}}^A \otimes \mathbb{1} + \mathbb{1} \otimes \widehat{\mathfrak{J}}^A + \frac{1}{2} f_{BC}^A \mathfrak{J}^B \otimes \mathfrak{J}^C. \tag{2.19}$$

Antipode and counit are easily derived from the Hopf algebra axioms.

2.4. Drinfeld's second realisation. Drinfeld's second realisation of $\mathcal{Y}(\mathfrak{g})$ is given in terms of generators $\kappa_{i,m}, \xi^{\pm}_{i,m}$, $i = 1, \dots, \mathrm{rank}\mathfrak{g}$, $m = 0, 1, 2, \dots$, and relations

$$\begin{gathered}
[\kappa_{i,m}, \kappa_{j,n}] = 0, \quad [\kappa_{i,0}, \xi^{\pm}_{j,m}] = \pm a_{ij}\, \xi^{+}_{j,m}, \quad [\xi^{+}_{j,m}, \xi^{-}_{j,n}] = \delta_{i,j}\, \kappa_{j,n+m},\\
[\kappa_{i,m+1}, \xi^{\pm}_{j,n}] - [\kappa_{i,m}, \xi^{\pm}_{j,n+1}] = \pm\tfrac{1}{2} a_{ij} \{\kappa_{i,m}, \xi^{\pm}_{j,n}\},\\
[\xi^{\pm}_{i,m+1}, \xi^{\pm}_{j,n}] - [\xi^{\pm}_{i,m}, \xi^{\pm}_{j,n+1}] = \pm\tfrac{1}{2} a_{ij} \{\xi^{\pm}_{i,m}, \xi^{\pm}_{j,n}\},\\
i \neq j, \;\; n_{ij} = 1 + |a_{ij}|, \quad Sym_{\{k\}} [\xi^{\pm}_{i,k_1}, [\xi^{\pm}_{i,k_2}, \dots [\xi^{\pm}_{i,k_{n_{ij}}}, \xi^{\pm}_{j,l}] \dots]] = 0.
\end{gathered} \tag{2.20}$$

In the above, a_{ij} is the Cartan matrix, which we assume to be symmetric and integer valued.

In the second realisation, one can very easily generate higher levels via a systematic procedure in any specific representation. One can take for instance the relations

$$\begin{gathered}
[\kappa_{i,m}, \kappa_{j,n}] = 0, \qquad [\kappa_{i,0}, \xi^{\pm}_{j,m}] = \pm a_{ij}\, \xi^{\pm}_{j,m},\\
[\kappa_{i,m+1}, \xi^{\pm}_{j,n}] - [\kappa_{i,m}, \xi^{\pm}_{j,n+1}] = \pm\frac{1}{2} a_{ij} \{\kappa_{i,m}, \xi^{\pm}_{j,n}\},
\end{gathered} \tag{2.21}$$

and assume one knows the entire level 0 and 1 representation. By choosing i and j such that $a_{ij} \neq 0$, one then obtains

$$\begin{aligned}
[\kappa_{i,0+1}, \xi^{\pm}_{j,1}] - [\kappa_{i,0}, \xi^{\pm}_{j,1+1}] &= \pm\frac{1}{2} a_{ij} \{\kappa_{i,0}, \xi^{\pm}_{j,1}\} \;=\; [\kappa_{i,1}, \xi^{\pm}_{j,1}] \mp a_{ij}\, \xi^{\pm}_{j,2}\\
&= \pm\frac{1}{2} a_{ij} \{\kappa_{i,0}, \xi^{\pm}_{j,1}\},
\end{aligned}$$

from which it is straightforward to extract $\xi^{\pm}_{j,2}$.

2.5. The Yangian of the AdS/CFT S-matrix. The $\mathfrak{psl}(2|2)_c$ Yangian symmetry of the AdS/CFT S-matrix in the fundamental representation has been obtained by explicit computation in [**13**]:

$$\Delta^{op}(\widehat{\mathbb{J}})\, R = R\, \Delta(\widehat{\mathbb{J}}). \tag{2.22}$$

To be a Lie algebra homomorphism, the coproduct has to respect (2.16) and therefore has to take into account the deformation in (2.8):

$$\Delta(\widehat{\mathfrak{J}}^A) = \widehat{\mathfrak{J}}^A \otimes \mathbb{1} + U^{[[A]]} \otimes \widehat{\mathfrak{J}}^A + \frac{1}{2} f^A_{BC}\, \mathfrak{J}^B\, U^{[[C]]} \otimes \mathfrak{J}^C, \qquad U = e^{ip}. \tag{2.23}$$

The representation for $\widehat{\mathfrak{J}}^A$ is the *evaluation* representation, obtained by multiplying the level-zero generators by a *spectral* parameter u:

$$\widehat{\mathfrak{J}}^A = u\, \mathfrak{J}^A = \frac{g}{4i}\left(x^+ + \frac{1}{x^+} + x^- + \frac{1}{x^-}\right) \mathfrak{J}^A, \tag{2.24}$$

where g depends on the Yang-Mills coupling and $x^\pm$ parameterise the central charges in the fundamental representation. The reason for the relationship (2.24) is that the level-one central charges also possess a central coproduct, therefore have to satisfy $\Delta^{op}(\widehat{\mathbb{C}}) = \Delta(\widehat{\mathbb{C}})$, *etc.* This fixes the dependence of u on the $\mathfrak{psl}(2|2)_c$ representation labels, up to an additive constant which we have disregarded thanks to the the shift automorphism (2.18):

$$u \longrightarrow u + c. \tag{2.25}$$

A special comment concerns the dual structure constants f^A_{BC} in (2.23), which are in principle ill-defined as the Killing form of $\mathfrak{psl}(2|2)_c$ is identically zero. In [**13**], the Yangian coproducts are explicitly given without relying on any particular index-lowering procedure[3].

In the case of simple Lie algebras, the spectral parameter is an unconstrained variable. The shift-automorphism $u \to u+c$ then implies that the Yangian-invariant S-matrix must depends only on the difference of the spectral parameters:

$$R = R(u_1 - u_2).$$

The dependence of u on the level-zero representation variables produces instead a lack of difference form in the fundamental AdS/CFT S-matrix.

Drinfeld's second realisation of the AdS/CFT Yangian was found in [**62**]:

$$\begin{aligned}
&[\kappa_{i,m}, \kappa_{j,n}] = 0, \quad [\kappa_{i,0}, \xi^\pm_{j,m}] = \pm a_{ij}\, \xi^\pm_{j,m}, \quad [\xi^+_{i,m}, \xi^-_{j,n}] = \delta_{ij}\, \kappa_{j,m+n},\\
&[\kappa_{i,m+1}, \xi^\pm_{j,n}] - [\kappa_{i,m}, \xi^\pm_{j,n+1}] = \pm\frac{1}{2} a_{ij}\, \{\kappa_{i,m}, \xi^\pm_{j,n}\},\\
&[\xi^\pm_{i,m+1}, \xi^\pm_{j,n}] - [\xi^\pm_{i,m}, \xi^\pm_{j,n+1}] = \pm\frac{1}{2} a_{ij}\, \{\xi^\pm_{i,m}, \xi^\pm_{j,n}\}, \qquad [\xi^\pm_{2,m}, \xi^\pm_{3,n}] = C^\pm_{m+n},
\end{aligned} \tag{2.26}$$

where $C^\pm_n$ are an infinite tower of central extensions.

3. The Yangian of $\mathfrak{psu}(2,2|4)$

Yangians have appeared in several other shapes within the AdS/CFT spectral problem. One instance is the perturbative expansion of the gauge theory anomalous dimensions, when expressed in terms of spin-chains [**1**, **34**, **35**, **70**] and before considering magnonic excitations. The level-zero generators act as local charges

$$\mathfrak{J}^A = \sum_k \mathfrak{J}^A(k), \tag{3.1}$$

[3]See [**13**, **61**] for an interpretation of these coproducts as arising from the dual structure constants of a bigger non-degenerate algebra, equipped with an invertible bilinear form.

where k runs over the spin-chain sites, while the level-one Yangian generators are bilocal combinations (on infinite chains):

$$\widehat{\mathfrak{J}}^A = \sum_{k<n} f^A_{BC}\, \mathfrak{J}^B(k)\, \mathfrak{J}^C(n). \tag{3.2}$$

The full integrable structure of the $\mathcal{N} = 4$ SYM is not entirely understood, as one lacks an explicit expression for the (long-range) interaction Hamiltonian[4].

On the other side of the correspondence, [**21**] established the classical integrability of the Green-Schwarz superstring sigma model, see also [**3**, **7**, **27**, **28**, **45**, **64**]. The semiclassical soliton argument [**51**, **52**] we presented before for the non-trivial deformation of the level zero coproduct applies to the non-local expression of the level one classical sigma model charges:

$$\begin{aligned}\widehat{\mathfrak{J}}^A_{|profile} = &\int_{-\infty}^0 dx\, J_1^A + \frac{1}{2} f^A_{BC} \int_{-\infty}^0 dx\, J_0^B(x) \int_{-\infty}^x dy\, J_0^C(y) + \int_0^\infty dx\, J_1^A \\ &+ \frac{1}{2} f^A_{BC} \int_0^\infty dx\, J_0^B(x) \int_0^x dy\, J_0^C(y) \\ &+ \frac{1}{2} f^A_{BC} \int_0^\infty dx\, J_0^B(x) \int_{-\infty}^0 dy\, J_0^C(y),\end{aligned} \tag{3.3}$$

schematically reproducing (2.19). In the absence of anomalies, one can quantise this action to one on the Hilbert space of the asymptotic states.

4. Secret symmetry

In this section we focus only on a few forms in which an additional *secret* symmetry manifests itself in the AdS/CFT correspondence. For a more thorough review, including secret symmetry in boundary problems [**59**] and quantum affine deformations [**32**], see [**31**].

4.1. Secret symmetry of the R-matrix. The fundamental R-matrix we have been discussing admits the extra symmetry [**20**, **54**]

$$\begin{aligned}&\Delta^{op}(\widehat{\mathbb{B}})\, R = R\, \Delta(\widehat{\mathbb{B}}), \\ &\Delta(\widehat{\mathbb{B}}) = \widehat{\mathbb{B}} \otimes \mathbb{1} + \mathbb{1} \otimes \widehat{\mathbb{B}} + \left(e^{\frac{i}{2}p}\, \mathbb{G}_a^{\ \alpha} \otimes \mathbb{Q}_\alpha^{\ a} + e^{-\frac{i}{2}p}\, \mathbb{Q}_\alpha^{\ a} \otimes \mathbb{G}_a^{\ \alpha}\right), \\ &\widehat{\mathbb{B}} = B_0\; diag(1, 1, -1, -1).\end{aligned} \tag{4.1}$$

This was generalised to bound states by [**29**]. As $\widehat{\mathbb{B}}$ is not supertraceless, one might postulate an enlarged Yangian symmetry algebra isomorphic to $Y\big(\mathfrak{gl}(2|2)\big)$. However, the level zero partner of $\mathbb{B}$, namely

$$\Delta(\mathbb{B}) = \mathbb{B} \otimes \mathbb{1} + \mathbb{1} \otimes \mathbb{B}, \qquad \mathbb{B} \propto diag(1, 1, -1, -1), \tag{4.2}$$

is not a symmetry of the fundamental R-matrix. One has therefore a very peculiar "indented" Yangian, with one generator at level one admitting no level zero partners.

---

[4] See however [**12**, **18**, **71**], where progress is driven by the fact that Yangian symmetry can provide a proof of integrability order by order in perturbation theory. See also [**2**, **70**, **72**].

4.2. Secret symmetry in Amplitudes. In [**19**] an analogous mechanism was noticed for tree-level planar n-particle color-ordered amplitudes $\mathcal{A}_n$. These amplitudes are functions of spinor-helicity variables $(\lambda_k, \tilde{\lambda}_k, \eta_k)$, $k = 1, \ldots, n$, with $\lambda_k, \tilde{\lambda}_k \in \mathbb{C}^2$ complex conjugate spinors and $\eta_k \in \mathbb{C}^{0|4}$ a Grassmann variable. The individual particles' light-like momentum is given by $p_k = \lambda_k \tilde{\lambda}_k$.

The $\mathfrak{psu}(2,2|4)$ superconformal generators $\mathfrak{J}^A$ act as differential operators $\mathfrak{J}^A_k$ annihilating the amplitudes. In [**40**], an additional Yangian symmetry $\widehat{\mathfrak{J}}^A$ was found:

$$\mathfrak{J}^A = \sum_{i=1}^{n} \mathfrak{J}^A_i \, , \qquad \widehat{\mathfrak{J}}^A = f^A_{BC} \sum_{j<k=1}^{n} \mathfrak{J}^B_j \mathfrak{J}^C_k \, , \tag{4.3}$$

with f^A_{BC} the $\mathfrak{psu}(2,2|4)$ structure constants. Cyclicity of color-ordered amplitudes is guaranteed by the vanishing of the dual Coxeter number of $\mathfrak{psu}(2,2|4)$.

The extra *bonus* symmetry [**19**], once again without a level zero partner, is given by

$$\widehat{\mathfrak{B}} = \sum_{k=1}^{n-1} \sum_{j=k+1}^{n} \left(\mathfrak{Q}^{\alpha b}_k \mathfrak{S}_{j,\alpha b} - \bar{\mathfrak{Q}}^{\dot{\alpha}}_{k,b} \bar{\mathfrak{S}}^b_{j,\dot{\alpha}} - \mathfrak{Q}^{\alpha b}_j \mathfrak{S}_{k,\alpha b} + \bar{\mathfrak{Q}}^{\dot{\alpha}}_{j,b} \bar{\mathfrak{S}}^b_{k,\dot{\alpha}} \right) . \tag{4.4}$$

The would-be level zero $\mathfrak{B} = \sum_i \eta^A_i \partial / \partial \eta^A_i$ counts the total helicity, which is not conserved for MHV amplitudes.

The authors of [**19**] proved cyclicity of (4.4) and invariance under $\widehat{\mathfrak{B}}$ of the Grassmannian integral formula [**4**] for the leading singularities in tree amplitudes. Moreover, they also showed that the secret generator can be consistently corrected by adding suitable length-changing operators to ensure that the conformal anomaly is properly taken into account, as for the other Yangian charges. It is fascinating to notice the same effect of length-changing in the symmetry action on amplitudes, as it occurs in the spin-chain picture of the spectral problem [**41, 58**].

4.3. Secret symmetry in the pure spinor formalism. In [**22**], a secret symmetry present at odd Yangian levels was found within the pure-spinor formulation of the $AdS_5 \times S^5$ superstring. There, one introduces a group variable $g \in PSU(2,2|4)$ with an action written in terms of the right-invariant current

$$J = - dg \, g^{-1} \, . \tag{4.5}$$

The existence of a Lax connection $J_{\pm}(z)$ [**55**]

$$[\partial_+ + J_+(z) \, , \, \partial_- + J_-(z)] = 0 \tag{4.6}$$

guarantees classical integrability, with the non-local conserved charges being generated by the transfer matrix

$$T(z) = g(+\infty)^{-1} \left(P \, \exp \int_C \left(-J_+(z) d\tau^+ - J_-(z) d\tau^- \right) \right) g(-\infty) \tag{4.7}$$

for a choice of the contour C. In [**22**] it is argued that $T(z)$ is lifted to $SU(2,2|4)$ by lifting the element g to the same supergroup. By tracing with the hypercharge

$$s = \begin{pmatrix} \mathbb{1}_{4\times 4} & 0 \\ 0 & -\mathbb{1}_{4\times 4} \end{pmatrix}$$

one obtains an infinite family of *bonus* charges at odd levels, the first one corresponding to the first order in the Taylor expansion of the transfer matrix:

$$\int\int_{\sigma_1<\sigma_2} [j(\sigma_1)\ ,\ j(\sigma_2)] - \int k\ . \tag{4.8}$$

The density j defines the level zero charges, while k is suitably chosen to enforce the necessary consistency requirements for the bonus charges.

5. Acknowledgements

The author would like to thank the organizers of the String-Math 2012 conference in Bonn for the invitation to speak at such a remarkable event, and for providing a fantastic environment for scientific exchange. The author also acknowledges useful conversations with the participants of the ESF and STFC supported workshop "Permutations and Gauge String duality (STFC- 4070083442)" (Queen Mary U. of London, July 2014). He is most indebted to his collaborators, and to the many people who shared ideas and contributed to this work with stimulating and insightful discussions. No new data has been produced as a result of this research.

References

[1] A. Agarwal and S. G. Rajeev, *Yangian symmetries of matrix models and spin chains: the dilatation operator of* $\mathcal{N}=4$ *SYM*, Internat. J. Modern Phys. A **20** (2005), no. 23, 5453–5490, DOI 10.1142/S0217751X05022822. MR2174934 (2006h:81265)

[2] A. Agarwal and S. G. Rajeev, *The dilatation operator of* $\mathcal{N}=4$ *SYM and classical limits of spin chains and matrix models*, Modern Phys. Lett. A **19** (2004), no. 34, 2549–2568, DOI 10.1142/S0217732304015725. MR2097407 (2005e:81221)

[3] L. F. Alday, *Non-local charges on* $\mathrm{AdS}_5\times S^5$ *and PP-waves*, J. High Energy Phys. **12** (2003), 033, 19 pp. (electronic), DOI 10.1088/1126-6708/2003/12/033. MR2041047 (2005a:83112)

[4] N. Arkani-Hamed, F. Cachazo, and C. Cheung, *The Grassmannian origin of dual superconformal invariance*, J. High Energy Phys. **3** (2010), 036, 9, DOI 10.1007/JHEP03(2010)036. MR2653476 (2012b:81167)

[5] G. Arutyunov, M. de Leeuw, R. Suzuki, and A. Torrielli, *Bound State Transfer Matrix for AdS5 x S5 Superstring*, JHEP **10** (2009), 025.

[6] G. Arutyunov, M. de Leeuw, and A. Torrielli, *The Bound State S-Matrix for AdS5 x S5 Superstring*, Nucl. Phys. **B819** (2009), 319.

[7] G. Arutyunov and S. Frolov, *Superstrings on* $\mathrm{AdS}_4\times\mathbb{CP}^3$ *as a coset sigma-model*, J. High Energy Phys. **9** (2008), 129, 27, DOI 10.1088/1126-6708/2008/09/129. MR2447661 (2010a:81179)

[8] G. Arutyunov and S. Frolov, *Foundations of the* AdS_5xS^5 *Superstring. Part I*, J. Phys. **A42** (2009), 254003.

[9] G. Arutyunov and S. Frolov, *Thermodynamic Bethe Ansatz for the AdS5 x S5 Mirror Model*, JHEP **05** (2009), 068.

[10] G. Arutyunov, S. Frolov, J. Plefka, and M. Zamaklar, *The off-shell symmetry algebra of the light-cone* $\mathrm{AdS}_5\times S^5$ *superstring*, J. Phys. A **40** (2007), no. 13, 3583–3605, DOI 10.1088/1751-8113/40/13/018. MR2325063 (2008h:81121)

[11] Z. Bajnok, *Review of AdS/CFT integrability, Chapter III.6: Thermodynamic Bethe ansatz*, Lett. Math. Phys. **99** (2012), no. 1-3, 299–320, DOI 10.1007/s11005-011-0512-y. MR2886424

[12] N. Beisert and T. Klose, *Long-range* $\mathfrak{gl}(n)$ *integrable spin chains and plane-wave matrix theory*, J. Stat. Mech. Theory Exp. **7** (2006), P07006, 21 pp. (electronic). MR2244326 (2007e:81039)

[13] Niklas Beisert, *The S-Matrix of AdS/CFT and Yangian Symmetry*, PoS **SOLVAY** (2006), 002.

[14] N. Beisert, *The Analytic Bethe Ansatz for a Chain with Centrally Extended* $su(2|2)$ *Symmetry*, J. Stat. Mech. **0701** (2007), P017.

[15] N. Beisert, *The* $\mathfrak{su}(2|2)$ *dynamic S-matrix*, Adv. Theor. Math. Phys. **12** (2008), no. 5, 945–979, DOI 10.4310/ATMP.2008.v12.n5.a1. MR2437847 (2010f:81265)

[16] N. Beisert, *On Yangian Symmetry in Planar N=4 SYM*, Trieste 2010, Gribov-80 Memorial Volume (2010), 413.

[17] N. Beisert, C. Ahn, L. F. Alday, and et al., *Review of AdS/CFT integrability: an overview*, Lett. Math. Phys. **99** (2012), no. 1-3, 3–32, DOI 10.1007/s11005-011-0529-2. MR2886411

[18] N. Beisert and D. Erkal, *Yangian symmetry of long-range* $\mathfrak{gl}(N)$ *integrable spin chains*, J. Stat. Mech. Theory Exp. **3** (2008), P03001, 27. MR2391305 (2009e:82015)

[19] N. Beisert and B. U. W. Schwab, *Bonus Yangian Symmetry for the Planar S-Matrix of N=4 Super Yang-Mills*, (2011).

[20] N. Beisert and F. Spill, *The classical r-matrix of AdS/CFT and its Lie bialgebra structure*, Comm. Math. Phys. **285** (2009), no. 2, 537–565, DOI 10.1007/s00220-008-0578-2. MR2461987 (2010a:81222)

[21] I. Bena, J. Polchinski, and R. Roiban, *Hidden symmetries of the* $\mathrm{AdS}_5 \times S^5$ *superstring*, Phys. Rev. D (3) **69** (2004), no. 4, 046002, 7, DOI 10.1103/PhysRevD.69.046002. MR2093115 (2005g:83115)

[22] N. Berkovits and A. Mikhailov, *Nonlocal charges for bonus Yangian symmetries of super-Yang-Mills*, J. High Energy Phys. **7** (2011), 125, 14, DOI 10.1007/JHEP07(2011)125. MR2875933

[23] D. Bernard, *An introduction to Yangian symmetries*, Internat. J. Modern Phys. B **7** (1993), no. 20-21, 3517–3530, DOI 10.1142/S0217979293003371. Yang-Baxter equations in Paris (1992). MR1241457

[24] D. Bombardelli, D. Fioravanti, and R. Tateo, *Thermodynamic Bethe ansatz for planar AdS/CFT: a proposal*, J. Phys. A **42** (2009), no. 37, 375401, 20, DOI 10.1088/1751-8113/42/37/375401. MR2540424 (2010i:81315)

[25] T. Creutzig, *Yangian superalgebras in conformal field theory*, Nuclear Phys. B **849** (2011), no. 3, 636–653, DOI 10.1016/j.nuclphysb.2011.04.004. MR2800449 (2012e:81168)

[26] T. Curtright and C. Zachos, *Supersymmetry and the nonlocal Yangian deformation symmetry*, Nuclear Phys. B **402** (1993), no. 3, 604–612, DOI 10.1016/0550-3213(93)90120-E. MR1236190 (94j:81213)

[27] A. K. Das, J. Maharana, A. Melikyan, and M. Sato, *The algebra of transition matrices for the AdS(5) x S**5 superstring*, JHEP **12** (2004), 055.

[28] A. K. Das, A. Melikyan, and M. Sato, *The algebra of flat currents for the string on AdS(5) x S**5 in the light-cone gauge*, JHEP **11** (2005), 015.

[29] M. de Leeuw, *Bound States, Yangian Symmetry and Classical r-matrix for the* $AdS_5 \times S^5$ *Superstring*, JHEP **06** (2008), 085.

[30] M. de Leeuw, *The Bethe Ansatz for* $AdS_5 \times S^5$ *Bound States*, JHEP **01** (2009), 005.

[31] M. de Leeuw, Takuya Matsumoto, Sanefumi Moriyama, Vidas Regelskis, and Alessandro Torrielli, *Secret Symmetries in AdS/CFT*, Phys.Scripta **02** (2012), 028502.

[32] M. de Leeuw, V. Regelskis, and A. Torrielli, *The quantum affine origin of the AdS/CFT secret symmetry*, J. Phys. A **45** (2012), no. 17, 175202, 20, DOI 10.1088/1751-8113/45/17/175202. MR2913472

[33] G. W. Delius, *Exact S-matrices with affine quantum group symmetry*, Nuclear Phys. B **451** (1995), no. 1-2, 445–465, DOI 10.1016/0550-3213(95)00336-Q. MR1352421 (96g:81158)

[34] L. Dolan, C. R. Nappi, and E. Witten, *A relation between approaches to integrability in superconformal Yang-Mills theory*, J. High Energy Phys. **10** (2003), 017, 17 pp. (electronic), DOI 10.1088/1126-6708/2003/10/017. MR2030769 (2005a:81141)

[35] L. Dolan, C. R. Nappi, and E. Witten, *Yangian symmetry in* $D = 4$ *superconformal Yang-Mills theory*, Quantum theory and symmetries, World Sci. Publ., Hackensack, NJ, 2004, pp. 300–315, DOI 10.1142/9789812702340_0036. MR2170727

[36] P. Dorey, *Exact S-matrices*, Conformal field theories and integrable models (Budapest, 1996), Lecture Notes in Phys., vol. 498, Springer, Berlin, 1997, pp. 85–125, DOI 10.1007/BFb0105279. MR1636799 (99k:81333)

[37] V. G. Drinfel′d, *Quantum groups*, Proceedings of the International Congress of Mathematicians, Vol. 1, 2 (Berkeley, Calif., 1986), Amer. Math. Soc., Providence, RI, 1987, pp. 798–820. MR934283 (89f:17017)

[38] V. G. Drinfel′d, *A new realization of Yangians and of quantum affine algebras* (Russian), Dokl. Akad. Nauk SSSR **296** (1987), no. 1, 13–17; English transl., Soviet Math. Dokl. **36** (1988), no. 2, 212–216. MR914215 (88j:17020)

[39] J. M. Drummond, *Hidden simplicity of gauge theory amplitudes*, Classical Quantum Gravity **27** (2010), no. 21, 214001, 38, DOI 10.1088/0264-9381/27/21/214001. MR2734064 (2012g:81151)

[40] J. Drummond, J. Henn, and J. Plefka, *Yangian symmetry of scattering amplitudes in $\mathcal{N}=4$ super Yang-Mills theory*, J. High Energy Phys. **5** (2009), 046, 23, DOI 10.1088/1126-6708/2009/05/046. MR2511396 (2011c:81187)

[41] C. Gómez and R. Hernández, *The magnon kinematics of the AdS/CFT correspondence*, J. High Energy Phys. **11** (2006), 021, 18, DOI 10.1088/1126-6708/2006/11/021. MR2270427 (2007k:81191)

[42] L. Gow, *Gauss decomposition of the Yangian $Y(\mathfrak{gl}_{m|n})$*, Comm. Math. Phys. **276** (2007), no. 3, 799–825, DOI 10.1007/s00220-007-0349-5. MR2350438 (2008h:17013)

[43] N. Gromov and V. Kazakov, *Review of AdS/CFT integrability, Chapter III.7: Hirota dynamics for quantum integrability*, Lett. Math. Phys. **99** (2012), no. 1-3, 321–347, DOI 10.1007/s11005-011-0513-x. MR2886425

[44] N. Gromov, V. Kazakov, and P. Vieira, *Exact spectrum of anomalous dimensions of planar $N=4$ supersymmetric Yang-Mills theory*, Phys. Rev. Lett. **103** (2009), no. 13, 131601, 4, DOI 10.1103/PhysRevLett.103.131601. MR2551684 (2010k:81270)

[45] M. Hatsuda and K. Yoshida, *Classical integrability and super Yangian of superstring on* $\mathrm{AdS}_5\times S^5$, Adv. Theor. Math. Phys. **9** (2005), no. 5, 703–728. MR2202289 (2006j:81140)

[46] K. Iohara and Y. Koga, *Central extensions of Lie superalgebras*, Comment. Math. Helv. **76** (2001), no. 1, 110–154, DOI 10.1007/s000140050152. MR1819663 (2002g:17009)

[47] R. A. Janik, *The* $\mathrm{AdS}_5\times S^5$ *superstring worldsheet S matrix and crossing symmetry*, Phys. Rev. D (3) **73** (2006), no. 8, 086006, 10, DOI 10.1103/PhysRevD.73.086006. MR2221398 (2006m:81230)

[48] S. M. Khoroshkin and V. N. Tolstoy, *Yangian double*, Lett. Math. Phys. **36** (1996), no. 4, 373–402, DOI 10.1007/BF00714404. MR1384643 (97e:81063)

[49] T. Klose, T. McLoughlin, R. Roiban, and K. Zarembo, *Worldsheet scattering in* $\mathrm{AdS}_5\times S^5$, J. High Energy Phys. **3** (2007), 094, 49 pp. (electronic), DOI 10.1088/1126-6708/2007/03/094. MR2313963 (2008b:81229)

[50] L. N. Lipatov, *High-energy asymptotics of multicolor QCD and exactly solvable lattice models*, JETP Lett (1994), 596.

[51] M. Lüscher and K. Pohlmeyer, *Scattering of massless lumps and non-local charges in the two-dimensional classical non-linear σ-model*, Nuclear Phys. B **137** (1978), no. 1-2, 46–54. MR0496052 (58 #14675)

[52] N. J. MacKay, *On the classical origins of Yangian symmetry in integrable field theory*, Phys. Lett. B **281** (1992), no. 1-2, 90–97, DOI 10.1016/0370-2693(92)90280-H. MR1163794 (94d:81077a)

[53] N. J. MacKay, *Introduction to Yangian symmetry in integrable field theory*, Internat. J. Modern Phys. A **20** (2005), no. 30, 7189–7217, DOI 10.1142/S0217751X05022317. MR2192353 (2006j:81092)

[54] T. Matsumoto, S. Moriyama, and A. Torrielli, *A secret symmetry of the AdS/CFT S-matrix*, J. High Energy Phys. **9** (2007), 099, 12, DOI 10.1088/1126-6708/2007/09/099. MR2342332 (2008k:81319)

[55] A. Mikhailov and S. Schäfer-Nameki, *Perturbative study of the transfer matrix on the string worldsheet in* $\mathrm{AdS}_5\times S^5$, Adv. Theor. Math. Phys. **15** (2011), no. 4, 913–971. MR2929679

[56] J. A. Minahan and K. Zarembo, *The Bethe-ansatz for $\mathcal{N}=4$ super Yang-Mills*, J. High Energy Phys. **3** (2003), 013, 29, DOI 10.1088/1126-6708/2003/03/013. MR1976005 (2004d:81131)

[57] A. Molev, *Yangians and classical Lie algebras*, Mathematical Surveys and Monographs, vol. 143, American Mathematical Society, Providence, RI, 2007. MR2355506 (2008m:17033)

[58] J. Plefka, F. Spill, and A. Torrielli, *Hopf algebra structure of the AdS/CFT S-matrix*, Phys. Rev. D (3) **74** (2006), no. 6, 066008, 6, DOI 10.1103/PhysRevD.74.066008. MR2276401 (2007j:81174)

[59] V. Regelskis, *The secret symmetries of the AdS/CFT reflection matrices*, J. High Energy Phys. **8** (2011), 006, 10, DOI 10.1007/JHEP08(2011)006. MR2876082

[60] C. Sieg, *Review of AdS/CFT integrability, Chapter I.2: The spectrum from perturbative gauge theory*, Lett. Math. Phys. **99** (2012), no. 1-3, 59–84, DOI 10.1007/s11005-011-0508-7. MR2886413

[61] F. Spill, *Hopf Algebras in the AdS/CFT Correspondence*, Diploma Thesis, Humboldt University of Berlin (2007).

[62] F. Spill and A. Torrielli, *On Drinfeld's second realization of the AdS/CFT* $\mathfrak{su}(2|2)$ *Yangian*, J. Geom. Phys. **59** (2009), no. 4, 489–502, DOI 10.1016/j.geomphys.2009.01.001. MR2504470 (2010c:81143)

[63] M. Staudacher, *The factorized S-matrix of CFT/AdS*, J. High Energy Phys. **5** (2005), 054, 35 pp. (electronic), DOI 10.1088/1126-6708/2005/05/054. MR2155052 (2006g:81191)

[64] B. Stefański Jr., *Green-Schwarz action for Type IIA strings on* $\mathrm{AdS}_4 \times \mathbf{C}\mathrm{P}^3$, Nuclear Phys. B **808** (2009), no. 1-2, 80–87, DOI 10.1016/j.nuclphysb.2008.09.015. MR2474085 (2010c:81216)

[65] V. Stukopin, *Yangians of classical Lie superalgebras: basic constructions, quantum double and universal R-matrix*, Symmetry in nonlinear mathematical physics. Part 1, 2, 3, Pr. Inst. Mat. Nats. Akad. Nauk Ukr. Mat. Zastos., 50, Part 1, vol. 2, Natsīonal. Akad. Nauk Ukraïni, Īnst. Mat., Kiev, 2004, pp. 1195–1201. MR2077938

[66] A. Torrielli, *The Hopf superalgebra of AdS/CFT*, J. Geom. Phys. **61** (2011), no. 1, 230–236, DOI 10.1016/j.geomphys.2010.10.003. MR2746994 (2012b:81104)

[67] A. Torrielli, *Yangians, S-matrices and AdS/CFT*, J. Phys. **A44** (2011), 263001.

[68] A. Torrielli, *Review of AdS/CFT integrability, Chapter VI.2: Yangian algebra*, Lett. Math. Phys. **99** (2012), no. 1-3, 547–565, DOI 10.1007/s11005-011-0491-z. MR2886434

[69] Y.-Z. Zhang and M. D. Gould, *Quasi-Hopf superalgebras and elliptic quantum supergroups*, J. Math. Phys. **40** (1999), no. 10, 5264–5282, DOI 10.1063/1.533029. MR1715313 (2001i:17020)

[70] B. I. Zwiebel, *Yangian symmetry at two loops for the* $\mathfrak{su}(2|1)$ *sector of* $\mathcal{N}=4$ *SYM*, J. Phys. A **40** (2007), no. 5, 1141–1151, DOI 10.1088/1751-8113/40/5/018. MR2307736 (2007m:81220)

[71] B. I. Zwiebel, *Iterative structure of the* $\mathcal{N}=4$ *SYM spin chain*, J. High Energy Phys. **7** (2008), 114, 49, DOI 10.1088/1126-6708/2008/07/114. MR2430070 (2009h:81308)

[72] B. I. Zwiebel, *Two-loop integrability of planar* $\mathcal{N}=6$ *superconformal Chern-Simons theory*, J. Phys. A **42** (2009), no. 49, 495402, 37, DOI 10.1088/1751-8113/42/49/495402. MR2566279 (2011e:81266)

Department of Mathematics, University of Surrey, Guildford, Surrey, GU2 7XH, United Kingdom

E-mail address: a.torrielli@surrey.ac.uk

Contributed talks

Proceedings of Symposia in Pure Mathematics
Volume **90**, 2015
http://dx.doi.org/10.1090/pspum/090/01519

On the Marginal Deformations of General (0,2) Non-Linear Sigma-Models

Ido Adam

ABSTRACT. In this note we explore the possible marginal deformations of general (0,2) non-linear sigma-models, which arise as descriptions of the weakly-coupled (large radius) limits of four-dimensional $\mathcal{N} = 1$ compactifications of the heterotic string, to lowest order in α' and first order in conformal perturbation theory. The results shed light from the world-sheet perspective on the classical moduli space of such compactifications. This is a contribution to the proceedings of String-Math 2012.

1. Introduction

One possible way of obtaining gauge theories in dimensions lower than ten from superstring theory is to consider the heterotic string on $\mathbb{R}^{d-1,1} \times M_{10-d}$, where $\mathbb{R}^{d-1,1}$ is d-dimensional Minkowski space and M_{10-d} is a compact $(10-d)$-dimensional manifold. For such compactifications one has also to specify the background gauge fields on M_{10-d} (a vector bundle V), which will break the large heterotic SO(32) or $\mathrm{E}_8 \times \mathrm{E}_8$ gauge group into smaller gauge groups more suitable for a realistic description of nature (for example getting an SU(5) GUT).

There is particular interest in compactifications with $\mathcal{N} = 1$ supersymmetry, since in that case there exist powerful non-renormalization theorems protecting the superpotential from perturbative α' corrections and in many cases ensuring the existence of the vacuum in string perturbation theory. The non-renormalization theorem can be violated by instanton effects, but in some favorable cases these can be shown to be absent (see, for example, [**4**, **6**, **10**]).

It is a well known result that $\mathcal{N} = 1$ space-time supersymmetry in four dimensions requires that the local (0,1) superconformal symmetry of the world-sheet description of the Ramond-Neveu-Schwarz superstring be enhanced to a global (0,2) superconformal symmetry (in which the local one is embedded) and that extended space-time supersymmetry leads to even higher superconformal symmetry on the world-sheet [**2**, **3**, **7**]. When there is a geometric description of the compactification (i.e., as a manifold and vector bundle) the compact CFT on M can be described in the large radius limit as a (0,2) superconformal non-linear sigma-model. The

2010 *Mathematics Subject Classification.* Primary 83E30.
Key words and phrases. Sigma models, conformal field models in string theory, flux compactifications.
The author's research was supported by FAPESP post-doctoral fellowship 2010/07439-0.

marginal deformations preserving the world-sheet supersymmetry (and hence the target-space one) would then correspond to massless modes parameterizing the moduli space of the compactification near the point represented by the world-sheet non-linear sigma-model. This led the authors of [**9**] to use an application of the methods of [**8**] to unitary two-dimensional (0,2) SCFTs [**1**] in order to determine those moduli in the case of compactifications possessing $G = G' \times \mathrm{E}_8$ space-time gauge symmetry where G' contains a non-anomalous $\mathrm{U}(1)_\mathrm{L}$ symmetry.

In this publication we will examine to lowest order in α' and to first order in conformal perturbation theory the general case of a gauge group G without requiring the existence of $\mathrm{U}(1)_\mathrm{L}$ factors.

The paper is organized as follows. In section 2 we describe the general (0,2) non-linear sigma-model, in section 3 we find the marginal deformations preserving the (0,2) superconformal symmetry to lowest order in α' and in conformal perturbation theory.

2. The general (0,2) non-linear sigma-model

In this section the non-linear sigma-model describing the weakly-coupled limit of a compactification of the heterotic string on a complex manifold M with a complex vector bundle $V \to M$ is constructed. For the background to be consistent, it has to satisfy the Bianchi identity $d\tilde{H} = \mathrm{ch}_2(\mathrm{T}M) - \mathrm{ch}_2(V)$, where $\tilde{H} = dB - \omega(\mathrm{T}M) - \omega(V)$ is the torsion field shifted by the Chern-Simons three-forms constructed from the connection of the gauge vector bundle and the tangent bundle. We will further assume that the compactification has $\mathcal{N} = 1$ target space supersymmetry, so the sigma-model has (0,2) global supersymmetry [**2, 3, 7**]. Target-space supersymmetry to one-loop order also implies the Hermitian Yang-Mills equations $\mathcal{F}_{ij} = \mathcal{F}_{\bar{i}\bar{j}} = 0$, $g^{\bar{j}i}\mathcal{F}_{i\bar{j}} = 0$, where $\mathcal{F}$ is the curvature of the bundle, so in particular, V is a holomorphic vector bundle.

2.1. (0,2) superspace conventions. The easiest way to write the most-general (0,2) non-linear sigma-model is in (0,2) superspace. We use mostly the conventions of [**13**]. In particular the right-moving supercharges and super-derivatives are given by

$$Q = \frac{\partial}{\partial\theta} + i\bar{\theta}\bar{\partial}\ , \quad \bar{Q} = -\frac{\partial}{\partial\bar{\theta}} - i\theta\bar{\partial}\ , \quad \mathcal{D} = \frac{\partial}{\partial\theta} - i\bar{\theta}\bar{\partial}\ , \quad \bar{\mathcal{D}} = -\frac{\partial}{\partial\bar{\theta}} + i\theta\bar{\partial} \tag{1}$$

A generic (0,2) superfield is of the form

$$\Phi = \phi + \sqrt{2}\theta\psi + \sqrt{2}\bar{\theta}\bar{\psi} + i\theta\bar{\theta}F\ . \tag{2}$$

A chiral superfield is constrained to satisfy $\bar{\mathcal{D}}\Phi = 0$, while an anti-chiral superfield satisfies $\mathcal{D}\bar{\Phi} = 0$. They are of the form

$$\Phi = \phi + \sqrt{2}\theta\psi - i\theta\bar{\theta}\bar{\partial}\phi\ , \quad \bar{\Phi} = \bar{\phi} - \sqrt{2}\bar{\theta}\bar{\psi} + i\theta\bar{\theta}\bar{\partial}\bar{\phi}\ . \tag{3}$$

The model also includes chiral and anti-chiral Fermi superfields satisfying $\bar{\mathcal{D}}\Gamma = 0$ and $\mathcal{D}\bar{\Gamma} = 0$. Their form is

$$\Gamma = \gamma - \sqrt{2}\theta G - i\theta\bar{\theta}\bar{\partial}\gamma\ , \quad \bar{\Gamma} = \bar{\gamma} - \sqrt{2}\bar{\theta}\bar{G} + i\theta\bar{\theta}\bar{\partial}\bar{\gamma}\ . \tag{4}$$

The Hermiticity conditions relating the two kinds of fields are the trivial ones $\theta^\dagger = \bar{\theta}$, $\phi^\dagger = \bar{\phi}$, $\psi^\dagger = \bar{\psi}$ and $\gamma^\dagger = \bar{\gamma}$.

TABLE 1. Conformal weights and right R-charges of the various superfields

	h	$\bar{h}$	$\bar{q}$
θ	0	$-\frac{1}{2}$	$+1$
$\bar{\theta}$	0	$-\frac{1}{2}$	-1
$\mathcal{D}$, Q	0	$\frac{1}{2}$	-1
$\bar{\mathcal{D}}$, $\bar{Q}$	0	$\frac{1}{2}$	$+1$
Φ^i, $\bar{\Phi}^{\bar{i}}$	0	0	0
Γ^α, $\bar{\Gamma}^{\bar{\alpha}}$	$\frac{1}{2}$	0	0

2.2. The (0,2) non-linear sigma-model. The general (0,2) non-linear sigma-model has been written by [**5**]. Its field content is comprised of the chiral and anti-chiral superfields Φ^i, $\bar{\Phi}^{\bar{i}}$ $(i, \bar{i} = 1, \ldots, n)$, which are coordinates of a complex manifold M of dimension n (for a Calabi-Yau three-fold, which is the case of most interest, $n = 3$), and the chiral and anti-chiral Fermi superfields Γ^α and $\bar{\Gamma}^{\bar{\alpha}}$, which take values in the vector bundle V over M. Their conformal weights and $\mathrm{U}(1)_\mathrm{R}$ charges are given in Table 1.

The integrand of the superspace integral must be of conformal weight (1,0) and have no $\mathrm{U}(1)_\mathrm{R}$ charge. The most general such action is of the form

$$
\begin{aligned}
S &= -\frac{i}{8\pi\alpha'}\int d^2x d^2\theta \Big[\mathcal{K}_i(\Phi,\bar{\Phi})\partial\Phi^i - \bar{\mathcal{K}}_{\bar{i}}(\Phi,\bar{\Phi})\partial\bar{\Phi}^{\bar{i}} + \\
&\quad + i\left(\mathcal{H}_{\alpha\beta}(\Phi,\bar{\Phi})\Gamma^\alpha\Gamma^\beta + 2\mathcal{H}_{\alpha\bar{\beta}}(\Phi,\bar{\Phi})\Gamma^\alpha\bar{\Gamma}^{\bar{\beta}} + \mathcal{H}_{\bar{\alpha}\bar{\beta}}(\Phi,\bar{\Phi})\bar{\Gamma}^{\bar{\alpha}}\bar{\Gamma}^{\bar{\beta}}\right)\Big], \qquad (5)
\end{aligned}
$$

where $\mathcal{K}_i(\Phi,\bar{\Phi})$ and $\bar{\mathcal{K}}_{\bar{i}}(\Phi,\bar{\Phi})$ can be regarded as the (1,0)- and (0,1)-forms $\mathcal{K} = \mathcal{K}_i d\Phi^i$ and $\bar{\mathcal{K}} = \bar{\mathcal{K}}_{\bar{i}} d\bar{\Phi}^{\bar{i}}$ on the target manifold and $\mathcal{H}_{\alpha\bar{\beta}}$ is a Hermitian structure on the vector bundle.

In the absence of world-sheet boundaries, the action is invariant under the transformations $\mathcal{K} \to \mathcal{K} + \partial f$, $\bar{\mathcal{K}} \to \bar{\mathcal{K}} + \bar{\partial} f$ for any (0,0)-form f, since these would shift the action by a total divergence. Similarly, shifting $\mathcal{K} \to \mathcal{K} + \omega$, $\bar{\mathcal{K}} \to \bar{\mathcal{K}} + \bar{\omega}$, where ω and $\bar{\omega}$ are a holomorphic (1,0)-form and an anti-holomorphic (0,1)-form, respectively, again only shifts the integral world-sheet integral by a total derivative, because ω depends only on the target-space-holomorphic fields Φ^i so

$$
\int_\Sigma \int d^2\theta\, \omega(\Phi) = i\int_\Sigma \frac{\partial}{\partial\phi^i}\omega(\phi)\bar{\partial}\phi^i = i\int_\Sigma \bar{\partial}\omega(\phi) = 0\,.
$$

Requiring that the action S be real leads to the Hermiticity conditions $\mathcal{K}_i^\dagger = \bar{\mathcal{K}}_{\bar{i}}$, $\mathcal{H}^\dagger_{\bar{\alpha}\bar{\beta}} = \mathcal{H}_{\beta\alpha}$ and $\mathcal{H}^\dagger_{\alpha\bar{\beta}} = \mathcal{H}_{\beta\bar{\alpha}}$.

In order to get some intuition about the meaning of the various background superfields, we also write the action in component form (dropping total derivative terms which vanish in the absence of world-sheet boundaries):

$$\begin{aligned} S &= -\frac{1}{2\pi\alpha'}\int d^2z\Big[\frac{1}{2}g_{i\bar{j}}(\partial\phi^i\bar{\partial}\bar{\phi}^{\bar{j}}+\partial\bar{\phi}^{\bar{j}}\bar{\partial}\phi^i)+\frac{1}{2}B_{i\bar{j}}(\partial\phi^i\bar{\partial}\bar{\phi}^{\bar{j}}-\partial\bar{\phi}^{\bar{j}}\bar{\partial}\phi^i) \\ &+ig_{i\bar{j}}\bar{\psi}^{\bar{j}}\partial\psi^i-i\bar{\psi}^{\bar{j}}\left(\Omega^-_{\bar{j}ki}\partial\phi^k+\Omega^-_{\bar{j}\bar{k}i}\partial\bar{\phi}^{\bar{k}}\right)\psi^i-i\bar{\gamma}_\alpha(\bar{\partial}\gamma^\alpha+\mathcal{A}^\alpha_{i\gamma}\bar{\partial}\phi^i\gamma^\gamma) \\ &-\frac{i}{2}\bar{\mathcal{A}}^{\bar{\alpha}}_{\bar{i}\beta}\bar{\partial}\bar{\phi}^{\bar{i}}\gamma_{\bar{\alpha}}\gamma^\beta-\frac{i}{2}\mathcal{A}^\alpha_{i\bar{\beta}}\bar{\partial}\phi^i\bar{\gamma}_\alpha\bar{\gamma}^{\bar{\beta}}-\frac{1}{2}\tilde{\mathcal{F}}^{\bar{\alpha}}_{i\bar{j}\beta}\psi^i\bar{\psi}^{\bar{j}}\gamma_{\bar{\alpha}}\gamma^\beta+\tilde{\mathcal{F}}^\alpha_{\bar{j}i\beta}\psi^i\bar{\psi}^{\bar{j}}\bar{\gamma}_\alpha\gamma^\beta \\ &+\frac{1}{2}\tilde{\mathcal{F}}^\alpha_{\bar{j}i\bar{\beta}}\psi^i\bar{\psi}^{\bar{j}}\bar{\gamma}_\alpha\bar{\gamma}^{\bar{\beta}}\Big]\,, \end{aligned} \tag{6}$$

where the metric and the B-fields are given by

$$g_{i\bar{j}}=\frac{1}{2}(\partial_{\bar{j}}\mathcal{K}_i+\partial_i\bar{\mathcal{K}}_{\bar{j}})\,,\quad B_{i\bar{j}}=\frac{1}{2}(\partial_{\bar{j}}\mathcal{K}_i-\partial_i\bar{\mathcal{K}}_{\bar{j}})\,, \tag{7}$$

and the torsion-twisted connection is

$$\Omega^-_{\bar{j}ki}=\Gamma_{\bar{j}ki}-\frac{1}{2}H_{\bar{j}ki}\,,\quad \Omega^-_{\bar{j}\bar{k}i}=\Gamma_{\bar{j}\bar{k}i}-\frac{1}{2}H_{\bar{j}\bar{k}i}\,, \tag{8}$$

with Γ being the Christoffel symbol of the first kind associated with the metric $g_{i\bar{j}}$ and $H=dB$ is the H-field. Furthermore, for brevity we define the holomorphic and anti-holomorphic connections on the vector bundle

$$\mathcal{A}^\alpha_{i\beta}\equiv\partial_i\mathcal{H}_{\beta\bar{\gamma}}\mathcal{H}^{\bar{\gamma}\alpha}\,,\quad \bar{\mathcal{A}}^{\bar{\alpha}}_{\bar{i}\bar{\beta}}\equiv\mathcal{H}^{\bar{\alpha}\gamma}\partial_{\bar{i}}\mathcal{H}_{\gamma\bar{\beta}} \tag{9}$$

(we use the notation that $\mathcal{H}^{\bar{\alpha}\beta}$ is the inverse Hermitian metric of $\mathcal{H}_{\alpha\bar{\beta}}$) as well as the symbols

$$\begin{aligned} \bar{\mathcal{A}}^{\bar{\alpha}}_{\bar{i}\beta} &\equiv \mathcal{H}^{\bar{\alpha}\gamma}\partial_{\bar{i}}\mathcal{H}_{\gamma\beta}\,,\quad \mathcal{A}^\alpha_{i\bar{\beta}}\equiv\partial_i\mathcal{H}_{\bar{\beta}\bar{\gamma}}\mathcal{H}^{\bar{\gamma}\alpha}\,, \\ \tilde{\mathcal{F}}^{\bar{\alpha}}_{j\bar{k}\beta} &\equiv \partial_j\bar{\mathcal{A}}^{\bar{\alpha}}_{\bar{k}\beta}-\mathcal{A}^\gamma_{j\beta}\bar{\mathcal{A}}^{\bar{\alpha}}_{\bar{k}\gamma}\,,\quad \tilde{\mathcal{F}}^\alpha_{\bar{k}j\beta}\equiv\partial_{\bar{k}}\mathcal{A}^\alpha_{j\beta}-\mathcal{A}^\alpha_{j\bar{\gamma}}\bar{\mathcal{A}}^{\bar{\gamma}}_{\bar{k}\beta}\,, \\ \tilde{\mathcal{F}}^\alpha_{\bar{k}j\bar{\beta}} &\equiv \partial_{\bar{k}}\mathcal{A}^\alpha_{j\bar{\beta}}-\bar{\mathcal{A}}^{\bar{\gamma}}_{\bar{k}\bar{\beta}}\mathcal{A}^\alpha_{j\bar{\gamma}}\,. \end{aligned} \tag{10}$$

Note that these are not connections and curvatures of the vector bundle (the curvature is $\mathcal{F}^\alpha_{\bar{j}i\beta}=\partial_{\bar{j}}\mathcal{A}^\alpha_{i\beta}$). We hope that the use of the letters $\mathcal{A}$ and $\mathcal{F}$ will not cause any confusion.

In the sequel we will require the equations of motion for the various superfields. Since the fields are either chiral or anti-chiral, we need their variations to obey the chirality/anti-chirality constraints. This is easily done by writing the variations as

$$\delta\Phi^i=\bar{\mathcal{D}}\delta X^i\,,\quad \delta\bar{\Phi}^{\bar{i}}=\mathcal{D}\delta\bar{X}^{\bar{i}}\,,\quad \delta\Gamma^\alpha=\bar{\mathcal{D}}\delta\Lambda^\alpha\,,\quad \delta\bar{\Gamma}^{\bar{\alpha}}=\mathcal{D}\delta\bar{\Lambda}^{\bar{\alpha}}\,. \tag{11}$$

The equations of motion thus obtained are

$$\begin{aligned} E^\Phi_i &= 2H_{j\bar{k}i}\bar{\mathcal{D}}\bar{\Phi}^{\bar{k}}\partial\Phi^j-2\bar{\mathcal{D}}(g_{i\bar{j}}\partial\bar{\Phi}^{\bar{j}})+i\bar{\mathcal{D}}\partial_i\mathcal{H}_{\alpha\beta}\Gamma^\alpha\Gamma^\beta- \\ &\quad -2i\bar{\mathcal{D}}(\partial_i\mathcal{H}_{\alpha\bar{\beta}}\bar{\Gamma}^{\bar{\beta}})\Gamma^\alpha+i\bar{\mathcal{D}}(\partial_i\mathcal{H}_{\bar{\alpha}\bar{\beta}}\bar{\Gamma}^{\bar{\alpha}}\bar{\Gamma}^{\bar{\beta}})=0\,, \end{aligned} \tag{12}$$

$$\begin{aligned} E^{\bar{\Phi}}_{\bar{i}} &= 2\mathcal{D}(g_{j\bar{i}}\partial\Phi^j)+2H_{k\bar{j}\bar{i}}\mathcal{D}\Phi^k\partial\bar{\Phi}^{\bar{j}}+i\mathcal{D}(\partial_{\bar{i}}\mathcal{H}_{\alpha\beta}\Gamma^\alpha\Gamma^\beta)+ \\ &\quad +2i\mathcal{D}(\partial_{\bar{i}}\mathcal{H}_{\alpha\bar{\beta}}\Gamma^\alpha\bar{\Gamma}^{\bar{\beta}})+i\mathcal{D}\partial_{\bar{i}}\mathcal{H}_{\bar{\alpha}\bar{\beta}}\bar{\Gamma}^{\bar{\alpha}}\bar{\Gamma}^{\bar{\beta}}=0\,, \end{aligned} \tag{13}$$

$$E^\Gamma_\alpha = \bar{\mathcal{D}}\mathcal{H}_{\alpha\beta}\Gamma^\beta+\bar{\mathcal{D}}(\mathcal{H}_{\alpha\bar{\beta}}\bar{\Gamma}^{\bar{\beta}})=0\,, \tag{14}$$

$$E^{\bar{\Gamma}}_{\bar{\alpha}} = \mathcal{D}(\mathcal{H}_{\beta\bar{\alpha}}\Gamma^\beta)-\mathcal{D}\mathcal{H}_{\bar{\alpha}\bar{\beta}}\bar{\Gamma}^{\bar{\beta}}=0\,. \tag{15}$$

3. Marginal deformations

In this section we consider the marginal deformations to the lowest order in α' and first order in conformal perturbation theory.

It can be seen that much like in the four-dimensional case [**8**] there are no Kähler deformations of the form

$$\int d^2z \mathcal{D}\bar{\mathcal{D}}X \ .$$

The only type of marginal deformations are of the form

$$S_{\mathcal{W}} = -\frac{i}{8\pi\alpha'}\int d^2z\, \mathcal{D}\mathcal{W} + \text{h.c.}\ , \tag{16}$$

where $\mathcal{W}$ must be a chiral primary of weights $(1,\, \frac{1}{2})$ and $\mathrm{U}(1)_{\mathrm{R}}$ charge $+1$. (For $S_{\mathcal{W}}$ to be truly marginal, these conditions should hold to any order in conformal perturbation theory and in α'.)

Since $\mathcal{D}^2 = 0$, $S_{\mathcal{W}}$ clearly remains unmodified under $\mathcal{W} \to \mathcal{W} + \mathcal{D}Y$, where Y has conformal weight $(1,\, 0)$ and R-charge $+2$. In the absence of world-sheet boundaries, it is also invariant under $\mathcal{W} \to \mathcal{W} + \partial\mathcal{Z}$ with $\mathcal{Z}$ being a superfield of weight $(0,\, \frac{1}{2})$ and R-charge $+1$, and $\partial\mathcal{Z}$ is required to be chiral. Finally, if we deform using $\mathcal{W}' = \mathcal{W} + \bar{\mathcal{D}}X$ one obtains an equivalent deformation $S_{\mathcal{W}'}$ because

$$S_{\mathcal{W}'} - S_{\mathcal{W}} = \int d^2z\, \mathcal{D}\bar{\mathcal{D}}X, \tag{17}$$

which is a trivial deformation.

A short note about the condition of chirality is in order. Working in conformal perturbation theory, we should expand the deformed action around the undeformed conformal theory. Therefore, we should treat the deformation $S_{\mathcal{W}}$ as a series of operator insertions in the undeformed correlation function evaluated at the conformal point. Insertions in a path integral satisfy the equations of motion of the undeformed action (up to possible contact terms with other insertions). Another point of view is that terms in the action that are proportional to the equations of motion can be removed by a field redefinition. Henceforth, *on-shell* will always mean on-shell with respect to the undeformed equations of motion.

Since our analysis is done at the first order in conformal perturbation theory and at tree-level in α', all the fields have their classical dimensions and we can treat the deformation as a classical object. The most general deformation with the required (1,1) conformal weight and R-charge $+1$ is

$$\begin{aligned} \mathcal{W} \;=\;& (\Lambda^{\bar\alpha}_{\bar i\beta}(\Phi,\bar\Phi)\Gamma_{\bar\alpha}\Gamma^{\beta} + \Lambda^{\alpha}_{\bar i\beta}\bar\Gamma_{\alpha}\Gamma^{\beta} + \Lambda^{\alpha}_{\bar i\bar\beta}(\Phi,\bar\Phi)\bar\Gamma_{\alpha}\bar\Gamma^{\bar\beta})\bar{\mathcal{D}}\Phi^{\bar i} + \\ & + (Y_{i\bar j}(\Phi,\bar\Phi)\partial\Phi^{i} + g_{i\bar k}Z_{\bar j}{}^{i}(\Phi,\bar\Phi)\partial\bar\Phi^{\bar k})\bar{\mathcal{D}}\bar\Phi^{\bar j}\ , \end{aligned} \tag{18}$$

where $\Lambda^{\bar\alpha}_{\bar i\beta} = \mathcal{H}^{\bar\alpha\gamma}\Lambda_{\bar i\gamma\beta}$ $(\Lambda_{\bar i\alpha\beta} = -\Lambda_{\bar i\beta\alpha})$ and $\Lambda^{\alpha}_{\bar i\bar\beta} = \mathcal{H}^{\bar\gamma\alpha}\Lambda_{\bar i\bar\gamma\bar\beta}$ $(\Lambda_{\bar i\bar\alpha\bar\beta} = -\Lambda_{\bar i\bar\beta\bar\alpha})$. A term of the form

$$\tilde{\mathcal{W}} = (\Omega_{\alpha\beta}(\Phi,\bar\Phi)\Gamma^{\alpha} + \Omega_{\bar\alpha\bar\beta}(\Phi,\bar\Phi)\bar\Gamma^{\bar\alpha})(\bar{\mathcal{D}}\bar\Gamma^{\bar\beta} + \bar{\mathcal{A}}^{\bar\beta}_{\bar i\bar\gamma}\bar{\mathcal{D}}\bar\Phi^{\bar i}\bar\Gamma^{\bar\gamma})$$

(where the derivative has been replaced by a gauge-covariant derivative to maintain gauge-invariance) does not appear because it can be absorbed in the deformation (18) by using the undeformed equations of motion.

For the theory to be well defined on the entire compact space, the deformation parameters must be sections of the appropriate bundles:

$$\Lambda \in \Gamma(\Omega^{0,1} \otimes \mathrm{End}\, V)\ , \quad Y \in \Gamma(\Omega^{1,1})\ , \quad Z \in \Gamma(\Omega^{0,1} \otimes \mathrm{T}^{1,0} M)\ .$$

The deformation (18) is not manifestly chiral as it depends on anti-chiral fields as well as chiral ones. However, as discussed above, it needs only be chiral on-shell in order to preserve (0,2) supersymmetry in conformal perturbation theory. On-shell

$$\begin{aligned}
\bar{\mathcal{D}}\mathcal{W} \;=\;& \left(\partial_{\bar{j}}\Lambda^{\bar{\alpha}}_{\bar{i}\beta} + \bar{\mathcal{A}}^{\bar{\alpha}}_{\bar{j}\bar{\gamma}}\Lambda^{\bar{\gamma}}_{\bar{i}\beta} + \frac{i}{2}Z_{\bar{i}}{}^{k}\tilde{\mathcal{F}}^{\bar{\alpha}}_{k\bar{j}\beta}\right)\Gamma_{\bar{\alpha}}\Gamma^{\beta}\bar{\mathcal{D}}\bar{\Phi}^{\bar{j}}\bar{\mathcal{D}}\bar{\Phi}^{\bar{i}} + \\
&+ \left(\partial_{\bar{j}}\Lambda^{\alpha}_{\bar{j}\beta} - \Lambda^{\alpha}_{\bar{i}\bar{\gamma}}\bar{\mathcal{A}}^{\bar{\gamma}}_{\bar{j}\beta} - \Lambda^{\alpha}_{\bar{i}\bar{\gamma}}\bar{\mathcal{A}}^{\bar{\gamma}}_{\bar{j}\beta} - iZ_{\bar{i}}{}^{k}\tilde{\mathcal{F}}^{\alpha}_{\bar{i}k\beta}\right)\bar{\Gamma}_{\alpha}\Gamma^{\beta}\bar{\mathcal{D}}\bar{\Phi}^{\bar{j}}\bar{\mathcal{D}}\bar{\Phi}^{\bar{i}} + \\
&+ \left(\partial_{\bar{j}}\Lambda^{\alpha}_{\bar{i}\bar{\beta}} - \bar{\mathcal{A}}^{\bar{\gamma}}_{\bar{j}\bar{\beta}}\Lambda^{\alpha}_{\bar{i}\bar{\gamma}} + \frac{i}{2}Z_{\bar{j}}{}^{k}\tilde{\mathcal{F}}^{\alpha}_{\bar{i}k\bar{\beta}}\right)\bar{\Gamma}_{\alpha}\bar{\Gamma}^{\bar{\beta}}\bar{\mathcal{D}}\bar{\Phi}^{\bar{j}}\bar{\mathcal{D}}\bar{\Phi}^{\bar{i}} + \\
&+ (\partial_{\bar{k}}Y_{i\bar{j}} + Z_{\bar{j}}{}^{l}H_{i\bar{k}l})\partial\Phi^{i}\bar{\mathcal{D}}\bar{\Phi}^{\bar{j}}\bar{\mathcal{D}}\Phi^{\bar{k}} + g_{i\bar{k}}\partial_{\bar{l}}Z_{\bar{j}}{}^{i}\partial\bar{\Phi}^{\bar{k}}\bar{\mathcal{D}}\bar{\Phi}^{\bar{l}}\bar{\mathcal{D}}\bar{\Phi}^{\bar{j}}\ .
\end{aligned}$$

Requiring that $\bar{\mathcal{D}}\mathcal{W} = 0$ yields the following constraints of the deformation parameters

$$\begin{aligned}
\mathcal{H}_{\alpha\bar{\gamma}}\left(\partial_{[\bar{j}}\Lambda^{\bar{\gamma}}_{\bar{i}]\beta} + \bar{\mathcal{A}}^{\bar{\gamma}}_{[\bar{j}\bar{\delta}}\Lambda^{\bar{\delta}}_{\bar{i}]\beta} + \frac{i}{2}Z_{[\bar{i}}{}^{k}\tilde{\mathcal{F}}^{\bar{\gamma}}_{|k|\bar{j}]\beta}\right) - (\alpha \leftrightarrow \beta) &= 0\ , \\
\partial_{[\bar{j}}\Lambda^{\alpha}_{\bar{i}]\beta} - \Lambda^{\alpha}_{[\bar{i}\bar{\gamma}}\bar{\mathcal{A}}^{\bar{\gamma}}_{\bar{j}]\beta} - iZ_{[\bar{i}}{}^{k}\tilde{\mathcal{F}}^{\alpha}_{\bar{j}]k\beta} &= 0\ , \\
\mathcal{H}_{\gamma\bar{\alpha}}\left(\partial_{[\bar{j}}\Lambda^{\gamma}_{\bar{i}]\bar{\beta}} - \bar{\mathcal{A}}^{\bar{\delta}}_{[\bar{j}\bar{\beta}}\Lambda^{\gamma}_{\bar{i}]\bar{\delta}} + \frac{i}{2}Z_{[\bar{j}}{}^{k}\tilde{\mathcal{F}}^{\gamma}_{\bar{i}]k\bar{\beta}}\right) - (\bar{\alpha} \leftrightarrow \bar{\beta}) &= 0\ , \\
\partial_{[\bar{k}}Y_{|i|\bar{j}]} - H_{i[\bar{k}|l|}Z_{\bar{j}]}{}^{l} &= 0\ , \\
\partial_{[\bar{l}}Z_{\bar{j}]}{}^{i} &= 0\ ,
\end{aligned} \tag{19}$$

where $[\dots]$ denotes anti-symmetrization with respect to space indices only and indices between bars are excluded from the anti-symmetrization.

As discussed earlier, deformations are subject to the equivalence relation $\mathcal{W} \simeq \mathcal{W} + \bar{\mathcal{D}}X + \partial\mathcal{Z}$. The most general X of weight (1,0) and R-charge 0 is (again at the classical level)

$$\begin{aligned}
X \;=\;& \lambda_{\alpha\beta}(\Phi,\bar{\Phi})\Gamma^{\alpha}\Gamma^{\beta} + \lambda^{\alpha}{}_{\beta}(\Phi,\bar{\Phi})\bar{\Gamma}_{\alpha}\Gamma^{\beta} + \lambda_{\bar{\alpha}\bar{\beta}}(\Phi,\bar{\Phi})\bar{\Gamma}^{\bar{\alpha}}\bar{\Gamma}^{\bar{\beta}} + \\
&+ \mu_{i}(\Phi,\bar{\Phi})\partial\Phi^{i} + g_{i\bar{j}}\zeta^{i}(\Phi,\bar{\Phi})\partial\bar{\Phi}^{\bar{j}}\ ,
\end{aligned} \tag{20}$$

where

$$\lambda^{\alpha}{}_{\beta}, \lambda^{\alpha}{}_{\bar{\beta}}, \lambda^{\bar{\alpha}}{}_{\beta} \in \Gamma(\mathrm{End}\, V)\ , \quad \mu \in \Gamma(\Omega^{1,0}(M))\ , \quad \zeta \in \Gamma(\mathrm{T}^{1,0}M)\ .$$

The most general $\mathcal{Z}$ of weight $(0,\frac{1}{2})$ and R-charge +1 is

$$\mathcal{Z} = \xi_{\bar{j}}(\Phi,\bar{\Phi})\bar{\mathcal{D}}\bar{\Phi}^{\bar{j}}\ . \tag{21}$$

$\partial\mathcal{Z}$ is chiral on-shell provided

$$\begin{aligned}
\nabla^{-}_{k}\partial_{[\bar{j}}\xi_{\bar{i}]} = 0\ , \quad \nabla^{-}_{\bar{k}}\partial_{[\bar{j}}\xi_{\bar{i}]} &= 0\ , \\
g^{\bar{i}j}(\partial_{[\bar{i}}\xi_{\bar{l}]}\tilde{\mathcal{F}}^{\bar{\alpha}}_{j\bar{k}\beta} - \partial_{[\bar{i}}\xi_{\bar{k}]}\tilde{\mathcal{F}}^{\bar{\alpha}}_{\bar{l}j\beta}) &= 0\ , \\
g^{\bar{i}j}(\partial_{[\bar{i}}\xi_{\bar{l}]}\tilde{\mathcal{F}}^{\alpha}_{\bar{k}j\beta} - \partial_{[\bar{i}}\xi_{\bar{k}]}\tilde{\mathcal{F}}^{\alpha}_{\bar{l}j\beta}) &= 0\ , \\
g^{\bar{i}j}(\partial_{[\bar{i}}\xi_{\bar{l}]}\tilde{\mathcal{F}}^{\alpha}_{\bar{k}j\bar{\beta}} - \partial_{[\bar{i}}\xi_{\bar{k}]}\tilde{\mathcal{F}}^{\alpha}_{\bar{l}j\bar{\beta}}) &= 0\ .
\end{aligned} \tag{22}$$

(The relations [**12**]

$$\Gamma_{ki\bar{j}} = \frac{1}{2}H_{ki\bar{j}} = -\frac{1}{2}(\partial_k g_{i\bar{j}} - \partial_i g_{k\bar{j}}) \,, \tag{23}$$

$$\Gamma_{l\bar{i}\bar{j}} = \frac{1}{2}(\partial_{\bar{i}} g_{l\bar{j}} + \partial_{\bar{j}} g_{l\bar{j}}) \,, \quad H_{l\bar{i}\bar{j}} = \partial_{\bar{i}} g_{l\bar{j}} - \partial_{\bar{j}} g_{l\bar{i}} \tag{24}$$

were used to rewrite the result in terms of the H-twisted connection ∇^-.)

Putting all these together, the equivalence relation $\mathcal{W} \simeq \mathcal{W} + \bar{\mathcal{D}}X + \partial\mathcal{Z}$ in component form are

$$\Lambda^{\bar{\alpha}}_{\bar{i}\beta} \simeq \Lambda^{\bar{\alpha}}_{\bar{i}\beta} + \mathcal{H}^{\bar{\alpha}\gamma}\partial_{\bar{i}}\lambda_{\gamma\beta} + \bar{\mathcal{A}}^{\bar{\alpha}}_{\bar{i}\gamma}\lambda^{\gamma}{}_{\beta} + \frac{i}{2}(\zeta^j + g^{\bar{k}j}\xi_{\bar{k}})\tilde{\mathcal{F}}^{\bar{\alpha}}_{j\bar{i}\beta} \,, \tag{25}$$

$$\Lambda^{\alpha}_{\bar{i}\beta} \simeq \Lambda^{\alpha}_{\bar{i}\beta} + \partial_{\bar{i}}\lambda^{\alpha}{}_{\beta} - 2\lambda^{\alpha}{}_{\bar{\gamma}}\bar{\mathcal{A}}^{\bar{\gamma}}_{\bar{i}\beta} - i(\zeta^j + g^{\bar{k}j}\xi_{\bar{k}})\tilde{\mathcal{F}}^{\alpha}_{\bar{i}j\beta} \,, \tag{26}$$

$$\Lambda^{\alpha}_{\bar{i}\bar{\beta}} \simeq \Lambda^{\alpha}_{\bar{i}\bar{\beta}} + \partial_{\bar{i}}\lambda^{\alpha}{}_{\bar{\beta}} - \bar{\mathcal{A}}^{\bar{\gamma}}_{\bar{i}\bar{\beta}}\lambda^{\alpha}{}_{\bar{\gamma}} - \frac{i}{2}(\zeta^j + g^{\bar{k}j}\xi_{\bar{k}})\tilde{\mathcal{F}}^{\alpha}_{\bar{i}j\bar{\beta}} \,, \tag{27}$$

$$Y_{i\bar{j}} \simeq Y_{i\bar{j}} + \partial_{\bar{j}}\mu_i + \partial_i\xi_{\bar{j}} + H_{i\bar{j}k}(\zeta^k + g^{\bar{l}k}\xi_{\bar{l}}) \,, \tag{28}$$

$$Z_{\bar{j}}{}^i \simeq Z_{\bar{j}}{}^i + \partial_{\bar{j}}(\zeta^i + g^{\bar{k}i}\xi_{\bar{k}}) + g^{\bar{k}i}(\partial_{\bar{k}}\xi_{\bar{j}} - \partial_{\bar{j}}\xi_{\bar{k}}) \,. \tag{29}$$

4. An example: $G = G' \times \mathrm{E}_8$, $\mathrm{U}(1)_{\mathrm{L}} \subset G'$

In this section we reconsider the case in which the bundle's surviving structure group is $G = G' \times \mathrm{E}_8$ with G' containing a $\mathrm{U}(1)_{\mathrm{L}}$ factor [**9**]. The analysis here differs from that in [**9**] by the inclusion of bundle deformations which break the $\mathrm{U}(1)_{\mathrm{L}}$ symmetry.

In this case

$$\mathcal{H}_{\alpha\beta} = \mathcal{H}_{\bar{\alpha}\bar{\beta}} = 0 \,, \tag{30}$$

from which it follows that

$$\mathcal{A}^{\alpha}_{i\bar{\beta}} = \bar{\mathcal{A}}^{\bar{\alpha}}_{\bar{i}\beta} = 0 \,, \quad \tilde{\mathcal{F}}^{\bar{\alpha}}_{j\bar{k}\beta} = \tilde{\mathcal{F}}^{\alpha}_{\bar{k}j\bar{\beta}} = 0, \quad \tilde{\mathcal{F}}^{\alpha}_{\bar{k}j\beta} = \mathcal{F}^{\alpha}_{\bar{k}j\beta} \,. \tag{31}$$

Thus, the constraints on the deformations (19) become

$$\begin{aligned}
\mathcal{H}_{\alpha\bar{\gamma}}(\partial_{[\bar{j}}\Lambda^{\bar{\gamma}}_{\bar{i}]\beta} + \bar{\mathcal{A}}^{\bar{\gamma}}_{[\bar{j}\bar{\delta}}\Lambda^{\bar{\delta}}_{\bar{i}]\beta}) - (\alpha \leftrightarrow \beta) &= 0 \,, \\
\partial_{[\bar{j}}\Lambda^{\alpha}_{\bar{i}]\beta} - iZ_{[\bar{i}}{}^k\mathcal{F}^{\alpha}_{\bar{j}]k\beta} &= 0 \,, \\
\mathcal{H}_{\gamma\bar{\alpha}}(\partial_{[\bar{j}}\Lambda^{\gamma}_{\bar{i}]\bar{\beta}} - \bar{\mathcal{A}}^{\bar{\delta}}_{[\bar{j}\bar{\beta}}\Lambda^{\gamma}_{\bar{i}]\bar{\delta}}) - (\alpha \leftrightarrow \beta) &= 0 \,, \\
\partial_{\bar{k}}Y_{i\bar{j}} - H_{i\bar{k}l}Z_{\bar{j}}{}^l &= \partial_{\bar{j}}Y_{i\bar{k}} - H_{i\bar{j}l}Z_{\bar{k}}{}^l \,, \\
\partial_{[\bar{k}}Z_{\bar{j}]}{}^i &= 0 \,.
\end{aligned} \tag{32}$$

The components of $\mathcal{Z}$ should satisfy

$$\begin{aligned}
\nabla^-_k\partial_{[\bar{i}}\xi_{\bar{j}]} &= 0 \,, \\
\nabla^-_{\bar{k}}\partial_{[\bar{i}}\xi_{\bar{j}]} &= 0 \,, \\
g^{\bar{i}j}(\partial_{[\bar{i}}\xi_{\bar{m}]}\mathcal{F}^{\alpha}_{\bar{k}j\beta} - \partial_{[ib}\xi_{\bar{k}]}\mathcal{F}^{\alpha}_{\bar{m}j\beta}) &= 0 \,.
\end{aligned} \tag{33}$$

The equivalence relations (29) are then reduced to

$$\Lambda^{\bar\alpha}_{\bar i\beta} \simeq \Lambda^{\bar\alpha}_{\bar i\beta} + \mathcal{H}^{\bar\alpha\gamma}\partial_{\bar i}\lambda_{\gamma\beta} \,, \tag{34}$$

$$\Lambda^{\alpha}_{\bar i\beta} \simeq \Lambda^{\alpha}_{\bar i\beta} + \partial_{\bar i}\lambda^{\alpha}{}_{\beta} - i(\zeta^j + g^{\bar k j}\xi_{\bar k})\mathcal{F}^{\alpha}_{\bar i j\beta} \,, \tag{35}$$

$$\Lambda^{\alpha}_{\bar i\bar\beta} \simeq \Lambda^{\alpha}_{\bar i\bar\beta} + \partial_{\bar i}\lambda^{\alpha}{}_{\bar\beta} - \bar{\mathcal{A}}^{\bar\gamma}_{\bar i\bar\beta}\lambda^{\alpha}{}_{\bar\gamma} \,, \tag{36}$$

$$Y_{i\bar j} \simeq Y_{i\bar j} + \partial_{\bar j}\mu_i + \partial_i\xi_{\bar j} + H_{i\bar j k}(\zeta^k + g^{\bar l k}\xi_{\bar l}) \,, \tag{37}$$

$$Z_{\bar j}{}^{i} \simeq Z_{\bar j}{}^{i} + \partial_{\bar j}(\zeta^i + g^{\bar k i}\xi_{\bar k}) + g^{\bar k i}(\partial_{\bar k}\xi_{\bar j} - \partial_{\bar j}\xi_{\bar k}) \,. \tag{38}$$

These are the same as the results obtained in [**9**] with the addition of deformations which break the $\mathrm{U}(1)_{\mathrm{L}}$ symmetry.

We can bring the extra deformation parameterized by $\Lambda^{\bar\alpha}_{\bar i\beta}$ and $\Lambda^{\alpha}_{\bar i\bar\beta}$ to a nicer form. Doing a little algebra yields

$$\begin{aligned} \mathcal{H}_{\alpha\bar\gamma}\partial_{\bar j}\Lambda^{\bar\gamma}_{\bar i\beta} &= \partial_{\bar j}\Lambda_{\bar i\alpha\beta} - \partial_{\bar j}\mathcal{H}_{\alpha\bar\gamma}\Lambda^{\bar\gamma}_{\bar i\beta} \,, \\ \mathcal{H}_{\alpha\bar\gamma}\bar{\mathcal{A}}^{\bar\gamma}_{\bar j\bar\delta}\Gamma^{\bar\delta}_{\bar i\beta} &= \partial_{\bar j}\mathcal{H}_{\alpha\bar\gamma}\Lambda^{\bar\gamma}_{\bar i\beta} \,, \end{aligned} \tag{39}$$

so we can rewrite (34) and its associated equivalence relation as

$$\partial_{[\bar j}\Lambda_{\bar i]\alpha\beta} = 0 \,, \quad \Lambda_{\bar i\alpha\beta} \simeq \Lambda_{\bar i\alpha\beta} + \partial_{\bar i}\lambda_{\alpha\beta} \,. \tag{40}$$

Hence, these extra deformations are elements of $H^1(M, V^* \wedge V^*)$. A similar manipulation of (36) gives

$$\partial_{[\bar i}\Lambda^{\alpha\beta}_{\bar j]} = 0 \,, \quad \Lambda^{\alpha\beta}_{\bar i} \simeq \Lambda^{\alpha\beta} + \partial_{\bar i}\lambda^{\alpha\beta}, \tag{41}$$

which are elements of $H^1(M, V \wedge V)$.

In particular, for the heterotic string compactified on a Calabi-Yau three-fold and the standard embedding of the tangent bundle in the vector bundle, which breaks the first E_8 into E_6, these new deformations will be in $H^1(M, \mathrm{T}^*M \wedge \mathrm{T}^*M) \simeq H^{2,1}(M)$ and $H^1(M, \mathrm{T}M \wedge \mathrm{T}M) \simeq H^{1,1}(M)$.

These are the same as the results obtained in [**9**] with the addition of deformations that break the $\mathrm{U}(1)_{\mathrm{L}}$ symmetry. To our knowledge, these deformations are new.[1] A possible application of these new deformations is breaking the E_6 gauge group of the standard embedding to SO(10). In this case the **78** adjoint representation of E_6 is decomposed under its $\mathrm{SO}(10) \times \mathrm{U}(1)_{\mathrm{L}}$ subgroup into $\mathbf{45}_0 \oplus \mathbf{16}_{-3} \oplus \overline{\mathbf{16}}_3 \oplus \mathbf{1}_0$ [**11**]. A deformation breaking the $\mathrm{U}(1)_{\mathrm{L}}$ should Higgs all but the **45** of SO(10). The **1** is clearly lifted and the two spinor representations must become massive as well for the consistency of the low-energy effective theory.

Acknowledgments

The author would like to thank Ilarion Melnikov and Ronen Plesser for collaboration on a related but yet unpublished project on marginal deformations of (0,2) SCFTs, whose intermediate results proved to be useful for this paper. The author would also like to thank Ilarion Melnikov for many useful discussions, sharing some of his notes and for commenting on an initial draft. This research was supported by FAPESP fellowship 2010/07439-0.

---

[1]The author would like to thank I. V. Melnikov for discussions on this point as well as for sharing some results obtained by him and by J. McOrist.

References

[1] Ido Adam, Ilarion V. Melnikov, and M. Ronen Plesser, *On marginal deformations of (0,2) SCFTs*, unpublished.

[2] Tom Banks and Lance J. Dixon, *Constraints on String Vacua with Space-Time Supersymmetry*, Nucl.Phys. **B307** (1988), 93–108.

[3] Thomas Banks, Lance J. Dixon, Daniel Friedan, and Emil Martinec, *Phenomenology and conformal field theory or can string theory predict the weak mixing angle?*, Nuclear Phys. B **299** (1988), no. 3, 613–626, DOI 10.1016/0550-3213(88)90551-2. MR930976 (89b:81244)

[4] Chris Beasley and Edward Witten, *Residues and world-sheet instantons*, J. High Energy Phys. **10** (2003), 065, 39 pp. (electronic), DOI 10.1088/1126-6708/2003/10/065. MR2030598 (2005f:81223)

[5] Michael Dine and Nathan Seiberg, $(2,0)$ *superspace*, Phys. Lett. B **180** (1986), no. 4, 364–369, DOI 10.1016/0370-2693(86)91203-7. MR867261 (88b:81143)

[6] Jacques Distler and Brian Greene, *Aspects of* $(2,0)$ *string compactifications*, Nuclear Phys. B **304** (1988), no. 1, 1–62, DOI 10.1016/0550-3213(88)90619-0. MR953711 (89m:81135)

[7] S. Ferrara, D. Lüst, A. Shapere, and S. Theisen, *Modular invariance in supersymmetric field theories*, Phys. Lett. B **225** (1989), no. 4, 363–366, DOI 10.1016/0370-2693(89)90583-2. MR1009091 (90k:83097)

[8] Daniel Green, Zohar Komargodski, Nathan Seiberg, Yuji Tachikawa, and Brian Wecht, *Exactly marginal deformations and global symmetries*, J. High Energy Phys. **6** (2010), 106, 19, DOI 10.1007/JHEP06(2010)106. MR2680311 (2011g:81284)

[9] Ilarion V. Melnikov and Eric Sharpe, *On marginal deformations of* $(0,2)$ *non-linear sigma models*, Phys. Lett. B **705** (2011), no. 5, 529–534, DOI 10.1016/j.physletb.2011.10.055. MR2860523

[10] Eva Silverstein and Edward Witten, *Criteria for conformal invariance of* $(0,2)$ *models*, Nuclear Phys. B **444** (1995), no. 1-2, 161–190, DOI 10.1016/0550-3213(95)00186-V. MR1344416 (96j:81106)

[11] Richard Slansky, *Group theory for unified model building*, Phys. Rep. **79** (1981), no. 1, 1–128, DOI 10.1016/0370-1573(81)90092-2. MR639396 (83d:81112)

[12] Andrew Strominger, *Superstrings with torsion*, Nuclear Phys. B **274** (1986), no. 2, 253–284, DOI 10.1016/0550-3213(86)90286-5. MR851702 (87m:81177)

[13] Edward Witten, *Phases of* $N = 2$ *theories in two dimensions*, Nuclear Phys. B **403** (1993), no. 1-2, 159–222, DOI 10.1016/0550-3213(93)90033-L. MR1232617 (95a:81261)

Instituto de Física Teórica, Universidade Estadual Paulista, R. Dr. Bento T. Ferraz 271, Bloco II, 01140-070, São Paulo, SP, Brasil

E-mail address: idoadam.physics@yahoo.com

Proceedings of Symposia in Pure Mathematics
Volume **90**, 2015
http://dx.doi.org/10.1090/pspum/090/01523

Quantum Hypermultiplet Moduli Spaces in $\mathcal{N}=2$ String Vacua: A Review

Sergei Alexandrov, Jan Manschot, Daniel Persson, and Boris Pioline

Abstract. The hypermultiplet moduli space $\mathcal{M}_H$ in type II string theories compactified on a Calabi-Yau threefold X is largely constrained by supersymmetry (which demands quaternion-Kählerity), S-duality (which requires an isometric action of $SL(2,\mathbb{Z})$) and regularity. Mathematically, $\mathcal{M}_H$ ought to encode all generalized Donaldson-Thomas invariants on X consistently with wall-crossing, modularity and homological mirror symmetry. We review recent progress towards computing the exact metric on $\mathcal{M}_H$, or rather the exact complex contact structure on its twistor space.

Contents

1. Introduction

String vacua with $N=2$ supersymmetry in four dimensions offer a unique opportunity to investigate non-perturbative aspects of the low energy effective action and of the spectrum of black hole bound states. Unlike in vacua with higher supersymmetry, the two-derivative effective action in general receives non-trivial quantum corrections, while degeneracies of BPS black holes depend non trivially on the value of the moduli at spatial infinity. Both issues are in fact related, since BPS black holes in 4 dimensions yield BPS instantons upon reduction on a circle, and the resulting instanton corrections to the three-dimensional effective action can sometimes (after T-duality along the circle) lift back to 4 dimensions.

For ungauged $\mathcal{N}=2$ vacua, the complete two-derivative effective action is encoded in the Riemannian metric on the moduli space, which famously factorizes as the product $\mathcal{M}_V \times \mathcal{M}_H$ of the vector multiplet (VM) and hypermultiplet (HM)

2010 *Mathematics Subject Classification.* Primary 11F03, 53D37, 81T30, 83E50.
Prepared for the proceedings of String Math 2012, Bonn.

moduli spaces, respectively [**1**, **2**]. A complete understanding of the former was achieved in the 90s, leading to deep connections with algebraic geometry, most notably the discovery of classical mirror symmetry. By contrast, our understanding of the latter has long remained rudimentary, mainly due to the difficulty of parametrizing quaternion-Kähler (QK) metrics on $\mathcal{M}_H$, as required by supersymmetry. The situation has considerably improved in recent years, as twistorial techniques [**3**–**9**] were used to reformulate this problem analytically, in terms of the complex contact structure on the twistor space $\mathcal{Z}$ (a $\mathbb{P}^1$-bundle over $\mathcal{M}_H$), and a suitable set of complex Darboux coordinates on $\mathcal{Z}$.

The purpose of this contribution is to give a survey of recent progress towards determining the exact hypermultiplet moduli space metric (see [**10**] for a review with different emphasis). We focus on type II strings compactified on Calabi-Yau (CY) threefolds , although other dual formulations of the same vacua (see [**11**–**13**] for recent progress on $K3 \times T^2$ heterotic vacua) may eventually be useful for achieving the stated goal. In §2, we summarize the structure of the perturbative hypermultiplet moduli space in type IIA and type IIB vacua, emphasizing its twistorial description. In §3, we discuss instanton corrections from Euclidean D-branes wrapped on supersymmetric cycles inside the CY threefold X, and provide a twistorial construction of these corrections parametrized by the Donaldson-Thomas invariants of X. We explain the consistency of this construction with wall-crossing using the so called QK/HK correspondence which allows to reformulate the resulting corrections to the QK metric on $\mathcal{M}_H$ in terms of corrections to an auxiliary, or "dual", hyperkähler (HK) space $\mathcal{M}'_H$. In §4, we discuss the implications of the modular symmetry of type IIB strings (known as S-duality) for these D-instanton corrections, with particular emphasis on D3-brane instantons (corresponding to divisors in X). We show that in the large volume, one-instanton approximation, S-duality holds thanks to special modular properties of the DT invariants for divisors, and of the indefinite theta series which sum up D1-D(-1) instanton effects at fixed D3-brane charge. In §5, we use the same duality to obtain Neveu-Schwarz (NS) fivebrane instantons from D5-instantons, and relate these contributions to the topological string amplitude on X. We also discuss some conjectural relations between NS5-branes and quantum integrable systems.

Acknowledgments

We are grateful to P. Roche, F. Saueressig and S. Vandoren for collaboration on some of the material presented here, and to N. Hitchin, A. Kleinschmidt, S. Monnier, R. Minasian, G. Moore, A. Neitzke, B. Nilsson, and Y. Soibelman for related discussions. DP and BP wish to thank the organizers of String Math 2012 for the opportunity to report on some of this work. DP also thanks the organizers and participants of the mini-workshop on "Hypers" in Hamburg, March 2013, where a series of lectures on part of this work was given. JM is supported in part by a Krupp fellowship.

Note added in proof, Jan 2015. This review was updated to incorporate references to relevant works which have appeared since the first release in April 2013.

2. Perturbative moduli space

In this section we discuss the one-loop corrected hypermultiplet moduli space in type IIA and type IIB string theories compactified on CY threefolds X and $\hat{X}$, respectively. Higher loop corrections are expected to vanish after suitable field redefinitions [**14**–**16**]. If $(X, \hat{X})$ is a mirror pair, then the moduli spaces are isometric as a consequence of classical mirror symmetry [**17**].

2.1. Type IIA.

2.1.1. *Topology.* Type IIA string theory associates to each compact CY threefold X a real $4(h_{2,1}(X)+1)$-dimensional quaternion-Kähler space $\mathcal{M}_H = \mathcal{M}_H(X)$. Topologically, $\mathcal{M}_H$ is a $\mathbb{C}^\times$-bundle

$$\mathbb{C}^\times \longrightarrow \mathcal{M}_H(X) \longrightarrow \mathcal{J}_W(X) \longrightarrow \mathcal{M}_C(X) \tag{2.1}$$

over the Weil intermediate Jacobian $\mathcal{J}_W(X)$. The latter is a torus bundle over the complex structure moduli space $\mathcal{M}_C(X)$, with generic fiber $\mathcal{T} = H^3(X,\mathbb{R})/H^3(X,\mathbb{Z})$, endowed with the Weil complex structure where $H^{3,0} \oplus H^{1,2}$ generate the holomorphic tangent space.

To see how this arises from physics, consider the $\mathbb{C}^\times$-bundle $\mathcal{L}_X \to \mathcal{M}_C(X)$ with fibre the space of nowhere vanishing holomorphic 3-forms $\Omega^{3,0}$ on X. Fixing a symplectic basis (A^Λ, B_Λ), $\Lambda = 0, \ldots, h_{2,1}(X)$ of $\Gamma = H_3(X,\mathbb{Z})$, the period integrals

$$X^\Lambda = \int_{A^\Lambda} \Omega^{3,0}, \qquad F_\Lambda = \int_{B_\Lambda} \Omega^{3,0} \tag{2.2}$$

realize $\mathcal{L}_X$ as a complex Lagrangian cone in $H^3(X,\mathbb{C})$. Locally, the B-periods F_Λ can be expressed in terms of the A-periods X^Λ as derivatives of a holomorphic prepotential $F(X^\Lambda)$ homogeneous of degree 2. The ratios $z^a = X^a/X^0$, $a = 1, \ldots, h_{2,1}$, parametrize the moduli space of complex structures $\mathcal{M}_C(X)$, and describe the scalar degrees of freedom in type IIA/X originating from the metric in 10 dimensions. The periods (X^Λ, F_Λ) are valued in the Hodge bundle $\mathcal{L}_X$ (times a symplectic vector bundle associated to changes of the symplectic basis).

In addition, the periods of the ten-dimensional Ramond-Ramond (RR) three-form C

$$\zeta^\Lambda = \int_{A^\Lambda} C, \qquad \tilde{\zeta}_\Lambda = \int_{B_\Lambda} C \tag{2.3}$$

yield scalar moduli valued in $H^3(X,\mathbb{R})$. Invariance under large gauge transformations $C \to C + H$ with $H \in \Gamma$ imply that $(\zeta^\Lambda, \tilde{\zeta}_\Lambda)$ are periodic with integer periods, hence live in the torus $\mathcal{T}$. Sometimes we abuse notation and write $\Omega^{3,0} = (X^\Lambda, F_\Lambda)$, $C = (\zeta^\Lambda, \tilde{\zeta}_\Lambda)$. Just like $\Omega^{3,0}$, the vector C transforms by a symplectic rotation under monodromies in $\mathcal{M}_C(X)$, implying that $\mathcal{T}$ is non-trivially fibered over $\mathcal{M}_X$. The total space of this bundle is the intermediate Jacobian $\mathcal{J}_W(X)$.

Finally, the four-dimensional dilaton e^ϕ and the Poincaré dual σ to the B-field in four dimensions provide an additional complex scalar degree of freedom in four dimensions, corresponding to the $\mathbb{C}^\times$ fiber in (2.1). Large gauge transformations of the B-field identify $\sigma \mapsto \sigma + 2\kappa$ with κ integer (for a suitable normalization), while the afore-mentioned large gauge transformations also act on the axion σ by a shift [**18**, **19**],

$$(C, \sigma) \longmapsto \Big(C + H, \sigma + 2\kappa + \langle C, H\rangle + 2c(H) \Big). \tag{2.4}$$

Here $c(H)$ provides the quadratic refinement $\lambda(H) \equiv (-1)^{2c(H)}$ of the intersection form $\langle\, ,\, \rangle$ on Γ satisfying

$$\lambda(H + H') = (-1)^{\langle H, H'\rangle} \lambda(H)\, \lambda(H'). \tag{2.5}$$

Given a choice of symplectic basis of $H^3(X,\mathbb{Z})$, any quadratic refinement can be parametrized as

$$\lambda(H) = (-1)^{2c(H)} = e^{-i\pi m_\Lambda n^\Lambda + 2\pi i(m_\Lambda \theta^\Lambda - n^\Lambda \phi_\Lambda)}, \tag{2.6}$$

where $H = (n^\Lambda, m_\Lambda)$ are the components of H along (A^Λ, B_Λ) and $\Theta \equiv (\theta^\Lambda, \phi_\Lambda)$ are a choice of characteristics in $\mathcal{T}$. Note that (2.6) defines $c(H)$ only modulo integers, but the corresponding ambiguity in (2.4) can be absorbed in κ. The extra shift $2c(H)$ of σ in (2.4) is needed to ensure the closure of the group action. Altogether, the large gauge transformations of the B and C fields define a discrete Heisenberg group action $\mathrm{H}(\mathbb{Z})$ which will play a central role in the discussion of NS5-brane instanton effects in §5. Eq. (2.4) completely specifies the restriction of the $\mathbb{C}^\times$-bundle (or rather, its unit circle bundle $\mathscr{C}_\sigma$, with S^1-fiber parametrized by the axion σ) in (2.1) to the torus $\mathcal{T}$. The topology of the bundle over the full intermediate Jacobian $\mathcal{J}_W(X)$ will be discussed in the next paragraph after discussing the one-loop corrected metric and in more detail in §5.1.

2.1.2. *Perturbative metric.* At tree-level, the metric on $\mathcal{M}_H$ belongs to the class of 'semi-flat' QK metrics discovered in [**20, 21**]. In particular, its restriction to $\mathcal{M}_C(X)$ is the special Kähler metric $g_{\mathcal{M}_C(X)}$ deduced from the prepotential $F(X^\Lambda)$, with Kähler potential [**22**]

$$\mathcal{K} = -\log\left[\mathrm{i}\int_X \Omega^{3,0}\wedge\overline{\Omega^{3,0}}\right] = -\log\left[\mathrm{i}(\bar{X}^\Lambda F_\Lambda - X^\Lambda \bar{F}_\Lambda)\right], \tag{2.7}$$

while along $\mathcal{T}\ltimes S^1$ it has continuous isometries. As shown in [**14, 15, 23, 24**], the one-loop correction takes the metric outside the above class, still preserving flatness along $\mathcal{T}\ltimes S^1$. The resulting metric on $\mathcal{M}_H$ can be written as

$$g_{\mathrm{pert}} = \frac{r+2c}{r^2(r+c)}\,\mathrm{d}r^2 + \frac{4(r+c)}{r}\,g_{\mathcal{M}_C(X)} + \frac{1}{r}\,g_{\mathcal{T}}(c) + \frac{r+c}{16r^2(r+2c)}\left(\mathrm{d}\sigma + \mathcal{A}(c)\right)^2, \tag{2.8}$$

where $r = e^\phi$ and the parameter $c = -\chi(X)/(192\pi)$ encodes the one-loop correction, governed solely by the Euler number of X. Here $g_{\mathcal{T}}(c)$ denotes a deformation of the standard Weil metric on the torus $\mathcal{T}$ which can be found in [**24**]. Most importantly, the connection $\mathcal{A}(c)$ on the circle bundle $\mathscr{C}_\sigma$ is given by

$$\mathcal{A}(c) = \tilde{\zeta}_\Lambda \mathrm{d}\zeta^\Lambda - \zeta^\Lambda \mathrm{d}\tilde{\zeta}_\Lambda + 8c\mathcal{A}_K, \qquad \mathcal{A}_K = \frac{\mathrm{i}}{2}(\partial_{z^a}\mathcal{K}\mathrm{d}z^a - \partial_{\bar{z}^{\bar{a}}}\mathcal{K}\mathrm{d}\bar{z}^{\bar{a}}). \tag{2.9}$$

Here $\mathcal{A}_K$ is the Kähler connection on the Hodge bundle $\mathcal{L}_X$ with Kähler potential (2.7). The second term in (2.9) follows by reducing the topological coupling $B\wedge I_8$ in the ten-dimensional type IIA action and dualizing B into σ (see [**19**] for details). The tree-level metric is recovered by setting $c=0$.

The connection (2.9) implies that the circle bundle $\mathscr{C}_\sigma$ has non-trivial curvature both along the torus $\mathcal{T}$, in accordance with (2.4), but also along the base $\mathcal{M}_C(X)$ of the intermediate Jacobian $\mathcal{J}_W(X)$; the first Chern class is given by

$$c_1(\mathscr{C}_\sigma) = \omega_{\mathcal{T}} + \frac{\chi(X)}{24}\,\omega_C, \qquad \omega_{\mathcal{T}} \equiv \mathrm{d}\tilde{\zeta}_\Lambda\wedge\mathrm{d}\zeta^\Lambda, \qquad \omega_C \equiv -\frac{1}{2\pi}\,\mathrm{d}\mathcal{A}_K. \tag{2.10}$$

We return to the topology of the axion circle bundle in relation to NS5-instantons in §5. For now, notice that putative higher loop corrections would in general induce corrections to $c_1(\mathscr{C}_\sigma)$ suppressed by inverse powers of r, contradicting the requirement that $c_1(\mathscr{C}_\sigma)\in H^2(\mathcal{M}_H,\mathbb{Z})$. Note that for $\chi(X)>0$, the metric (2.8) has a curvature singularity at $r=-2c$ (while $r=0$, $r=-c$ are coordinate singularities).

This singularity is expected to be resolved once the full set of non-perturbative corrections is included [**25**].

2.1.3. *Twistor space description.* The most convenient way of describing a QK manifold $\mathcal{M}$ is via its twistor space $\mathcal{Z}$ [**3**]. Recall that a quaternion-Kähler manifold of real dimension $4n$ has holonomy group contained in $USp(n)\times SU(2)\subset SO(4n)$. In particular, it has a triplet of almost complex structures $\vec{J}$ (defined locally up to $SU(2)$ rotations) satisfying the quaternion algebra, corresponding two-forms $\vec{\omega}$ and a globally defined closed 4-form $\vec{\omega}\wedge\vec{\omega}$. The J_i's are not integrable unless the scalar curvature of $\mathcal{M}$ vanishes, in which case $\mathcal{M}$ is hyperkähler. Nevertheless, it is possible to encode the geometry of $\mathcal{M}$ complex analytically, by passing to its twistor space $\mathcal{Z}$, the total space of a canonical $\mathbb{P}^1$-bundle over $\mathcal{M}$. $\mathcal{Z}$ carries a canonical complex contact structure, given by the kernel of the $\mathcal{O}(2)$-twisted, (1,0)-form

$$Dt=\mathrm{d}t+p_+-\mathrm{i}p_3t+p_-t^2, \tag{2.11}$$

where t is a stereographic coordinate on $\mathbb{P}^1$ and $(p_\pm=-\frac{1}{2}(p_1\mp\mathrm{i}p_2),p_3)$ denotes the $SU(2)$-part of the Levi-Civita connection on $\mathcal{M}$.

Locally on an open patch $\mathcal{U}_i\subset\mathcal{Z}$ there exists a function $\Phi_{[i]}$, the 'contact potential', which is holomorphic along the twistor lines (i.e. the fibers of $\mathcal{Z}\longrightarrow\mathcal{M}$) and such that the product

$$\mathcal{X}^{[i]}=-4\mathrm{i}\,e^{\Phi_{[i]}}\,Dt/t \tag{2.12}$$

is a holomorphic (i.e. $\bar\partial$-closed) one-form. The nowhere vanishing holomorphic top-form $\mathcal{X}\wedge(\mathrm{d}\mathcal{X})^n$ defines the complex contact structure on $\mathcal{Z}$. Locally, by a complex–contact analogue of the Darboux theorem, one can always choose complex coordinates $(\xi^\Lambda_{[i]},\tilde\xi^{[i]}_\Lambda,\alpha^{[i]})$ in $\mathcal{U}_i$ such that the contact one-form (2.12) takes the canonical form [**26, 27**]

$$\mathcal{X}^{[i]}=\mathrm{d}\alpha^{[i]}+\xi^\Lambda_{[i]}\mathrm{d}\tilde\xi^{[i]}_\Lambda. \tag{2.13}$$

In what follows it will often be convenient to combine ξ^Λ and $\tilde\xi_\Lambda$ into a symplectic vector $\Xi=(\xi^\Lambda,\tilde\xi_\Lambda)$, and to define a variant $\tilde\alpha=-2\alpha-\tilde\xi_\Lambda\xi^\Lambda$ of the coordinate α such that

$$\mathcal{X}^{[i]}=-\frac{1}{2}\left(\mathrm{d}\tilde\alpha^{[i]}+\tilde\xi^{[i]}_\Lambda\mathrm{d}\xi^\Lambda_{[i]}-\xi^\Lambda_{[i]}\mathrm{d}\tilde\xi^{[i]}_\Lambda\right)=-\frac{1}{2}\left(\mathrm{d}\tilde\alpha^{[i]}+\langle\Xi^{[i]},\mathrm{d}\Xi^{[i]}\rangle\right). \tag{2.14}$$

The global complex contact structure on $\mathcal{Z}$ is then encoded into the set of complex contact transformations between overlapping Darboux coordinate systems on $\mathcal{U}_i\cap\mathcal{U}_j$. A convenient way of specifying these contact transformations is via a set of Hamilton generating functions $H^{[ij]}(\xi,\tilde\xi,\alpha)\in H^1(\mathcal{Z},\mathcal{O}(2))$, as explained in detail in [**16, 27**]. The QK metric can be reconstructed by (i) parametrizing the twistor lines, i.e. expressing the complex Darboux coordinates (Ξ,α) in terms of the local coordinates (t,x^μ) on $\mathbb{P}^1\times\mathcal{M}$; (ii) evaluating the contact one-form (2.13) and matching the result with (2.11) so as to extract the $SU(2)$ connection $\vec{p}$; (iii) computing the quaternionic 2-forms $\vec{\omega}$ via $\mathrm{d}\vec{p}+\frac{1}{2}\vec{p}\wedge\vec{p}=\frac{\nu}{2}\vec{\omega}$ (ν is the constant curvature of $\mathcal{M}_H$); (iv) constructing the space of $(1,0)$-forms with respect to J_3, by expanding the differentials $(\mathrm{d}\Xi,\mathrm{d}\alpha)$ around $t=0$, and finally, (v) contracting the Kähler form ω_3 with the complex structure J_3.

In this framework, the perturbative metric (2.8) is captured by the following Darboux coordinates in the patch $\mathcal{U}_0=\mathbb{P}^1\backslash\{0,\infty\}$ [**27**] (building on earlier work

[**24**, **26**, **28**]):

$$\begin{aligned} \xi^\Lambda &= \zeta^\Lambda + 2\sqrt{r+c}\, e^{\mathcal{K}/2}\big(t^{-1}X^\Lambda - t\,\bar{X}^\Lambda\big), \\ \tilde{\xi}_\Lambda &= \tilde{\zeta}_\Lambda + 2\sqrt{r+c}\, e^{\mathcal{K}/2}\big(t^{-1}F_\Lambda - t\,\bar{F}_\Lambda\big), \\ \tilde{\alpha} &= \sigma + 2\sqrt{r+c}\, e^{\mathcal{K}/2}\big(t^{-1}W - t\,\bar{W}\big) - 8\mathrm{i}\, c\log t, \end{aligned} \tag{2.15}$$

where $W \equiv F_\Lambda\zeta^\Lambda - X^\Lambda\tilde{\zeta}_\Lambda$, whereas the contact potential coincides with the dilaton, $\Phi = \phi$. The last term in the expression for $\tilde{\alpha}$ is the sole effect of the one-loop correction in this framework (except for a field redefinition $r \to r + c$). Under a holomorphic rescaling $\Omega^{3,0} \to e^f\Omega^{3,0}$, the Kähler potential $\mathcal{K}$ and coordinates t, σ vary according to $\mathcal{K} \to \mathcal{K} - f - \bar{f}$, $t \mapsto e^{\mathrm{i}\,\mathrm{Im}\, f}t$, $\sigma \mapsto \sigma - 8c\,\mathrm{Im}\, f$, leaving (2.15) invariant.

For our purposes, it is important to note two key properties of the twistorial approach. First, quaternionic isometries of $\mathcal{M}$ (i.e. preserving the 4-form $\vec{\omega}\wedge\vec{\omega}$) are classified by the Cech cohomology group $H^0(\mathcal{Z}, \mathcal{O}(2))$ via the moment map construction, and therefore lift to holomorphic actions on $\mathcal{Z}$ [**29**]. In particular, the action of the Heisenberg group (2.4) on $\mathcal{M}_H$ lifts to a holomorphic action on the Darboux coordinates (2.15) as

$$\big(\Xi, \tilde{\alpha}\big) \longmapsto \big(\Xi + H,\ \tilde{\alpha} + 2\kappa + \langle \Xi, H\rangle + 2c(H)\big). \tag{2.16}$$

The second property is that linear deformations of a QK space $\mathcal{M}$ are classified by sections of $H^1(\mathcal{Z}, \mathcal{O}(2))$ [**30**].

2.2. Type IIB. We now turn to the perturbative HM moduli space $\hat{\mathcal{M}}_H(\hat{X})$ in type IIB string theory compactified on a CY threefold $\hat{X}$. Mirror symmetry requires that it should be isometric to the previously discussed type IIA HM moduli space $\mathcal{M}_H(X)$ whenever $(X, \hat{X})$ form a dual pair. The two spaces however come with different natural coordinates, and it is important to determine the 'mirror map' between the two sides.

On the type IIB side, the HM moduli space has a similar fibration structure as in (2.1),

$$\mathbb{C}^\times \longrightarrow \hat{\mathcal{M}}_H(\hat{X}) \longrightarrow \mathcal{J}_K(\hat{X}) \longrightarrow \mathcal{M}_K(\hat{X}), \tag{2.17}$$

where the 'even Jacobian' $\mathcal{J}_K(\hat{X})$ is a torus bundle over $\mathcal{M}_K(\hat{X})$, the moduli space of complexified Kähler structures on $\hat{X}$, with fiber $\hat{\mathcal{T}} = H^{\mathrm{even}}(\hat{X}, \mathbb{R})/\hat{\Gamma}$ where $\hat{\Gamma}$ is a lattice which will be specified below. Physically, the $\mathbb{C}^\times$-fiber is parametrized by the type IIB dilaton $\tau_2 = 1/g_s$ and the NS-axion ψ, while $\hat{\mathcal{T}}$ corresponds to the periods of the ten-dimensional RR form $C^{\mathrm{even}} = C^{(0)} + C^{(2)} + C^{(4)} + C^{(6)} = H^{\mathrm{even}}(\hat{X}, \mathbb{R})$. A convenient set of coordinates is given by [**16**, **31**]

(2.18)

$$\begin{aligned} &c^0 = C^{(0)}, \qquad c^a = \int_{\gamma^a} C^{(2)}, \qquad \tilde{c}_a = -\int_{\gamma_a}\Big(C^{(4)} - \tfrac{1}{2}B\wedge C^{(2)}\Big), \\ &\tilde{c}_0 = -\int_{\hat{X}}\Big(C^{(6)} - B\wedge C^{(4)} + \tfrac{1}{3}B\wedge B\wedge C^{(2)}\Big), \qquad b^a + \mathrm{i}t^a = \int_{\gamma^a}(B + \mathrm{i}J). \end{aligned}$$

where γ_a, $a = 1, \ldots, h^{1,1}$, is a basis of 4-cycles in $H_4(\hat{X}, \mathbb{R})$, and γ^a is the dual basis of 2-cycles in $H_2(\hat{X}, \mathbb{R})$.

By classical mirror symmetry, $\mathcal{M}_K(\hat{X}) = \mathcal{M}_C(X)$, for a suitable map between the complex structure moduli $z^a = X^a/X^0$ and the Kähler moduli (b^a, t^a). In the large volume limit on the type IIB side, the prepotential on $\mathcal{M}_K(\hat{X})$ is given by

$$F^{\rm cl} = -\kappa_{abc}\,\frac{X^a X^b X^c}{6X^0} + \frac{1}{2}\,A_{\Lambda\Sigma}X^\Lambda X^\Sigma\,, \tag{2.19}$$

where κ_{abc} is the triple intersection product on $H^2(\hat{X},\mathbb{Z})$ and $A_{\Lambda\Sigma}$ is a real symmetric matrix which does not affect the Kähler potential, but which is important for consistency with charge quantization [**19**]. In this limit, the mirror map reduces to $z^a = b^a + it^a$ for a suitable choice of symplectic basis on the type IIA side adapted to the point of maximal unipotent monodromy. The classical metric on $\hat{\mathcal{M}}_H(\hat{X})$ then takes the semi-flat form (2.8) with $c = 0$ and the prepotential (2.19), provided $\mathcal{M}_K(\hat{X})$ is identified with $\mathcal{M}_C(X)$ and the natural coordinates on the type IIB side are related to the coordinates (2.3) on the type IIA side by [**32**]:

(2.20)
$$\begin{aligned}
&r = \frac{\tau_2^2}{2}\,\mathcal{V}\,, \qquad X^a/X^0 = z^a = b^a + \mathrm{i}t^a\,, \qquad \zeta^0 = \tau_1\,, \qquad \zeta^a = -(c^a - \tau_1 b^a)\,,\\
&\tilde\zeta'_a = \tilde c_a + \frac{1}{2}\,\kappa_{abc}\,b^b(c^c - \tau_1 b^c)\,, \qquad \tilde\zeta'_0 = \tilde c_0 - \frac{1}{6}\,\kappa_{abc}\,b^a b^b(c^c - \tau_1 b^c)\,,\\
&\sigma = -2(\psi + \frac{1}{2}\tau_1\tilde c_0) + \tilde c_a(c^a - \tau_1 b^a) - \frac{1}{6}\,\kappa_{abc}\,b^a c^b(c^c - \tau_1 b^c)\,,
\end{aligned}$$

where $\mathcal{V} = \frac{1}{6}\kappa_{abc}t^a t^b t^c$ denotes the volume of $\hat{X}$ and the prime denotes fields obtained by the symplectic transformation removing the quadratic term in (2.19), namely,

$$\tilde\zeta'_\Lambda = \tilde\zeta_\Lambda - A_{\Lambda\Sigma}\zeta^\Sigma. \tag{2.21}$$

By mirror symmetry, the lattice[1] $\hat\Gamma \subset H^{\rm even}(\hat{X},\mathbb{R})$ must be (indeed, is) the image of the lattice $\Gamma \subset H^3(X,\mathbb{R})$ under the map (2.20) between the type IIA RR fields $\zeta^\Lambda, \tilde\zeta_\Lambda$ and the type IIB RR fields $c^0, c^a, \tilde c_a, \tilde c_0$. Beyond the large volume limit, the prepotential (2.19) and mirror map (2.20) acquire worldsheet instanton corrections, governed by the genus zero Gopakumar-Vafa invariants $n^{(0)}_{q_a}$ of $\hat{X}$ [**16, 33–35**]. Together with the one-loop correction proportional to $\chi(\hat{X})$, this produces the same metric (2.8) as on the type IIA side.

Besides establishing mirror symmetry, the mirror map (2.20) has another virtue: it exposes the invariance of the HM moduli space $\mathcal{M}_H(X) = \hat{\mathcal{M}}_H(\hat{X})$, in the large volume/weak coupling limit where it holds, under the action of $SL(2,\mathbb{R})$, corresponding to the continuous S-duality symmetry of ten-dimensional type IIB supergravity. Of course, this continuous symmetry is broken by quantum corrections, but there is overwhelming evidence that a discrete $SL(2,\mathbb{Z})$ subgroup remains unbroken, providing a strong constraint on possible non-perturbative effects [**36**]. The action of $g = \left(\begin{smallmatrix} a & b \\ c & d\end{smallmatrix}\right) \in SL(2,\mathbb{Z})$ is simplest in type IIB variables [**14, 32**]:

$$\begin{aligned}
&\tau \mapsto \frac{a\tau + b}{c\tau + d}\,, \qquad t^a \mapsto t^a|c\tau + d|\,, \qquad \tilde c_a \mapsto \tilde c_a - c_{2,a}\varepsilon(g)\,,\\
&\begin{pmatrix} c^a \\ b^a \end{pmatrix} \mapsto \begin{pmatrix} a & b \\ c & d \end{pmatrix}\begin{pmatrix} c^a \\ b^a \end{pmatrix}, \qquad \begin{pmatrix} \tilde c_0 \\ \psi \end{pmatrix} \mapsto \begin{pmatrix} d & -c \\ -b & a \end{pmatrix}\begin{pmatrix} \tilde c_0 \\ \psi \end{pmatrix}.
\end{aligned} \tag{2.22}$$

[1] More precisely, Γ and $\hat\Gamma$ are local system of lattices over $\mathcal{M}_C(X)$ and $\mathcal{M}_K(\hat{X})$, due to monodromies.

In this action, we have included a shift of the RR coordinate $\tilde{c}_a$, overlooked in early studies but crucial for maintaining S-duality invariance under D3 and D5-NS5 instanton corrections [**19, 37**], as we shall see in §4 and §5. Here, $c_{2,a}$ and $\varepsilon(g) \in \mathbb{Q}$ are defined by

$$c_{2,a} \equiv \int_{\gamma_a} c_2(\hat{X}), \qquad \eta\left(\frac{a\tau+b}{c\tau+d}\right)/\eta(\tau) = \mathbf{E}(\varepsilon(g))\,(c\tau+d)^{-1/2}, \tag{2.23}$$

where $\eta(\tau)$ is the Dedekind eta-function and $\mathbf{E}(x) = \exp(2\pi\mathrm{i}\,x)$. We stress that as defined so far, the metric on $\hat{\mathcal{M}}_H(\hat{X})$ is only invariant under $SL(2,\mathbb{Z})$ in the strict infinite volume, zero string coupling limit, where it is actually enhanced to $SL(2,\mathbb{R})$. Both worldsheet instanton corrections to the classical prepotential (2.19) and the one-loop correction break this symmetry, and it is necessary to include non-perturbative effects in order to recover it. The holomorphic action of S-duality in twistor space will be described in §4.1.

3. D-instantons, wall-crossing and the QK/HK correspondence

The perturbative metric (2.8), while being valid to all orders at small string coupling g_s, is expected to receive non-perturbative corrections of order e^{-1/g_s}, due to Euclidean D-branes wrapping supersymmetric cycles in X (or $\hat{X}$). In this section we discuss some general aspects of D-instantons and their relation with Donaldson-Thomas (DT) invariants, describe how they modify the twistorial description of §2.1.3, and how they result in a smooth quantum corrected metric, despite discontinuities of the DT invariants across certain walls in complex structure (or Kähler) moduli space. For this purpose, a new duality between quaternion-Kähler and hyperkähler manifolds will turn out to be useful.

3.1. D-instantons, Donaldson-Thomas invariants and wall-crossing.

3.1.1. *Derived category of D-instantons.* On the type IIA side, the leading corrections to the perturbative hypermultiplet metric g_{pert} described in §2.1.2 come from Euclidean D2-branes wrapping Lagrangian 3-cycles (sLags) in X endowed with a flat $U(1)$ connection. On the type IIB side, they correspond to superpositions of D(-1)-D1-D3-D5 instantons wrapping complex even-dimensional cycles, or more generally coherent sheaves (holomorphic vector bundles supported on (singular) submanifolds). Most generally, D-instantons are objects in a bounded derived Fukaya category $\mathrm{D^bFuk}(X)$ on the type IIA side, or the derived category of coherent sheaves $\mathrm{D^bCoh}(\hat{X})$ on the type IIB side [**38, 39**]. Each of them is graded by the Grothendieck group, an extension of the lattice $\Gamma = H_3(X,\mathbb{Z})$ or $\hat{\Gamma} \subset H_{\text{even}}(\hat{X},\mathbb{Z})$ of electromagnetic charges. The fact that the same categories also govern the spectrum of BPS states in type IIB on X and type IIA on $\hat{X}$, respectively, can be understood by compactifying on a circle down to 3 space-time dimensions: T-duality along the circle exchanges 4-dimensional D-instantons with 4-dimensional BPS states whose worldline winds around the circle [**16**]. Kontsevich's homological mirror symmetry conjecture [**40**], the mathematical counterpart of non-perturbative mirror symmetry [**41**], asserts that these two categories are isomorphic when $(X, \hat{X})$ is a dual pair, in particular $\Gamma \simeq \hat{\Gamma}$.

3.1.2. *Stability and DT invariants.* Among all the objects in the derived category, those which correspond to supersymmetric, elementary D-instantons (or dually, one-particle BPS states) are the semi-stable ones [**39**]. Stability can be assessed using the central charge Z, a homomorphism $Z : \Gamma \to \mathbb{C}$ which varies holomorphically over $\mathcal{B}$ ($= \mathcal{M}_C(X)$ or $\mathcal{M}_K(\hat{X})$) and which determines the classical action (or dually, the mass) of the instanton. In type IIA we have

$$Z_\gamma(z) = e^{\mathcal{K}/2} \int_\gamma \Omega^{3,0}, \tag{3.1}$$

while in the large-volume limit in type IIB

$$Z_\gamma(z) = \int_{\hat{X}} e^{B+\mathrm{i}J} \mathrm{ch}(\mathscr{E}) \sqrt{\mathrm{Td}(\hat{X})}, \tag{3.2}$$

where $\mathscr{E}$ is a coherent sheaf, whose Mukai vector $\mathrm{ch}(\mathscr{E})\sqrt{\mathrm{Td}(\hat{X})} \in H^{\mathrm{even}}(\hat{X}, \mathbb{Q})$ is identified with the charge vector γ of E. Semi-stability is most easily defined for Abelian categories as follows. An object F with charge γ is called semi-stable if for every subobject $F' \subset F$ with charge γ', $\varphi(\gamma') \leq \varphi(\gamma)$, where $\varphi(\gamma)$ is the argument of the central charge $Z_\gamma(z)$. F is called stable if the inequality is strict for strict subobjects. This notion can be extended to derived categories, by considering an Abelian subcategory of the derived category, the "heart of the t-structure" [**42**, **43**]. On the IIA side, semi-stable objects of $\mathrm{D^b Fuk}(X)$ are *special* (or calibrated) Lagrangian cycles L, i.e. such that the phase of $\Omega|_L/\mathrm{d}V_L$ is constant, where $\mathrm{d}V_L$ is the volume form on L [**44**]. On the IIB side in the infinite volume limit, semi-stable objects are the semi-stable coherent sheaves in the classical sense of Gieseker stability.

An important property of semi-stable objects is that their space of deformations is finite-dimensional, although it can be singular. The generalized DT invariant $\Omega(\gamma; z)$ is defined as the (weighted) Euler number of this moduli space. It is the mathematical counterpart of the BPS index, which counts BPS black holes or instantons of charge γ. It is a locally constant function of the moduli z (through the central charge Z_γ), away from certain walls of marginal stability described below. It is also monodromy invariant, in the sense that $\Omega(M \cdot \gamma; M \cdot z) = \Omega(\gamma; z)$, where $M \in Sp(m; \mathbb{Z})$ is the symplectic rotation induced by a monodromy along a loop in $\mathcal{B}$. Homological mirror symmetry implies that the DT invariants $\Omega(\gamma; z)$ associated to $\mathrm{D^b Fuk}(X)$ and $\mathrm{D^b Coh}(\hat{X})$ are the same, provided the charges γ and moduli z are related according to the classical mirror map. As we shall see, this guarantees that the D-instanton corrected HM moduli spaces $\mathcal{M}_H(X)$ and $\hat{\mathcal{M}}_H(\hat{X})$ are isometric.

3.1.3. *Wall-crossing.* Physically, the jump of the DT invariants $\Omega(\gamma, z)$ across codimension one walls in $\mathcal{B}$ corresponds to the decay of bound states into more elementary stable constituents. For any pair of charge vectors (γ_1, γ_2), the decay of a D-brane of charge $\gamma = M\gamma_1 + N\gamma_2$ into constituents of charges $M_i\gamma_1 + N_i\gamma_2$ with $\sum(M_i, N_i) = (M, N)$ is energetically possible only if the phases of the central charges align, i.e. $\varphi(\gamma_1) = \varphi(\gamma_2)$, which defines the wall of marginal stability $W(\gamma_1, \gamma_2) \subset \mathcal{B}$. Let $z_\pm \in \mathcal{B}$ denote two points infinitesimally displaced on either side of such a wall. We can always choose the basis γ_1, γ_2 of the two-dimensional lattice $\mathbb{Z}\gamma_1 + \mathbb{Z}\gamma_2$ such that only the first and third quadrants are populated on either side of the wall, $\Omega^\pm(M\gamma_1 + N\gamma_2) = 0$ if $MN \leq 0$ [**45**].

Several formulae exist in the mathematics and physics literature for how to compute the jump of $\Omega(M\gamma_1 + N\gamma_2; z)$ across the wall $W(\gamma_1, \gamma_2)$ (see, e.g., [**46**] for a review). The one relevant here is the Kontsevich-Soibelman (KS) formula [**47**], which has a clear geometric interpretation. To write their formula, one introduces the Lie algebra of (twisted) infinitesimal symplectomorphisms of the complex torus $\Gamma \otimes_{\mathbb{Z}} \mathbb{C}^\times$ generated by vector fields $(e_\gamma)_{\gamma\in\Gamma}$ satisfying

$$[e_\gamma, e_{\gamma'}] = (-1)^{\langle\gamma,\gamma'\rangle} \langle\gamma,\gamma'\rangle\, e_{\gamma+\gamma'}. \tag{3.3}$$

For any $\gamma \in \Gamma$, $z \in \mathcal{B}$ we also define the group element

$$U_\gamma(z) = \exp\left(\Omega(\gamma; z) \sum_{n=1}^{\infty} \frac{e_{n\gamma}}{n^2}\right). \tag{3.4}$$

The KS wall-crossing formula then asserts the following equality between oppositely ordered infinite products of symplectomorphisms [**47**]

$$\prod^{\curvearrowleft}_{\substack{\gamma=m\gamma_1+n\gamma_2\\ m\geq 0, n\geq 0}} U_\gamma(z_+) = \prod^{\curvearrowright}_{\substack{\gamma=m\gamma_1+n\gamma_2\\ m\geq 0, n\geq 0}} U_\gamma(z_-). \tag{3.5}$$

By projecting this equality on finite dimensional quotients, one can determine $\Delta\Omega = \Omega(\gamma; z_+) - \Omega(\gamma; z_-)$ for any $\gamma = M\gamma_1 + N\gamma_2$. This formula was interpreted physically in the context of $\mathcal{N} = 2$ gauge theory à la Seiberg-Witten in [**48**], as ensuring the smoothness of the hyperkähler metric on the Coulomb branch of the gauge theory on $\mathbb{R}^3 \times S^1$. From this point of view, the U_γ's are complex symplectomorphisms relating different Darboux coordinate systems on the twistor space. Below we will show that an extension of (3.5), where the U_γ's now are complex contact transformations, also ensures the smoothness of the HM moduli space $\mathcal{M}_H$ in $\mathcal{N} = 2$ string vacua across walls in $\mathcal{B}$.

3.2. D-instantons in twistor space. Away from the zero-coupling limit, the perturbative hypermultiplet metric (2.8) receives non-perturbative corrections due to D-brane instantons. In the one-instanton approximation these corrections take the schematic form [**16**]

$$g_D \sim \sum_{\gamma\in\Gamma} \lambda_{\rm D}(\gamma)\, \bar\Omega(\gamma; z) \exp\Big(- 8\pi\sqrt{r}\, |Z_\gamma| - 2\pi \mathrm{i}\, \langle\gamma, C\rangle \Big), \tag{3.6}$$

where the exponential is the classical action of the D-instanton, with $\sqrt{r} \sim 1/g_s$. The prefactor in principle originates from integrating the fluctuation determinant around the classical solution over collective coordinates. It is natural to expect that it is proportional to the DT invariant $\Omega(\gamma, z)$ introduced in §3.1.2, however consistency with wall-crossing will dictate the less obvious product of a quadratic refinement $\lambda_{\rm D}(\gamma)$, analogous to the one in (2.5), with the 'rational DT invariant' [**47, 49, 50**]

$$\overline{\Omega}(\gamma; z) = \sum_{d|\gamma} \frac{\Omega(\gamma/d; z)}{d^2}. \tag{3.7}$$

When γ is a primitive charge vector, $\overline{\Omega}(\gamma; z) = \Omega(\gamma; z)$.

In order to incorporate these corrections to the metric while maintaining its quaternion-Kähler structure, it is best to do this at the level of the twistor space $\mathcal{Z}$. As explained in [**16, 51, 52**], in close analogy with the field theory construction in [**48**], the instanton corrections modify the contact structure on $\mathcal{Z}$, presented in §2.1.3, by replacing the patch $\mathcal{U}_0$ around the equator of $\mathbb{P}^1$ with an infinite set of angular sectors separated by so called "BPS-rays"

$$\ell_\gamma = \{t \in \mathbb{P}^1 \;:\; Z_\gamma(z)\, t^{-1} \in \mathrm{i}\mathbb{R}_-\}, \tag{3.8}$$

where $Z_\gamma(z)$ is the central charge function (3.1). Across ℓ_γ the Darboux coordinates $(\xi^\Lambda, \tilde\xi_\Lambda, \tilde\alpha)$ must jump by a complex contact transformation. We postulate that the jump of the holomorphic Fourier modes on the torus $\Gamma\otimes_{\mathbb{Z}}\mathbb{C}^\times$,

$$\mathcal{X}_\gamma = \mathbf{E}(-\langle\gamma,\Xi\rangle) = e^{-2\pi\mathrm{i}\left(q_\Lambda\xi^\Lambda - p^\Lambda\tilde\xi_\Lambda\right)}, \tag{3.9}$$

is the standard KS symplectomorphism

$$U_\gamma\,:\,\mathcal{X}_{\gamma'} \longmapsto \mathcal{X}_{\gamma'}\,(1-\lambda_{\mathrm{D}}(\gamma)\mathcal{X}_\gamma)^{\Omega(\gamma)\langle\gamma,\gamma'\rangle}\,. \tag{3.10}$$

Requiring that the contact one-form (2.13) is preserved determines the discontinuity in the remaining Darboux coordinate $\tilde\alpha$. As a result, the full contact transformation is given by

$$V_\gamma\,:\,(\mathcal{X}_{\gamma'},\,\tilde\alpha)\longmapsto\left(\mathcal{X}_{\gamma'}\,(1-\lambda_{\mathrm{D}}(\gamma)\mathcal{X}_\gamma)^{\Omega(\gamma)\langle\gamma,\gamma'\rangle}\,,\,\tilde\alpha+\frac{\Omega(\gamma)}{2\pi^2}\,L_{\lambda_{\mathrm{D}}(\gamma)}\left(\lambda_{\mathrm{D}}(\gamma)\mathcal{X}_\gamma\right)\right), \tag{3.11}$$

where $L_\epsilon(x)$ is a variant of the Rogers dilogarithm,

$$L_\epsilon(x)\equiv \mathrm{Li}_2(x)+\tfrac12\log(\epsilon^{-1}x)\log(1-x). \tag{3.12}$$

Requiring further that the Darboux coordinates reduce to the uncorrected ones (2.15) near $t=0$ and $t=\infty$, one may recast the gluing conditions (3.11) across the BPS rays ℓ_γ as a system of integral equations for the Fourier modes $\mathcal{X}_\gamma$ [**48**, **51**],

$$\mathcal{X}_\gamma(t)=\mathcal{X}^{\mathrm{sf}}_\gamma(t)\,\exp\left[\frac{1}{4\pi\mathrm{i}}\sum_{\gamma'}\Omega(\gamma')\,\langle\gamma,\gamma'\rangle\int_{\ell_{\gamma'}}\frac{\mathrm{d}t'}{t'}\frac{t+t'}{t-t'}\log\left(1-\lambda_{\mathrm{D}}(\gamma')\mathcal{X}_{\gamma'}(t')\right)\right], \tag{3.13}$$

where $\mathcal{X}^{\mathrm{sf}}_\gamma$ are the 'semi-flat' Fourier modes obtained from (2.15)

$$\mathcal{X}^{\mathrm{sf}}_\gamma(t)=\exp\left[-2\pi\mathrm{i}\left(\langle\gamma,C\rangle+\frac{\tau_2}{2}\,e^{-\mathcal{K}/2}\left(t^{-1}Z_\gamma-t\bar Z_\gamma\right)\right)\right]. \tag{3.14}$$

The remaining coordinate $\tilde\alpha$ and the contact potential, which in this case is globally defined, independent of t and can be identified with the 4-dimensional dilaton $\Phi=\phi$, are then obtained from the solutions of (3.13) via [**51**, **52**]

$$\tilde\alpha = \sigma+t^{-1}\mathcal{W}-t\bar{\mathcal{W}}+\frac{\mathrm{i}\chi(X)}{24\pi}\log t+\frac{\mathrm{i}}{8\pi^3}\sum_\gamma\Omega(\gamma)\int_{\ell_\gamma}\frac{\mathrm{d}t'}{t'}\frac{t+t'}{t-t'}\,L_{\lambda_{\mathrm{D}}(\gamma)}\left(\lambda_{\mathrm{D}}(\gamma)\mathcal{X}_\gamma\right), \tag{3.15}$$

$$e^\phi = \frac{\tau_2^2}{16}\,e^{-\mathcal{K}}+\frac{\chi(X)}{192\pi}-\frac{\mathrm{i}\tau_2}{64\pi^2}\,e^{-\mathcal{K}/2}\sum_\gamma\Omega(\gamma)\int_{\ell_\gamma}\frac{\mathrm{d}t}{t}\left(t^{-1}Z_\gamma-t\bar Z_\gamma\right)\log\left(1-\lambda_{\mathrm{D}}(\gamma)\mathcal{X}_\gamma\right), \tag{3.16}$$

where the expression for $\mathcal{W}$ can be found in [**52**]. While (3.13) cannot be solved exactly in general, it can be solved approximately by first plugging in $\mathcal{X}_\gamma=\mathcal{X}^{\mathrm{sf}}_\gamma$ on the r.h.s., computing the l.h.s. and iterating. This generates a formal infinite series of terms labelled by decorated rooted trees, interpreted as multi-instanton corrections [**48**, **53**]. The first term in this expansion, known as the one-instanton approximation, is an integral governed by a saddle point at $t=\mathrm{i}\sqrt{Z_\gamma/\bar Z_\gamma}$, which produces a result of the expected form (3.6). Ref. [**54**] argued that the third term on the right hand side in (3.16) corresponds to the contribution of multi-particle states to the Witten index (whereas $\Omega(\gamma)$ counts single-particle BPS states).

It should be stressed that the gluing conditions (3.11) apply only in an open set on $\mathcal{Z}$ away from any wall of marginal stability. Across such a wall, the DT invariants $\Omega(\gamma)$ will jump, but so will the ordering of the BPS rays ℓ_γ. By the same reasoning as in [**48**], the consistency of the construction and the smoothness of the instanton corrected metric (including all multi-instanton corrections) requires an analogue of the KS wall-crossing formula (3.5), where the U_γ's are replaced by V_γ's, and equality holds modulo the axion periodicity $\tilde{\alpha} \to \tilde{\alpha} + 2\kappa$, $\kappa \in \mathbb{Z}$. On the other hand, unlike the HK situation in [**48**], the twistor space $\mathcal{Z}$ is not a trivial fibration $\mathcal{M}_H \to \mathcal{Z} \to \mathbb{P}^1$, but rather an opposite, non-trivial fibration $\mathbb{P}^1 \to \mathcal{Z} \to \mathcal{M}_H$, and the above construction does not address the global structure of the twistor space. To circumvent this problem, it will be convenient to relate the twistor space of the D-instanton corrected QK manifold to the twistor space of a 'dual' HK manifold, using a general correspondence between QK and HK manifolds with isometries, which we now explain.

3.3. The QK/HK correspondence. Let us first recall the notion of *hyperholomorphic line bundle* on a HK manifold $\mathcal{M}'$: A line bundle $\mathscr{L} \to \mathcal{M}'$ is hyperholomorphic if its first Chern class $c_1(\mathscr{L})$ is of type $(1,1)$ with respect to the whole S^2 of complex structures on $\mathcal{M}'$ [**55–57**]. In real dimension 4 this reduces to the notion of self-dual curvature. A hyperholomorphic connection is a one-form λ whose curvature $\mathrm{d}\lambda$ satisfies the same condition. It is the curvature of a hyperholomorphic line bundle if and only if $\mathrm{d}\lambda \in H^2(\mathcal{M}',\mathbb{Z})$.

3.3.1. *Theorem* [**52, 58, 59**]. Given a quaternion-Kähler manifold $\mathcal{M}$ with a quaternionic circle action generated by a Killing vector κ, there exists a 'dual' hyperkähler manifold $\mathcal{M}'$ of the same dimension, equipped with a circle action generated by the Killing vector κ', that fixes one of the complex structures, J_3', and rotates J_1', J_2'. Choosing coordinates θ, θ' adapted to the circle actions on both sides, such that $\kappa = \partial_\theta$, $\kappa' = \partial_{\theta'}$, the QK metric on $\mathcal{M}$ and HK metric on $\mathcal{M}'$ are related by

$$\mathrm{d}s^2_{\mathcal{M}} = \tau\,(\mathrm{d}\theta + \Theta)^2 + \mathrm{d}s^2_{\mathcal{M}/\partial_\theta}, \qquad \mathrm{d}s^2_{\mathcal{M}'} = (\nu+\rho)\,(\mathrm{d}\theta' + \Theta')^2 + \mathrm{d}s^2_{\mathcal{M}'/\partial_{\theta'}} \tag{3.17}$$

with

$$\mathrm{d}s^2_{\mathcal{M}'/\partial_{\theta'}} = \frac{\mathrm{d}\rho^2}{\rho} + 4\rho\,|p_+|^2 - 2\rho\,\mathrm{d}s^2_{\mathcal{M}/\partial_\theta}. \tag{3.18}$$

Here, ρ is a function on $\mathcal{M}'/\partial_{\theta'}$ defined as the moment map of κ' with respect to J_3'; it is identified with the function $1/(2|\vec{\mu}|)$ on $\mathcal{M}/\partial_\theta$, where $\vec{\mu}$ is quaternionic moment map of κ on $\mathcal{M}$. τ and ν are functions on $\mathcal{M}/\partial_\theta$ and $\mathcal{M}'/\partial_{\theta'}$, respectively, related by $\tau = \frac{\nu+\rho}{2\rho^2\nu}$. Finally, p_+ is the $+$ component of the $SU(2)$ connection on $\mathcal{M}$, related to the same component of the $SU(2)$ connection on $\mathcal{M}'$ via $p_+' = \rho\, e^{\mathrm{i}\theta'} p_+$, while the one-forms Θ and Θ' appear in the decomposition of the third component, $p_3 = -\frac{1}{\rho}(\mathrm{d}\theta + \Theta) + \Theta'$, $p_3' = \rho(\mathrm{d}\theta' + \Theta') - \Theta$.

Under this correspondence, the HK manifold $\mathcal{M}'$ is naturally endowed with a hyperholomorphic connection[2]

$$\lambda = \nu\,(\mathrm{d}\theta' + \Theta') + \Theta, \qquad \mathrm{d}\lambda = 2\mathrm{i}\,\partial\bar{\partial}\rho - w_3' \in H^2(\mathcal{M}',\mathbb{Z}), \tag{3.19}$$

where w_3' is the Kähler form on $\mathcal{M}'$ associated to J_3' and ∂ is the Dolbeault derivative in the same complex structure. Conversely, given a HK manifold with a

[2]Hyperholomorphicity is guaranteed by the second equation in (3.19).

hyperholomorphic line bundle, there exists a one-parameter family of QK metrics given by the same formula. The one-parameter ambiguity stems from the fact that the moment map ρ of κ' is defined up to an additive constant, which can be absorbed in a shift of ν. This affects the hyperholomorphic connection λ but not its curvature. Examples of such dual pairs are provided by the rigid and (one-loop deformed) local c-map spaces associated to the same prepotential $F(X)$ [**52, 60**].

3.3.2. *Twistor space realization.* The QK/HK correspondence is most easily understood by using Swann's relation between QK manifolds and hyperkähler cones [**7**]. Indeed, the total space $\mathcal{S}$ of the $\mathcal{O}(-2)$ line bundle over the twistor space $\mathcal{Z}$ of a QK manifold $\mathcal{M}$ carries a canonical HK cone metric. Any quaternionic circle action of $\mathcal{M}$ lifts to a triholomorphic circle action on $\mathcal{S}$. Taking the hyperkähler quotient of $\mathcal{S}$ with respect to this circle action then produces the dual hyperkähler manifold $\mathcal{M}'$, equipped with a natural hyperholomorphic circle bundle [**55, 56**].

For our purposes it will be more useful to realize the QK/HK correspondence directly at the level of the twistor spaces $\mathcal{Z}$ and $\mathcal{Z}'$, without invoking the Swann bundle. For this purpose, choose local contact Darboux coordinates $(\xi^\Lambda, \tilde\xi_\Lambda, \alpha)$ on the QK side, such that the Killing vector κ globally lifts to ∂_α (the Reeb vector for the contact one-form (2.12)). This implies that contact transformations between different patches must reduce to symplectomorphisms of $(\xi^\Lambda, \tilde\xi_\Lambda)$, supplemented by a suitable, Ξ-dependent shift of α. On the HK side, we choose local (symplectic) Darboux coordinates $(\eta^\Lambda, \mu_\Lambda)$ on $\mathcal{Z}'$ such that the holomorphic symplectic form on $\mathcal{Z}'$ is $\mathrm{d}\eta^\Lambda \wedge \mathrm{d}\mu_\Lambda$. We further choose coordinates x^μ on $\mathcal{M}'/\partial_{\theta'}$, ζ on the $\mathbb{P}^1$ base on the HK side and t on the $\mathbb{P}^1$ fiber on the QK side, such that $\zeta = 0, \infty$ correspond to the complex structure J_3' preserved by κ'. The fact that κ' rotates J_1' into J_2' means that the Darboux coordinates $(\eta^\Lambda, \mu_\Lambda)$ on $\mathcal{Z}'$ depend only on x^μ and $\zeta e^{-\mathrm{i}\theta'}$. Moreover, it implies that transition functions between different patches must be complex symplectomorphisms of $(\eta^\Lambda, \mu_\Lambda)$, independent of ζ. Then the correspondence shows that the Darboux coordinates $(\xi^\Lambda, \tilde\xi_\Lambda)$ on $\mathcal{Z}$, as functions of (x^μ, θ, t), can be identified with $(\eta^\Lambda, \mu_\Lambda)$ for $t = \zeta e^{-\mathrm{i}\theta'}$. In particular, the complex contact structure on $\mathcal{Z}$ and symplectic structure on $\mathcal{Z}'$ can be described globally by the same symplectomorphisms, supplemented on the QK side by a suitable shift of α. In fact, the Darboux coordinate α provides, on the dual HK side, a holomorphic section

$$\Upsilon^{[i]} \equiv e^{2\pi i \alpha^{[i]}} \tag{3.20}$$

of a line bundle $\mathscr{L}'_{\mathcal{Z}}$ over $\mathcal{Z}'$, with holomorphic connection given by the contact one-form (2.13). This section is non-zero along each twistor line, and hence by the Atiyah-Ward twistor correspondence (see [**61**]) yields a hyperholomorphic line bundle $\mathscr{L}$ over $\mathcal{M}'$ with connection

$$\lambda = \frac{1}{4}\left(\bar\partial^{(\zeta)}\alpha + \partial^{(\zeta)}\bar\alpha\right), \tag{3.21}$$

which can be shown to agree with (3.19). It is also worth noting that the contact potential e^ϕ on the QK side is identified with the moment map ρ on the HK side.

3.4. Wall-crossing revisited. After this digression on the QK/HK correspondence, we now return to the problem of wall-crossing on the D-instanton corrected HM moduli space $\mathcal{M}_H$. Since D-instanton corrections are independent of the NS-axion σ, they preserve the quaternionic Killing vector ∂_σ. Therefore, by the QK/HK correspondence, we can trade the construction of the QK space $\mathcal{M}_H$ with that of a HK space $\mathcal{M}'_H$ equipped with a hyperholomorphic connection λ. Twistorially, this is equivalent to constructing the twistor space $\mathcal{Z}'$ and complex line bundle $\mathscr{L}_{\mathcal{Z}'}$. The space $\mathcal{Z}'$, parametrized by the Darboux coordinates $(\xi^\Lambda, \tilde\xi_\Lambda)$, is defined by the same gluing conditions (3.10) as in [**48**], specialized to the case where the prepotential is homogeneous. To construct[3] $\mathscr{L}_{\mathcal{Z}'}$, one must lift the symplectomorphisms U_γ to complex gauge transformations V_γ preserving the holomorphic connection (2.13), which we identify with the contact transformations (3.11). The consistency of these gluing conditions across walls of marginal stability require that the KS formula (3.5) lifts to

$$\prod_{\substack{\gamma=m\gamma_1+n\gamma_2\\ m\geq 0, n\geq 0}}^{\curvearrowleft} V_\gamma(z_+) = \prod_{\substack{\gamma=m\gamma_1+n\gamma_2\\ m\geq 0, n\geq 0}}^{\curvearrowright} V_\gamma(z_-). \tag{3.22}$$

Clearly, assuming (3.5) is satisfied, (3.22) could fail at most by a translation $\tilde\alpha \to \tilde\alpha + \Delta\tilde\alpha$ along the $\mathbb{C}^\times$-fiber. The global existence of $\mathscr{L}_{\mathcal{Z}'}$ requires $\Delta\tilde\alpha \in 2\mathbb{Z}$, the natural ambiguity in the coordinate $\tilde\alpha$. To see why this is so, let us rewrite (3.22) as an identity $\prod_s V_{\gamma_s}^{\epsilon_s} = 1$ by assembling all operators on one side. Here, ϵ_s is a sign that changes from $+1$ on the right side of the product (corresponding to the r.h.s. of (3.22)) to -1 on the left side (corresponding to the inverse of the l.h.s. of (3.22)). The total shift of $\tilde\alpha$ so obtained can be written as

$$\Delta\tilde\alpha = \frac{1}{2\pi^2}\sum_s \epsilon_s \Omega(\gamma_s) L_{\lambda_{\rm D}(\gamma_s)}\left(\mathcal{X}_{\gamma_s}(s)\right), \tag{3.23}$$

where $\mathcal{X}_{\gamma_s}(s) = U_{\gamma_{s-1}} \circ U_{\gamma_{s-2}} \circ \cdots \circ U_{\gamma_1} \cdot \mathcal{X}_{\gamma_s}$. In [**52**] it was shown that the quantization property $\Delta\tilde\alpha \in 2\mathbb{Z}$ follows from the motivic wall-crossing formula of Kontsevich and Soibelman [**47**] in an appropriate classical limit. This formula generalizes various known and conjectural identities for the Rogers dilogarithm associated with cluster algebras of Dynkin quivers (see, e.g., [**63, 64**]).

Let us end this discussion with an important remark. Although the above construction formally gives a satisfactory solution to the wall-crossing problem in the hypermultiplet sector of type IIA string theory on a CY threefold X, it ignores a crucial problem, namely the exponential growth $\Omega(\gamma) \sim e^{S(\gamma)}$ of the DT invariants for large charges, where $S(\gamma)$ is the entropy of a 4D BPS black hole of charge γ. Since $S(\gamma)$ scales quadratically in γ for large classical black holes, any generating function of the form $\sum_\gamma \Omega(\gamma) q^\gamma$ is divergent, hence making the integral equation (3.13) ill-defined. It was observed in [**65**] that the ambiguity in such divergent sums is of the same order e^{-1/g_s^2} as NS5-instanton effects, which might therefore cure this problem. We return to NS5-instantons in §5, after discussing S-duality in the presence of D3-instantons.

4. S-duality, D3-instantons and mock theta series

The invariance of ten-dimensional type IIB string theory under S-duality is a well supported fact (see e.g. [**36**] and much subsequent work). It is important to test whether it continues to hold in vacua with less supersymmetry, in particular in type IIB compactified on a CY threefold $\hat X$. Assuming that it does, one may e.g. deduce D1-D(-1) instanton corrections from worldsheet instantons [**33, 34**],

[3] The construction of a canonical hyperholomorphic connection on the Coulomb branch of $\mathcal{N}=2$ gauge theories was independently given in [**62**].

obtaining the first hint of the form of D-instanton corrections to the HM metric, or NS5-instantons from D5 [**19**], as we discuss in §5. Here, we focus on the intermediate case of D3-instantons, which are singlets of $SL(2,\mathbb{Z})$ and should therefore preserve S-duality by themselves [**35, 37**]. We shall show that this is indeed the case thanks to special modular properties of D3-D1-D(-1) Donaldson-Thomas invariants and of certain indefinite theta series.

4.1. S-duality in twistor space. We start by discussing general constraints imposed by S-duality on the twistor space construction. By a suitable choice of coordinates, we can assume that, even after the inclusion of instanton corrections, $SL(2,\mathbb{Z})$ acts on $\hat{\mathcal{M}}_H(\hat{X})$ by (2.22). This action must lift to a holomorphic action on the twistor space $\mathcal{Z}$. At the classical level, using the Darboux coordinates (2.15) with $F=F^{\rm cl}$ (2.19), $r=\frac{\tau_2^2}{16}\,e^{-\mathcal{K}}$ and $c=0$, one can check that (2.22) lifts to the complex contact transformation [**16**]

(4.1)
$$\xi^0\mapsto\frac{a\xi^0+b}{c\xi^0+d},\qquad \xi^a\mapsto\frac{\xi^a}{c\xi^0+d},\qquad \tilde\xi'_a\mapsto\tilde\xi'_a+\frac{c}{2(c\xi^0+d)}\kappa_{abc}\xi^b\xi^c-c_{2,a}\,\varepsilon(g),$$
$$\begin{pmatrix}\tilde\xi'_0\\ \alpha'\end{pmatrix}\mapsto\begin{pmatrix}d & -c\\ -b & a\end{pmatrix}\cdot\begin{pmatrix}\tilde\xi'_0\\ \alpha'\end{pmatrix}+\frac{1}{6}\,\kappa_{abc}\xi^a\xi^b\xi^c\begin{pmatrix}c^2/(c\xi^0+d)\\ -[c^2(a\xi^0+b)+2c]/(c\xi^0+d)^2\end{pmatrix},$$

provided the action (2.22) on $\mathcal{M}_H$ is supplemented by a suitable action on the $\mathbb{P}^1$ fiber, e.g. in the gauge $X^0=1$,

$$z\mapsto\frac{c\bar\tau+d}{|c\tau+d|}\,z,\qquad z\equiv\frac{t+\mathrm{i}}{t-\mathrm{i}}. \tag{4.2}$$

Beyond the classical limit, the Darboux coordinates are no longer given by the simple formulae (2.15), however, they should still transform as in (4.1) up to local contact transformations, if S-duality is to remain unbroken. This in turn constrains the transformations of the transition functions on overlapping patches. The constraint can be easily formulated by considering a covering by an infinite set of open patches $\mathcal{U}_{m,n}$, which are mapped to each other by S-duality, including a S-duality invariant patch $\mathcal{U}_0\equiv\mathcal{U}_{0,0}$ [**34, 35**]. Then at the linearized level S-duality requires that the generating functions $H_{m,n}$ of the contact transformations from $\mathcal{U}_0$ to $\mathcal{U}_{m,n}$ transform as

$$H_{m,n}\mapsto\frac{H_{m',n'}}{c\xi^0+d}+\text{reg.},\qquad \begin{pmatrix}m'\\ n'\end{pmatrix}=\begin{pmatrix}a & c\\ b & d\end{pmatrix}\begin{pmatrix}m\\ n\end{pmatrix}. \tag{4.3}$$

Heuristically, ignoring the fact that the functions $H_{m,n}$ are attached to different (pairs of) patches, the constraint (4.3) says that the formal sum $\sum H_{m,n}$ should transform as a holomorphic modular form of weight -1. In [**35**] this constraint was promoted to the full non-linear level under the assumption that the QK space has two commuting continuous isometries, which holds for the D3-instanton corrected HM moduli space having two continuous isometries along fivebrane axions. How to drop this assumption and describe an arbitrary QK space carrying an isometric action of $SL(2,\mathbb{Z})$ was understood in [**66, 67**].

4.2. Modularity of DT invariants. Before discussing the S-duality invariance of the D3-instanton corrected metric, let us first recall the modular properties of DT invariants associated to dimension 2 sheaves. The same invariants control D4-D2-D0 and M5-brane black holes, which have been the subject of much research

[**68**–**73**]. As discussed in §3.1.2, given a coherent sheaf $\mathscr{E}$ on $\hat{X}$, D-brane charges are components of the generalized Mukai charge vector γ:

$$\gamma = \mathrm{ch}(\mathscr{E})\sqrt{\mathrm{Td}\,\hat{X}} = p^0 + p^a\omega_a - q'_a\omega^a + q'_0\,\omega_{\hat{X}}\,, \tag{4.4}$$

where $\{\omega_a\}$, $\{\omega^a\}$ and $\omega_{\hat{X}}$ are respectively a basis of 2-forms, 4-forms and the volume form of $\hat{X}$, and $q'_\Lambda = q_\Lambda - A_{\Lambda\Sigma}p^\Sigma$ are in general non-integral [**19**, **37**]. We shall denote the corresponding DT invariants by $\Omega(p^0, p^a, q'_a, q'_0; z)$. Dimension-one or zero sheaves ($p^a = p^0 = 0$) correspond to D1-D(-1) instantons. As shown in [**33**], the D1-D(-1)-instanton corrected metric is invariant under $SL(2,\mathbb{Z})$ provided

$$\Omega(0,0,0,q_0) = -\chi(\hat{X})\,, \qquad \Omega(0,0,q_a,q_0) = n^{(0)}_{q_a}\,, \tag{4.5}$$

where $n^{(0)}_{q_a}$ are the genus 0 Gopakumar-Vafa invariants governing the worldsheet instanton corrections. D3-brane instantons correspond to dimension-two sheaves ($p^0 = 0, p^a \neq 0$), supported on a divisor $\mathcal{D} \subset \hat{X}$. Dimension-three sheaves will be discussed in §5.2.1.

We assume that $\mathcal{D}$ is an ample divisor, i.e. that $[\mathcal{D}]$ belongs to the Kähler cone. The intersection matrix of 2-cycles of an ample divisor provides a natural quadratic form $\kappa_{abc}p^c$ on $\Lambda = H_4(\hat{X},\mathbb{Z})$, with signature $(1, b_2(\hat{X}) - 1)$. This also provides a quadratic form $\kappa^{ab} = (\kappa_{abc}p^c)^{-1}$ on Λ^*. In the following, we shall use this quadratic form to identify Λ as a sublattice of Λ^*, $k^a \mapsto k_a \equiv \kappa_{abc}p^bk^c$. In the following we shall denote the vector $(k^1, \dots, k^{b_2(\hat{X})})$ as $\boldsymbol{k}$. For a general $\boldsymbol{k} \in \Lambda$, the vectors $\boldsymbol{k}_\pm \in \Lambda \otimes \mathbb{R}$ are projections of $\boldsymbol{k}$ onto the positive and negative definite subspaces of $\Lambda \otimes \mathbb{R}$ defined by the magnetic charge vector $\boldsymbol{p}$ and the Kähler moduli $\boldsymbol{t}$:

$$\boldsymbol{k}_+ = \frac{\boldsymbol{k}\cdot\boldsymbol{t}}{\boldsymbol{p}\cdot\boldsymbol{t}^2}\,\boldsymbol{t}, \qquad \boldsymbol{k}_- = \boldsymbol{k} - \boldsymbol{k}_+, \qquad \boldsymbol{k}^2 = \boldsymbol{k}_+^2 + \boldsymbol{k}_-^2, \tag{4.6}$$

which satisfy $\boldsymbol{k}_+^2 > 0$, $\boldsymbol{k}_-^2 < 0$ for all $\boldsymbol{k} \neq \boldsymbol{0}$. We also use the notation k_+ to denote the modulus of the vector $\boldsymbol{k}_+$.

As explained in §3.1.2, the DT invariants $\Omega(\gamma; \boldsymbol{z})$ are piecewise constant in Kähler moduli, but can be discontinuous across walls of marginal stability. This moduli dependence persists in the large volume limit, and complicates the analysis of the modular properties of $\Omega(\gamma; \boldsymbol{z})$ [**50**, **74**]. To deal with this problem, we express the DT invariants $\Omega(\gamma; \boldsymbol{z})$ in terms of the 'MSW invariants' $\Omega_{\boldsymbol{p}}(\boldsymbol{q}', q'_0)$. The latter coincide with the DT invariants at the so-called 'large volume attractor point' [**75**][4]

$$\Omega_{\boldsymbol{p}}(\boldsymbol{q}', q'_0) = \Omega\left(0, \boldsymbol{p}, \boldsymbol{q}', q'_0; \boldsymbol{z}_\infty(\gamma)\right), \qquad \boldsymbol{z}_\infty(\gamma) = \lim_{\lambda\to+\infty}\left(\boldsymbol{b}(\gamma) + \mathrm{i}\lambda\,\boldsymbol{t}(\gamma)\right), \tag{4.7}$$

where $\boldsymbol{z}(\gamma) = \boldsymbol{b}(\gamma) + \mathrm{i}\boldsymbol{t}(\gamma)$ is the standard attractor point. Away from the large volume attractor point $\boldsymbol{z}_\infty(\gamma)$ (but still at large volume), the DT invariant $\Omega(0, \boldsymbol{p}, \boldsymbol{q}', q'_0; \boldsymbol{z})$ differs from the MSW invariant $\Omega_{\boldsymbol{p}}(\boldsymbol{q}', q'_0)$ by terms of higher order in the MSW invariants [**74**, **76**]. The higher order terms can be thought of as describing bound states of the MSW constituents, which exist away from the large volume attractor point $\boldsymbol{z}_\infty(\gamma)$. This decomposition is analogous to the decomposition of the index in terms of the multi-centered black hole bound states. As shown in [**74**], the 'two-centered' contribution leads to a modular invariant partition function with the same modular properties as the elliptic genus, and it is expected that modularity

---

[4] Note that for CY threefolds with $b_2(\hat{X}) = 1$, the walls of marginal stability for D3-instantons do not extend to large volume regime, hence the MSW and DT invariants coincide.

persists to all orders in the MSW invariants. We stress that the expansion in MSW invariants is not a Taylor expansion in a small parameter, rather it is a finite sum in any chamber separated by a finite number of walls from the large volume attractor chamber. In the remainder of this work, we shall be concerned with only the first term in this expansion, which we call the 'one-instanton approximation' or 'dilute instanton approximation'.

As the name suggests, the MSW invariants are the BPS indices of the MSW $(0,4)$ superconformal field theory (SCFT) describing D4-brane or M5-branes wrapped on the divisor $\mathcal{D}$ [**68**]. They are unchanged by the 'spectral flow' transformations of the charges:

$$\boldsymbol{p}\mapsto\boldsymbol{p},\qquad \boldsymbol{q}'\mapsto\boldsymbol{q}'-\boldsymbol{\epsilon},\qquad q_0'\mapsto q_0'-\boldsymbol{\epsilon}\cdot\boldsymbol{q}'+\frac{1}{2}\,\boldsymbol{p}\cdot\boldsymbol{\epsilon}^2, \tag{4.8}$$

which are induced by monodromies around the large volume point of $\mathcal{M}_H(\hat{X})$, $\boldsymbol{z}\to\boldsymbol{z}+\boldsymbol{\epsilon}$, and leave

$$\hat{q}_0\equiv q_0'-\frac{1}{2}\,\boldsymbol{q}'^2 \tag{4.9}$$

invariant. Decomposing

$$\boldsymbol{q}'=\boldsymbol{\mu}+\boldsymbol{\epsilon}+\tfrac{1}{2}\,\boldsymbol{p}, \tag{4.10}$$

where $\boldsymbol{\mu}\in\Lambda^*/\Lambda$ is the residue class of $\boldsymbol{q}'-\frac{1}{2}\boldsymbol{p}$ modulo $\boldsymbol{\epsilon}$, it follows that the MSW invariant $\Omega_{\boldsymbol{p}}(\boldsymbol{q}',q_0')\equiv\Omega_{\boldsymbol{p},\boldsymbol{\mu}}(\hat{q}_0)$ depends only on $\boldsymbol{p}$, $\boldsymbol{\mu}$ and $\hat{q}_0$. The partition function of MSW invariants for fixed divisor $\mathcal{D}$ is the elliptic genus of the SCFT,

$$\mathcal{Z}_{\boldsymbol{p}}(\tau,\boldsymbol{y}) = \sum_{\boldsymbol{q}',q_0'}(-1)^{\boldsymbol{p}\cdot\boldsymbol{q}}\,\overline{\Omega}_{\boldsymbol{p}}(\boldsymbol{q}',q_0')\,\mathbf{E}\Big(-\hat{q}_0\tau-\frac{1}{2}\,\boldsymbol{q}_-'^2\tau-\frac{1}{2}\,\boldsymbol{q}_+'^2\bar{\tau}+\boldsymbol{q}'\cdot\boldsymbol{y}\Big) \tag{4.11}$$

with $\boldsymbol{y}\in\Lambda\otimes\mathbb{C}$ and $\overline{\Omega}_{\boldsymbol{p}}(\boldsymbol{q}',q_0')$ the rational MSW invariant defined analogously to (3.7). When γ is primitive, it follows from general properties of the MSW SCFT that the elliptic genus (4.11) is a multi-variable Jacobi form of weight $(-\frac{3}{2},\frac{1}{2})$ under $SL(2,\mathbb{Z})$, with multiplier system $M_{\mathcal{Z}}=\mathbf{E}(\varepsilon(g)\,\boldsymbol{c_2}\cdot\boldsymbol{p})$, where $\varepsilon(g)$ is as in (2.23). If $\boldsymbol{p}$ is not a primitive vector, the BPS indices might not be related to a proper CFT due to states at threshold stability. However, wall-crossing arguments [**50**] and explicit calculations [**77**] suggest that nevertheless the generating function of $\overline{\Omega}_{\boldsymbol{p}}(\boldsymbol{q}',q_0')$ exhibits good modular properties under $SL(2,\mathbb{Z})$.

The invariance of $\overline{\Omega}_{\boldsymbol{p}}(\boldsymbol{q}',q_0')$ under the spectral flow (4.8) implies that the elliptic genus has a theta function decomposition:

$$\mathcal{Z}_{\boldsymbol{p}}(\tau,\boldsymbol{y},\boldsymbol{t})=\sum_{\boldsymbol{\mu}\in\Lambda^*/\Lambda}h_{\boldsymbol{p},\boldsymbol{\mu}}(\tau)\,\overline{\theta_{\boldsymbol{p},\boldsymbol{\mu}}(\tau,\boldsymbol{y},\boldsymbol{t})}\,, \tag{4.12}$$

where $\theta_{\boldsymbol{p},\boldsymbol{\mu}}$ is the Siegel-Narain theta series associated to the lattice Λ equipped with the quadratic form κ_{ab} of signature $(1,b_2(\hat{X})-1)$,

$$\theta_{\boldsymbol{p},\boldsymbol{\mu}}(\tau,\boldsymbol{y},\boldsymbol{t})=\sum_{\boldsymbol{k}\in\Lambda+\boldsymbol{\mu}+\frac{1}{2}\boldsymbol{p}}(-1)^{\boldsymbol{k}\cdot\boldsymbol{p}}\,\mathbf{E}\Big(\frac{1}{2}\,\boldsymbol{k}_+^2\tau+\frac{1}{2}\,\boldsymbol{k}_-^2\bar{\tau}+\boldsymbol{k}\cdot\boldsymbol{y}\Big)\,, \tag{4.13}$$

and the $\boldsymbol{y}$-independent coefficients are given by

$$h_{\boldsymbol{p},\boldsymbol{\mu}}(\tau)=\sum_{\hat{q}_0\le r\chi(\mathcal{D})/24}\overline{\Omega}_{\boldsymbol{p},\boldsymbol{\mu}}(\hat{q}_0)\,\mathbf{E}(-\hat{q}_0\tau)\,. \tag{4.14}$$

Since the theta series (4.13) is a vector valued Jacobi form of modular weight $(\frac{1}{2}, \frac{b_2(\hat{X})-1}{2})$ and multiplier system M_θ under $SL(2,\mathbb{Z})$, it follows that $h_{\boldsymbol{p},\boldsymbol{\mu}}$ must transform as a vector-valued holomorphic[5] modular form of negative weight $(-\frac{b_2(\hat{X})}{2} - 1, 0)$ and multiplier system $M(g) = M_{\mathcal{Z}} \times \overline{M_\theta}^{-1}$ under the full modular group $SL(2,\mathbb{Z})$. The latter is equivalent to $M(g) = M_{\mathcal{Z}} \times M_\theta$ since M_θ is unitary.

4.3. Dilute D3-instantons. Let us now explain why D3-D1-D(-1) instanton corrections are consistent with S-duality, at least in the dilute instanton approximation. To this end it is sufficient to consider the contribution of a single homology class $\boldsymbol{p}$, but to include the sum over embedded classes $\boldsymbol{q}$ and q_0. This avoids the complications arising from the divergent sum over $\boldsymbol{p} \in H^4(\hat{X}, \mathbb{Z})$, as mentioned at the end of §3.2.

4.3.1. *Transition functions.* First, we observe that the twistor description of D3-instantons satisfies the requirement of §4.1, namely that the formal sum of transition functions transforms as a holomorphic modular form of weight -1. The transition functions generating the contact transformations (3.11) are

$$H_\gamma = \frac{\Omega(\gamma)}{(2\pi)^2} \operatorname{Li}_2(\lambda_{\mathrm{D}}(\gamma)\mathcal{X}_\gamma) + \cdots , \tag{4.15}$$

where $\gamma = (0, \boldsymbol{p}, \boldsymbol{q}, q_0)$ and $\cdots$ denote terms non-linear in the DT invariants whose explicit form can be found in [51]. In the leading term one can replace $\operatorname{Li}_2(x)$ by x at the cost of replacing the integer DT invariants by their rational counterparts (3.7). Furthermore, in the dilute instanton approximation the latter can be replaced by the MSW invariants $\overline{\Omega}_{\boldsymbol{p}}(\boldsymbol{q}', q_0')$ introduced in the previous subsection. Choosing $\lambda_{\mathrm{D}}(\gamma) = (-1)^{\boldsymbol{p}\cdot\boldsymbol{q}}$, we arrive at the formal sum

$$H_{\boldsymbol{p}} = \sum_{\boldsymbol{q}', q_0'} \overline{H}_\gamma, \qquad \overline{H}_\gamma = \frac{(-1)^{\boldsymbol{p}\cdot\boldsymbol{q}'}}{(2\pi)^2} \overline{\Omega}_{\boldsymbol{p},\boldsymbol{\mu}}(\hat{q}_0)\, \mathbf{E}\Big(\boldsymbol{p}\cdot\tilde{\boldsymbol{\xi}}' - \boldsymbol{q}'\cdot\boldsymbol{\xi} - q_0'\xi^0\Big). \tag{4.16}$$

Similarly to (4.12), using monodromy invariance the sum can be rewritten as

$$H_{\boldsymbol{p}} = \frac{1}{(2\pi)^2} \mathbf{E}\Big(\boldsymbol{p}\cdot\tilde{\boldsymbol{\xi}}'\Big) \sum_{\boldsymbol{\mu}\in\Lambda^*/\Lambda} h_{\boldsymbol{p},\boldsymbol{\mu}}(\xi^0)\, \Xi_{\boldsymbol{p},\boldsymbol{\mu}}(\xi^0, \boldsymbol{\xi}), \tag{4.17}$$

where $h_{\boldsymbol{p},\boldsymbol{\mu}}(\xi^0)$ is the modular function defined in (4.14), now evaluated at $\tau = \xi^0$, and $\Xi_{\boldsymbol{p},\boldsymbol{\mu}}(\xi^0, \boldsymbol{\xi})$ is a holomorphic theta series defined by the quadratic form $-\kappa_{ab}$ which transforms formally as a holomorphic Jacobi form of weight $\frac{b_2}{2}$, multiplier system M_θ^{-1} and index $m_{ab} = -\frac{1}{2}\kappa_{ab}$ [37]. Finally, thanks to the term proportional to $c_{2,a}$ in the transformation of $\tilde{\boldsymbol{\xi}}'$, the exponential prefactor in (4.17) transforms as the automorphy factor of a multi-variable holomorphic Jacobi theta series with the index $m_{ab} = \frac{1}{2}\kappa_{ab}$ and multiplier system $M_{\mathcal{Z}}^{-1}$. Combining the transformation properties of all factors, we conclude that, under the action (4.1), the formal sum (4.16) indeed transforms as a holomorphic Jacobi form of weight -1 and trivial multiplier system, as required by S-duality.

---

[5]Non-compact directions in the target space of the CFT could potentially lead to mock modular forms and thus holomorphic anomalies [78]. For local CY manifolds, e.g. $\mathcal{O}(-K_{\mathbb{P}^2}) \to \mathbb{P}^2$, it is known that the holomorphic generating series of DT-invariants for sheaves with rank > 1 requires a non-holomorphic addition in order to transform as a modular form [79]. We assume that this issue does not arise here.

However, this analysis overlooks the important fact that the quadratic form κ_{ab} has indefinite signature $(1, b_2-1)$, and therefore the theta series $\Xi_{\boldsymbol{p},\boldsymbol{\mu}}(\xi^0,\boldsymbol{\xi})$ is divergent. Fortunately, it never actually arises as such in the computation of the metric, rather each of the terms in (4.16) must be integrated along a different contour, which renders the resulting series convergent. To put the above heuristic argument on solid ground, one should analyze the transformation properties of the quantities which enter the computation of the metric, namely the Darboux coordinates and the contact potential. Below we restrict to the contact potential and Darboux coordinate $\boldsymbol{\xi}$, which exhibit the key mechanism, referring the reader to [**37**] for a complete discussion.

4.3.2. *Contact potential.* The contact potential e^{ϕ} is given in (3.16). The first term on the right hand side is the classical term, with $\frac{1}{8}e^{-\mathcal{K}(z,\bar z)}$ equal to the volume $\mathcal{V}=\frac{1}{6}\boldsymbol{t}^3$ of $\hat X$. Using the transformation properties of τ_2 and $\boldsymbol{t}$, one easily checks that the classical part transforms with weight $(-\frac12,-\frac12)$. The instanton corrections involve the Darboux coordinates $\mathcal{X}_\gamma$, which are solutions to the integral equations (3.13). In order to relate these corrections to the modular functions of the previous section, we start by writing the 3rd term on the right hand side of (3.16) as:

$$\delta_{\boldsymbol{p}}\, e^{\phi} = -\frac{\mathrm{i}\tau_2}{16}\sum_{q_\Lambda}\int_{\ell_\gamma}\frac{\mathrm{d}t}{t}\left(t^{-1}Z_\gamma - t\bar Z_\gamma\right)\overline{H}_\gamma(\xi^0,\boldsymbol{\xi},\tilde{\boldsymbol{\xi}}), \tag{4.18}$$

where $\overline{H}_\gamma$ is given in (4.16). In the dilute instanton approximation we can replace the Darboux coordinates appearing in the arguments of $\overline{H}_\gamma$ by their classical expressions (2.15) with the prepotential (2.19). Furthermore, we keep only the leading terms in the limit where $\boldsymbol{t}\to\infty$ and the product $z\,\boldsymbol{t}$ remains constant (where z is the Cayley-rotated coordinate on $\mathbb{P}^1$, see (4.2)). This is motivated by the fact that the saddle points of the integral lie at $z_\gamma = -\mathrm{i}(k+b)_+/\sqrt{\boldsymbol{p}\cdot\boldsymbol{t}^2}$ for $\boldsymbol{p}\cdot\boldsymbol{t}^2>0$. As a result, $\overline{H}_\gamma(\xi^0,\boldsymbol{\xi},\tilde{\boldsymbol{\xi}})$ simplifies to the following form [**37**]:

$$\overline{H}_\gamma = \frac{(-1)^{\boldsymbol{p}\cdot\boldsymbol{q}}}{(2\pi)^2}\,\overline{\Omega}_{\boldsymbol{p},\boldsymbol{\mu}}(\hat q_0)\,\mathbf{E}\Big(\mathrm{i}S_{\rm cl} - \frac12(\boldsymbol{q}+\boldsymbol{b})_+^2\,\bar\tau - \Big(\hat q_0 + \frac12(\boldsymbol{q}+\boldsymbol{b})_-^2\Big)\tau + \boldsymbol{c}\cdot(\boldsymbol{q}+\frac12\boldsymbol{b}) + \mathrm{i}Q_\gamma(z)\Big), \tag{4.19}$$

where $S_{\rm cl}$ is the leading part of the Euclidean D3-instanton action in the large volume limit, and $Q_\gamma(z)$ is the only part which depends on the fiber coordinate z,

$$S_{\rm cl} = \frac{\tau_2}{2}\,\boldsymbol{p}\cdot\boldsymbol{t}^2 - \mathrm{i}\,\tilde{\boldsymbol{c}}\cdot\boldsymbol{p}, \qquad Q_\gamma(z) = \tau_2\,\boldsymbol{p}\cdot\boldsymbol{t}^2\left(z + \mathrm{i}\,\frac{(k+b)_+}{\sqrt{\boldsymbol{p}\cdot\boldsymbol{t}^2}}\right)^2. \tag{4.20}$$

Keeping the leading contributions to the remaining terms in (4.18) leads to

$$\delta_{\boldsymbol{p}}\, e^{\phi} = -\frac{\tau_2}{4}\sum_{q_\Lambda}\int_{\ell_\gamma} dz\left[\hat q_0 + \frac12(\boldsymbol{k}+\boldsymbol{b}-\mathrm{i}z\boldsymbol{t})\cdot(\boldsymbol{k}+\boldsymbol{b}-3\mathrm{i}z\boldsymbol{t})\right]H_\gamma + \text{c.c.} \tag{4.21}$$

The integral over z is now Gaussian, leading to

$$\delta_{\boldsymbol{p}}\, e^{\phi} = \frac{\tau_2\, e^{-2\pi S_{\rm cl}}}{16\pi^2\sqrt{2\tau_2\,\boldsymbol{p}\cdot\boldsymbol{t}^2}}\,\mathcal{D}_{-\frac32}\sum_{\boldsymbol{\mu}\in\Lambda^*/\Lambda} h_{\boldsymbol{p},\boldsymbol{\mu}}(\tau)\,\overline{\theta_{\boldsymbol{p},\boldsymbol{\mu}}(\tau,\boldsymbol{t},\boldsymbol{b},\boldsymbol{c})} + \text{c.c.}, \tag{4.22}$$

where $\theta_{\boldsymbol{p},\boldsymbol{\mu}}(\tau,\boldsymbol{t},\boldsymbol{b},\boldsymbol{c})$ is a theta function similar to (4.13), which transforms as a vector-valued modular form with weight $\frac12(1,b_2(\hat X)-1)$ and multiplier system M_θ, see [**37**, Eq. (A.6)]. The action of $\mathcal{D}_{-\frac32} = \frac{1}{2\pi\mathrm{i}}\left(\partial_\tau - \frac{3/2}{2\mathrm{i}\tau_2}\right)$ raises the modular weight from $(-\frac{3}{2},\frac{1}{2})$ to $(\frac{1}{2},\frac{1}{2})$, while the overall factor of τ_2 reduces this to $(-\frac{1}{2},-\frac{1}{2})$. The transformation property of $\tilde{c}_a$ (2.22) is now seen to cancel the non-trivial phases due to the multiplier system $M_{\mathcal{Z}}$ of $\mathcal{Z}_{\boldsymbol{p}}(\tau,\boldsymbol{y},\boldsymbol{t})$, establishing that the instanton correction $\delta_{\boldsymbol{p}}\, e^{\phi}$ transforms correctly under S-duality.

4.3.3. *Darboux coordinates and Eichler integrals.* Our next example is the coordinate $\boldsymbol{\xi}$ on twistor space, which has the general form:

$$\boldsymbol{\xi} = \boldsymbol{\zeta} + \frac{\tau_2}{2}\left(\frac{\boldsymbol{z}}{t} - \bar{\boldsymbol{z}}t\right) + \frac{1}{8\pi^2}\sum_{\gamma}\Omega(\gamma;\boldsymbol{z})\,\boldsymbol{p}\int_{\ell_\gamma}\frac{\mathrm{d}t'}{t'}\frac{t+t'}{t-t'}\log\left[1-\sigma_\gamma\mathcal{X}_\gamma\right]. \tag{4.23}$$

As explained in [**35**], the S-duality transformations of such expressions can be greatly simplified by modifying the integration kernel by a t-independent term, at the cost of correcting the mirror map (2.20). After this change and in the large volume limit, $\delta_{\boldsymbol{p}}\,\boldsymbol{\xi}$ becomes

$$\delta_{\boldsymbol{p}}\,\boldsymbol{\xi} = \boldsymbol{p}\sum_{q_\Lambda}\int_{\ell_\gamma}\frac{dz'}{z'-z}\,\overline{H}_\gamma. \tag{4.24}$$

Unlike (4.21), the integral is no longer Gaussian but can be expressed as an Eichler integral

$$\delta_{\boldsymbol{p}}\,\boldsymbol{\xi} = -\frac{e^{-2\pi S_{\rm cl}}}{4\pi}\boldsymbol{p}\sum_{\boldsymbol{\mu}\in\Lambda^*/\Lambda}h_{\boldsymbol{p},\boldsymbol{\mu}}(\tau)\int_{\bar\tau}^{-\mathrm{i}\infty}\frac{\overline{\Upsilon_{\boldsymbol{\mu}}(w,\bar\tau;\bar z)\,\mathrm{d}\bar w}}{\sqrt{\mathrm{i}(\bar w-\tau)}}, \tag{4.25}$$

where $\Upsilon_{\boldsymbol{\mu}}(w,\bar\tau;\bar z)$ (given [**37**, Eq. (4.24)]) is a modular form of weight $(\frac{3}{2},\frac{b_2-1}{2})$. Such integrals transform inhomogeneously under $SL(2,\mathbb{Z})$,

$$\delta_{\boldsymbol{p}}\,\boldsymbol{\xi} \mapsto (c\tau+d)^{-1}\left(\delta_{\boldsymbol{p}}\,\boldsymbol{\xi} + \frac{e^{-2\pi S_{\rm cl}}}{4\pi}\boldsymbol{p}\sum_{\boldsymbol{\mu}\in\Lambda^*/\Lambda}h_{\boldsymbol{p},\boldsymbol{\mu}}(\tau)\int_{-d/c}^{\mathrm{i}\infty}\frac{\overline{\Upsilon_{\boldsymbol{\mu}}(w,\bar\tau;\bar z)\,\mathrm{d}\bar w}}{[\mathrm{i}(\bar w-\tau)]^{1/2}}\right). \tag{4.26}$$

The overall weight $(-1,0)$ is in agreement with the classical transformation of $\boldsymbol{\xi}$ (4.1), however the period integral $\int_{-d/c}^{\mathrm{i}\infty}\mathrm{d}\bar w$ makes this transformation anomalous. Remarkably, using standard techniques for mock theta series [**80**] one can show that

$$\delta_{\boldsymbol{p}}\,\widehat{\boldsymbol{\xi}} = \delta_{\boldsymbol{p}}\,\boldsymbol{\xi} - 2\pi\mathrm{i}\,\boldsymbol{p}\,H_{\rm anom} \tag{4.27}$$

does transform as a modular form of weight $(-1,0)$. Here $H_{\rm anom}$ is an indefinite theta function:

$$H_{\rm anom} = \frac{1}{2}\sum_{q_\Lambda}\left[\mathrm{sgn}\left((\boldsymbol{k}+\boldsymbol{b})\cdot\boldsymbol{t}\right) - \mathrm{sgn}\left((\boldsymbol{k}+\boldsymbol{b})\cdot\boldsymbol{t}'\right)\right]\overline{H}_\gamma, \tag{4.28}$$

with $\boldsymbol{t}'$ lying on the boundary of the Kähler cone. $H_{\rm anom}$ transforms also by a period integral, precisely cancelling the one of $\delta_{\boldsymbol{p}}\,\boldsymbol{\xi}$. More generally, $H_{\rm anom}$ generates a contact transformation which precisely cancels the modular anomaly in all Darboux coordinates, thus establishing the S-duality invariance of the D3-instanton corrected HM metric. It would be very interesting to derive $H_{\rm anom}$ and the choice of $\boldsymbol{t}'$ from first principles.

5. Toward NS5-instanton effects

In addition to D-instanton effects, the metric on the HM moduli space $\mathcal{M}_H$ receives instanton corrections of order e^{-1/g_s^2} from Euclidean NS5-branes wrapping the whole threefold X [**81**]. On the type IIB side, it is clear that such corrections are necessary to restore S-duality, since D5 and NS5 branes transform as a doublet under $SL(2,\mathbb{Z})$. For k NS5-branes the expected correction is, schematically,

$$g_{\rm NS5} \sim e^{-4\pi|k|r-\pi i k\sigma}\,\mathscr{Z}_k, \tag{5.1}$$

where $\mathscr{Z}_k$ is the partition function for the degrees of freedom localized on the NS5-branes and $r\sim \mathcal{V}/g_s^2$. In particular, unlike D-instanton effects, these corrections break the continuous translational symmetry of the NS-axion to a discrete subgroup $\sigma\mapsto\sigma+2\kappa$, $\kappa\in\mathbb{Z}$, as anticipated in equation (2.4). Our goal in this section is to infer the form of $\mathscr{Z}_k$ (more precisely, the corresponding correction to the twistor space $\mathcal{Z}$) from the known D5-instanton corrections, at least in a linearized approximation. Before doing this, we discuss constraints on $\mathscr{Z}_k$ coming from the topology of the axion circle bundle $\mathscr{C}_\sigma$. We close with some speculative comments on relations between NS5-instantons and quantum integrable systems.

5.1. Topology of the axion circle bundle.

5.1.1. *NS5-partition function and theta series.* The fivebrane partition function $\mathscr{Z}_k$ is in general a function of the dilaton ϕ, complex structure moduli z^a and RR moduli C (in type IIA variables). Its dependence on C is strongly constrained by the fact that the corrected metric must stay invariant under the large gauge transformations (2.4). In view of (5.1), this implies that, under an integer shift $C\mapsto C+H$ with $H\in H^3(X,\mathbb{Z})$,

$$\mathscr{Z}_k(C+H)=\big(\lambda(H)\big)^k\,\mathbf{E}\big(\tfrac{k}{2}\langle C,H\rangle\big)\,\mathscr{Z}_k(C). \tag{5.2}$$

In words, $\mathscr{Z}_k$ must be a section of the theta line bundle $(\mathscr{L}_\Theta)^{\otimes k}$ over $\mathcal{T}$, with first Chern class $c_1(\mathscr{L}_\Theta)=\omega_{\mathcal{T}}$ (see (2.10)). In the Weil complex structure, $(\mathscr{L}_\Theta)^{\otimes k}$ is known to admit $|k|^{b_3(X)/2}$ holomorphic sections [**82**], corresponding to the Siegel theta series

(5.3)
$$\vartheta_{k,\mu}(C)=\sum_{n\in\Gamma_m+\mu+\theta}\mathbf{E}\Big(\tfrac{k}{2}(\zeta^\Lambda-n^\Lambda)\bar{\mathcal{N}}_{\Lambda\Sigma}(\zeta^\Sigma-n^\Sigma)+k(\tilde\zeta_\Lambda-\phi_\Lambda)n^\Lambda+\tfrac{k}{2}(\theta^\Lambda\phi_\Lambda-\zeta^\Lambda\tilde\zeta_\Lambda)\Big)$$

labelled by vectors $\mu\in\Gamma_m/|k|\Gamma_m$, where Γ_m is a Lagrangian sublattice of $H^3(X,\mathbb{Z})$, here spanned by A-cycles. Physically, the sum over n^Λ labels the topological sectors of the (imaginary) self-dual 3-form field strength $\mathcal{H}$ living on k fivebranes. Indeed, the Siegel theta series can be obtained by holomorphic factorization of the (non-holomorphic) partition function $\mathscr{Z}_{k,3-\text{form}}$ of a Gaussian 3-form on X [**82–86**]:

$$\mathscr{Z}_{k,3-\text{form}}\sim\sum_{\mu\in\Gamma_m/|k|\Gamma_m}\vartheta_{k,\mu}(C)\,\overline{\vartheta_{k,\mu}(C)}.$$

In general however, the chiral fivebrane worldvolume theory is non-Gaussian, and the only conclusion that can be drawn from (5.2) is that $\mathscr{Z}_k$ must be a linear combination of non-Gaussian theta series

(5.4)
$$\mathscr{Z}_k(C)=\sum_{\mu\in\Gamma_m/(|k|\Gamma_m)}\sum_{n\in\Gamma_m+\mu+\theta}\Psi_{k,\mu}(\zeta^\Lambda-n^\Lambda)\mathbf{E}\Big(k(\tilde\zeta_\Lambda-\phi_\Lambda)n^\Lambda+\tfrac{k}{2}(\theta^\Lambda\phi_\Lambda-\zeta^\Lambda\tilde\zeta_\Lambda)\Big),$$

where we only displayed the dependence on the C-field. The $\Psi_{k,\mu}$'s can be interpreted as wave-functions in the real polarization corresponding to the Lagrangian subspace of $H_3(X,\mathbb{R})$ spanned by periods along the A-cycles.

5.1.2. *Metric dependence.* More generally, for consistency of the correction (5.1), the partition function $\mathscr{Z}_k$ should be a (not necessarily holomorphic) section of $\mathscr{C}_\sigma^k$, where $\mathscr{C}_\sigma$ is the circle bundle where the NS axion $e^{\mathrm{i}\pi\sigma}$ is valued [**19**]. As indicated in (2.10), this circle bundle has curvature both over the fiber $\mathcal{T}$ and base $\mathcal{M}_C(X)$ of the intermediate Jacobian $\mathcal{J}_W(X)$. The curvature over $\mathcal{T}$ reflects the non-trivial behavior (2.4), (5.2) under large gauge transformations, while the curvature over $\mathcal{M}_C(X)$ shows that under a monodromy M in $\mathcal{M}_C(X)$, under which the holomorphic 3-form transforms as $\Omega_{3,0}\mapsto e^f\Omega_{3,0}$, the NS-axion and NS5 partition function transform by

$$\sigma\mapsto\sigma+\frac{\chi(X)}{24\pi}\,\mathrm{Im}\,f+2\kappa(M),\qquad \mathscr{Z}_k\mapsto e^{\mathrm{i}k\frac{\chi(X)}{24}\,\mathrm{Im}\,f+2\pi ik\kappa(M)}\,\mathscr{Z}_k, \tag{5.5}$$

where $\kappa(M)$ is the logarithm of a character of the monodromy group (see [**19**] for more details).

The cancellation of phases between $e^{-\mathrm{i}\pi k\sigma}$ and $\mathscr{Z}_k$ is expected by the general inflow mechanism for local anomalies [**87**]. A complication however is that the first Chern class (2.10) is not an integer cohomology class, rather $2c_1(\mathscr{C}_\sigma)\in H_2(\mathcal{M}_H,\mathbb{Z})$, since $\frac{1}{12}\chi(X)\omega_C$ can be identified as the first Chern class of the determinant line bundle of the Dirac operator for chiral spinors on X [**88, 89**], also known as the BCOV line bundle [**90**]. This means that σ is only defined up to a half-period, and $\mathscr{Z}_k$ has a sign ambiguity. This sign ambiguity corresponds to a choice of character $\kappa(M)$ in (5.5), and also appears to be related to the 'orientation data' in the theory of generalized Donaldson-Thomas invariants [**47, 91**]. We expect that global anomaly cancellation will ensure that these two ambiguities cancel.

5.2. Fivebrane corrections to the contact structure.

5.2.1. *NS5-branes from D5-branes.* Since (NS5,D5)-branes transform as a doublet under S-duality, the corrections to the complex contact structure on $\mathcal{Z}$ induced by NS5-instantons can in principle be inferred from the D5-instanton corrections described in §3.2. This procedure was implemented at the linearized level in [**19**] as follows. Start from the transition function

$$\overline{H}_\gamma=\frac{\overline{\Omega}(\gamma)}{(2\pi)^2}\,\lambda_{\mathrm{D}}(\gamma)\,\mathbf{E}\Big(p^\Lambda\tilde{\xi}'_\Lambda-q'_\Lambda\xi^\Lambda\Big) \tag{5.6}$$

generating the contact transformation across a BPS ray associated to an instanton with non-vanishing D5-brane charge p^0, and act on it by an $SL(2,\mathbb{Z})$ transformation $g=\left(\begin{smallmatrix}a&b\\c&d\end{smallmatrix}\right)$, choosing $(c,d)=(-k/p^0,p/p^0)$ such that $p^0=gcd(k,p)$. Such a transformation maps a D5-brane into a (k,p)-fivebrane, where k labels the NS5-brane charge. Using (4.1) and (4.3), one finds that the transition function associated to such a fivebrane takes the form

$$H_{k,p,\hat\gamma}=-\frac{\overline{\Omega}(\gamma)}{(2\pi)^2}\,\frac{k}{p^0}\,(\xi^0-n^0)\,\lambda_{\mathrm{D}}(\gamma)\mathbf{E}\left(kS_\alpha+\frac{p^0\left(k\hat{q}_a(\xi^a-n^a)+p^0\hat{q}_0\right)}{k^2(\xi^0-n^0)}+a\,\frac{p^0q'_0}{k}-c_{2,a}p^a\epsilon(g)\right), \tag{5.7}$$

where $\hat\gamma$ denotes the remaining D3-D1-D(-1)-charges and

$$S_\alpha=\alpha+n^\Lambda\tilde{\xi}_\Lambda+F^{\mathrm{cl}}(\xi-n)-\frac{1}{2}\,A_{\Lambda\Sigma}n^\Lambda n^\Sigma,\qquad (n^0,n^a)=\left(\frac{p}{k},\frac{p^a}{k}\right)\in\mathbb{Z}/k. \tag{5.8}$$

This transition function generates a discontinuity in the Darboux coordinates across the image of the BPS ray ℓ_γ under the same $SL(2,\mathbb{Z})$ transformation, namely a meridian $\ell_{k,p,\hat{\gamma}} \subset \mathbb{P}^1$ joining the two roots of the equation $\xi^0(t) = n^0$ where (5.7) has essential singularities.

One can perform two consistency checks on the above result. First, by evaluating the Penrose transform of (5.7)

$$\int_{\ell_{k,p,\hat{\gamma}}} \frac{\mathrm{d}t}{t}\, H_{k,p,\hat{\gamma}}(\xi(t), \tilde{\xi}(t), \alpha(t)) \tag{5.9}$$

in the small string coupling limit, which justifies the use of the saddle point approximation, it was checked [**19**] that (5.7) produces corrections with the correct semi-classical action known from the analysis of instanton solutions in $N=2$ supergravity [**92**]. The second check is to verify that (5.7) is consistent with large gauge transformations (2.4) and monodromy around the large volume point. This requires that the set of transition functions should be mapped to itself under the corresponding actions on twistor space. This is indeed the case, as shown in [**19**] and further clarified in [**93**]. The problem of going beyond the linearized regime and elevating the infinitesimal contact transformations generated by (5.7) to their finite counterparts was addressed in [**67, 93**].

5.2.2. *Relation to topological strings.* A general prediction of S-duality is that the partition function of a single NS5-brane ($k=1$) on a CY threefold X should be governed by the (ordinary) DT invariants of X with $p^0 = 1$, which are in turn related to higher genus Gromov-Witten invariants [**94**], and therefore to topological strings [**84, 95, 96**]. A precise relation of this sort follows immediately from (5.7). Indeed, the formal sum over all charges

$$H^{(1)}_{\mathrm{NS5}}(\xi, \tilde{\xi}, \alpha) = \sum_{p, p^a, q_\Lambda} H_{1,p,\hat{\gamma}}(\xi, \tilde{\xi}, \alpha) \tag{5.10}$$

is invariant under Heisenberg transformations (2.16). Thus it can be cast in the form (5.4) of a non-Gaussian theta series [**19**]

$$H^{(1)}_{\mathrm{NS5}}(\xi, \tilde{\xi}, \alpha) = \frac{1}{4\pi^2} \sum_{n^\Lambda} \mathcal{H}^{(1)}_{\mathrm{NS5}}\left(\xi^\Lambda - n^\Lambda\right) \mathbf{E}\Big(\alpha + n^\Lambda(\tilde{\xi}_\Lambda - \phi_\Lambda)\Big). \tag{5.11}$$

Remarkably, the wave-function $\mathcal{H}^{(1)}_{\mathrm{NS5}}$ turns to be proportional (up to ξ^0-dependent factors involving the Mac-Mahon function) to the topological A-model string amplitude on $\hat{X}$ in the real polarization,

$$\mathcal{H}^{(1)}_{\mathrm{NS5}}(\xi^\Lambda) \sim \Psi^{\mathrm{top}}_{\mathbb{R}}(\xi^\Lambda). \tag{5.12}$$

This relation indicates that the proper habitat for the topological string amplitude is the space $H^1(\mathcal{Z}_D, O(2))$ parametrizing deformations of the D-instanton corrected twistor space.

5.3. NS5-branes and quantization of cluster varieties. We conclude our discussion of quantum corrected HM moduli spaces by collecting several indications pointing to deep relations with quantum integrable systems. The general relation to integrable systems is of course built in from the fact that $\mathcal{M}'_H$, the HK manifold dual to the perturbative $\mathcal{M}_H$ by the QK/HK correspondence of §3.3, is a HK manifold fibered by algebraic tori, and therefore provides an example of a complex integrable system.[6]

More interestingly, this relation seems to extend in the presence of D-instanton corrections, in view of the fact [**48**, **98**] that the integral equations (3.13) governing the D-instanton corrections to the metric on $\mathcal{M}_H$ coincide with equations of Thermodynamic Bethe Ansatz (TBA) typically describing the spectrum of two-dimensional integrable models [**99**]. Moreover, the D-instanton corrected contact potential (3.16) is identified with the free energy of the system associated to this TBA, and the corresponding S-matrix can be shown to satisfy all axioms of integrability [**98**].

The TBA equations can often be rewritten as a system of discrete equations, known as Y-system, corresponding to gluing conditions (3.11) of our framework. In particular, the same dilogarithm identities which ensure consistency with wall-crossing are well-known to arise as a consequence of periodicity of Y-systems [**100**]. Their mathematical underpinning is Fomin and Zelevinsky's theory of *cluster algebras*[7] [**102**], and its generalization to *cluster varieties*, developed by Fock and Goncharov [**103**]. In particular, a structure very similar to NS5-instanton corrections arises when quantizing *cluster $\mathcal{A}$-varieties* [**104**]. An N-dimensional cluster $\mathcal{A}$-variety is a collection of complex tori $(\mathbb{C}^\times)^N$ glued together into a symplectic algebraic variety using cluster transformations. Geometric quantization produces a pre-quantum vector bundle $\mathscr{V}_\hbar \to \mathcal{A}$, depending on a rational parameter $\hbar = s/r$, where s is the first Chern class and r encodes the rank of the bundle. For $\hbar = 1$, the corresponding line bundle $\mathscr{V}_1$ is described by exactly the same gluing conditions (3.11) as for the holomorphic bundle $\mathscr{L}_{\mathcal{Z}'}$ entering the QK/HK-correspondence. Thus we may view the hyperholomorphic bundle $\mathscr{L}$ as a pre-quantum line bundle for the geometric quantization of $\mathcal{M}'$ (as also observed in [**59**]).

To explain the relation with NS5-branes, let us for simplicity restrict to the case of a two-dimensional cluster $\mathcal{A}$-variety, with local coordinates $(a,b) \in \mathbb{C}^\times \times \mathbb{C}^\times$. Sections of $\mathscr{V}_1$ are multivalued functions $F(a,b)$ on $\mathbb{C}^\times \times \mathbb{C}^\times$, and by a choice of polarization one may restrict to holomorphic 'wave functions' $\Psi(a)$ on $\mathbb{C}^\times$. Fock and Goncharov implement this restriction explicitly via Fourier expansion in one of the variables [**104**]:

$$F(a,b) = \sum_{n\in\mathbb{Z}} \Psi(\log a - 2\pi \mathrm{i} n) \exp\left[-\tfrac{\log a \,\log b}{4\pi \mathrm{i}}\right] b^n. \tag{5.13}$$

Identifying $a = e^{2\pi \mathrm{i}\xi}$ and $b = e^{2\pi\mathrm{i}\tilde\xi}$, up to the factor $e^{-\mathrm{i}\pi\tilde\alpha}$, this takes the same form as the NS5-partition function (5.11). Thus the twistor space partition function of a single NS5-brane ($k = 1$) may be thought of as a section of a pre-quantum line bundle over an $\mathcal{A}$-cluster variety. More generally, for $\hbar \neq 1$, sections of the vector bundle $\mathscr{V}_\hbar$ can be represented by vector-valued functions $\Psi_{\hbar^{-1},\ell}(\xi)$ via a generalization of the expansion (5.13). In the case $s = 1$ and $r = 1/\hbar \in \mathbb{Z}_+$, this becomes

$$F_\hbar(\xi,\tilde\xi) = \sum_{\ell\in\mathbb{Z}/(\mathbb{Z}|\hbar|^{-1})} \sum_{n\in\mathbb{Z}+\ell|\hbar|} \Psi_{\hbar^{-1},\ell}(\xi - n)\mathbf{E}\Big((2n\tilde\xi - \xi\tilde\xi)/2\hbar\Big), \tag{5.14}$$

[6]See [**97**] for a recent analysis of the relation between these integrable systems and wall-crossing in Donaldson-Thomas theory.

[7]The relation between wall-crossing and cluster transformations was noted already in [**47**], and further analyzed in the context of $\mathcal{N} = 2$ gauge theories [**101**] and $\mathcal{N} = 2$ supergravity [**52**].

which is recognized as the non-abelian Fourier expansion corresponding to k NS5-branes [**19**, **105**], provided we identify $\hbar = 1/k$ and set the characteristics (θ, ϕ) to zero. This identification of the NS5-brane charge k with the inverse of a quantization parameter $\hbar$ will be confirmed from a different perspective in §5.4 below.

Moreover, as explained in [**104**], the gluing conditions for wave functions $\Psi_{\hbar^{-1},\ell}$ involve a convolution with the quantum dilogarithm. Since the latter also governs the wall-crossing behavior of motivic DT invariants $\Omega_{\hbar^{-1}}(\gamma)$ [**47**], this suggests that NS5-instantons should be controlled by the same. On the other hand, S-duality implies that (p,k)5-brane instantons are determined by the ordinary DT invariants $\Omega(\gamma)$ suggesting an intriguing relation between ordinary and motivic DT invariants which it would be very interesting to spell out. Finally, while the analogy with quantum cluster varieties seems to open the way to the construction of a natural section of $H^1(\mathcal{Z}, \mathcal{O}(2))$, hence a linear correction to the D-instanton corrected metric on $\mathcal{M}_H$, it is as yet unclear how this could be lifted to a full non-linear deformation, consistent with S-duality and regularity.

5.4. Universal hypermultiplet and free fermions. Another interesting relation to quantum integrable systems can be seen in the special case of type IIA string theory compactified on a rigid CY threefold with $h^{2,1}(X) = 0$. The corresponding HM moduli space is a 4-dimensional QK space, often known as 'the universal hypermultiplet' (although it is hardly universal).

In the absence of NS5-brane corrections, $\mathcal{M}_H$ has a continuous isometry corresponding to shifts of the NS-axion σ. Four-dimensional QK spaces with an isometry are known to be described by solutions of the Toda equation [**106**]

$$\partial_z \partial_{\bar z} T + \partial_\rho^2 \, \mathrm{e}^T = 0, \tag{5.15}$$

which appears as the lowest equation of the dispersionless (i.e. classical) limit of Toda integrable hierarchy. In fact, one can show that the twistor framework is equivalent to the Lax formalism for this hierarchy [**107**]. In particular, the Darboux coordinates ξ and $\tilde\xi$ coincide with the two Lax operators, whereas the gluing conditions are identified with the so called string equations. To understand the role of the Darboux coordinate α, consider the perturbative HM moduli space (2.8) with $F = -\frac{\mathrm{i}}{4} X^2$. In this case the corresponding solution of (5.15) takes particularly simple form [**108**, **109**]

$$T = \log(\rho + c). \tag{5.16}$$

In [**110**] it was shown that the Darboux coordinate α given in (2.15) is related in a simple way to the WKB phase of the quasiclassical Baker-Akhiezer function Ψ associated with the solution (5.16). This, together with the known wave-function property of the Baker-Akhiezer function, suggests that Ψ might be related to NS5-brane effects, which typically have an exponential dependence on α, see (5.7).

Indeed, in [**111**] it was suggested that, for compactifications on rigid CY, NS5-brane instanton corrections to the contact structure on the twistor space are generated by the following holomorphic function

$$H_k^{\mathrm{NS5}} \sim (\xi - 2\mathrm{i}\tilde\xi)^{8\pi c k} \, e^{-\pi \mathrm{i} k \tilde\alpha - \pi k\left(\frac14 \xi^2 + \tilde\xi^2\right)}, \tag{5.17}$$

which is the type IIA counterpart of the type IIB description based on (5.7). One can easily verify that H_k^{NS5} is proportional to the Baker-Akhiezer function Ψ provided one identifies the NS5-brane charge with the inverse quantization parameter of the integrable hierarchy [**110**]

$$\hbar^{-1} = 8\pi k. \tag{5.18}$$

This nicely agrees with the relation found below (5.14), up to normalization.

Furthermore, this 4-dimensional example hints for a possible relation to free fermions. Indeed, the solution of Toda hierarchy based on (5.16) is known to describe non-critical $c = 1$ string theory in a non-trivial tachyon background compactified on a circle of the self-dual radius [**112**]. In turn, this theory is described by Matrix Quantum Mechanics, where integrability is manifest, and this matrix model is known to reduce to a system of free fermions. In this description, the Baker-Akhiezer function, shown above to encode NS5-brane effects, is just the one-fermion wave function. It would be exciting to use this idea to compute the exact quantum HM moduli space for rigid CY threefolds, and see whether S-duality or one of its extensions is indeed realized [**105, 113–115**].

References

[1] J. Bagger and E. Witten, *Matter couplings in $N = 2$ supergravity*, Nuclear Phys. B **222** (1983), no. 1, 1–10, DOI 10.1016/0550-3213(83)90605-3. MR708440 (84k:83050)

[2] B. de Wit, P. Lauwers, and A. Van Proeyen, "Lagrangians of N=2 Supergravity - Matter Systems," *Nucl.Phys.* **B255** (1985) 569.

[3] S. Salamon, *Quaternionic Kähler manifolds*, Invent. Math. **67** (1982), no. 1, 143–171, DOI 10.1007/BF01393378. MR664330 (83k:53054)

[4] A. Karlhede, U. Lindström, and M. Roček, *Self-interacting tensor multiplets in $N = 2$ superspace*, Phys. Lett. B **147** (1984), no. 4-5, 297–300, DOI 10.1016/0370-2693(84)90120-5. MR769049 (87c:81128)

[5] N. J. Hitchin, A. Karlhede, U. Lindström, and M. Roček, *Hyper-Kähler metrics and supersymmetry*, Comm. Math. Phys. **108** (1987), no. 4, 535–589. MR877637 (88g:53048)

[6] C. LeBrun, *Quaternionic-Kähler manifolds and conformal geometry*, Math. Ann. **284** (1989), no. 3, 353–376, DOI 10.1007/BF01442490. MR1001707 (90e:53062)

[7] A. Swann, *Hyper-Kähler and quaternionic Kähler geometry*, Math. Ann. **289** (1991), no. 3, 421–450, DOI 10.1007/BF01446581. MR1096180 (92c:53030)

[8] B. de Wit, M. Roček, and S. Vandoren, "Hypermultiplets, hyperkähler cones and quaternion-Kähler geometry," *JHEP* **02** (2001) 039, `hep-th/0101161`.

[9] U. Lindström and M. Roček, *Properties of hyperkähler manifolds and their twistor spaces*, Comm. Math. Phys. **293** (2010), no. 1, 257–278, DOI 10.1007/s00220-009-0923-0. MR2563806 (2011d:53094)

[10] S. Alexandrov, *Twistor approach to string compactifications: A review*, Phys. Rep. **522** (2013), no. 1, 1–57, DOI 10.1016/j.physrep.2012.09.005. MR3006184

[11] J. Louis and R. Valandro, "Heterotic-Type II Duality in the Hypermultiplet Sector," *JHEP* **1205** (2012) 016, `1112.3566`.

[12] S. Alexandrov and B. Pioline, *Heterotic-type II duality in twistor space*, J. High Energy Phys. **3** (2013), 085. MR3046713

[13] S. Alexandrov, J. Louis, B. Pioline, and R. Valandro, "$\mathcal{N} = 2$ Heterotic-Type II duality and bundle moduli," *JHEP* **1408** (2014) 092, `1405.4792`.

[14] H. Günther, C. Herrmann, and J. Louis, *Quantum corrections in the hypermultiplet moduli space*, Proceedings of the 32nd International Symposium Ahrenshoop on the Theory of Elementary Particles (Buckow, 1998), Fortschr. Phys. **48** (2000), no. 1-3, 119–123, DOI 10.1002/(SICI)1521-3978(20001)48:1/3⟨119::AID-PROP119⟩3.0.CO;2-#. MR1764059

[15] D. Robles Llana, F. Saueressig, and S. Vandoren, *String loop corrected hypermultiplet moduli spaces*, J. High Energy Phys. **3** (2006), 081, 35 pp. (electronic), DOI 10.1088/1126-6708/2006/03/081. MR2222955 (2006k:81319)

[16] S. Alexandrov, B. Pioline, F. Saueressig, and S. Vandoren, *D-instantons and twistors*, J. High Energy Phys. **3** (2009), 044, 40, DOI 10.1088/1126-6708/2009/03/044. MR2495672 (2011c:81241)

[17] P. Candelas, X. C. de la Ossa, P. S. Green, and L. Parkes, *A pair of Calabi-Yau manifolds as an exactly soluble superconformal theory*, Nuclear Phys. B **359** (1991), no. 1, 21–74, DOI 10.1016/0550-3213(91)90292-6. MR1115626 (93b:32029)

[18] S. Alexandrov, D. Persson, and B. Pioline, "On the topology of the hypermultiplet moduli space in type II/CY string vacua," *Phys.Rev.* **D83** (2011) 026001, 1009.3026.

[19] S. Alexandrov, D. Persson, and B. Pioline, *Fivebrane instantons, topological wave functions and hypermultiplet moduli spaces*, J. High Energy Phys. **3** (2011), 111, 73, DOI 10.1007/JHEP03(2011)111. MR2821100 (2012k:81228)

[20] S. Cecotti, S. Ferrara, and L. Girardello, *Geometry of type II superstrings and the moduli of superconformal field theories*, Internat. J. Modern Phys. A **4** (1989), no. 10, 2475–2529, DOI 10.1142/S0217751X89000972. MR1017548 (90j:81175)

[21] S. Ferrara and S. Sabharwal, *Quaternionic manifolds for type* II *superstring vacua of Calabi-Yau spaces*, Nuclear Phys. B **332** (1990), no. 2, 317–332, DOI 10.1016/0550-3213(90)90097-W. MR1046353 (91g:53051)

[22] A. Strominger, *Special geometry*, Comm. Math. Phys. **133** (1990), no. 1, 163–180. MR1071240 (92a:53068)

[23] I. Antoniadis, R. Minasian, S. Theisen, and P. Vanhove, *String loop corrections to the universal hypermultiplet*, Classical Quantum Gravity **20** (2003), no. 23, 5079–5102, DOI 10.1088/0264-9381/20/23/009. MR2024800 (2004k:81243)

[24] S. Alexandrov, *Quantum covariant c-map*, J. High Energy Phys. **5** (2007), 094, 18 pp. (electronic), DOI 10.1088/1126-6708/2007/05/094. MR2318145 (2008g:81178)

[25] S. Alexandrov and S. Banerjee, "Hypermultiplet metric and D-instantons," Dec 28, 2014. 36 pp. Published in J. High Energy Phys. 1502 (2015) 176

[26] A. Neitzke, B. Pioline, and S. Vandoren, *Twistors and black holes*, J. High Energy Phys. **4** (2007), 038, 41 pp. (electronic), DOI 10.1088/1126-6708/2007/04/038. MR2318751 (2008k:83188)

[27] S. Alexandrov, B. Pioline, F. Saueressig, and S. Vandoren, *Linear perturbations of quaternionic metrics*, Comm. Math. Phys. **296** (2010), no. 2, 353–403, DOI 10.1007/s00220-010-1022-y. MR2608119 (2011i:53058)

[28] M. Roček, C. Vafa, and S. Vandoren, *Hypermultiplets and topological strings*, J. High Energy Phys. **2** (2006), 062, 20 pp. (electronic), DOI 10.1088/1126-6708/2006/02/062. MR2219443 (2007m:81210)

[29] K. Galicki, *A generalization of the momentum mapping construction for quaternionic Kähler manifolds*, Comm. Math. Phys. **108** (1987), no. 1, 117–138. MR872143 (88f:53088)

[30] C. LeBrun and S. Salamon, *Strong rigidity of positive quaternion-Kähler manifolds*, Invent. Math. **118** (1994), no. 1, 109–132, DOI 10.1007/BF01231528. MR1288469 (95k:53059)

[31] J. Louis and A. Micu, *Type II theories compactified on Calabi-Yau threefolds in the presence of background fluxes*, Nuclear Phys. B **635** (2002), no. 1-2, 395–431, DOI 10.1016/S0550-3213(02)00338-3. MR1914159 (2003g:81202)

[32] R. Böhm, H. Günther, C. Herrmann, and J. Louis, *Compactification of type IIB string theory on Calabi-Yau threefolds*, Nuclear Phys. B **569** (2000), no. 1-3, 229–246, DOI 10.1016/S0550-3213(99)00796-8. MR1746656 (2001d:81063)

[33] D. Robles-Llana, M. Roček, F. Saueressig, U. Theis, and S. Vandoren, *Nonperturbative corrections to 4D string theory effective actions from* $\mathrm{SL}(2,\mathbb{Z})$ *duality and supersymmetry*, Phys. Rev. Lett. **98** (2007), no. 21, 211602, 4, DOI 10.1103/PhysRevLett.98.211602. MR2308929 (2008b:81242)

[34] S. Alexandrov and F. Saueressig, *Quantum mirror symmetry and twistors*, J. High Energy Phys. **9** (2009), 108, 36, DOI 10.1088/1126-6708/2009/09/108. MR2580696 (2011f:53085)

[35] S. Alexandrov and B. Pioline, *S-duality in twistor space*, J. High Energy Phys. **8** (2012), 112. MR3006925

[36] M. B. Green and M. Gutperle, *Effects of D-instantons*, Nuclear Phys. B **498** (1997), no. 1-2, 195–227, DOI 10.1016/S0550-3213(97)00269-1. MR1459079 (99e:81193)

[37] S. Alexandrov, J. Manschot, and B. Pioline, *D3-instantons, mock theta series and twistors*, J. High Energy Phys. **4** (2013), 002. MR3065975

[38] E. Sharpe, *D-branes, derived categories, and Grothendieck groups*, Nuclear Phys. B **561** (1999), no. 3, 433–450, DOI 10.1016/S0550-3213(99)00535-0. MR1726156 (2001k:14024)

[39] M. R. Douglas, B. Fiol, and C. Römelsberger, *Stability and BPS branes*, J. High Energy Phys. **9** (2005), 006, 15 pp. (electronic), DOI 10.1088/1126-6708/2005/09/006. MR2171358 (2006e:81207)

[40] M. Kontsevich, *Homological algebra of mirror symmetry*, Proceedings of the International Congress of Mathematicians, Vol. 1, 2 (Zürich, 1994), Birkhäuser, Basel, 1995, pp. 120–139. MR1403918 (97f:32040)

[41] S. Ferrara, J. A. Harvey, A. Strominger, and C. Vafa, *Second-quantized mirror symmetry*, Phys. Lett. B **361** (1995), no. 1-4, 59–65, DOI 10.1016/0370-2693(95)01074-Z. MR1362446 (97b:81097)

[42] M. R. Douglas, *Dirichlet branes, homological mirror symmetry, and stability*, Proceedings of the International Congress of Mathematicians, Vol. III (Beijing, 2002), Higher Ed. Press, Beijing, 2002, pp. 395–408. MR1957548 (2004c:81200)

[43] T. Bridgeland, *Stability conditions on triangulated categories*, Ann. of Math. (2) **166** (2007), no. 2, 317–345, DOI 10.4007/annals.2007.166.317. MR2373143 (2009c:14026)

[44] R. Harvey and H. B. Lawson Jr., *Calibrated geometries*, Acta Math. **148** (1982), 47–157, DOI 10.1007/BF02392726. MR666108 (85i:53058)

[45] E. Andriyash, F. Denef, D. L. Jafferis, and G. W. Moore, *Wall-crossing from supersymmetric galaxies*, J. High Energy Phys. **1** (2012), 115, 15. MR2949278

[46] B. Pioline, "Four ways across the wall," *J.Phys.Conf.Ser.* **346** (2012) 012017, `1103.0261`.

[47] M. Kontsevich and Y. Soibelman, "Stability structures, motivic Donaldson-Thomas invariants and cluster transformations," `0811.2435`.

[48] D. Gaiotto, G. W. Moore, and A. Neitzke, *Four-dimensional wall-crossing via three-dimensional field theory*, Comm. Math. Phys. **299** (2010), no. 1, 163–224, DOI 10.1007/s00220-010-1071-2. MR2672801 (2011m:81283)

[49] D. Joyce and Y. Song, *A theory of generalized Donaldson-Thomas invariants*, Mem. Amer. Math. Soc. **217** (2012), no. 1020, iv+199, DOI 10.1090/S0065-9266-2011-00630-1. MR2951762

[50] J. Manschot, B. Pioline, and A. Sen, *Wall crossing from Boltzmann black hole halos*, J. High Energy Phys. **7** (2011), 059, 73, DOI 10.1007/JHEP07(2011)059. MR2875965

[51] S. Alexandrov, *D-instantons and twistors: some exact results*, J. Phys. A **42** (2009), no. 33, 335402, 26, DOI 10.1088/1751-8113/42/33/335402. MR2525885 (2011b:81190)

[52] S. Alexandrov, D. Persson, and B. Pioline, *Wall-crossing, Rogers dilogarithm, and the QK/HK correspondence*, J. High Energy Phys. **12** (2011), 027, i, 64. MR2935636

[53] J. Stoppa, *Joyce-Song wall crossing as an asymptotic expansion*, Kyoto J. Math. **54** (2014), no. 1, 103–156, DOI 10.1215/21562261-2400292. MR3178548

[54] S. Alexandrov, G. W. Moore, A. Neitzke and B. Pioline, "$\mathbb{R}3$ Index for Four-Dimensional $N = 2$ Field Theories," Jun 9, 2014. 7 pp. Published in Phys. Rev. Lett. 114 (2015) 121601, DOI: 10.1103/PhysRevLett.114.121601. e-Print: arXiv:1406.2360.

[55] M. M. Capria and S. M. Salamon, *Yang-Mills fields on quaternionic spaces*, Nonlinearity **1** (1988), no. 4, 517–530. MR967469 (89k:58064)

[56] T. Gocho and H. Nakajima, *Einstein-Hermitian connections on hyper-Kähler quotients*, J. Math. Soc. Japan **44** (1992), no. 1, 43–51, DOI 10.2969/jmsj/04410043. MR1139657 (92k:53079)

[57] M. Verbitsky, *Hyperholomorphic bundles over a hyper-Kähler manifold*, J. Algebraic Geom. **5** (1996), no. 4, 633–669. MR1486984 (2000a:32051)

[58] A. Haydys, *HyperKähler and quaternionic Kähler manifolds with S^1-symmetries*, J. Geom. Phys. **58** (2008), no. 3, 293–306, DOI 10.1016/j.geomphys.2007.11.004. MR2394039 (2008m:53113)

[59] N. Hitchin, *On the hyperkähler/quaternion Kähler correspondence*, Comm. Math. Phys. **324** (2013), no. 1, 77–106, DOI 10.1007/s00220-013-1689-y. MR3116317

[60] D. V. Alekseevsky, V. Cortés, and T. Mohaupt, *Conification of Kähler and hyper-Kähler manifolds*, Comm. Math. Phys. **324** (2013), no. 2, 637–655, DOI 10.1007/s00220-013-1812-0. MR3117523

[61] R. S. Ward and R. O. Wells, *Twistor geometry and field theory*. Cambridge Monographs on Mathematical Physics. Cambridge University Press, Cambridge, 1990.

[62] A. Neitzke, "On a hyperholomorphic line bundle over the Coulomb branch," `1110.1619`.

[63] F. Gliozzi and R. Tateo, *ADE functional dilogarithm identities and integrable models*, Phys. Lett. B **348** (1995), no. 1-2, 84–88, DOI 10.1016/0370-2693(95)00125-5. MR1325407 (96e:33001)

[64] T. Nakanishi, *Dilogarithm identities for conformal field theories and cluster algebras: simply laced case*, Nagoya Math. J. **202** (2011), 23–43. MR2804544

[65] B. Pioline and S. Vandoren, *Large D-instanton effects in string theory*, J. High Energy Phys. **7** (2009), 008, 11, DOI 10.1088/1126-6708/2009/07/008. MR2545090 (2010j:81265)

[66] S. Alexandrov and S. Banerjee, *Modularity, quaternion-Kähler spaces, and mirror symmetry*, J. Math. Phys. **54** (2013), no. 10, 102301, 23, DOI 10.1063/1.4826603. MR3134587

[67] S. Alexandrov and S. Banerjee, "Fivebrane instantons in Calabi-Yau compactifications," *Phys.Rev.* **D90** (2014) 041902, 1403.1265.

[68] J. Maldacena, A. Strominger, and E. Witten, *Black hole entropy in M-theory*, J. High Energy Phys. **12** (1997), 16, DOI 10.1088/1126-6708/1997/12/002. MR1608487 (99d:83097)

[69] D. Gaiotto, A. Strominger, and X. Yin, "From AdS(3)/CFT(2) to black holes/topological strings,", Feb 2006. 11 pp. Published in J. High Energy Phys. 0709 (2007) 050, DOI: 10.1088/1126-6708/2007/09/050, e-Print: hep-th/0602046

[70] D. Gaiotto, A. Strominger, and X. Yin, *The M5-brane elliptic genus: modularity and BPS states*, J. High Energy Phys. **8** (2007), 070, 18 pp. (electronic), DOI 10.1088/1126-6708/2007/08/070. MR2342043 (2009f:81194)

[71] J. de Boer, M. C. N. Cheng, R. Dijkgraaf, J. Manschot, and E. Verlinde, *A Farey tail for attractor black holes*, J. High Energy Phys. **11** (2006), 024, 28, DOI 10.1088/1126-6708/2006/11/024. MR2270426 (2008f:83125)

[72] F. Denef and G. W. Moore, *Split states, entropy enigmas, holes and halos*, J. High Energy Phys. **11** (2011), 129, i, 152, DOI 10.1007/JHEP11(2011)129. MR2913216

[73] A. Dabholkar, F. Denef, G. W. Moore, and B. Pioline, *Precision counting of small black holes*, J. High Energy Phys. **10** (2005), 096, 90 pp. (electronic), DOI 10.1088/1126-6708/2005/10/096. MR2180982 (2007b:81232)

[74] J. Manschot, *Stability and duality in $\mathcal{N} = 2$ supergravity*, Comm. Math. Phys. **299** (2010), no. 3, 651–676, DOI 10.1007/s00220-010-1104-x. MR2718927 (2012a:81276)

[75] J. de Boer, F. Denef, S. El-Showk, I. Messamah, and D. Van den Bleeken, "Black hole bound states in $AdS_3 \times S^2$," *JHEP* **0811** (2008) 050, 0802.2257.

[76] J. Manschot, *Wall-crossing of D4-branes using flow trees*, Adv. Theor. Math. Phys. **15** (2011), no. 1, 1–42. MR2888006

[77] J. Manschot, *BPS invariants of $\mathcal{N} = 4$ gauge theory on Hirzebruch surfaces*, Commun. Number Theory Phys. **6** (2012), no. 2, 497–516, DOI 10.4310/CNTP.2012.v6.n2.a4. MR2993122

[78] J. Troost, *The non-compact elliptic genus: mock or modular*, J. High Energy Phys. **6** (2010), 104, 18, DOI 10.1007/JHEP06(2010)104. MR2680313 (2011h:81217)

[79] C. Vafa and E. Witten, *A strong coupling test of S-duality*, Nuclear Phys. B **431** (1994), no. 1-2, 3–77, DOI 10.1016/0550-3213(94)90097-3. MR1305096 (95k:81138)

[80] S. Zwegers, "Mock theta functions." PhD dissertation, 2002, Utrecht.

[81] K. Becker, M. Becker, and A. Strominger, "Five-branes, membranes and nonperturbative string theory," *Nucl. Phys.* **B456** (1995) 130–152, hep-th/9507158.

[82] E. Witten, *Five-brane effective action in M-theory*, J. Geom. Phys. **22** (1997), no. 2, 103–133, DOI 10.1016/S0393-0440(97)80160-X. MR1451550 (98e:81150)

[83] M. Henningson, B. E. W. Nilsson, and P. Salomonson, *Holomorphic factorization of correlation functions in $(4k+2)$-dimensional $(2k)$-form gauge theory*, J. High Energy Phys. **9** (1999), Paper 8, 13 pp. (electronic), DOI 10.1088/1126-6708/1999/09/008. MR1720225 (2001a:81141)

[84] R. Dijkgraaf, E. P. Verlinde, and M. Vonk, "On the partition sum of the NS five-brane," hep-th/0205281.

[85] G. W. Moore, *Anomalies, Gauss laws, and Page charges in M-theory* (English, with English and French summaries), C. R. Phys. **6** (2005), no. 2, 251–259, DOI 10.1016/j.crhy.2004.12.005. Strings 04. Part II. MR2121726 (2006i:81180)

[86] D. Belov and G. W. Moore, "Holographic action for the self-dual field," hep-th/0605038.

[87] M. J. Duff, J. T. Liu, and R. Minasian, *Eleven-dimensional origin of string/string duality: a one-loop test*, Nuclear Phys. B **452** (1995), no. 1-2, 261–282, DOI 10.1016/0550-3213(95)00368-3. MR1356403 (97c:81144)

[88] E. Witten, *Global gravitational anomalies*, Comm. Math. Phys. **100** (1985), no. 2, 197–229. MR804460 (87k:58282)

[89] S. Monnier, *Geometric quantization and the metric dependence of the self-dual field theory*, Comm. Math. Phys. **314** (2012), no. 2, 305–328, DOI 10.1007/s00220-012-1525-9. MR2958954

[90] H. Fang, Z. Lu, and K.-I. Yoshikawa, *Analytic torsion for Calabi-Yau threefolds*, J. Differential Geom. **80** (2008), no. 2, 175–259. MR2454893 (2009k:58069)

[91] Z. Hua, "Orientation data on moduli space of sheaves on Calabi-Yau threefold", http://arxiv.org/abs/1212.3790

[92] H. Looyestijn and S. Vandoren, *On NS5-brane instantons and volume stabilization*, J. High Energy Phys. **4** (2008), 024, 25, DOI 10.1088/1126-6708/2008/04/024. MR2425279 (2009f:81175)

[93] S. Alexandrov and S. Banerjee, *Dualities and fivebrane instantons*, J. High Energy Phys. **11** (2014), 040. MR3290808

[94] D. Maulik, N. Nekrasov, A. Okounkov, and R. Pandharipande, *Gromov-Witten theory and Donaldson-Thomas theory. I*, Compos. Math. **142** (2006), no. 5, 1263–1285, DOI 10.1112/S0010437X06002302. MR2264664 (2007i:14061)

[95] N. Nekrasov, H. Ooguri, and C. Vafa, *S-duality and topological strings*, J. High Energy Phys. **10** (2004), 009, 11 pp. (electronic), DOI 10.1088/1126-6708/2004/10/009. MR2116038 (2006b:81289)

[96] A. Kapustin, *Gauge theory, topological strings, and S-duality*, J. High Energy Phys. **9** (2004), 034, 8 pp. (electronic), DOI 10.1088/1126-6708/2004/09/034. MR2108759 (2005k:81288)

[97] M. Kontsevich and Y. Soibelman, "Wall-crossing structures in Donaldson-Thomas invariants, integrable systems and Mirror Symmetry," *ArXiv e-prints* (Mar., 2013) 1303.3253.

[98] S. Alexandrov and P. Roche, *TBA for non-perturbative moduli spaces*, J. High Energy Phys. **6** (2010), 066, 25, DOI 10.1007/JHEP06(2010)066. MR2680332 (2011m:81276)

[99] Al. B. Zamolodchikov, *Thermodynamic Bethe ansatz in relativistic models: scaling 3-state Potts and Lee-Yang models*, Nuclear Phys. B **342** (1990), no. 3, 695–720, DOI 10.1016/0550-3213(90)90333-9. MR1072737 (91g:81106)

[100] F. Ravanini, A. Valleriani, and R. Tateo, *Dynkin TBAs*, Internat. J. Modern Phys. A **8** (1993), no. 10, 1707–1727, DOI 10.1142/S0217751X93000709. MR1216231 (94h:81149)

[101] D. Gaiotto, G. W. Moore, and A. Neitzke, *Wall-crossing, Hitchin systems, and the WKB approximation*, Adv. Math. **234** (2013), 239–403, DOI 10.1016/j.aim.2012.09.027. MR3003931

[102] S. Fomin and A. Zelevinsky, *Cluster algebras. I. Foundations*, J. Amer. Math. Soc. **15** (2002), no. 2, 497–529 (electronic), DOI 10.1090/S0894-0347-01-00385-X. MR1887642 (2003f:16050)

[103] V. V. Fock and A. B. Goncharov, *The quantum dilogarithm and representations of quantum cluster varieties*, Invent. Math. **175** (2009), no. 2, 223–286, DOI 10.1007/s00222-008-0149-3. MR2470108 (2010e:14021)

[104] V. Fock and A. Goncharov, "unpublished." talk by V. Fock at the CQGM meeting in Aarhus, summer 2012.

[105] B. Pioline and D. Persson, *The automorphic NS5-brane*, Commun. Number Theory Phys. **3** (2009), no. 4, 697–754, DOI 10.4310/CNTP.2009.v3.n4.a5. MR2610484 (2011k:81227)

[106] M. Przanowski, *Killing vector fields in self-dual, Euclidean Einstein spaces with* $\Lambda \neq 0$, J. Math. Phys. **32** (1991), no. 4, 1004–1010, DOI 10.1063/1.529375. MR1097787 (91m:83029)

[107] P. Guha and K. Takasaki, *Dispersionless hierarchies, Hamilton-Jacobi theory and twistor correspondences*, J. Geom. Phys. **25** (1998), no. 3-4, 326–340, DOI 10.1016/S0393-0440(97)00034-X. MR1619849 (99e:58099)

[108] M. Davidse, F. Saueressig, U. Theis, and S. Vandoren, *Membrane instantons and de Sitter vacua*, J. High Energy Phys. **9** (2005), 065, 37 pp. (electronic), DOI 10.1088/1126-6708/2005/09/065. MR2174150 (2006d:83116)

[109] S. Alexandrov, F. Saueressig, and S. Vandoren, *Membrane and fivebrane instantons from quaternionic geometry*, J. High Energy Phys. **9** (2006), 040, 32 pp. (electronic), DOI 10.1088/1126-6708/2006/09/040. MR2257491 (2007i:81177)

[110] S. Alexandrov, *c-map as* $c = 1$ *string*, Nuclear Phys. B **863** (2012), no. 1, 329–346, DOI 10.1016/j.nuclphysb.2012.05.018. MR2930617

[111] S. Alexandrov, B. Pioline, and S. Vandoren, *Self-dual Einstein spaces, heavenly metrics, and twistors*, J. Math. Phys. **51** (2010), no. 7, 073510, 31, DOI 10.1063/1.3430574. MR2681102 (2012b:53084)

[112] S. Alexandrov, "Matrix quantum mechanics and two-dimensional string theory in non-trivial backgrounds," hep-th/0311273.

[113] L. Bao, A. Kleinschmidt, B. E. W. Nilsson, D. Persson, and B. Pioline, *Instanton corrections to the universal hypermultiplet and automorphic forms on* SU(2, 1), Commun. Number Theory Phys. **4** (2010), no. 1, 187–266, DOI 10.4310/CNTP.2010.v4.n1.a5. MR2679380 (2012b:81185)

[114] L. Bao, A. Kleinschmidt, B. E. W. Nilsson, D. Persson, and B. Pioline, "Rigid Calabi-Yau threefolds, Picard Eisenstein series and instantons," *J. Phys. Conf. Ser.* **462** (2013) 012026, 1005.4848.

[115] D. Persson, "Automorphic Instanton Partition Functions on Calabi-Yau Threefolds," *J.Phys.Conf.Ser.* **346** (2012) 012016, 1103.1014.

UNIVERSIÉ MONTPELLIER 2, LABORATOIRE CHARLES COULOMB, F-34095, MONTPELLIER, FRANCE
E-mail address: salexand@univ-montp2.fr

BETHE CENTER FOR THEORETICAL PHYSICS, BONN UNIVERSITY, NUSSALLEE 12, 53115 BONN, GERMANY

MAX PLANCK INSTITUTE FOR MATHEMATICS, VIVATSGASSE 7, 53111 BONN, GERMANY
E-mail address: manschot@uni-bonn.de

FUNDAMENTAL PHYSICS, CHALMERS UNIVERSITY OF TECHNOLOGY, 412 96, GOTHENBURG, SWEDEN
E-mail address: daniel.persson@chalmers.se

CERN PH-TH, CASE C01600, CERN, CH-1211 GENEVA 23, SWITZERLAND

LABORATOIRE DE PHYSIQUE THÉORIQUE ET HAUTES ENERGIES, CNRS UMR 7589, AND UNIVERSITÉ PIERRE ET MARIE CURIE - PARIS 6, 4 PLACE JUSSIEU, 75252 PARIS CEDEX 05, FRANCE
E-mail address: boris.pioline@cern.ch, pioline@lpthe.jussieu.fr

Proceedings of Symposia in Pure Mathematics
Volume **90**, 2015
http://dx.doi.org/10.1090/pspum/090/01518

Non-Geometric Fluxes Versus (Non)-Geometry

David Andriot

ABSTRACT. Non-geometry has been introduced when considering a new type of string backgrounds, for which stringy symmetries serve as transition functions between patches of the target space. Then, some terms in the potential of four-dimensional gauged supergravities, generated by so-called non-geometric fluxes, have been argued to find a higher-dimensional origin in these backgrounds, even if a standard compactification on those cannot be made. We present here recent results clarifying the relation between these two settings. Thanks to a field redefinition, we reformulate the NSNS Lagrangian in such a way that the non-geometric fluxes appear in ten dimensions. In addition, if an NSNS field configuration is non-geometric, its reformulation in terms of the new fields can restore a standard geometry. A dimensional reduction is then possible, and leads to the non-geometric terms in the four-dimensional potential. Reformulating similarly doubled field theory, we get a better understanding of the role of the non-geometric fluxes, and rewrite the Lagrangian in a manifestly diffeomorphism-covariant manner. We finally discuss the relevance of the field redefinition and the non-geometric fluxes when studying the non-commutativity of string coordinates. This paper is based on a talk given at String-Math 2012 in Bonn, Germany, and contributes to the proceedings of this conference.

1. Introduction

String backgrounds with non-trivial fluxes on an internal space are crucial for phenomenology. Solutions of four-dimensional supergravity with non-geometric fluxes are in that respect rather promising. Some of them are indeed among the few examples of metastable de Sitter solutions [**5**], and others allow to achieve a full moduli stabilisation [**6**]. Unfortunately, the uplift to ten dimensions of such solutions has been so far rather unclear, as it should involve a ten-dimensional non-geometry. One aim of the results [**1**–**4**] presented here is to understand better the relation between these two different settings. Before going any further, let us give a brief account on these ten- and four-dimensional perspectives. We restrict here the discussion to supergravity, and only consider the NSNS sector.

- In ten dimensions

  We first consider the target space of a string, divided locally into patches. If some fields are living on each of these patches, they can be defined globally by "gluing" them from one patch to the other using transition functions. The latter, for a standard differential geometry, are the

2010 *Mathematics Subject Classification.* Primary 51-06, 53-06, 81-06, 83-06.

diffeomorphisms (completed with gauge transformations), i.e. the usual symmetries of a point-like field theory. String theory has actually more symmetries, for instance T-duality on some backgrounds. The essential idea of non-geometry is then to use these more stringy symmetries to glue the fields from one patch to the other [**7**–**9**]. From the string theory point of view, the resulting geometry is equally fine, and could serve as a background. However, such stringy transition functions take us away from standard differential geometry, hence the name of "non-geometry". From the point of view of an effective theory in the target space, this situation can be more problematic. In practice, the ten-dimensional supergravity fields would look ill-defined, as they would not be single-valued in the usual sense. One typically faces global issues with such field configurations.

- In four dimensions

  Some four-dimensional gauged supergravities have in their super- or scalar potential specific terms generated by the so-called non-geometric fluxes. The latter are quantized objects $Q_p{}^{mn}$ and R^{mnp}, and can be identified with some structure constants of the gauge algebra; in other words they correspond to specific gaugings of the four-dimensional supergravity [**8**, **10**, **11**]. As discussed above, these non-geometric terms of a four-dimensional potential can be of phenomenological interest.

One reason to make a link between these two settings can be found in the study of a simple non-geometric field configuration [**12**] that we call the toroidal example. It has the property of being related to geometric configurations when applying standard T-duality transformations along isometries. Geometric backgrounds lead after compactification to specific terms in the four-dimensional potential, such as those generated by the NSNS H-flux or by the so-called "geometric flux" (related to the curvature of the internal manifold). One can then determine what is the corresponding transformation of these type of terms when performing a T-duality at the ten-dimensional level. The result is that they can transform into the terms generated by Q or R (these rules are described by the so-called T-duality chain). It is in particular the case for the toroidal example, where its non-geometric configuration should then correspond to a four-dimensional Q-flux (hence a "non-geometric" flux). However, this correspondence is not really constructive, as there is no direct determination of Q given the non-geometry. For more details on this discussion, and for a review on non-geometry, we refer the reader to [**1**].

A typical relation between a ten- and a four-dimensional supergravity would be a compactification, so can one get the non-geometric terms of the potential from such a process? To start with, the specific dependence of these terms in the scalar fields cannot be reproduced from a standard ten-dimensional supergravity. In addition, the index structure of Q and R is very special, and can a priori not be found in a flux or a field in ten dimensions; their origin is thus unclear. Finally, we mentioned that ten-dimensional non-geometric field configurations have been argued to be related to four-dimensional non-geometric potential terms. It is therefore tempting to consider a compactification on those. However, the global issues, characteristic of these field configurations, make a standard dimensional reduction not possible; in particular one cannot integrate the fields properly.

Here, we present recent progress [**1**–**3**] in relating ten-dimensional non-geometry and four-dimensional non-geometric fluxes. This is made possible thanks to a reformulation of standard supergravity. More precisely, we consider a field redefinition to be performed on the NSNS fields. The NSNS Lagrangian can this way be rewritten into a new ten-dimensional Lagrangian, in which the non-geometric Q- and R-fluxes appear. In addition, starting with a standard non-geometric field configuration, the redefined fields and new Lagrangian can be globally well-defined. One can then perform the dimensional reduction, and doing so gives precisely the non-geometric terms in the scalar potential. This way, the ten-dimensional and four-dimensional perspectives are finally related.

While these ideas are presented in section 2, we discuss in section 3 the interesting roles played by the field redefinition and the non-geometric fluxes in broader contexts. In double field theory (DFT) [**13**], they help to reformulate the DFT Lagrangian in a manifestly diffeomorphism-covariant manner. We then get a better understanding of the non-geometric fluxes: the R-flux is a tensor, while the Q-flux serves more as a connection; this is analogous to the NSNS H-flux, and the geometric flux. Another topic is the non-commutativity of string coordinates, that we studied in [**4**]: interesting relations occur between non-geometry, non-geometric fluxes and non-commutativity. We come back to them in more details.

Note: Due to huge delays in the publication process, some results presented in the initial version of this paper have been a little outdated. The necessary updates are now provided in three addenda at the end of the paper, to which we refer in the main text.

2. Reformulation of the NSNS sector of supergravity

2.1. The field redefinition. An important object in the reformulation is the field β^{mn}, which is an antisymmetric bivector. Our first motivations to consider this field came from [**14**–**16**], where generalized complex geometry tools are used in supergravity to study ten-dimensional non-geometry. Several arguments are then put forward to indicate that β is a good object to characterise the presence of non-geometry. Relations between β and the non-geometric Q- and R-fluxes are even proposed (see [**1**] for a more detailed account on these ideas). We concluded that making β appear could lead to a reformulation of ten-dimensional supergravity which would provide an origin to the four-dimensional non-geometric fluxes. A way of introducing β is by considering a different generalized vielbein than the standard NSNS one [**16**]. The idea used in [**1**] is then that β would appear through a reparameterization of the generalized metric, associated to the choice of this different generalized vielbein. More precisely, the generalized metric $\mathcal{H}$ usually depends on the standard NSNS metric g_{mn} and Kalb-Ramond field b_{mn}, and we considered the following reparametrization

$$\mathcal{H} = \begin{pmatrix} g - bg^{-1}b & bg^{-1} \\ -g^{-1}b & g^{-1} \end{pmatrix} = \begin{pmatrix} \tilde{g} & -\tilde{g}\beta \\ \beta\tilde{g} & \tilde{g}^{-1} - \beta\tilde{g}\beta \end{pmatrix} , \tag{2.1}$$

where $\tilde{g}_{mn}$ is a new metric. The above can be rewritten in several manners, among which

$$\left. \begin{array}{l} g = (\tilde{g}^{-1} + \beta)^{-1} \tilde{g}^{-1} (\tilde{g}^{-1} - \beta)^{-1} \\ b = -(\tilde{g}^{-1} + \beta)^{-1} \beta (\tilde{g}^{-1} - \beta)^{-1} \end{array} \right| \quad \Leftrightarrow \quad (g + b) = (\tilde{g}^{-1} + \beta)^{-1} , \tag{2.2}$$

where the last equality is useful in DFT (see section 3.1.1). The standard NSNS dilaton ϕ also needs a counterpart, so we introduced the new dilaton $\tilde{\phi}$, defined as follows

$$e^{-2\tilde{\phi}}\sqrt{|\tilde{g}|} = e^{-2\phi}\sqrt{|g|}\ , \tag{2.3}$$

so that the supergravity measure gets preserved.

To summarize, we first have a field redefinition from the standard NSNS fields (g, b, ϕ) to the new set of fields $(\tilde{g}, \beta, \tilde{\phi})$. A natural question is then that of the fate of the standard NSNS Lagrangian under this redefinition. As β has been proposed to be related to the non-geometric fluxes, they could appear in the resulting ten-dimensional Lagrangian. This is indeed the result obtained in [**1**, **3**], that we now detail. For completeness, let us indicate the existence of an alternative field redefinition, proposed in [**17**].

2.2. Rewriting the NSNS Lagrangian. The standard NSNS Lagrangian is given by

$$\mathcal{L} = e^{-2\phi}\sqrt{|g|}\left(\mathcal{R}(g) + 4(\partial\phi)^2 - \frac{1}{12}H_{mnp}H_{qrs}\ g^{mq}g^{nr}g^{ps}\right)\ , \tag{2.4}$$

where $\mathcal{R}(g)$ is the Ricci scalar associated to the metric g for a Levi-Civita connection, the H-flux is given by $H_{mnp} = 3\ \partial_{[m}b_{np]}$, and we refer to [**3**] for more conventions. Performing the field redefinition directly in $\mathcal{L}$ is a rather tedious computation. For instance, the Ricci scalar $\mathcal{R}(g)$ is given by the one, $\mathcal{R}(\tilde{g})$, associated to $\tilde{g}$, together with eleven lines of other terms (see equation (B.15) in [**3**]). It is therefore remarkable that most of these terms get in the end cancelled by others, coming from the two remaining terms of (2.4) once they are also rewritten. This leaves us with, in comparison, a surprisingly simple resulting Lagrangian $\tilde{\mathcal{L}}$, given by

$$\begin{aligned} \tilde{\mathcal{L}} &= \mathcal{L} - \partial(..) \\ &= e^{-2\tilde{\phi}}\sqrt{|\tilde{g}|}\left(\mathcal{R}(\tilde{g}) + 4(\partial\tilde{\phi})^2 - \frac{1}{4}Q_p{}^{mn}Q_q{}^{rs}\ \tilde{g}^{pq}\tilde{g}_{mr}\tilde{g}_{ns} + \dots \right. \\ &\qquad\qquad \left. - \frac{1}{12}R^{mnp}R^{qrs}\ \tilde{g}_{mq}\tilde{g}_{nr}\tilde{g}_{ps}\right) \end{aligned} \tag{2.5}$$

where we introduce ten-dimensional Q- and R-fluxes as

$$Q_p{}^{mn} = \partial_p\beta^{mn}\ ,\ R^{mnp} = 3\ \beta^{q[m}\partial_q\beta^{np]}\ . \tag{2.6}$$

Another, more efficient, method to derive this result is to use DFT as a tool [**2**, **3**]. We give more details about it in section 3.1.1. Note as well that we first computed $\tilde{\mathcal{L}}$ in [**1**] using a simplifying assumption, given by $\beta^{mn}\partial_n\cdot = 0$ where the $\cdot$ is a placeholder for any field. In that case, the dots in (2.5) and the R-flux term vanish. Without this assumption, the terms in the dots should be thought of as going together with the Q-flux term (see [**3**]), leading to a possible redefinition of the actual Q-flux (see addendum 1).

Applying the field redefinition to the NSNS Lagrangian thus gives, up to a total derivative, a new ten-dimensional Lagrangian $\tilde{\mathcal{L}}$. The latter depends on new objects denoted Q and R, which have the same index structure as their four-dimensional counterparts. Is there a more precise relation to the four-dimensional non-geometric fluxes? To verify this, one should perform a dimensional reduction on $\tilde{\mathcal{L}}$. This was done in [**1**, **3**] by studying the dependence of the four-dimensional scalar potential on two moduli: the volume and the four-dimensional dilaton. This is enough to see that the dimensional reduction of $\tilde{\mathcal{L}}$ gives the expected non-geometric terms of the potential (defined in [**18**]), while the standard NSNS Lagrangian $\mathcal{L}$ does not lead to such terms. We conclude that for the first time, these terms get this way a ten-dimensional supergravity origin. In addition, we know that the ten- and four-dimensional non-geometric fluxes are related, but the precise relation remains to be established (see addendum 1).

An important aspect of the dimensional reduction is the global behaviour of the background. The reformulation we have done so far is somehow made formally. In addition, the quantities involved (fields and Lagrangian) are local. But when performing a dimensional reduction, one needs to consider the action, and integrate over the background fields. This implies to consider the global behaviour of an actual background configuration of fields. This is where the discussion on non-geometry becomes crucial.

2.3. The global aspects. For a geometric background, the fields are by definition globally well-defined, and so is the NSNS Lagrangian $\mathcal{L}$. One can therefore perform a dimensional reduction on it, and this gives the standard terms due to the H-flux and the internal curvature. Let us now consider a non-geometric NSNS field configuration: as explained already, the global issues of the fields, and consequently of $\mathcal{L}$, make the integration of the Lagrangian over some internal directions not possible. It is therefore unclear how to derive a four-dimensional potential from such a background. What we have shown though in (2.5) is that $\mathcal{L}$ can be rewritten as

$$\mathcal{L}(g, b, \phi) = \tilde{\mathcal{L}}(\tilde{g}, \beta, \tilde{\phi}) + \partial(\dots) \ . \tag{2.7}$$

A nice feature of this reformulation is that despite the ill-definedness of $\mathcal{L}$, the new fields and new Lagrangian $\tilde{\mathcal{L}}$ can be globally well-defined. This is precisely what happens for the toroidal example mentioned in the introduction, and we expect it to hold for other examples as well [**1**] (see addendum 3). Given the equality (2.7), such a situation can only occur if the total derivative $\partial(\dots)$ is also ill-defined. For the toroidal example, it is not single-valued and therefore does not integrate to zero.

In that case, we propose to consider $\tilde{\mathcal{L}}$ as the correct effective description of string theory. In other words, for string theory on a non-geometric background, one should not consider the NSNS *action* to start with, as it does not make sense, but rather the action given by a well-defined $\tilde{\mathcal{L}}$ whenever it exists. The dimensional reduction of $\tilde{\mathcal{L}}$ is then allowed, and as we explained above, it gives the non-geometric terms of the four-dimensional potential. Let us emphasise once more that the global aspects are the whole subtlety of the discussion: one would naively think that a field redefinition does not change much; but on a non-geometric configuration, it can restore a standard geometry, at the cost of introducing new fluxes, and then leads a previously unreached four-dimensional theory.

Thanks to this procedure, we establish a relation between the ten-dimensional non-geometry and the four-dimensional non-geometric fluxes: given a non-geometric field configuration, first perform a field redefinition to a well-defined set of fields, and then dimensionally reduce to get the four-dimensional non-geometric terms of the potential. For the toroidal example, this procedure works perfectly, and the ten-dimensional Q-flux (2.6) matches its four-dimensional counterpart, given by the T-duality chain. We discuss in [**1**, **3**] the extension of such a situation to other examples (see addendum 3).

3. The non-geometric fluxes in broader contexts

3.1. Reformulation of double field theory.

3.1.1. *First properties and field redefinition.* Double field theory [**13**] is defined on a doubled space, where to the standard coordinates x^m are added so-called dual coordinates $\tilde{x}_m$; the resulting space has twice the usual number d of dimensions. Coordinates are gathered into $X^M = (\tilde{x}_m, x^m)$ and one introduces the associated derivative $\partial_M = (\tilde{\partial}^m, \partial_m)$. All the fields of DFT depend at first on both sets of coordinates. We consider here only the NSNS sector; one formulation of DFT is then given in terms of the combination $\mathcal{E}_{mn}(x,\tilde{x}) = (g+b)_{mn}(x,\tilde{x})$ and the dilaton.

A first property of DFT is its $O(d,d)$ invariance, where transformations are encoded as

$$\text{For } h = \begin{pmatrix} A & B \\ C & D \end{pmatrix} \in O(d,d) \ , \quad X' = hX \ , \ \mathcal{E}'(X') = (A\mathcal{E}(X) + B)(C\mathcal{E}(X) + D)^{-1} \ . \tag{3.1}$$

These reproduce the standard fractional linear transformation of T-duality, but also go beyond it. Indeed, acting along non-isometry directions is allowed here, when transforming in particular the coordinates, so this $O(d,d)$ action is more general than T-duality.

Another property is the requirement of the strong constraint, applied on all fields and their products. A consequence is that the fields depend locally only on half of the coordinates in X^M. We consider in the following this constraint in the form of $\tilde{\partial}^m = 0$, i.e. by loosing any dependence in the dual coordinates. Applying this to the DFT Lagrangian $\mathcal{L}_{\text{DFT}}$ gives the standard NSNS Lagrangian up to a total derivative

$$\mathcal{L}_{\text{DFT}}(\mathcal{E},\phi) \overset{\tilde{\partial}^m=0}{=\!=\!=\!=} \mathcal{L}(g,b,\phi) + \partial(\dots) \ . \tag{3.2}$$

These two properties are used to obtain the equality of Lagrangians (2.5). To start with, performing the field redefinition within DFT is simple, as it essentially involves the quantity $\mathcal{E}$. Indeed, as indicated above, the field redefinition (2.2) can take the form $\mathcal{E} = (g+b) = (\tilde{g}^{-1} + \beta)^{-1}$. The last inverse power makes things a little more involved though, and one uses in addition the $O(d,d)$ invariance to get the result (we refer the reader to [**3**] for the details of the procedure). We finally obtain the DFT Lagrangian expressed in terms of the new fields. Using again the strong constraint, the resulting Lagrangian is precisely $\tilde{\mathcal{L}}$ given in (2.5) up to a total derivative

$$\mathcal{L}_{\text{DFT}}(\tilde{g},\beta,\tilde{\phi}) \overset{\tilde{\partial}^m=0}{=\!=\!=\!=} \tilde{\mathcal{L}}(\tilde{g},\beta,\tilde{\phi}) + \partial(\dots) \ . \tag{3.3}$$

Thanks to (3.2) and (3.3), we recover the equality (2.5) relating the standard NSNS Lagrangian $\mathcal{L}$ to the new Lagrangian $\tilde{\mathcal{L}}$.

3.1.2. *Diffeomorphism covariance.* Another property of DFT is its gauge invariance under double diffeomorphisms. Those are the transformations of the two sets of coordinates that we parameterize as $x^m \to x^m - \xi^m(x,\tilde{x})$, $\tilde{x}_m \to$

$\tilde{x}_m - \tilde{\xi}_m(x, \tilde{x})$. The DFT Lagrangian is in most of its formulations not manifestly covariant under these transformations. Let us illustrate this point with $\mathcal{L}_{\rm DFT}$ depending on the new fields, which can be expressed as

$$(3.4)\quad \mathcal{L}_{\rm DFT}(\tilde{g}, \beta, \tilde{\phi}) = e^{-2\tilde{\phi}}\sqrt{|\tilde{g}|}\left(\mathcal{R}(\tilde{g}, \partial) + 4(\partial\tilde{\phi})^2 + \mathcal{R}(\tilde{g}^{-1}, \tilde{\partial}) + 4(\tilde{\partial}\tilde{\phi})^2 - \frac{1}{2}Q^2 - \frac{1}{2}R^2 + \dots\right)$$

where $\mathcal{R}(\tilde{g}, \partial)$ is the standard Ricci scalar already mentioned, and $\mathcal{R}(\tilde{g}^{-1}, \tilde{\partial})$ has exactly the same expression with all up and down indices exchanged. In addition, the Q^2 and R^2 denote the same expressions as in (2.5) with indices contracted by metrics. The DFT R-flux has however an additional term with respect to (2.6), as first proposed in [**19**]

$$(3.5)\qquad R^{mnp} = 3\big(\tilde{\partial}^{[m}\beta^{np]} + \beta^{q[m}\partial_q\beta^{np]}\big)\ .$$

Under the standard diffeomorphisms $x^m \to x^m - \xi^m(x)$, resp. the dual diffeomorphisms $\tilde{x}_m \to \tilde{x}_m - \tilde{\xi}_m(\tilde{x})$, the terms $\mathcal{R}(\tilde{g}, \partial) + 4(\partial\tilde{\phi})^2$, resp. $\mathcal{R}(\tilde{g}^{-1}, \tilde{\partial}) + 4(\tilde{\partial}\tilde{\phi})^2$, transform covariantly. This behaviour is however lost if one considers the dependence in the other coordinate, and looks at the transformation under the "unnatural" diffeomorphism. For instance, the terms $\mathcal{R}(\tilde{g}^{-1}, \tilde{\partial}) + 4(\tilde{\partial}\tilde{\phi})^2$ a priori do not transform covariantly under the ξ-diffeomorphisms $x^m \to x^m - \xi^m(x, \tilde{x})$. So we constructed in [**2, 3**] a covariant derivative $\tilde{\nabla}^m$ completing $\tilde{\partial}^m$ in order to restore covariance under the "unnatural" ξ-diffeomorphisms. The non-geometric fluxes were an inspiration for the structures involved: indeed, while the Q-flux, defined as in (2.6), is not a tensor, the DFT R-flux (3.5) is a tensor under the ξ-diffeomorphisms. The covariant derivative $\tilde{\nabla}^m$ is defined by the following action on a vector and a co-vector

$$(3.6)\qquad \tilde{\nabla}^m V^n = \tilde{D}^m V^n - \check{\Gamma}_p{}^{mn} V^p\ ,\quad \tilde{\nabla}^m V_n = \tilde{D}^m V_n + \check{\Gamma}_n{}^{mp} V_p\ ,$$

where its two building blocks are given by

$$(3.7)\qquad \tilde{D}^m = \tilde{\partial}^m + \beta^{pm}\partial_p\ ,$$
$$\check{\Gamma}_p{}^{mn} = \frac{1}{2}\tilde{g}_{pq}\left(\tilde{D}^m\tilde{g}^{nq} + \tilde{D}^n\tilde{g}^{mq} - \tilde{D}^q\tilde{g}^{mn}\right) + \tilde{g}_{pq}\tilde{g}^{r(m}Q_r{}^{n)q} - \frac{1}{2}Q_p{}^{mn}\ .$$

The derivative $\tilde{D}^m$ is tied to the R-flux (3.5), since $R^{mnp} = 3\tilde{D}^{[m}\beta^{np]}$. The Q-flux also enters these buildings blocks, as providing the non-standard part of the connection $\check{\Gamma}_p{}^{mn}$ (the first part of this connection is very analogous to the Levi-Civita connection); see addendum 2.

From these objects, we constructed a Riemann tensor $\check{\mathcal{R}}^{mn}{}_p{}^q$, a Ricci tensor $\check{\mathcal{R}}^{mn}$ and finally a Ricci scalar $\check{\mathcal{R}}$. We used the latter to rewrite the DFT Lagrangian as

$$(3.8)\quad \mathcal{L}_{\rm DFT}(\tilde{g}, \beta, \tilde{\phi}) - \partial(\dots) = e^{-2\tilde{\phi}}\sqrt{|\tilde{g}|}\left(\mathcal{R}(\tilde{g}, \partial) + 4(\partial\tilde{\phi})^2 + \check{\mathcal{R}}(\tilde{D}, \check{\Gamma}) + 4\big(\tilde{D}^m\tilde{\phi} + \check{\Gamma}_p{}^{pm}\big)^2 - \frac{1}{2}R^2\right)$$

where each term is, at last, manifestly covariant under the ξ-diffeomorphisms.

We get a better understanding of the role of the non-geometric fluxes through (3.8). The R-flux always appears as a tensor, and is then analogous to the standard NSNS H-flux. The Q-flux on the contrary is not a tensor, but rather plays the role of a connection, analogously to the geometric flux. In addition, all the Q-flux terms in (3.4), and in some sense, in (2.5), are absorbed within the connection $\check{\Gamma}_p{}^{mn}$, essentially appearing through $\check{\mathcal{R}}$. This provides some structure to these terms (see addendum 2).

3.2. Non-commutativity of the string coordinates. Non-commutativity in string theory first appeared in the context of the open string. By considering open strings ending on a D-brane with a constant b-field, their coordinates were found not to commute, as given by $[x^m(\tau), x^n(\tau)] = i\theta^{mn}$. The non-commutativity parameter θ^{mn} was defined [**20**], together with the so-called open string metric G_{mn}, by

$$G^{mn} = \left(\frac{1}{g + 2\pi\alpha' b}\, g\, \frac{1}{g - 2\pi\alpha' b}\right)^{mn} , \tag{3.9}$$
$$\theta^{mn} = -(2\pi\alpha')^2 \left(\frac{1}{g + 2\pi\alpha' b}\, b\, \frac{1}{g - 2\pi\alpha' b}\right)^{mn} .$$

As noticed in [**3**] and other references therein, it is striking to compare these formulas with the field redefinition (2.2), that we repeat here in a slightly rewritten form

$$\tilde{g}^{-1} = (g+b)^{-1}\, g\, (g-b)^{-1} \ , \ \beta = -(g+b)^{-1}\, b\, (g-b)^{-1} . \tag{3.10}$$

Up to conventions on α', they perfectly match. We are however working in a different context, involving closed strings and non-geometry. It would nevertheless be interesting to investigate further this relation.

Non-commutativity for closed strings has been studied more recently. In the serie of papers [**4**, **21**, **22**], a closed string, on various examples of non-geometric backgrounds, was shown to have some coordinates not commuting: $[\mathcal{X}^m(\tau,\sigma), \mathcal{X}^n(\tau,\sigma)] \neq 0$. Drawing the analogy with the open string, one would think of β as the non-commutativity parameter. However, in the examples studied, the non-vanishing commutator was rather parameterized by the Q-flux times a winding number N^p

$$[\mathcal{X}^m(\tau,\sigma), \mathcal{X}^n(\tau,\sigma)] \sim iN^p\, Q_p{}^{mn} . \tag{3.11}$$

Before we comment on this result, let us indicate that the R-flux as well served as a parameter, in the context of non-associativity [**23**]. More precisely, it parameterizes a non-vanishing jacobiator of closed string coordinates. So the new fields introduced, in particular β, and the non-geometric fluxes seem to play a role in characterising non-commutative and non-associative structures in string theory.

The right-hand side of (3.11) was discussed at length in [**4**], where the non-geometric background considered is that of the toroidal example. For this background, the space is given locally by a two-torus fibered over a base circle. The non-geometry occurs when going around the base: the fiber then requires a T-duality as a transition function. The Q-flux characterising this non-geometry, in the sense of section 2.3, is simply given by (2.6) and is constant. We discuss in [**4**] several arguments explaining why the non-commutativity of the fiber coordinates is related to the non-geometry; one of them being that the fiber is somehow fuzzy, so a precise determination of the position (in the point particle sense) is not possible. This relation justifies why the non-geometric Q-flux parameterizes the non-vanishing commutator. In addition, one can understand for this background the presence in (3.11) of the string winding number along the base circle. Such a wound closed string probes directly the non-geometry of the fiber, as it goes through its non-trivial monodromy when wrapping the base circle. On the contrary, a non-wound string would only probe the local geometry. Thanks to these arguments, the formula (3.11), or the more precise one given in [**4**], may be as well valid for a closed string on another non-geometric background of a similar type (those which can be viewed as a T-fold [**24**]).

4. Conclusion and outlook

We have presented a reformulation of the NSNS sector of supergravity in terms of new fields and fluxes. This reformulation is made through a field redefinition from the standard (g_{mn}, b_{mn}, ϕ) to new fields $(\tilde{g}_{mn}, \beta^{mn}, \tilde{\phi})$. Rewriting the standard NSNS Lagrangian $\mathcal{L}$ in terms of the new fields gives, up to a total derivative, the new Lagrangian $\tilde{\mathcal{L}}$. The latter depends on new objects that we identify as ten-dimensional non-geometric fluxes, namely the Q-flux given by ${Q_p}^{mn} = \partial_p \beta^{mn}$ (at least for $\beta^{mn}\partial_n \cdot = 0$) and the R-flux $R^{mnp} = 3\ \beta^{q[m}\partial_q \beta^{np]}$.

The new ten-dimensional Lagrangian $\tilde{\mathcal{L}}$ is proposed to provide an uplift to some four-dimensional gauged supergravities, because its dimensional reduction leads generically to non-geometric terms in the scalar potential. A concrete compactification can though only be done on a given background, and the global aspects of the latter are then crucial. For the non-geometric field configuration of the toroidal example, the global issues are cured when using the new fields; a standard differential geometry is also restored. $\tilde{\mathcal{L}}$ can then be dimensionally reduced in the usual way, while it was not possible with the standard NSNS Lagrangian. For this example, ten-dimensional and four-dimensional Q-fluxes match. This procedure establishes a relation between a ten-dimensional non-geometry, and four-dimensional non-geometric fluxes.

Double field theory can also be reformulated in terms of the new fields. While the resulting Lagrangian is not manifestly diffeomorphism-covariant, the non-geometric fluxes exhibit inspiring structures to rewrite it in such a manner. Doing so, the R-flux appears to always behave as a tensor, while the Q-flux does not, but rather plays the role of a connection.

The new fields and the non-geometric fluxes also appear as relevant quantities to study non-commutativity of string coordinates. In particular, for a closed string on a non-geometric background, the Q-flux, together with a winding number, are argued to be the parameters of the non-vanishing commutator.

It would be interesting develop more the reformulation of ten-dimensional supergravity. To do so, one should first get a better handle on the Q-flux terms that appear in $\tilde{\mathcal{L}}$, and clarify the relation to the four-dimensional Q-flux (see addenda 1 and 2). Then, one could consider extensions to other sectors, such as the Ramond-Ramond sector, where the fluxes are known to have non-geometric counterparts [**25**], or to the gauge fields of heterotic string. For the latter, introducing a generalized metric as in [**26**] should be helpful. Finally, there should as well exist non-geometric counterparts to the D-brane and O-plane sources, possibly related to the recently studied exotic branes (see [**27**] and references therein).

This reformulation of ten-dimensional supergravity is based on a field redefinition, so we believe that the symmetries should be preserved in the reformulation, even if possibly appearing in a more complicated fashion. This is also an interesting point to study. The resulting new supergravity would provide an uplift to some four-dimensional gauged supergravities, that did not have any before. It would then be nice to find new interesting solutions directly from ten dimensions, in particular de Sitter solutions. Due to the non-triviality of the global aspects, some new physics could be accessed that way. Indeed, suppose we have a solution of $\tilde{\mathcal{L}}$, which satisfies the usual compactification ansatz. It can be that its expression in terms of the standard NSNS fields is too complicated to have been considered before (especially because of its global properties, for instance if it is non-geometric). If in addition it is not T-dual to any geometric solution, this solution can fairly be thought of as new. As a solution of $\tilde{\mathcal{L}}$, it then provides new interesting physics (see addendum 3). We hope to come back to these ideas in future work.

Addendum 1: The reformulation of the standard ten-dimensional NSNS Lagrangian thanks to the field redefinition, discussed in sections 2.1 and 2.2, was clarified in [**28**]. There, the resulting Lagrangian $\tilde{\mathcal{L}}_\beta$ (it differs from $\tilde{\mathcal{L}}$ here by a total derivative) was said to correspond to the NSNS sector of a theory called for convenience β-supergravity. The form of $\tilde{\mathcal{L}}_\beta$ was simplified with respect to $\tilde{\mathcal{L}}$ thanks to a better identification of the fluxes: those are better defined in flat (tangent space) indices as

$$Q_c{}^{ab} = \partial_c \beta^{ab} - 2\beta^{d[a} f^{b]}{}_{cd} \ , \quad R^{abc} = 3\beta^{d[a}\nabla_d \beta^{bc]} \ , \quad f^a{}_{bc} = 2\tilde{e}^a{}_m \partial_{[b} \tilde{e}^m{}_{c]} \ . \tag{4.1}$$

The R-flux being a tensor, its formula (4.1) is the same the one in (2.6) multiplied by vielbeins $\tilde{e}^a{}_m$ (∇ is the usual covariant derivative with Levi-Civita connection). The Q-flux in (4.1) is however not a tensor, and is in general not directly related to the one in (2.6). This difference in the Q-flux, suggested several times in the present paper, was the key point to simplify the Lagrangian. The DFT completion of β-supergravity (in the form of [**28**]) is achieved in [**29**], building on [**19**], and the fluxes there match (4.1) upon the strong constraint. This DFT formulation is related to the Generalized Geometry one, derived in [**28**]. In that paper and in [**29**] is also discussed in more details the matching with four-dimensional gauged supergravity and its potential.

Addendum 2: The objects presented in section 3.1.2 were further developed in [**28**], although only at the supergravity level (meaning with $\tilde{\partial}^m = 0$). In particular, the derivative $\tilde{\nabla}$, denoted there $\check{\nabla}$, was shown to play a crucial role. We considered its flat indices version with a corresponding spin connection ω_Q. Then, the Q-flux (4.1) turned out to play precisely the same role in ω_Q as the geometric flux f usually plays in the standard spin connection of ∇. So the intuition of the Q-flux not being a tensor, but rather playing the role of a connection, here presented for DFT and with a different formula for this flux, actually still held in the subsequent work. The other objects presented here, such as $\check{\mathcal{R}}$, also got further developments and interpretations in [**28**].

Addendum 3: The discussion in sections 2.3 and 4 on global aspects and novelty of the backgrounds described got refined in later work. Because of (2.7), the symmetries of β-supergravity are generically those of standard supergravity, and are therefore restricted for the NSNS sector to diffeomorphisms and b-field gauge transformations. As stressed in [**30**], the symmetries of the theory are in principle the only transformations one should use to patch the fields. Therefore, in general, the description of a non-geometric background would not be improved by going from standard supergravity to β-supergravity. This statement can however be refined as follows, as pointed out in [**28, 31**]: if one focuses rather on subcase by making a further assumption or implementing a further constraint, then the symmetries can change and in particular get enhanced. This is precisely what happens here when considering the presence of n isometries: the T-duality group $O(n,n)$ then becomes part of the available symmetries. Of special interest are the subgroup of β-transforms, under which the Lagrangian of β-supergravity is manifestly invariant. We proved [**31**] that the whole class of backgrounds of standard supergravity patching with those (together with possible diffeomorphisms) would very likely be non-geometric, while their reformulation in β-supergravity would be geometric there (in particular the fluxes (4.1) are left invariant). This class of backgrounds thus realises the scenario described in section 2.3, and the toroidal example falls precisely in this class. Considering isometries is thus an important subcase. Unfortunately, we showed as well that the whole class was T-dual to geometric backgrounds of standard supergravity, preventing from getting new physics from them. It is for now unclear whether a different situation could be obtained from β-supergravity, by for instance considering another subcase, adding other sectors, etc.

References

[1] D. Andriot, M. Larfors, D. Lüst, and P. Patalong, *A ten-dimensional action for non-geometric fluxes*, J. High Energy Phys. **9** (2011), 134, 49, DOI 10.1007/JHEP09(2011)134. MR2889830

[2] D. Andriot, O. Hohm, M. Larfors, D. Lüst and P. Patalong, *A geometric action for non-geometric fluxes*, Phys. Rev. Lett. **108** (2012), 261602.

[3] D. Andriot, O. Hohm, M. Larfors, D. Lüst, and P. Patalong, *Non-geometric fluxes in supergravity and double field theory*, Fortschr. Phys. **60** (2012), no. 11-12, 1150–1186, DOI 10.1002/prop.201200085. MR2994452

[4] D. Andriot, M. Larfors, D. Lüst, and P. Patalong, *(Non-)commutative closed string on T-dual toroidal backgrounds*, J. High Energy Phys. **6** (2013), 021, front matter+50. MR3083389

[5] B. de Carlos, A. Guarino and J. M. Moreno, *Complete classification of Minkowski vacua in generalised flux models*, J. High Energy Phys. **02** (2010), 076 MR2672717 (2011i:81160); U. Danielsson and G. Dibitetto, *On the distribution of stable de Sitter vacua*, J. High Energy Phys. **03** (2013), 018; J. Blaback, U. Danielsson and G. Dibitetto, *Fully stable dS vacua from generalised fluxes*, J. High Energy Phys. **08** (2013), 054 MR3106384; C. Damian, O. Loaiza-Brito, L. Rey and M. Sabido, *Slow-Roll Inflation in Non-geometric Flux Compactification*, J. High Energy Phys. **06** (2013), 109 MR3083301.

[6] J. Shelton, W. Taylor and B. Wecht, *Generalized Flux Vacua*, J. High Energy Physics **02** (2007), 095, 28 pp. DOI 10.1088/1126-6708/2007/02/095 MR2318047 (2008g:81220); A. Micu, E. Palti and G. Tasinato, *Towards Minkowski Vacua in Type II String Compactifications*, J. High Energy Physics **03** (2007), 104, 34 pp. MR2313972 (2008g:81211)

[7] S. Hellerman, J. McGreevy, and B. Williams, *Geometric constructions of nongeometric string theories*, J. High Energy Phys. **1** (2004), 024, 39 pp. (electronic), DOI 10.1088/1126-6708/2004/01/024. MR2045882 (2005j:81144)

[8] A. Dabholkar and C. Hull, *Duality twists, orbifolds, and fluxes*, J. High Energy Phys. **9** (2003), 054, 25 pp. (electronic), DOI 10.1088/1126-6708/2003/09/054. MR2026959 (2005e:81169)

[9] A. Flournoy, B. Wecht, and B. Williams, *Constructing nongeometric vacua in string theory*, Nuclear Phys. B **706** (2005), no. 1-2, 127–149, DOI 10.1016/j.nuclphysb.2004.11.005. MR2112480 (2006e:81212)

[10] J. Shelton, W. Taylor, and B. Wecht, *Nongeometric flux compactifications*, J. High Energy Phys. **10** (2005), 085, 24 pp. (electronic), DOI 10.1088/1126-6708/2005/10/085. MR2180609 (2006m:81255)

[11] A. Dabholkar and C. Hull, *Generalised T-duality and non-geometric backgrounds*, J. High Energy Phys. **5** (2006), 009, 27 pp. (electronic), DOI 10.1088/1126-6708/2006/05/009. MR2231521 (2007d:81180)

[12] S. Kachru, M. B. Schulz, P. K. Tripathy and S. P. Trivedi, *New supersymmetric string compactifications*, J. High Energy Physics **03** (2003), 061, DOI 10.1088/1126-6708/2003/03/061. MR1975671 (2004k:81287; D. A. Lowe, H. Nastase and S. Ramgoolam, *Massive IIA string theory and matrix theory compactification*, Nuclear Phys. **B 667** (2003), 55–89. MR2000717 (2004m:81201)

[13] C. Hull and B. Zwiebach, *Double Field Theory*, J. High Energy Phys. **09** (2009), 099 MR2580704; C. Hull and B. Zwiebach, *The Gauge algebra of double field theory and Courant brackets*, J. High Energy Phys. **09** (2009), 090 MR2580711 (2011e:81217); O. Hohm, C. Hull and B. Zwiebach, *Background independent action for double field theory*, J. High Energy Phys. **07** (2010), 016 MR2720017 (2011j:81267); O. Hohm, C. Hull and B. Zwiebach, *Generalized metric formulation of double field theory*, J. High Energy Phys. **08** (2010), 008 MR2756111 (2012b:81201).

[14] P. Grange and S. Schäfer-Nameki, *T-duality with H-flux: non-commutativity, T-folds and $G \times G$ structure*, Nuclear Phys. B **770** (2007), no. 1-2, 123–144, DOI 10.1016/j.nuclphysb.2007.02.003. MR2316459 (2008k:81251)

[15] P. Grange and S. Schäfer-Nameki, *Towards mirror symmetry à la SYZ for generalized Calabi-Yau manifolds*, J. High Energy Phys. **10** (2007), 052, 31, DOI 10.1088/1126-6708/2007/10/052. MR2357926 (2009b:53071)

[16] M. Graña, R. Minasian, M. Petrini, and D. Waldram, *T-duality, generalized geometry and non-geometric backgrounds*, J. High Energy Phys. **4** (2009), 075, 39, DOI 10.1088/1126-6708/2009/04/075. MR2505954 (2010i:81323)

[17] R. Blumenhagen, A. Deser, E. Plauschinn and F. Rennecke, *A bi-invariant Einstein-Hilbert action for the non-geometric string*, Phys. Lett. **B 720** (2013), 215 MR3028102; R. Blumenhagen, A. Deser, E. Plauschinn and F. Rennecke, *Non-geometric strings, symplectic gravity and differential geometry of Lie algebroids*, J. High Energy Phys. **02** (2013), 122. MR3046548

[18] M. P. Hertzberg, S. Kachru, W. Taylor, and M. Tegmark, *Inflationary constraints on type IIA string theory*, J. High Energy Phys. **12** (2007), 095, 21, DOI 10.1088/1126-6708/2007/12/095. MR2366461 (2009b:83132)

[19] G. Aldazabal, W. Baron, D. Marqués, and C. Núñez, *The effective action of double field theory*, J. High Energy Phys. **11** (2011), 052, 34, DOI 10.1007/JHEP11(2011)052. MR2913284

[20] N. Seiberg and E. Witten, *String theory and noncommutative geometry*, J. High Energy Phys. **9** (1999), Paper 32, 93 pp. (electronic), DOI 10.1088/1126-6708/1999/09/032. MR1720697 (2001i:81237)

[21] D. Lüst, *T-duality and closed string non-commutative (doubled), geometry*, J. High Energy Phys. **12** (2010), 084.

[22] C. Condeescu, I. Florakis, and D. Lüst, *Asymmetric orbifolds, non-geometric fluxes and non-commutativity in closed string theory*, J. High Energy Phys. **4** (2012), 121, front matter+28. MR2967590

[23] R. Blumenhagen and E. Plauschinn, *Nonassociative Gravity in String Theory?*, J.Phys. **A 44** (2011), 015401 MR2749110 (2012c:81155); R. Blumenhagen, A. Deser, D. Lüst, E. Plauschinn and F. Rennecke, *Non-geometric Fluxes, Asymmetric Strings and Nonassociative Geometry*, J. Phys. **A 44** (2011), 385401 MR2835158; D. Mylonas, P. Schupp and R. J. Szabo, *Membrane Sigma-Models and Quantization of Non-Geometric Flux Backgrounds*, J. High Energy Phys. **09** 2012), 012 MR3044978.

[24] C. M. Hull, *A geometry for non-geometric string backgrounds*, J. High Energy Phys. **10** (2005), 065, 30 pp. (electronic), DOI 10.1088/1126-6708/2005/10/065. MR2180623 (2007d:81188)

[25] G. Aldazabal, P. G. Camara, A. Font and L. Ibáñez, *More dual fluxes and moduli fixing*, J. High Energy Phys. **05** (2006), 070 MR2231476 (2007f:81189); G. Aldazabal, P. G. Camara and J. Rosabal, *Flux algebra, Bianchi identities and Freed-Witten anomalies in F-theory compactifications*, Nucl. Phys. **B 814** (2009), 21 MR2504677; G. Dibitetto, R. Linares and

D. Roest, *Flux Compactifications, Gauge Algebras and De Sitter*, Phys. Lett. **B 688** (2010), 96 MR2629870 (2011d:81231).

[26] D. Andriot, *Heterotic string from a higher dimensional perspective*, Nuclear Phys. B **855** (2012), no. 2, 222–267, DOI 10.1016/j.nuclphysb.2011.10.007. MR2854537

[27] J. de Boer and M. Shigemori, *Exotic branes in string theory*, Phys. Rep. **532** (2013), no. 3, 65–118, DOI 10.1016/j.physrep.2013.07.003. MR3120691

[28] D. Andriot and A. Betz, *β-supergravity: a ten-dimensional theory with non-geometric fluxes, and its geometric framework*, J. High Energy Phys. **12** (2013), 083.

[29] D. Geissbühler, D. Marqués, C. Núñez, and V. Penas, *Exploring double field theory*, J. High Energy Phys. **6** (2013), 101, front matter+46. MR3083309

[30] R. Blumenhagen, A. Deser, E. Plauschinn, F. Rennecke, and C. Schmid, *The intriguing structure of non-geometric frames in string theory*, Fortschr. Phys. **61** (2013), no. 10, 893–925. MR3118856

[31] D. Andriot and A. Betz, *NS-branes, source corrected Bianchi identities, and more on backgrounds with non-geometric fluxes*, J. High Energy Phys. **07** (2014), 059.

Max-Planck-Institut für Physik, Föhringer Ring 6, 80805 München, Germany

Current address: Max-Planck-Institut für Gravitationsphysik, Albert-Einstein-Institut, Am Mühlenberg 1, 14467 Potsdam-Golm, Germany

E-mail address: `david.andriot@aei.mpg.de`

Proceedings of Symposia in Pure Mathematics
Volume **90**, 2015
http://dx.doi.org/10.1090/pspum/090/01529

The Geometric Algebra of Supersymmetric Backgrounds

C. I. Lazaroiu, E. M. Babalic, and I. A. Coman

ABSTRACT. Supersymmetry-preserving backgrounds in supergravity and string theory can be studied using a powerful framework based on a natural realization of Clifford bundles. We explain the geometric origin of this framework and show how it can be used to formulate a theory of 'constrained generalized Killing forms', which gives a useful geometric translation of supersymmetry conditions in the presence of fluxes.

1. The geometric algebra approach to spinors

Let (M, g) be a paracompact, smooth, connected and oriented pseudo-Riemannian manifold of dimension $d = p + q$, where p and q denote the numbers of positive and negative eigenvalues of the metric. Inhomogeneous differential forms on M constitute a $\mathbb{Z}$-graded $\mathcal{C}^\infty(M, \mathbb{R})$-module $\Omega(M) \stackrel{\text{def.}}{=} \Gamma(M, \wedge T^*M)$, whose fixed rank components we denote by $\Omega^k(M) = \Gamma(M, \wedge^k T^*M)$. One approach to the spin geometry [**1**] of (M, g) starts with the Clifford bundle $\mathrm{Cl}(T^*M)$ of the pseudo-Euclidean vector bundle T^*M, where the latter is endowed with the pairing $\hat{g}$ induced by g. Then a spinor bundle S over (M, g) is [1] a bundle of modules over the even sub-bundle $\mathrm{Cl}^{\mathrm{ev}}(T^*M)$ while a pinor bundle is a bundle of modules over $\mathrm{Cl}(T^*M)$. Sections of these bundles are called *spinors* and *pinors*, respectively and correspond in physics to particles of arbitrary spin; of course, a pinor is a spinor in a natural fashion. The case of spin 1/2 arises when S is a bundle of *simple* modules over $\mathrm{Cl}^{\mathrm{ev}}(T^*M)$ or $\mathrm{Cl}(T^*M)$, in which case we say that S is a spin or pin bundle, respectively.

The Kähler-Atiyah bundle and Kähler-Atiyah algebra. The Clifford bundle is, conceptually, not the most 'rigid' representative of its isomorphism class as a bundle of algebras. This can be cured by using a particular isomorphic realization of the Clifford bundle (which in some ways goes back to Chevalley [**3**] and Riesz

2010 *Mathematics Subject Classification.* Primary 53C27, 53Z05, 53C10.

The work of C.I.L was financed by the research grant IBS-R003-G1 and by the CNCS-UEFISCDI grant PN-II-ID-PCE 50/2011.

The work of E.M.B. was supported by the CNCS-UEFISCDI grants PN-II-RU-TE 77/2010 and PN-II-ID-PCE 121/2011.

The work of I.A.C. was supported by the CNCS-UEFISCDI grant PN-II-RU-TE 77/2010. I.A.C. also acknowledges her *Open Horizons* scholarship, which financed part of her studies.

[1]See [**2**] for the relation with the construction based on vector bundles associated with spin structures, i.e. covers of the principal bundle of pseudo-orthonormal frames.

[**4**]) known as the *Kähler-Atiyah bundle* of (M, g). The Chevalley-Riesz construction identifies the underlying vector bundle of $\mathrm{Cl}(T^*M)$ with the exterior bundle $\wedge T^*M$ of M using a very special isomorphism of vector bundles which is determined by the metric g, while transporting the Clifford product of the former to a unital, associative but non-commutative fiberwise binary operation on the latter which we denote by $\diamond$ and call the *geometric product* of (M, g). By definition, the Kähler-Atiyah bundle of (M, g) is the bundle of unital associative algebras $(\wedge T^*M, \diamond)$, while the Kähler-Atiyah algebra is its $\mathcal{C}^\infty(M, \mathbb{R})$-algebra of global sections $(\Omega(M), \diamond)$, where we denote the operation induced on global sections through the same symbol. The geometric product is uniquely determined by the differential and metric structure of (M, g). Though it is inhomogeneous with respect to the natural $\mathbb{Z}$-grading of the exterior bundle, it does preserve rank parity and hence it is even with respect to the induced $\mathbb{Z}_2$-grading. This construction shows that the Clifford bundle itself is unnecessary, since the only information which it adds to the natural geometry of (M, g) is the geometric product on the exterior bundle, which itself is entirely determined by the metric g through an expansion involving wedge products and contractions. This point of view allows one to interpret the geometric product as a deformation of the wedge product which admits a finite expansion in powers of the inverse of the radius of (M, g), an interpretation which is related to deformation quantization.

Generalized products. The Chevalley-Riesz construction implies that the geometric product admits an expansion into a finite sum of binary fiberwise-bilinear operations $\triangle_k : \wedge T^*M \times_M \wedge T^*M \to \wedge T^*M$ $(k = 0 \ldots d)$ which are homogeneous of degree $-2k$ with respect to the rank grading, being known as the *generalized products* determined by the metric g. This expansion takes the form:

$$\diamond = \sum_{k=0}^{[\frac{d}{2}]} (-1)^k \triangle_{2k} + \sum_{k=0}^{[\frac{d-1}{2}]} (-1)^{k+1} \triangle_{2k+1} \circ (\pi \otimes \mathrm{id}_{\wedge T^*M}) , \tag{1.1}$$

where π is a certain unital automorphism of the Kähler-Atiyah bundle (known as the *parity automorphism*), which is defined through:

$$\pi \stackrel{\text{def.}}{=} \oplus_{k=0}^{d} (-1)^k \mathrm{id}_{\wedge^k T^*M} .$$

Connection to 'partial quantization' and to the large radius expansion. The expression (1.1) can be viewed as the semiclassical expansion of the geometric product when the latter is identified with the star product arising in a certain 'vertical' partial quantization procedure in which the role of the Planck constant is played by the inverse of the overall scale of the metric g. In the classical limit $g \to \infty$ (i.e. when M is of 'infinite size', or 'large radius' when measured using g), the geometric product reduces to $\triangle_0$, which coincides with the wedge product $\wedge$. The corrections to this limit occur as powers in the inverse of the size of M, thereby providing a natural realization of 'compactification/decompactification' limit arguments which are sometimes used in supergravity and string theory.

Recursive construction. The higher generalized products $\triangle_k$ $(k = 1 \ldots d)$ depend on g, their action on inhomogeneous differential forms (=sections of $\wedge T^*M$)

being determined recursively through:

$$\omega \triangle_{k+1} \eta = \frac{1}{k+1} g^{ab}(e_a \lrcorner \omega) \triangle_k (e_b \lrcorner \eta) = \frac{1}{k+1} g_{ab}(\iota_{e^a}\omega) \triangle_k (\iota_{e^b}\eta) ,$$

where ι denotes the so-called *interior product*, which is defined as the adjoint of the wedge product with respect to the pairing induced by g on $\wedge T^*M$ (see [**5**] for details). In the formulas above, $(e_a)_{a=1\ldots d}$ denotes a local frame of TM and $(e^a)_{a=1\ldots d}$ its dual local coframe (which satisfies $e^a(e_b) = \delta^a_b$ and $g(e^a, e^b) = g^{ab}$, where (g^{ab}) is the inverse of the matrix (g_{ab})). For latter reference, recall that the corresponding contragradient frame $(e^a)^\#$ and coframe $(e_a)_\#$ satisfy $(e^a)^\# = g^{ab}e_b$ and $(e_a)_\# = g_{ab}e^b$, where the $\#$ subscript and superscript denote the (mutually-inverse) musical isomorphisms between TM and T^*M, given respectively by lowering and raising of indices with the metric g. For latter reference, the *main antiautomorphism* (also known as *reversion*) of the Kähler-Atiyah bundle is the involutive antiautomorphism which acts on global sections through:

$$\tau(\omega) = (-1)^{\frac{k(k-1)}{2}}\omega \ , \ \forall \omega \in \Omega^k(M) \ .$$

The generalized products satisfy various identities which are consequences of associativity and unitality of the geometric product and of the fact that the volume form $\nu = \mathrm{vol}_M \in \Omega^d(M)$ of (M, g) satisfies:

$$\begin{aligned} \nu \diamond \nu &= (-1)^{q+\left[\frac{d}{2}\right]} 1_M = \begin{cases} (-1)^{\frac{p-q}{2}} 1_M \ , & \text{if } d = \text{even} \\ (-1)^{\frac{p-q-1}{2}} 1_M \ , & \text{if } d = \text{odd} \end{cases} , \\ \nu \diamond \omega &= \pi^{d-1}(\omega) \diamond \nu \ , \ \forall \omega \in \Omega(M) \ . \end{aligned}$$

Hence ν is central in the Kähler-Atiyah algebra $(\Omega(M), \diamond)$ when d is odd and twisted central (i.e., $\nu \diamond \omega = \pi(\omega) \diamond \nu$) when d is even. In Table 1, we indicate the values of $p - q$ (mod 8) for which the volume form ν has the corresponding properties. Various aspects of the geometric algebra formalism are discussed in detail in [**5**–**7**].

	$\nu \diamond \nu = +1$	$\nu \diamond \nu = -1$
ν is central	$\mathbf{1}(\mathbb{R}), \mathbf{5}(\mathbb{H})$	$\mathbf{3}(\mathbb{C}), \mathbf{7}(\mathbb{C})$
ν is twisted central	$\mathbf{0}(\mathbb{R}), \mathbf{4}(\mathbb{H})$	$\mathbf{2}(\mathbb{R}), \mathbf{6}(\mathbb{H})$

Table 1: Properties of the volume form.

Reconsidering (s)pinor bundles. A pinor bundle S can be viewed as a bundle of modules over the Kähler-Atiyah bundle of (M, g), the module structure being defined by a morphism $\gamma : (\wedge T^*M, \diamond) \to (\mathrm{End}(S), \circ)$ of bundles of algebras. Since we are interested in pinors of spin 1/2, we assume that γ is fiberwise irreducible. Since $\mathrm{Cl}^{\mathrm{ev}}(T^*M)$ identifies with the sub-bundle of algebras $(\wedge^{\mathrm{ev}}T^*M, \diamond)$, a spinor bundle is a bundle of modules over the latter, being called a spin bundle when its fibers are simple modules. A particularly important role in the study of (s)pin bundles is played by the endomorphism $\gamma(\nu) \in \Gamma(M, \mathrm{End}(S))$, which is central or twisted central in the algebra $(\Gamma(M, \mathrm{End}(S)), \circ)$ depending on the value of $p - q$ (mod 8). Notice that γ provides a globally meaningful interpretation of physicists' 'gamma matrices' (which are ordinarily defined only locally). In the interpretation through vertical quantization, the space of sections $\Gamma(M, S)$ plays a role akin to that of the Hilbert space while γ is the quantization map which translates between the deformation quantization description (provided by the Kähler-Atiyah algebra) and the operator quantization provided by the algebra of operators $\Gamma(M, \mathrm{End}(S)) \simeq \mathrm{End}_{\mathcal{C}^\infty(M,\mathbb{R})}(\Gamma(M,S))$. This interpretation can be exploited in various dimensions and signatures, providing new points of view on various geometric structures.

Local expressions. Given a local pseudo-orthonormal frame (e_a) of (M, g) with dual local coframe (e^a), a general inhomogeneous form $\omega \in \Omega(M)$ expands as:

$$\omega = \sum_{k=0}^{d} \omega^{(k)} =_U \sum_{k=0}^{d} \frac{1}{k!} \omega^{(k)}_{a_1 \ldots a_k} e^{a_1 \ldots a_k} \text{ with } \omega^{(k)} \in \Omega^k(M) \ , \tag{1.2}$$

where $e^{a_1 \ldots a_k} \stackrel{\text{def.}}{=} e^{a_1} \wedge \ldots \wedge e^{a_k}$ and the symbol $=_U$ means that equality holds only after restriction of ω to U . We let $\gamma^a \stackrel{\text{def.}}{=} \gamma(e^a) \in \Gamma(U, \mathrm{End}(S))$ and $\gamma_a \stackrel{\text{def.}}{=} g_{ab}\gamma^b \in \Gamma(U, \mathrm{End}(S))$ be the contravariant and covariant 'gamma matrices' associated with the given local orthonormal (co)frame and $\gamma_{a_1 \ldots a_k}$ denote the complete antisymmetrization of the composition $\gamma_{a_1} \circ \ldots \circ \gamma_{a_k}$.

Spin projectors and spin bundles. Giving a direct sum bundle decomposition $S = S_+ \oplus S_-$ amounts to giving a *product structure* on S, i.e. a bundle endomorphism $\mathcal{R} \in \Gamma(M, \mathrm{End}(S)) \setminus \{-\mathrm{id}_S, \mathrm{id}_S\}$ satisfying:

$$\mathcal{R}^2 = \mathrm{id}_S \ .$$

A product structure shall be called a *spin endomorphism* if it also satisfies:

$$[\mathcal{R}, \gamma(\omega)]_{-,\circ} = 0 \ , \ \forall \omega \in \Omega^{\mathrm{ev}}(M) \ .$$

A spin endomorphism exists only when $p - q \equiv_8 0, 4, 6, 7$. When S is a pin bundle, the restriction $\gamma_{\mathrm{ev}} \stackrel{\text{def.}}{=} \gamma|_{\wedge^{\mathrm{ev}} T^*M} : (\wedge^{\mathrm{ev}} T^*M, \diamond) \to (\mathrm{End}(S), \circ)$ is fiberwise reducible iff. S admits a spin endomorphism, in which case we define the *spin projectors* determined by $\mathcal{R}$ to be the globally-defined endomorphisms $\mathcal{P}^{\mathcal{R}}_{\pm} \stackrel{\text{def.}}{=} \frac{1}{2}(\mathrm{id}_S \pm \mathcal{R})$, which are complementary idempotents in $\Gamma(M, \mathrm{End}(S))$. The eigen-subbundles $S^{\pm} \stackrel{\text{def.}}{=} \mathcal{P}^{\mathcal{R}}_{\pm}(S)$ corresponding to the eigenvalues ± 1 of $\mathcal{R}$ are complementary in S and $\mathcal{R}$ determines a nontrivial direct sum decomposition $\gamma_{\mathrm{ev}} = \gamma^+ \oplus \gamma^-$.

The effective domain of definition of γ. We let $\wedge^{\pm} T^*M$ denote the bundle of twisted (anti-)selfdual forms [**5**]. Its space $\Omega^{\pm}(M) \stackrel{\text{def.}}{=} \Gamma(M, \wedge^{\pm} T^*M)$ of smooth global sections is the $\mathcal{C}^\infty(M, \mathbb{R})$-module consisting of those forms $\omega \in \Omega(M)$ which satisfy the condition $\omega \diamond \nu = \pm\omega$. Defining:

$$\wedge^{\gamma} T^*M \stackrel{\text{def.}}{=} \begin{cases} \wedge T^*M \ , & \text{if } \gamma \text{ is fiberwise injective (simple case)} \ , \\ \wedge^{\epsilon_\gamma} T^*M \ , & \text{if } \gamma \text{ is not fiberwise injective (non} - \text{simple case)} \ , \end{cases}$$

$$\wedge^{-\gamma} T^*M \stackrel{\text{def.}}{=} \begin{cases} 0 \ , & \text{if } \gamma \text{ is fiberwise injective (simple case)} \ , \\ \wedge^{-\epsilon_\gamma} T^*M \ , & \text{if } \gamma \text{ is not fiberwise injective (non} - \text{simple case)} \ , \end{cases}$$

one finds that γ restricts to zero on $\wedge^{-\gamma} T^*M$ and to a monomorphism of vector bundles on $\wedge^{\gamma} T^*M$. Due to this fact, we say that $\wedge^{\gamma} T^*M$ is the *effective domain of definition* of γ.

Schur algebras and representation types. Let S be a pin bundle of (M, g) and x be any point of M. The *Schur algebra* of γ_x is the unital subalgebra $\Sigma_{\gamma,x}$ of

$(\mathrm{End}(S_x), \circ)$ defined through:

$$\Sigma_{\gamma,x} \stackrel{\text{def.}}{=} \{T_x \in \mathrm{End}(S_x) \mid [T_x, \gamma_x(\omega_x)]_{-,\circ} = 0 \ , \ \forall \omega_x \in \wedge T_x^* M\} \ .$$

The subset $\Sigma_\gamma = \{(x, T_x) \mid x \in M \ , \ T_x \in \Sigma_{\gamma,x}\} = \sqcup_{x\in M}\Sigma_{\gamma,x}$ is a sub-bundle of unital algebras of the bundle of algebras $(\mathrm{End}(S), \circ)$, called the *Schur bundle* of γ. The isomorphism type of the fiber $(\Sigma_{\gamma,x}, \circ_x)$ is denoted by $\mathbb{S}$, being called the *Schur algebra* of γ. Real pin bundles S fall into three classes: *normal*, *almost complex* or *quaternionic*, depending on whether their Schur algebra is isomorphic with $\mathbb{R}$, $\mathbb{C}$ or $\mathbb{H}$. Some of the properties of these types are summarized in the tables below. When γ is fiberwise irreducible (i.e. in the case of pin bundles), the Schur algebra depends only on $p - q$ (mod 8), being indicated in parantheses in Tables 1 and 3. The real Clifford algebra $\mathrm{Cl}(p, q)$ (which, up to isomorphism, coincides with any fiber of the Kähler-Atiyah bundle) is non-simple iff. $p - q \equiv_8 1, 5$.

$\mathbb{S}$	$p-q$ mod 8	$\wedge T_x^* M \approx \mathrm{Cl}(p,q)$	Δ	N	Number of choices for γ	$\gamma_x(\wedge T_x^* M)$	Fiberwise injectivity of γ
$\mathbb{R}$	**0, 2**	$\mathrm{Mat}(\Delta, \mathbb{R})$	$2^{[\frac{d}{2}]} = 2^{\frac{d}{2}}$	$2^{[\frac{d}{2}]}$	1	$\mathrm{Mat}(\Delta, \mathbb{R})$	injective
$\mathbb{H}$	**4, 6**	$\mathrm{Mat}(\Delta, \mathbb{H})$	$2^{[\frac{d}{2}]-1} = 2^{\frac{d}{2}-1}$	$2^{[\frac{d}{2}]+1}$	1	$\mathrm{Mat}(\Delta, \mathbb{H})$	injective
$\mathbb{C}$	**3, 7**	$\mathrm{Mat}(\Delta, \mathbb{C})$	$2^{[\frac{d}{2}]} = 2^{\frac{d-1}{2}}$	$2^{[\frac{d}{2}]+1}$	1	$\mathrm{Mat}(\Delta, \mathbb{C})$	injective
$\mathbb{H}$	**5**	$\mathrm{Mat}(\Delta, \mathbb{H})^{\oplus 2}$	$2^{[\frac{d}{2}]-1} = 2^{\frac{d-3}{2}}$	$2^{[\frac{d}{2}]+1}$	2 ($\epsilon_\gamma = \pm 1$)	$\mathrm{Mat}(\Delta, \mathbb{H})$	non-injective
$\mathbb{R}$	**1**	$\mathrm{Mat}(\Delta, \mathbb{R})^{\oplus 2}$	$2^{[\frac{d}{2}]} = 2^{\frac{d-1}{2}}$	$2^{[\frac{d}{2}]}$	2 ($\epsilon_\gamma = \pm 1$)	$\mathrm{Mat}(\Delta, \mathbb{R})$	non-injective

Table 2: Summary of pin bundle types. $N \stackrel{\text{def.}}{=} \mathrm{rk}_{\mathbb{R}} S$ is the real rank of S while $\Delta \stackrel{\text{def.}}{=} \mathrm{rk}_{\Sigma_\gamma} S$ is the Schur rank of S.

	injective	non-injective
surjective	$\mathbf{0}(\mathbb{R}), \mathbf{2}(\mathbb{R})$	$\mathbf{1}(\mathbb{R})$
non-surjective	$\mathbf{3}(\mathbb{C}), \mathbf{7}(\mathbb{C}), \mathbf{4}(\mathbb{H}), \mathbf{6}(\mathbb{H})$	$\mathbf{5}(\mathbb{H})$

Table 3: Fiberwise character of real pin representations γ.

Fiberwise injectivity and surjectivity of γ. Basic facts from the representation theory of Clifford algebras imply:

1. γ is *fiberwise injective* iff. $\mathrm{Cl}(p, q)$ is simple as an associative $\mathbb{R}$-algebra, i.e. iff. $p - q \not\equiv_8 1, 5$ (the so-called *simple case*).

2. When γ is fiberwise *non-injective* (i.e. when $p - q \equiv_8 1, 5$, the so-called *non-simple case*), we have $\gamma(\nu) = \epsilon_\gamma \mathrm{id}_S$, where the sign factor $\epsilon_\gamma \in \{-1, 1\}$ is called the *signature* of γ. The two choices for ϵ_γ lead to two inequivalent choices for γ. The fiberwise injectivity and surjectivity of γ are summarized in Table3.

Using the approach outlined above, one can reformulate numerous constructions which are common in spin geometry and its applications to gravitational physics (in particular, to supergravity and string theory). For example, one can give [**7**] a unified, systematic and computationally efficient approach to a certain class of Fierz identities which are central in the study of supergravity backgrounds. One can also use this approach to develop [**5, 6**] certain aspects of spin geometry in a manner which allows progress in the analysis and classification of flux backgrounds [**14, 15**] as well as in the analysis of effective actions in string theory. Since a full treatment of each of these directions is quite technical and involved, we shall merely illustrate this with an example.

2. Application to general $\mathcal{N}=2$ flux compactifications of eleven-dimensional supergravity on eight-manifolds

Consider eleven-dimensional supergravity on a connected, oriented eleven-manifold $\tilde{M}$ admitting a spin structure. The physical fields are the metric $\tilde{g}$ (taken to be of mostly plus Lorentzian signature), the three-form potential $\tilde{C}$ with non-trivial four-form field strength $\tilde{G}$ and the gravitino $\tilde{\Psi}_M$. The pin bundle $\tilde{S}$ of $\tilde{M}$ can be viewed as a bundle of simple modules over the Clifford bundle of $T^*\tilde{M}$. The supersymmetry generator is a section of $\tilde{S}$. Vanishing of the supersymmetry variation of the gravitino requires:

$$\delta_{\tilde{\eta}}\tilde{\Psi}_M \stackrel{\text{def.}}{=} \tilde{\mathcal{D}}_M\tilde{\eta}=0 \ . \tag{2.1}$$

The 'supercovariant derivative' $\tilde{\mathcal{D}}_M$ takes the form: $T\tilde{M}$,

$$\tilde{\mathcal{D}}_M \stackrel{\text{def.}}{=} \tilde{\nabla}^{\text{spin}}_M - \frac{1}{288}\left(\tilde{G}_{NPQR}\tilde{\Gamma}^{NPQR}{}_M - 8\tilde{G}_{MNPQ}\tilde{\Gamma}^{NPQ}\right) \tag{2.2}$$

in a local orthonormal frame $(\tilde{e}_M)_{M=0\ldots10}$, where $\tilde{\nabla}^{\text{spin}}_M = \partial_M + \frac{1}{4}\tilde{\omega}_{MNP}\tilde{\Gamma}^{NP}$ is the connection induced on $\tilde{S}$ by the Levi-Civita connection of $\tilde{M}$, $\tilde{\omega}_{MNP}$ are the totally covariant spin connection coefficients and $\tilde{\Gamma}_M$ are the gamma 'matrices' of $\mathrm{Cl}(10,1)$ in the irreducible representation characterized by $\tilde{\Gamma}_{11} = +\mathrm{id}_{\tilde{S}}$. Setting $\tilde{M} = M_3 \times M$ with usual warped product ansatz for $\tilde{g}$ (see [**5, 8–10**]), condition (2.1) reduces to the following two conditions (known as the *constrained generalized Killing (CGK) pinor equations* [**5**]) for the internal part $\xi \in \Gamma(M,S)$ of $\tilde{\eta}$, which is a section of the pin bundle S of the internal manifold M:

(2.3)
$$D_m\xi = 0 \text{ with } D_m \stackrel{\text{def.}}{=} \nabla^{\text{spin}}_m + A_m \ , \ A_m = -\frac{1}{4}f_n\gamma^n{}_m\gamma_9 + \frac{1}{24}F_{mpqr}\gamma^{pqr} + \kappa\gamma_m\gamma_9 \ ,$$

(2.4)
$$Q\xi = 0 \text{ with } Q = \frac{1}{2}\gamma^m\partial_m\Delta - \frac{1}{288}F_{mnpq}\gamma^{mnpq} - \frac{1}{6}f_m\gamma^m\gamma_9 - \kappa\gamma_9 \ , \ \forall m,n,p,q = 1\ldots 8 \ .$$

We refer to conditions (2.3) and (2.4) above as the differential and algebraic constraints, respectively; they have two independent global solutions (ξ_1, ξ_2) when the background preserves $\mathcal{N}=2$ supersymmetry (see [**5, 6**]). In (2.3) and (2.4), the gamma 'matrices' γ^m transform in the representations of the real Clifford algebra $\mathrm{Cl}(8,0)$, while κ is a positive real number which is proportional to the square root of the cosmological constant of the external AdS_3 space.

The Fierz isomorphism. When (M,g) has Euclidean signature with $d \equiv_8 0, 1$, the morphism $\gamma : (\wedge T^*M, \diamond) \to (\mathrm{End}(S), \circ)$ is surjective and has a partial inverse [**5**] which allows one to translate the CGK pinor conditions into conditions on differential forms on M constructed as bilinear combinations of sections of S. The process is encoded by the so-called *Fierz isomorphism* [**5**], which identifies the bundle of bispinors $S \otimes S$ with a sub-bundle $\wedge^\gamma T^*M$ of the exterior bundle playing the role of 'effective domain of definition' [2] of γ. As explained in detail in [**5, 7, 11, 12**], one can construct *admissible bilinear pairings* $\mathcal{B}$ on the pin bundle satisfying certain properties[3], using which one defines a $\mathcal{B}$-dependent bundle isomorphism:

$$E : S \otimes S \to \mathrm{End}(S) \text{ where } E_{\xi,\xi'}(\xi'') \stackrel{\text{def.}}{=} \mathcal{B}(\xi'', \xi')\xi \text{ such that}$$

$$E_{\xi_1,\xi_2} \circ E_{\xi_3,\xi_4} = \mathcal{B}(\xi_3, \xi_2) E_{\xi_1,\xi_4} ,$$

which induces the Fierz isomorphism $\check{E} = \gamma^{-1} \circ E : S \otimes S \to \wedge^\gamma T^*M$ mentioned above.

Lift to a nine-manifold. To investigate general supersymmetric compactifications on eight-manifolds, one must analyze the CGK pinor equations (2.3), (2.4). The requirement of $\mathcal{N} = 2$ supersymmetry means that these equations must have two linearly independent global solutions $\xi_1, \xi_2 \in \Gamma(M, S)$ — whose values $\xi_1(x)$, $\xi_2(x) \in S_x$ then are [**6**] also linearly independent at any point x of M. To simplify the analysis, one can lift [**6**] the problem to the nine-dimensional metric cone $\hat{M}$ over M. The solutions (ξ_1, ξ_2) determine a point on the second Stiefel manifold $V_2(S_x)$ of each fiber S_x of S. Thus solutions can be classified according to the orbit of the representation of Spin(8) on S_x, which induces a corresponding action on $V_2(S_x)$. Since the latter action fails to be transitive, a generic basis (ξ_1, ξ_2) of solutions of the CGK pinor equations does not determine a global reduction of the structure group of M. Using the natural embedding $\mathrm{Spin}(8) \subset \mathrm{Spin}(9) \subset \mathrm{Cl}(9, 0)$, the action of Spin(8) on $V_2(S_x)$ extends to an action of Spin(9), which turns out to be transitive — thereby suggesting that an interpretation in terms of reduction of structure group could be given upon passing to some Riemannian nine-manifold naturally associated with M. The action of Spin(9) can indeed be geometrized [**6**] by passing to the metric cone $\hat{M}$ over M, whose Clifford bundle is of non-simple type [**5–7, 13**]. Using the construction of [**6**], this allows one to provide a global description of 8-manifold compactifications preserving $\mathcal{N} = 2$ supersymmetry through reduction of the structure group of $\hat{M}$ (rather than of M itself). We therefore consider the (punctured) metric cone $\hat{M} = (0, +\infty) \times M$ (with squared line element given by $\mathrm{d}s^2_{\mathrm{cone}} = \mathrm{d}r^2 + r^2 \mathrm{d}s^2$), where the canonical normalized one-form $\theta = \mathrm{d}r$ is a special Killing-Yano form with respect to the metric g_{cone} of $\hat{M}$. We can pull back the Levi-Civita connection of (M, g) along the natural projection $\Pi : \hat{M} \to M$ and also lift the connection (2.3) from S to the pin bundle $\hat{S}$ of $\hat{M}$ – using the fact [**6**] that $\hat{S}$ can be identified with the Π-pullback of S. We define $\Pi^*(h) = h \circ \Pi$ for $h \in \mathcal{C}^\infty(M, \mathbb{R})$. It turns out that the morphism $\gamma_{\mathrm{cone}} : \wedge T^*\hat{M} \to \mathrm{End}(\hat{S})$ is determined by the morphism γ of the manifold M. The bundle morphism γ_{cone} is fiberwise surjective but not fiberwise injective, so there are two inequivalent choices distinguished by the property $\gamma_{\mathrm{cone}}(\nu_{\mathrm{cone}}) = \pm \mathrm{id}_{\hat{S}}$, where ν_{cone} is the volume form of $\hat{M}$. We choose to work with the representation of signature $+1$, which satisfies $\gamma_{\mathrm{cone}}(\nu_{\mathrm{cone}}) = +\mathrm{id}_{\hat{S}}$. Then the Kähler-Atiyah algebra $(\Omega(\hat{M}), \diamond^{\mathrm{cone}})$ of the cone can be realized through the truncated model $(\Omega^<(\hat{M}), \blacklozenge^{\mathrm{cone}})$ of [**6**], where

[2]When $d \equiv_8 0$ in Euclidean signature, the bundle morphism γ is fiberwise injective and we have $\wedge^\gamma T^*M = \wedge T^*M$. When $d \equiv_8 1$, the effective domain of γ is a proper sub-bundle of the exterior bundle since γ fails to be injective in that case.

[3]These properties are symmetry, type and — when applicable — isotropy

$\Omega^<(\hat{M}) = \oplus_{k=0}^4 \Omega^k(\hat{M})$ and $\blacklozenge^{\rm cone}$ is the reduced geometric product of the cone. Rescaling the metric through $g \to 2\kappa^2 g$, the connection induced on $\hat{S}$ by the Levi-Civita connection of $\hat{M}$ takes the form:

$$\nabla^{\hat{S}}_{e_m} = \nabla^{S}_{e_m} + \kappa\gamma_{9m} \ , \ \nabla^{\hat{S}}_{\partial_r} = \partial_r \ , \ \text{where } m = 1\dots 8 \ .$$

Defining the connection $\hat{D} \stackrel{\rm def.}{=} D^* = \nabla^{\hat{S},{\rm cone}} + A^{\rm cone}$ as the pullback of (2.3) to $\hat{M}$, the expression for $A^{\rm cone}$ becomes [**6**]:

$$A^{\rm cone} = \Pi^*(A) - \frac{1}{2r} e^m_{\rm cone} \otimes \gamma_{\rm cone}((e^{\rm cone}_m)_{\sharp_{\rm cone}} \wedge \theta) \ ,$$

where $\Pi^*(A)$ denotes the pull-back connection and $(e^{\rm cone}_m)_{\sharp_{\rm cone}}$ is the one-form dual to the base vector $e^{\rm cone}_m$ with respect to the metric on $\hat{M}$. Lifting to the cone and 'dequantizing' as in [**5**, **6**], the connection A_m of (2.3) and the endomorphism Q from (2.4) induce the following inhomogeneous differential forms on $\hat{M}$:

$$(2.5)\quad \check{A}_m = \frac{1}{4}\iota_{e^{\rm cone}_m} F + \frac{1}{4}(e^{\rm cone}_m)_\sharp \wedge f \wedge \theta \quad , \quad \check{Q} = \frac{1}{2} r({\rm d}\Delta) - \frac{1}{6} f \wedge \theta - \frac{1}{12} F - \kappa\theta \ .$$

Analysis of CGK pinor equations. We are interested in form-valued pinor bilinears written locally as follows (where the subscript '+' denotes the twisted self-dual part [**5**, **6**]) :

$$\check{E}^{(k)}_{\hat{\xi}_i,\hat{\xi}_j} \equiv \check{E}^{(k)}_{ij} = \frac{1}{k!}\mathcal{B}(\hat{\xi}_i, \gamma^{\rm cone}_{a_1\dots a_k}\hat{\xi}_j) e_+^{a_1\dots a_k} \ , \ \text{where } a_1 \dots a_k \in \{1\dots 9\} \ ,$$

with $i, j \in \{1, 2\}$ and in the weighted sums:

$$\check{E}_{ij} = \frac{N}{2^d} \sum_{k=0}^{d} \check{E}^{(k)}_{ij} \ ,$$

which are inhomogeneous differential forms generating the algebra $(\Omega^+(\hat{M}), \diamond)$. Here, $N = 2^{\left[\frac{d}{2}\right]}$ is the real rank of the pin bundle S, $d = 9$ is the dimension of $\hat{M}$ and we normalized the two pinors through $\mathcal{B}(\xi_i, \xi_j) = \delta_{ij}$. Using the properties of the admissible bilinear paring on $\hat{S}$, we find:

$$\mathcal{B}(\hat{\xi}_i, \gamma^{a_1\dots a_k}_{\rm cone}\hat{\xi}_j) = (-1)^{\frac{k(k-1)}{2}} \mathcal{B}(\hat{\xi}_j, \gamma^{a_1\dots a_k}_{\rm cone}\hat{\xi}_i) \ , \ \forall i, j = 1, 2 \ ,$$

which implies that the non-trivial form-valued bilinears of rank ≤ 4 are three 1-forms V_k dual to the vector fields with local coefficients given (after raising indices to avoid notational clutter) by:

$$(2.6)\qquad V_1^a = \mathcal{B}(\hat{\xi}_1, \gamma^a_{\rm cone}\hat{\xi}_1) \ , \ V_2^a = \mathcal{B}(\hat{\xi}_2, \gamma^a_{\rm cone}\hat{\xi}_2) \ , \ V_3^a = \mathcal{B}(\hat{\xi}_1, \gamma^a_{\rm cone}\hat{\xi}_2)$$

together with one 2-form K, one 3-form Ψ and three 4-forms ϕ_k with strict coefficients given similarly by:

$$(2.7)\qquad K^{ab} = \mathcal{B}(\hat{\xi}_1, \gamma^{ab}_{\rm cone}\hat{\xi}_2) \ , \ \psi^{abc} = \mathcal{B}(\hat{\xi}_1, \gamma^{abc}_{\rm cone}\hat{\xi}_2) \ , \ \phi_1^{abce} = \mathcal{B}(\hat{\xi}_1, \gamma^{abce}_{\rm cone}\hat{\xi})$$

The forms above are the definite rank components of the truncated inhomogeneous forms:

$$\check{E}^<_{11} = \frac{1}{32}(1+V_1+\phi_1) \ , \ \check{E}^<_{12} = \frac{1}{32}(V_3+K+\psi+\phi_3) \ , \ \check{E}^<_{21} = \frac{1}{32}(V_3-K-\psi+\phi_3) \ , \ \check{E}^<_{22} = \frac{1}{32}(1+V_2+\phi_2)$$

which are the basis elements of the truncated Fierz algebra of [**5**, **6**] and satisfy the truncated Fierz identities[4]:

$$\check{E}^{<}_{ij}\blacklozenge\check{E}^{<}_{kl} = \frac{1}{2}\delta_{jk}\check{E}^{<}_{il} \ , \ \forall i,j,k,l = 1,2 \ . \tag{2.8}$$

The full analysis of these equations is quite involved. We list only some of the relations implied by the truncated Fierz identities:

$$\iota_{V_1}V_3 = 0 \ , \ \iota_{V_1}\phi_3 - \psi + V_1 \wedge K = 0 \ , \ \iota_{V_3}\phi_1 + \psi - V_1 \wedge K = 0 \ . \tag{2.9}$$

Let us also discuss the constraints on these forms implied by the CGK pinor equations. As explained in [**5**, **6**], the differential constraints (2.3) imply:

$$\mathrm{d}\check{E}^{<}_{ij} = e^a \wedge \nabla_a \check{E}^{<}_{ij} \text{ where } \nabla_a \check{E}^{<}_{ij} = -[\check{A}_a, \check{E}^{<}_{ij}]_{-,\blacklozenge} \ , \ \forall i,j \in \{1,2\} \ , \tag{2.10}$$

whereas the algebraic constraints (2.4) reduce to:

$$\check{Q}\blacklozenge\check{E}^{<}_{ij} \mp \check{E}^{<}_{ij}\blacklozenge\hat{\tau}(\check{Q}) = 0 \ , \ \forall i,j \in \{1,2\} \ . \tag{2.11}$$

The complete set of conditions is rather involved, so only part of which will be reproduced here. The first equation in (2.11) (the one with the *minus* sign) for $\check{E}^{<}_{12}$ leads to the following constraints when separating into rank components:

$$\iota_{f\wedge\theta}K = 0 \ , \tag{2.12}$$

$$r\iota_{\mathrm{d}\Delta}K + \frac{1}{3}\iota_{f\wedge\theta}\psi - \frac{1}{6}\iota_{\psi}F - 2\kappa\iota_{\theta}K = 0 \ , \tag{2.13}$$

$$\frac{1}{3}\iota_{f\wedge\theta}\phi_3 - \frac{1}{6}F \bigtriangleup_3 \phi_3 + r(\mathrm{d}\Delta)\wedge V_3 + 2\kappa V_3 \wedge \theta = 0 \ , \tag{2.14}$$

while using the truncated inhomogeneous form $\check{E}^{<}_{11}$ amounts to:

$$-\frac{1}{3}f\wedge\theta + \frac{1}{3}\iota_{f\wedge\theta}\phi_1 - \frac{1}{6}F \bigtriangleup_3 \phi_1 + r(\mathrm{d}\Delta)\wedge V_1 + 2\kappa V_1\wedge\theta = 0 \ . \tag{2.15}$$

Expanding (2.10) for $\check{E}^{<}_{11}$ and $\check{E}^{<}_{12}$ gives the following expressions for the covariant derivatives of V_1 and V_3:

$$\nabla_m V_{1n} = f_s\theta_p{\phi_1}^{sp}{}_{mn} - \frac{1}{6}F_{spqm}{\phi_1}^{spq}{}_n \ , \ \nabla_m V_{3n} = f_s\theta_p{\phi_3}^{sp}{}_{mn} - \frac{1}{6}F_{spqm}{\phi_3}^{spq}{}_n \ . \tag{2.16}$$

Upon antisymmetrization, these give the differential constraints:

$$\mathrm{d}V_1 = 2\iota_{f\wedge\theta}\phi_1 - F \bigtriangleup_3 \phi_1 \ , \ \mathrm{d}V_3 = 2\iota_{f\wedge\theta}\phi_3 - F\bigtriangleup_3 \phi_3 \ .$$

Using the differential and algebraic constraints for the form-valued pinor bilinears, one can investigate the geometric implications of the $\mathcal{N} = 2$ supersymmetry condition. This analysis, as well as the a discussion of the physics implications, is rather involved and we shall not attempt to summarize it here.

3. Outlook

Throughout the years, there have been numerous attempts to find a unified mathematical perspective on the class of 'supersymmetric flux compactifications' of supergravity and string theories. One of the most elementary of these relies on the classical theory of G-structures, using the fact that a system of spinors which are locally linearly independent provides – modulo finite-covering issues – a local reduction of structure group of the special orthonormal frame bundle of (M, g). It

[4]Note that we will omit writing the 'cone' superscript on $\blacklozenge^{\mathrm{cone}}$ to simplify notation.

is immediately apparent that such reductions will rarely be global, which forces one to consider more complex – and less well-studied – frameworks such as a theory of 'generalized G-structures', which extends the theory of singular (or generalized) cosmooth Frobenius-like distributions used in certain problems related to dynamical systems. Another point of view, which is driven by the study of symmetries, is the approach through generalized geometry, which is appropriate in the perturbative framework of nonlinear sigma models and the more recent framework of exceptional generalized geometry, which is inspired by M-theory. The last of these encounters serious limitations when applied to the study of compactifications on internal spaces of high enough dimension. The limits of both of these frameworks are illustrated rather clearly by the class of supersymmetric compactifications of M-theory on eight-manifolds, where in general one has no global reduction of structure group of the special orthonormal frame bundle and no clear interpretation through exceptional generalized geometry. In this class of compactifications, detailed analysis leads to a rich geometric picture which involves the theory of singular foliations (in the sense of Haefliger structures), to various unsolved problems in noncommutative geometry and to an entire landscape of effective theories whose geometric interpretation is surprisingly subtle.

The point of view provided by the Kähler-Atiyah bundle is both more general and different in spirit when compared with other frameworks. Rather than pursuing symmetries or local structures, this approach makes a direct connection to non-commutative geometry and geometric quantization, while interpreting fluxes as higher order corrections which can be understood through deformation theory, using expansions which are controlled by the inverse of the radius of the compactification manifold. The central object of interest is not the algebroid structure arising from symmetries, but the associative structure arising from vertical quantization, which is the central feature of any supergravity theory since local supersymmetry mixes bosonic fields (which from this perspective are classical objects) with fermionic fields (which are quantum objects from the point of view of vertical quantization). Hence supersymmetry transformations can be understood as 'quantum symmetries' in the vertical quantization procedure which 'explains' how fermions arise geometrically. Following this point of view allows one to apply methods and ideas from the theory of quantization and from algebraic homotopy theory to the subject of flux compactifications, thereby providing novel perspectives on that subject. It also provides new perspectives on the theory of G-structures and on their cosmooth generalizations as well as a different starting point for ideas of generalized geometry, in that one does not look simply at symmetries of geometric structures which parameterize supersymmetric backgrounds, but at a certain class of vertical quantization procedures which leads to supersymmetric theories and which, in special situations, happens to preserve certain symmetries encoded by various types of generalized geometry.

References

[1] H. B. Lawson Jr. and M.-L. Michelsohn, *Spin geometry*, Princeton Mathematical Series, vol. 38, Princeton University Press, Princeton, NJ, 1989. MR1031992 (91g:53001)

[2] A. Trautman, *Connections and the Dirac operator on spinor bundles*, J. Geom. Phys. **58** (2008), no. 2, 238–252, DOI 10.1016/j.geomphys.2007.11.001. MR2384313 (2008k:58077)

[3] C. Chevalley, *The algebraic theory of spinors and Clifford algebras*, Springer-Verlag, Berlin, 1997. Collected works. Vol. 2; Edited and with a foreword by Pierre Cartier and Catherine Chevalley; With a postface by J.-P. Bourguignon. MR1636473 (99f:01028)

[4] M. Riesz, *Clifford numbers and spinors: With the author's private lectures to E. Folke Bolinder and a historical review by Pertti Lounesto*, Fundamental Theories of Physics, vol. 54, Kluwer Academic Publishers Group, Dordrecht, 1993. Edited by Bolinder and Pertti Lounesto and with a preface by Bolinder. MR1247961 (94i:15024)

[5] C. I. Lazaroiu, E. M. Babalic, and I. A. Coman, *"Geometric algebra techniques in flux compactifications (I)"*, arXiv:1212.6766 [hep-th].

[6] C.-I. Lazaroiu and E.-M. Babalic, *Geometric algebra technique in flux compactifications (II)*, J. High Energy Phys. **6** (2013), 054. MR3083356

[7] C. I. Lazaroiu, E. M. Babalic, and I. A. Coman, *The geometric algebra of Fierz identities in arbitrary dimensions and signatures*, J. High Energy Phys. **9** (2013), 156, 72, DOI 10.1007/JHEP09(2013)156. MR3108766

[8] K. Becker and M. Becker, *$\mathcal{M}$-theory on eight-manifolds*, Nuclear Phys. B **477** (1996), no. 1, 155–167, DOI 10.1016/0550-3213(96)00367-7. MR1413258 (98b:81178)

[9] D. Martelli and J. Sparks, *G structures, fluxes, and calibrations in M theory*, Phys. Rev. D (3) **68** (2003), no. 8, 085014, 19, DOI 10.1103/PhysRevD.68.085014. MR2039402 (2005b:53079)

[10] D. Tsimpis, *M-theory on eight-manifolds revisited: $\mathcal{N} = 1$ supersymmetry and generalized* Spin(7) *structures*, J. High Energy Phys. **4** (2006), 027, 26 pp. (electronic), DOI 10.1088/1126-6708/2006/04/027. MR2219075 (2007b:81218)

[11] D. V. Alekseevsky and V. Cortés, *Classification of N-(super)-extended Poincaré algebras and bilinear invariants of the spinor representation of* $\mathrm{Spin}(p,q)$, Comm. Math. Phys. **183** (1997), no. 3, 477–510, DOI 10.1007/s002200050039. MR1462223 (99b:17004)

[12] D. V. Alekseevsky, V. Cortés, C. Devchand, and A. Van Proeyen, *Polyvector super-Poincaré algebras*, Comm. Math. Phys. **253** (2005), no. 2, 385–422, DOI 10.1007/s00220-004-1155-y. MR2140254 (2006c:17008)

[13] S. Okubo, *Representations of Clifford algebras and its applications*, Math. Japon. **41** (1995), no. 1, 59–79. MR1317752 (95k:81044)

[14] E. M. Babalic and C. I. Lazaroiu, *Foliated eight-manifolds for M-theory compactification*, JHEP **01** (2015) 140, arXiv:1411.3148 [hep-th]

[15] E. M. Babalic and C. I. Lazaroiu, *Singular foliations for M-theory compactification*, JHEP **03** (2015) 116, arXiv:1411.3497 [hep-th].

Department of Theoretical Physics, National Institute of Physics and Nuclear Engineering, Str. Reactorului no.30, P.O. BOX MG-6, Postcode 077125, Bucharest-Magurele, Romania

Current address: Center for Geometry and Physics, Institute for Basic Science, Pohang 790-784, Republic of Korea

E-mail address: calin@ibs.re.kr

Department of Theoretical Physics, National Institute of Physics and Nuclear Engineering, Str. Reactorului no.30, P.O.BOX MG-6, Postcode 077125, Bucharest - Magurele, Romania

E-mail address: mbabalic@theory.nipne.ro

Department of Theoretical Physics, National Institute of Physics and Nuclear Engineering, Str. Reactorului no.30, P.O.BOX MG-6, Postcode 077125, Bucharest - Magurele, Romania

E-mail address: icoman@theory.nipne.ro

Proceedings of Symposia in Pure Mathematics
Volume **90**, 2015
http://dx.doi.org/10.1090/pspum/090/01517

A Toolkit for Defect Computations in Landau-Ginzburg Models

Nils Carqueville and Daniel Murfet

Abstract. We review the results of [**13**] on orientation reversal and duality for defects in topological Landau-Ginzburg models, with the intention of providing an easily accessible toolkit for computations. As an example we include a proof of the main result on adjunctions in a special case, using Pauli matrices. We also explain how to compute arbitrary correlators of defect-decorated planar worldsheets, and briefly discuss the relation to generalised orbifolds.

1. Introduction

There is a wide spectrum of reasons to study Landau-Ginzburg models, of which we name only but a few. A certain subclass of such models is believed to have infrared fixed points under renormalisation group flow that describe 'stringy' string vacua in the moduli space of Calabi-Yau compactifications [**23, 27, 41**]. On the one hand, this CY/LG correspondence makes Landau-Ginzburg models a central player in the world of (homological) mirror symmetry. On the other hand it embeds into the more general CFT/LG correspondence which states that many properties of the conformal fixed point can already be directly described on the level of Landau-Ginzburg models. By now much circumstantial evidence has been collected in the bulk, boundary and defect sectors, see e. g. [**2, 5, 14, 15, 20, 28, 35, 40**]. Out of this (together with an increasing understanding of matrix factorisations) emerges the hope of better control over non-rational conformal field theories, as Landau-Ginzburg models appear comparably indifferent towards additional symmetries. Finally, Landau-Ginzburg models also feature in the construction [**33**] of homological link invariants [**12, 25, 36**]; the results reported in this note may help to establish the precise connection to the work of [**24, 42**].

Regarding the subject of defects in two-dimensional (topological) field theories we refer e. g. to [**18, 29**] as excellent introductions. Generally speaking one should take the defect sector of a theory as seriously as (its special case) the boundary sector. The fact that the former's target space interpretation is currently much less clear than the latter's in terms of D-branes poses a challenge whose resolution may significantly advance our view on string theory. Among the numerous important applications of defects we single out their use for renormalisation group flow [**8, 22**], in deforming conformal field theories away from the rational point [**10, 39**], their

2010 *Mathematics Subject Classification.* Primary 18D05, 57R56.

role in the AGT correspondence [**1**], as well as their relevance for the refined link invariant constructions mentioned above.

The main part of this note is organised as follows. In Section 2 we start by recalling the description of defects in topological Landau-Ginzburg models in terms of matrix factorisations. Then we explain the origin and nature of the type of duality we wish to consider, to wit adjunctions between defects. Our main theorem states the existence of these adjunctions and provides explicit evaluation and coevaluation maps to satisfy the needs of the practically-minded. Throughout we use a rigorous diagrammatic language which lends itself to efficient computations. The fact that it also has a natural physical interpretation is the starting point for the applications discussed in Section 3: any planar defect-decorated correlator can be straightforwardly computed from our results. To illustrate this we first derive a residue expression for the action of defects on bulk fields; among its special cases we recover the Kapustin-Li disc correlator and a formula for boundary states. Next we consider a genus-one worldsheet and give a simple one-line proof of the Cardy condition. We end with a brief look on the generalised orbifolds of [**16**].

2. Adjoint defects in Landau-Ginzburg models

We recall from [**6, 7, 30, 34**] that oriented *defects* $X : W \to V$ between topological Landau-Ginzburg models with potentials $W \in \mathbb{C}[x] \equiv \mathbb{C}[x_1, \dots, x_n]$ and $V \in \mathbb{C}[z] \equiv \mathbb{C}[z_1, \dots, z_m]$ are described by matrix factorisations of $V - W$. This means that $X = X^0 \oplus X^1$ is a free $\mathbb{Z}_2$-graded $\mathbb{C}[z, x]$-module equipped with a twisted differential, i. e. an odd operator d_X which squares to $(V - W) \cdot 1_X$. *Defect changing fields* $\Psi \in \operatorname{Hom}(X, Y)$ take values in the cohomology of the BRST operator $\Psi \mapsto d_Y \Psi - (-1)^{|\Psi|} \Psi d_X$. Throughout this note we stick to the above notation of variable numbers n, m, potentials W, V, and defect X:

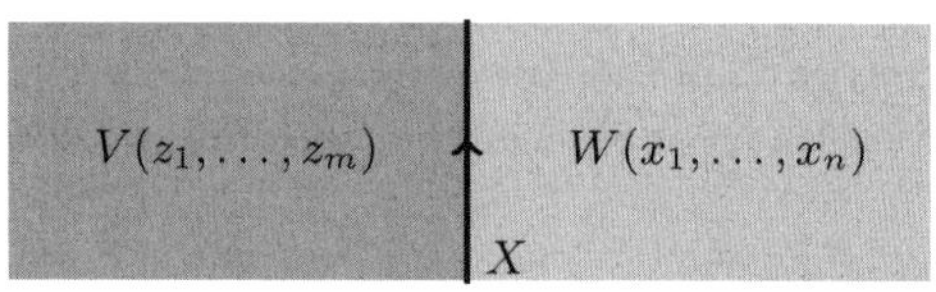

In the presence of topological defects there are two types of 'multiplications': the *operator product* of fields given by matrix multiplication, and the *fusion* of defects and their fields. The latter is given by the tensor product $X \otimes_{\mathbb{C}[x]} Z$ with twisted differential $d_X \otimes 1 + 1 \otimes d_Z$ where X is as before and $Z : U \to W$ for some potential U. The unit for the fusion product is the *invisible defect* $I_W : W(x') \to W(x)$, where we agree to always use primed notation for the copy of variables pertaining to the source of an endodefect. As a module it is given by $I_W = \bigwedge(\bigoplus_{i=1}^n \mathbb{C}[x, x'] \cdot \theta_i)$, i. e. by the exterior algebra generated by n anticommuting variables θ_i (sometimes interpreted as boundary fermions). For the twisted differential on I_W we have $d_{I_W} = \sum_{i=1}^n [(x_i - x'_i) \cdot \theta_i^* + \partial_{[i]}^{x,x'} W \cdot \theta_i]$ where we use the divided difference operators defined by

$$\partial_{[i]}^{x,x'} W = \frac{W(x'_1, \dots, x'_{i-1}, x_i, \dots, x_n) - W(x'_1, \dots, x'_i, x_{i+1}, \dots, x_n)}{x_i - x'_i} \tag{2.1}$$

and by definition $\theta_i^*(\theta_j) = \delta_{ij}$. It is straightforward to verify that $\mathrm{End}(I_W)$ is isomorphic to $\mathbb{C}[x]/(\partial_i W)$, in line with the intuition that fields living on I_W are nothing but bulk fields.

The invisible defect is the unit for fusion in the sense that there are natural isomorphisms (up to homotopy, i. e. up to BRST exact terms) $\lambda_X : I_V \otimes X \to X$ and $\rho_X : X \otimes I_W \to X$ that implement its λeft and ρight action on X. For example, λ_X is the projection $I_V \to \mathbb{C}[z, z']$ to θ-degree zero, followed by identifying $z' = z$, and analogously for ρ_X. We may depict these maps and their inverses as follows:

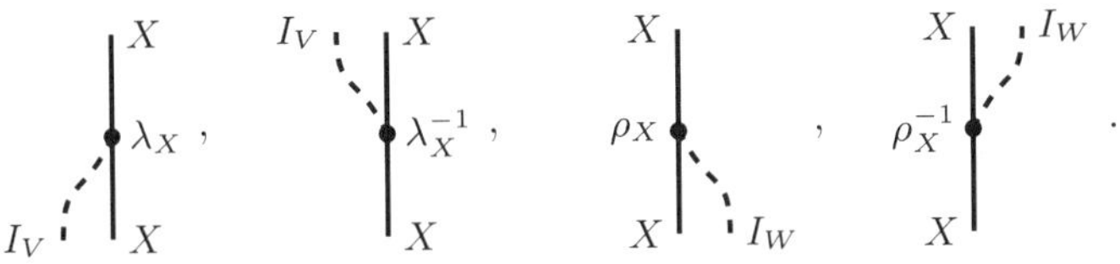

More generally, any collection of fields sitting on defects adorning some worldsheet is naturally described by such a diagram. To determine a given diagram's 'value', i. e. the effective field inserted at the junction of all in- and outgoing defects, we read it from bottom to top (operator product) and from right to left (fusion product).

Next we consider the defect $X^\dagger$ that has the opposite orientation but otherwise imposes the same constraints on bulk fields as X. For Landau-Ginzburg models we have $X^\dagger = X^\vee[n]$ where $X^\vee := \mathrm{Hom}_{\mathbb{C}[z,x]}(X, \mathbb{C}[z,x])$, the twisted differential is

$$d_{X^\dagger} = \begin{pmatrix} 0 & (d_X^0)^\vee \\ -(d_X^1)^\vee & 0 \end{pmatrix}[n], \quad \text{if} \quad d_X = \begin{pmatrix} 0 & d_X^1 \\ d_X^0 & 0 \end{pmatrix},$$

and $[n]$ denotes the n-fold shift functor (which basically exchanges d_X^0 and d_X^1 n times). For $\Psi \in \mathrm{Hom}(X, Y)$ the orientation reversed field is $\Psi^\dagger = \Psi^\vee[n] \in \mathrm{Hom}(Y^\dagger, X^\dagger)$.

Oriented defects may 'take a U-turn', so we expect canonical defect fields

(2.2)  $\equiv \widetilde{\mathrm{ev}}_X : X \otimes X^\dagger \longrightarrow I_V$,  $\equiv \widetilde{\mathrm{coev}}_X : I_W \longrightarrow X^\dagger \otimes X$

where up- and downwards oriented lines are implictly labelled by X and $X^\dagger$, respectively, and we no longer display dashed lines for the invisible defect. Furthermore, by the topological nature of X we expect the identities

(2.3) 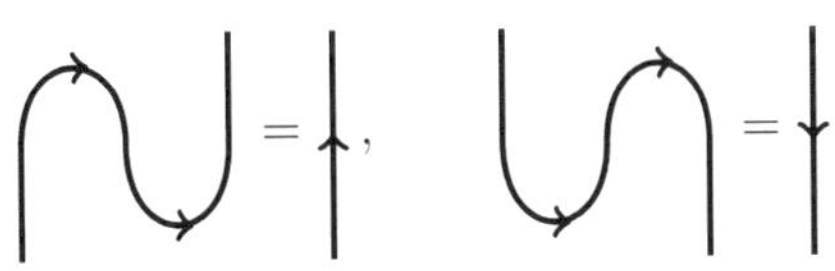

to hold. While the diagrams on either side of each equality are certainly isotopic, (2.3) does impose nontrivial conditions on the maps (2.2); e. g. the second identity reads $\rho_{X^\dagger} \circ (1_{X^\dagger} \otimes \widetilde{\mathrm{ev}}_X) \circ (\widetilde{\mathrm{coev}}_X \otimes 1_{X^\dagger}) \circ \lambda_{X^\dagger}^{-1} = 1_{X^\dagger}$ in clumsy diagram-free notation. We say that two defect fields $\widetilde{\mathrm{ev}}_X, \widetilde{\mathrm{coev}}_X$ exhibit $X^\dagger$ as the *right adjoint* of X if the *Zorro moves* (2.3) hold. This generalises the adjunction for a finite-dimensional vector space V and its dual $V^\dagger$ encoded in the standard evaluation and coevaluation maps defined by $\widetilde{\mathrm{ev}}_V(v \otimes \alpha) = \alpha(v)$ and $\widetilde{\mathrm{coev}}_V : 1 \mapsto \sum_i e_i^* \otimes e_i$ for any basis $\{e_i\}$ of V. Analogously, ${}^\dagger X = X^\vee[m]$ is *left adjoint* to X if there are defect fields

(2.4)  $\equiv \mathrm{ev}_X : {}^\dagger X \otimes X \to I_W$,  $\equiv \mathrm{coev}_X : I_V \to X \otimes {}^\dagger X$

satisfying their versions of Zorro moves:

(2.5) 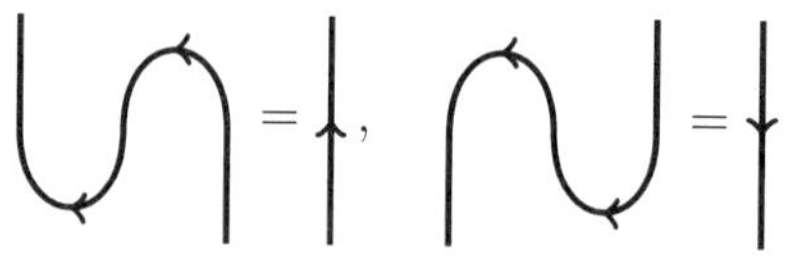

The main theorem, quoted below in its explicit form, gives expressions for the adjunction maps (2.2), (2.4) as well as the inverses $\lambda_X^{-1}, \rho_X^{-1}$. For practical purposes it is important to note that the formulas are all expressed concretely in terms of matrix multiplication, taking derivatives, and 'integrating' – where we recall that residues are computed by the rules

$$\mathrm{Res}\left[\frac{\mathrm{d}x}{x_1^{a_1},\dots,x_n^{a_n}}\right] = \delta_{a_1,1}\dots\delta_{a_n,1} \quad \text{and} \quad \mathrm{Res}\left[\frac{h\ \mathrm{d}x}{f_1,\dots,f_n}\right] = \mathrm{Res}\left[\frac{\det(C)h\ \mathrm{d}x}{g_1,\dots,g_n}\right]$$

if $g_i = \sum_{j=1}^n C_{ij} f_j$ for polynomials h, f_i, C_{ij}. The upshot is that the correlator of any planar defect-decorated worldsheet can be straightforwardly (and often gainfully) computed by viewing it as a diagram made up of the maps (2.2), (2.4), $\lambda^{\pm 1}, \rho^{\pm 1}$ and whatever other defect fields may be present. We will return to this point and discuss several examples in the next section.

THEOREM ([**13**]). *Any matrix factorisation X of $V(z_1,\dots,z_m)-W(x_1,\dots,x_n)$ has left and right adjoints. The associated structure maps have the following explicit presentations:*

$$\widetilde{\mathrm{ev}}_X(e_j\otimes e_i^*) = \sum_{l\geqslant 0}\sum_{a_1<\dots<a_l} (-1)^{l+(n+1)|e_j|}\,\theta_{a_1}\dots\theta_{a_l} \cdot \mathrm{Res}\left[\frac{\{\partial^{z,z'}_{[a_l]} d_X \dots \partial^{z,z'}_{[a_1]} d_X\, \Lambda^{(x)}\}_{ij}\ \mathrm{d}x}{\partial_{x_1}W,\dots,\partial_{x_n}W}\right],$$

$$\mathrm{ev}_X(e_i^*\otimes e_j) = \sum_{l\geqslant 0}\sum_{a_1<\dots<a_l} (-1)^{\binom{l}{2}+l|e_j|}\,\theta_{a_1}\dots\theta_{a_l} \cdot \mathrm{Res}\left[\frac{\{\Lambda^{(z)}\,\partial^{x,x'}_{[a_1]} d_X \dots \partial^{x,x'}_{[a_l]} d_X\}_{ij}\ \mathrm{d}z}{\partial_{z_1}V,\dots,\partial_{z_m}V}\right],$$

$$\widetilde{\mathrm{coev}}_X(\bar\gamma) = \sum_{i,j}(-1)^{(\bar r+1)|e_j|+s_n}\left\{\partial^{x,x'}_{[\bar b_{\bar r}]}(d_X)\dots\partial^{x,x'}_{[\bar b_1]}(d_X)\right\}_{ji} e_i^*\otimes e_j\,,$$

$$\mathrm{coev}_X(\gamma) = \sum_{i,j}(-1)^{\binom{r+1}{2}+mr+s_m}\left\{\partial^{z,z'}_{[b_1]} d_X\dots\partial^{z,z'}_{[b_r]} d_X\right\}_{ij} e_i\otimes e_j^*\,,$$

$$\lambda_X^{-1}(e_i) = \sum_{l\geqslant 0}\sum_{a_1<\dots<a_l}\sum_j \theta_{a_1}\dots\theta_{a_l}\left\{\partial^{z,z'}_{[a_l]} d_X\dots\partial^{z,z'}_{[a_1]} d_X\right\}_{ji}\otimes e_j\,,$$

$$\rho_X^{-1}(e_i) = \sum_{l\geqslant 0}\sum_{a_1<\dots<a_l}\sum_j (-1)^{\binom{l}{2}+l|e_i|} e_j\otimes\left\{\partial^{x,x'}_{[a_1]} d_X\dots\partial^{x,x'}_{[a_l]} d_X\right\}_{ji}\theta_{a_1}\dots\theta_{a_l}$$

where $\{e_i\}$ is a basis of X, $\Lambda^{(x)} := (-1)^n\partial_{x_1}d_X\dots\partial_{x_n}d_X$, $\Lambda^{(z)} := \partial_{z_1}d_X\dots\partial_{z_m}d_X$, and $b_i, \bar b_{\bar\jmath}$ and $s_n, s_m\in\mathbb{Z}_2$ are uniquely determined by requiring that $b_1<\dots<b_r$, $\bar b_1<\dots<\bar b_{\bar r}$, and $\bar\gamma\theta_{\bar b_1}\dots\theta_{\bar b_{\bar r}} = (-1)^{s_n}\theta_1\dots\theta_n$ and $\gamma\theta_{b_1}\dots\theta_{b_r} = (-1)^{s_m}\theta_1\dots\theta_m$.

Let us briefly comment on the proof. Its first ingredient is homological perturbation [**17**], which allows us to construct $\lambda_X^{-1}, \rho_X^{-1}$ by viewing parts of the twisted differential of, say, $I_V \otimes X$ as a perturbation [**13**, Sect. 4]. Then using noncommutative forms for another natural presentation of the invisible defect leads to a conceptually solid description of $\lambda_X^{-1}, \rho_X^{-1}$ in terms of 'associative Atiyah classes' [**13**, Sect. 3]. Computing them explicitly gives rise to the divided difference operators $\partial_{[i]}^{x,x'} d_X$ in the formulas above.

The coevaluation is the image of the identity under the map [**13**, Eq. (5.11)]

$$\operatorname{Hom}(X,X) \cong X^\vee \otimes X \xrightarrow{\rho'^{-1}_{X^\vee \otimes X}} X^\vee \otimes X \otimes I'_W \cong \operatorname{Hom}(I_W, X^\vee[n] \otimes X)$$

where I'_W denotes I_W but with twisted differential $\sum_{i=1}^n [(x_i - x'_i)\cdot \theta_i^* - \partial_{[i]}^{x,x'} W \cdot \theta_i]$. Note that the nontrivial part $\rho'^{-1}_{X^\vee \otimes X}$ is an isomorphism only up to homotopy. Similarly, the evaluation is constructed by lifting the Kapustin-Li pairing [**26, 31**] $X \otimes_{\mathbb{C}[x]} X^\dagger \to \mathbb{C}[z]$ using homological perturbation [**13**, Sect. 5.2].

Proving that the above adjunction maps really satisfy the Zorro moves (up to homotopy) is one of the central results of [**13**, Sect. 6]. To give a taste we work out a simple example that boils down to properties of Pauli matrices in Appendix A. The general case essentially amounts to artfully manipulating Atiyah classes inside the residue and supertrace.

Finally, we collect a few useful identities. Together with the Zorro moves and the expressions in our theorem they are all we need to compute any correlator of defect fields.

PROPOSITION ([**13**]). *For $\Psi \in \operatorname{Hom}(X,Y)$ and composable X, Z we have*

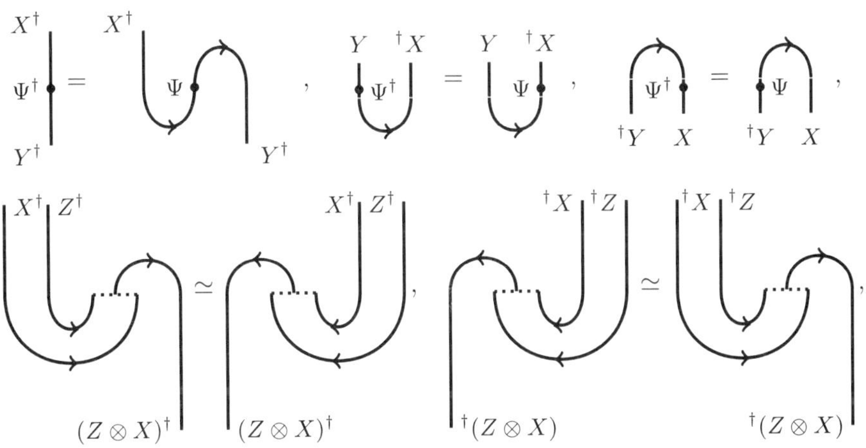

hiding certain signs (explained at length in [**13**, Sect. 7]*) in the symbol* $\simeq$.

3. Applications

We now illustrate how the general results collected in the previous section are put to use. As already stated, the explicit expressions in our theorem allow us to compute arbitrary correlators of planar worldsheets with defects, for which we give two examples. Another application is to the theory of generalised orbifolds.

Defect actions. Let us fix a defect $X : W \to V$, a defect field $\Psi \in \operatorname{End}(X)$ and a bulk field $\phi \in \operatorname{End}(I_W)$. We will explain that for the associated *defect action on bulk fields* we have

(3.1)

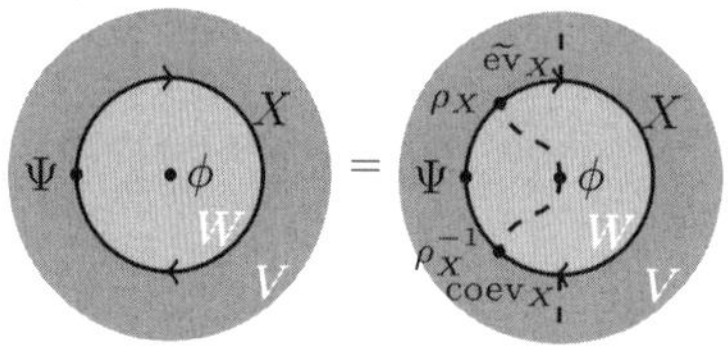

$$= (-1)^{\binom{m+1}{2}} \operatorname{Res}\left[\frac{\phi \operatorname{str}\left(\Psi\left(\prod_i \partial_{x_i} d_X\right)\left(\prod_j \partial_{z_j} d_X\right)\right)\mathrm{d}x}{\partial_{x_1} W, \ldots, \partial_{x_n} W}\right].$$

Here the left-hand side shows the physical picture of the defect X, decorated by Ψ, wrapping around the bulk field ϕ of the theory W. Because of the topological nature of the situation, X may wrap ϕ as tightly as we please; the limit of the defect line collapsing on the bulk field (of the inner theory W) is effectively described by a new bulk field (of the outer theory V) that we call $\mathcal{D}_X^\Psi(\phi)$. It is precisely this effective bulk field that is computed in (3.1).

The first step in determining the defect action in (3.1) is to translate the left-most physical picture into rigorous mathematical language. Given our diagrammatics this is straightforward: all we have to do is view bulk fields as endomorphisms of the invisible defect ($\phi \in \operatorname{End}(I_W)$, $\mathcal{D}_X^\Psi(\phi) \in \operatorname{End}(I_V)$), make the latter and its action on other defects visible (i. e. insert ρ_X, ρ_X^{-1} in our example), and label cups and caps of defect lines by appropriate adjunction maps ($\widetilde{\operatorname{ev}}_X$, coev_X). This is done in the first step of (3.1); reading the resulting diagram from bottom to top and from right to left produces the new bulk field $\mathcal{D}_X^\Psi(\phi)$ in theory V. To compute it explicitly, in the second step of (3.1) we simply plug in our expressions for $\rho^{\pm 1}, \widetilde{\operatorname{ev}}, \operatorname{coev}$ from the previous section. After a short calculation [**13**, Sect. 8] this leads to the residue formula on the right-hand side. Note that it gives a precise meaning to 'integrating out' the x-dependent degrees of freedom in theory W. It is clear that any planar configuration of defects with field insertions can be computed in this way.

The defect action (3.1) has several special cases:

- If we set both ϕ and Ψ equal to the respective identities, i. e. if we simply consider an empty defect bubble labelled by X, then by definition we obtain the *(right) quantum dimension* of X.
- If $V = 0$, i. e. if the outer theory is trivial and the defect becomes a boundary condition, then (3.1) recovers the Kapustin-Li *disc correlator* of [**26, 31**].
- If $W = 0$, i. e. if the inner theory is trivial, (3.1) describes the *boundary-bulk map* $\Psi \mapsto (-1)^{\binom{m+1}{2}} \operatorname{str}(\Psi\, \partial_{z_1} d_X \ldots \partial_{z_m} d_X)$ of [**32**]. If we further restrict to $\Psi = 1_X$, we obtain what is traditionally called the *boundary state* or *Chern character* of X.

Cardy condition. We just saw that disc correlators and boundary-bulk maps can be neatly formulated in defect language. Actually the complete structure of open/closed topological field theory (TFT) naturally fits into this framework [**13**, Sect. 9]. As a second example of our theorem's practical uses we now recall how it is used to give a 'one-line proof' of the Cardy condition.

The Cardy condition is most familiar in two-dimensional conformal field theory (CFT), where it is a consequence of open/closed duality and can be derived by evaluating a cylinder amplitude in two different ways, see e. g. [**3**]. It asserts that the overlap of two boundary states equals the associated open sector partition function which is computed as a certain trace. Similarly, the TFT version of the Cardy condition says that the two-point bulk correlator of two images under the boundary-bulk map is the same as a certain supertrace in the boundary sector. It can be considered the most 'quantum' gluing axiom as it originates from inspection of a genus-one worldsheet. Another incarnation of its importance is that in the case of B-twisted sigma models the Cardy condition manifests itself as a generalisation of the Hirzebruch-Riemann-Roch theorem [**11**].

We will now first formulate the Cardy condition for Landau-Ginzburg models (originally proven in [**9**, **19**, **38**]) and then explain how to effortlessly derive it from the defect perspective advocated in this note.

Let X, Y be two matrix factorisations of $W(x_1, \ldots, x_n)$ and $\Phi \in \mathrm{End}(X)$, $\Psi \in \mathrm{End}(Y)$. Then the Cardy condition states

(3.2)
$$(-1)^{\binom{n+1}{2}} \operatorname{Res}\left[\frac{\operatorname{str}(\Phi\, \partial_{x_1} d_X \ldots \partial_{x_n} d_X)\operatorname{str}(\Psi\, \partial_{x_1} d_Y \ldots \partial_{x_n} d_Y)\,\mathrm{d}x}{\partial_{x_1} W, \ldots, \partial_{x_n} W}\right] = \operatorname{str}({}_\Psi m_\Phi)$$

where ${}_\Psi m_\Phi$ is the operator on the open string space $\mathrm{Hom}(X, Y)$ that precomposes with Φ and postcomposes with Ψ (so in particular ${}_{1_Y} m_{1_X} = 1_{\mathrm{Hom}(X,Y)}$ and the supertrace becomes a simple index).

In order to argue for (3.2) we think of a cylinder with boundary conditions as an annulus correlator. Accordingly, we claim that the proof of the Cardy condition is contained in the following identities:

(3.3)

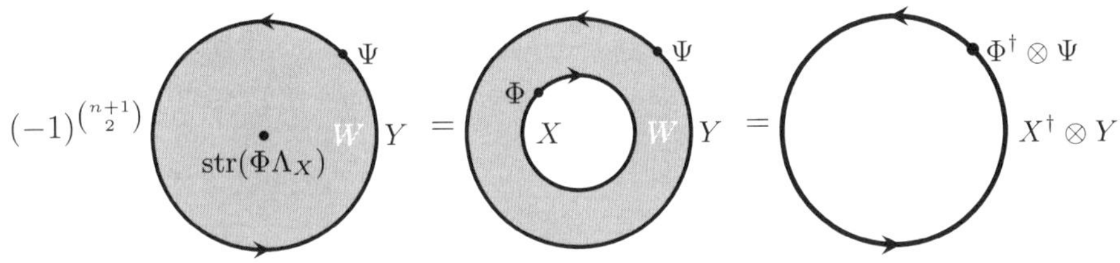

To make sense of this we start from the annulus diagram in the middle. Since we are dealing with a topological theory, the size of the inner X-boundary does not matter and may be shrunk to zero. This is the left equality, where we have used the knowledge from our first example, namely that the boundary-bulk map is given by $(-1)^{\binom{n+1}{2}} \operatorname{str}(\Phi\Lambda_X)$ with $\Lambda_X = \partial_{x_1} d_X \ldots \partial_{x_n} d_X$. Next we use the other special case of (3.1), to wit the one with trivial outer theory, to immediately find that the disc correlator on the left-hand side of (3.3) is precisely given by the left-hand side of (3.2). To see that the right-hand side of (3.2) matches the right-hand side of (3.3), we note that the second equality in the latter comes about by expanding the inner boundary until it fuses with the outer boundary (where on the technical level use has to be made of a property called 'pivotality', see [**13**, Sect. 7]). One can then convince oneself that the right-hand side of (3.3) is indeed given by $\operatorname{str}({}_\Psi m_\Phi)$, thus concluding our 'one-line proof' of the Cardy condition.

Generalised orbifolds. As a further application of our explicit control over adjunctions in Landau-Ginzburg models we mention generalised orbifolds. Recall that given a finite symmetry group G of some two-dimensional field theory T, correlators in the conventional orbifold theory T^G can be computed as correlators in the original theory T, but with a network of defects A_G (implementing the group action) covering the worldsheet. It was first realised in [**21**] (within the framework of rational CFT) that this construction can be generalised by allowing *any* defect A that shares certain crucial properties with A_G (to wit, A must be a 'separable symmetric Frobenius algebra', see e. g. [**16**, Sect. 2.2]). Among many other consequences, this means that any two rational CFTs with identical chiral algebras and central charges (that have a unique vacuum and nondegenerate two-point correlators) are generalised orbifolds of one another; this is in particular true for all minimal models with ADE-type modular invariants.

These results were developed into a general theory of *orbifold completion* for arbitrary two-dimensional TFTs with defects in [**16**]. Unorbifolded theories T are 'completed' by considering all pairs (T, A) where $A : T \to T$ is a separable symmetric Frobenius algebra. Such pairs are called *generalised orbifolds*. The original theory T is identified with (T, I_T), and ordinary orbifolds are the special cases $(T, A_G) \equiv T^G$.

Much can be said already on this general level; to keep the discussion brief we shall only mention one central result of [**16**]. Let $X : T \to T'$ be a defect between theories T, T' that has invertible quantum dimension. Then with $A_X := X^\dagger \otimes X$ we have an equivalence of theories

$$T' \equiv (T', I_{T'}) \cong (T, A_X)\,. \tag{3.4}$$

On the level of correlators, the idea behind this construction is illustrated by

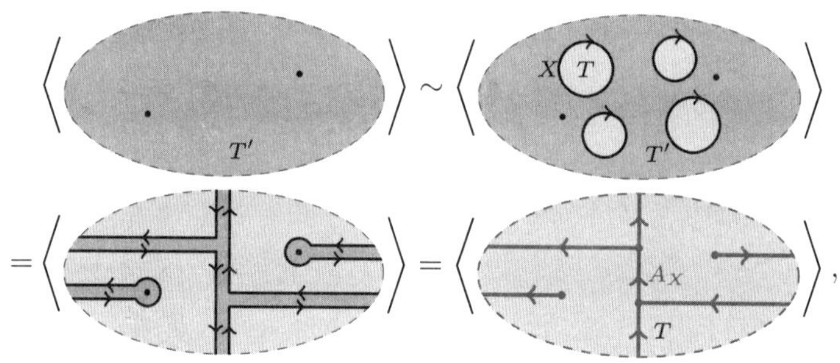

expressing a T'-correlator in terms of a T-correlator with a network of A_X-defects. As a special case (3.4) implies that boundary conditions of theory T' are in one-to-one correspondence with A_X-modules.

Naturally the above constructions can be applied to Landau-Ginzburg models. If G is a finite symmetry group of a potential W, then one finds [**16**, Thm. 7.2] that G-equivariant matrix factorisations of W are equivalent to modules over $A_G = \bigoplus_{g\in G}(I_W)_g$ where $(-)_g$ denotes twisting by g from the right. This recasts conventional Landau-Ginzburg orbifolds (including discrete torsion) in defect language, as worked out in detail in [**4**].

To construct other examples of generalised orbifolds between Landau-Ginzburg models with potentials W and V, it suffices to find a defect $X : W \to V$ with invertible quantum dimension. In this case, as an immediate consequence of (3.4), we have an equivalence between matrix factorisations of V and A_X-modules.

To some degree equivalences of this form come very cheaply: once a candidate X is identified, our explicit residue expression (3.1) makes computing $\dim(X)$

a straightforward exercise. While finding defects with invertible quantum dimension is still nontrivial, we nonetheless expect that many new equivalences can be produced in this way. For example, (generalised) orbifolds between A- and D-type minimal models were constructed in [**16**, Sect. 7.3], and the CFT/LG correspondence predicts generalised orbifolds relating them to exceptional minimal models as well. We further expect generalised orbifolds involving Calabi-Yau compactifications à la [**27, 37**], and eventually for them to play a role in a generalisation of homological mirror symmetry to the defect sector.

Acknowledgements. We thank Ilka Brunner, Daniel Plencner and Ingo Runkel for helpful comments on the manuscript. Nils Carqueville is grateful to Marc Lehmacher and Natina Zulficar.

Appendix A. Zorro proof for beginners

For a general defect $X : W \longrightarrow V$ the proof of the Zorro moves in [**13**] requires the introduction and careful study of Atiyah classes. Nevertheless the key idea is a simple one, and in this section we explain a special case where the proof reduces to a straightforward calculation with Pauli matrices $\sigma_1, \sigma_2, \sigma_3$.

We take $W = 0$, $V = z_1^2 + z_2^2$ and

$$d_X = \begin{pmatrix} 0 & z_1 - \mathrm{i}z_2 \\ z_1 + \mathrm{i}z_2 & 0 \end{pmatrix} = \sigma_1 z_1 + \sigma_2 z_2 \, .$$

Then X is a defect $W \to V$ and

$$\partial_{[1]}^{z,z'} d_X = \partial_{z_1} d_X = \sigma_1 = \begin{pmatrix} 0 & 1 \\ 1 & 0 \end{pmatrix}, \quad \partial_{[2]}^{z,z'} d_X = \partial_{z_2} d_X = \sigma_2 = \begin{pmatrix} 0 & -\mathrm{i} \\ \mathrm{i} & 0 \end{pmatrix}.$$

Let e_0, e_1 constitute a $\mathbb{C}[z_1, z_2]$-basis of X with $|e_i| = i$ and note that $\sigma_3 = -\mathrm{i}\sigma_1\sigma_2$ is the grading operator $\sigma_3(v) = (-1)^{|v|}v$. There are four Zorro moves to prove: the two in (2.3) are trivial, so we focus our attention on the first identity in (2.5) (the second is similar). Let $\mathcal{Z}$ denote the composite

$$X \xrightarrow{\lambda^{-1}} I_V \otimes_{\mathbb{C}[z]} X \xrightarrow{\mathrm{coev} \otimes 1} X \otimes_{\mathbb{C}} X^\vee \otimes_{\mathbb{C}[z]} X \xrightarrow{\rho \circ (1 \otimes \mathrm{ev})} X \, .$$

We prove that this is the identity map. In this case our explicit formulas give

$$\begin{aligned} \lambda^{-1}(e_i) &= \sum_{l \geqslant 0} \sum_{a_1 < \cdots < a_l} \sum_j \theta_{a_1} \ldots \theta_{a_l} \{\sigma_{a_l} \ldots \sigma_{a_1}\}_{ji} \otimes e_j \, , \\ \mathrm{ev}(\nu \otimes \eta) &= \frac{\mathrm{i}}{4} (-1)^{|\eta|} \nu(\eta)\big|_{z_1 = z_2 = 0} \, , \\ \mathrm{coev}(\theta_{a_1} \ldots \theta_{a_l}) &= \sum_{i,j} (-1)^{\binom{r+1}{2} + s} \{\sigma_{b_1} \ldots \sigma_{b_r}\}_{ij} \, e_i \otimes e_j^* \end{aligned}$$

where $\theta_{a_1}\dots\theta_{a_l}\theta_{b_1}\dots\theta_{b_r} = (-1)^s\theta_1\theta_2$. Substituting, we find

$$\begin{aligned}
\mathcal{Z}(e_i) &= (1\otimes \mathrm{ev})\sum_{l\geqslant 0}\sum_{a_1<\dots<a_l}\sum_{j}\mathrm{coev}(\theta_{a_1}\dots\theta_{a_l})\{\sigma_{a_l}\dots\sigma_{a_1}\}_{ji}\otimes e_j\\
&=\sum_{l\geqslant 0}\sum_{a_1<\dots<a_l}\sum_{j,i',j'}(-1)^{\binom{r+1}{2}+s}\{\sigma_{a_l}\dots\sigma_{a_1}\}_{ji}\{\sigma_{b_1}\dots\sigma_{b_r}\}_{i'j'}e_{i'}\otimes \mathrm{ev}(e^*_{j'}\otimes e_j)\\
&=\sum_{l\geqslant 0}\sum_{a_1<\dots<a_l}\sum_{i'}\frac{\mathrm{i}}{4}(-1)^{\binom{r+1}{2}+s+|e_i|+l}\{\sigma_{b_1}\dots\sigma_{b_r}\sigma_{a_l}\dots\sigma_{a_1}\}_{i'i}\,e_{i'}\\
&=\sum_{l\geqslant 0}\sum_{a_1<\dots<a_l}\sum_{i'}\frac{\mathrm{i}}{4}(-1)^{\binom{r+1}{2}+s+|e_i|+l+rl+\binom{l}{2}}\{\sigma_{a_1}\dots\sigma_{a_l}\sigma_{b_1}\dots\sigma_{b_r}\}_{i'i}\,e_{i'}\\
&=\sum_{l\geqslant 0}\sum_{a_1<\dots<a_l}\sum_{i'}\frac{\mathrm{i}}{4}(-1)^{1+|e_i|}\{\sigma_1\sigma_2\}_{i'i}\,e_{i'}=\sum_{l\geqslant 0}\sum_{a_1<\dots<a_l}\frac{1}{4}e_i=e_i\,.
\end{aligned}$$

In the last line we sum over the $2^m = 4$ basis elements of I_V, and this factor of 4 cancels with the $\frac{1}{4}$ coming from the residue denominator in the evaluation map. In the general case something analogous happens: the quantum dimension of I_V 'cancels' with the residue, and the subtle part of the proof lies in showing that λ^{-1} and the coevaluation map combine to produce the quantum dimension.

References

[1] L. F. Alday, D. Gaiotto, S. Gukov, Y. Tachikawa, and H. Verlinde, *Loop and surface operators in $\mathcal{N}=2$ gauge theory and Liouville modular geometry*, J. High Energy Phys. **1** (2010), 113, 50, DOI 10.1007/JHEP01(2010)113. MR2660780 (2011k:81250)

[2] S. K. Ashok, E. Dell'Aquila, and D.-E. Diaconescu, *Fractional branes in Landau-Ginzburg orbifolds*, Adv. Theor. Math. Phys. **8** (2004), no. 3, 461–513. MR2105188 (2006k:81278)

[3] R. Blumenhagen and E. Plauschinn, *Introduction to conformal field theory: With applications to string theory*, Lecture Notes in Physics, vol. 779, Springer, Dordrecht, 2009. MR2848105 (2012h:81238)

[4] I. Brunner, N. Carqueville, and D. Plencner, work in progress.

[5] I. Brunner and M. R. Gaberdiel, *The matrix factorizations of the D-model*, J. Phys. A **38** (2005), no. 36, 7901–7919, DOI 10.1088/0305-4470/38/36/008. MR2185420 (2006m:81213)

[6] I. Brunner, M. Herbst, W. Lerche, and B. Scheuner, *Landau-Ginzburg realization of open string TFT*, J. High Energy Phys. **11** (2006), 043, 28 pp. (electronic), DOI 10.1088/1126-6708/2006/11/043. MR2270412 (2008b:81267)

[7] I. Brunner and D. Roggenkamp, *B-type defects in Landau-Ginzburg models*, J. High Energy Phys. **8** (2007), 093, 46 pp. (electronic), DOI 10.1088/1126-6708/2007/08/093. MR2342020 (2009a:81178)

[8] I. Brunner and D. Roggenkamp, *Defects and bulk perturbations of boundary Landau-Ginzburg orbifolds*, J. High Energy Phys. **4** (2008), 001, 35, DOI 10.1088/1126-6708/2008/04/001. MR2425302 (2010d:81241)

[9] R.-O. Buchweitz and D. Van Straten, *An index theorem for modules on a hypersurface singularity* (English, with English and Russian summaries), Mosc. Math. J. **12** (2012), no. 2, 237–259, 459. MR2978754

[10] D. Bücher and I. Runkel, *Integrable perturbations of conformal field theories and Yetter-Drinfeld modules*, [arXiv:1211.4726].

[11] A. Căldăraru and S. Willerton, *The Mukai pairing. I. A categorical approach*, New York J. Math. **16** (2010), 61–98. MR2657369 (2011g:18012)

[12] N. Carqueville and D. Murfet, *Computing Khovanov-Rozansky homology and defect fusion*, [arXiv:1108.1081].

[13] N. Carqueville and D. Murfet, *Adjunctions and defects in Landau-Ginzburg models*, [arXiv:1208.1481].

[14] N. Carqueville and I. Runkel, *On the monoidal structure of matrix bi-factorizations*, J. Phys. A **43** (2010), no. 27, 275401, 33, DOI 10.1088/1751-8113/43/27/275401. MR2658288 (2012a:81243)

[15] N. Carqueville and I. Runkel, *Rigidity and defect actions in Landau-Ginzburg models*, Comm. Math. Phys. **310** (2012), no. 1, 135–179, DOI 10.1007/s00220-011-1403-x. MR2885616

[16] N. Carqueville and I. Runkel, *Orbifold completion of defect bicategories*, [arXiv:1210.6363].

[17] M. Crainic, *On the perturbation lemma, and deformations*, [math/0403266].

[18] A. Davydov, L. Kong, and I. Runkel, *Field theories with defects and the centre functor*, Mathematical foundations of quantum field theory and perturbative string theory, Proc. Sympos. Pure Math., vol. 83, Amer. Math. Soc., Providence, RI, 2011, pp. 71–128, DOI 10.1090/pspum/083/2742426. MR2742426 (2012j:57056)

[19] T. Dyckerhoff and D. Murfet, *Pushing forward matrix factorisations*, [arXiv:1102.2957].

[20] H. Enger, A. Recknagel, and D. Roggenkamp, *Permutation branes and linear matrix factorisations*, J. High Energy Phys. **1** (2006), 087, 37 pp. (electronic), DOI 10.1088/1126-6708/2006/01/087. MR2200799 (2007m:81202)

[21] J. Fröhlich, J. Fuchs, I. Runkel, and C. Schweigert, *Defect lines, dualities and generalised orbifolds*, XVIth International Congress on Mathematical Physics, World Sci. Publ., Hackensack, NJ, 2010, pp. 608–613, DOI 10.1142/9789814304634_0056. MR2730830 (2012f:81254)

[22] D. Gaiotto, *Domain walls for two-dimensional renormalization group flows*, J. High Energy Phys. **12** (2012), 103, front matter + 25. MR3045259

[23] B. R. Greene, C. Vafa, and N. P. Warner, *Calabi-Yau manifolds and renormalization group flows*, Nuclear Phys. B **324** (1989), no. 2, 371–390, DOI 10.1016/0550-3213(89)90471-9. MR1025421 (91c:14042)

[24] S. Gukov, A. Schwarz, and C. Vafa, *Khovanov-Rozansky homology and topological strings*, Lett. Math. Phys. **74** (2005), no. 1, 53–74, DOI 10.1007/s11005-005-0008-8. MR2193547 (2007a:57014)

[25] S. Gukov and J. Walcher, *Matrix Factorizations and Kauffman Homology*, [hep-th/0512298].

[26] M. Herbst and C.-I. Lazaroiu, *Localization and traces in open-closed topological Landau-Ginzburg models*, J. High Energy Phys. **5** (2005), 044, 31 pp. (electronic), DOI 10.1088/1126-6708/2005/05/044. MR2155540 (2007a:81134)

[27] M. Herbst, K. Hori, and D. Page, *Phases Of $\mathcal{N} = 2$ Theories In $1 + 1$ Dimensions With Boundary*, [arXiv:0803.2045].

[28] P. Howe and P. West, *$N = 2$ superconformal models, Landau-Ginsburg hamiltonians and the ϵ expansion*, Phys. Lett. B **223** (1989), 377–385; *Chiral Correlators In Landau-Ginzburg Theories And $N = 2$ Superconformal Models*, Phys. Lett. B **227** (1989), 397–405; *Fixed points in multifield Landau-Ginsburg models*, Phys. Lett. B **244** (1990), 270–274.

[29] A. Kapustin, *Topological Field Theory, Higher Categories, and Their Applications*, [arXiv:1004.2307].

[30] A. Kapustin and Y. Li, *D-branes in Landau-Ginzburg models and algebraic geometry*, J. High Energy Phys. **12** (2003), 005, 44 pp. (electronic), DOI 10.1088/1126-6708/2003/12/005. MR2041170 (2005b:81179b)

[31] A. Kapustin and Y. Li, *Topological correlators in Landau-Ginzburg models with boundaries*, Adv. Theor. Math. Phys. **7** (2003), no. 4, 727–749. MR2039036 (2005b:81179a)

[32] A. Kapustin and L. Rozansky, *On the relation between open and closed topological strings*, Comm. Math. Phys. **252** (2004), no. 1-3, 393–414, DOI 10.1007/s00220-004-1227-z. MR2104884 (2006a:81134)

[33] M. Khovanov and L. Rozansky, *Matrix factorizations and link homology*, Fund. Math. **199** (2008), no. 1, 1–91, DOI 10.4064/fm199-1-1. MR2391017 (2010a:57011)

[34] C. I. Lazaroiu, *On the boundary coupling of topological Landau-Ginzburg models*, J. High Energy Phys. **5** (2005), 037, 19 pp. (electronic), DOI 10.1088/1126-6708/2005/05/037. MR2155384 (2006f:81171)

[35] E. J. Martinec, *Algebraic geometry and effective Lagrangians*, Phys. Lett. B **217** (1989), no. 4, 431–437, DOI 10.1016/0370-2693(89)90074-9. MR981536 (90c:81129)

[36] D. Murfet and R. Rouquier, work in progress.

[37] D. Orlov, *Derived categories of coherent sheaves and triangulated categories of singularities*, Algebra, arithmetic, and geometry: in honor of Yu. I. Manin. Vol. II, Progr. Math., vol. 270, Birkhäuser Boston, Inc., Boston, MA, 2009, pp. 503–531, DOI 10.1007/978-0-8176-4747-6_16. MR2641200 (2011c:14050)

[38] A. Polishchuk and A. Vaintrob, *Chern characters and Hirzebruch-Riemann-Roch formula for matrix factorizations*, [arXiv:1002.2116].

[39] I. Runkel, *Non-local conserved charges from defects in perturbed conformal field theory*, J. Phys. A **43** (2010), no. 36, 365206, 27, DOI 10.1088/1751-8113/43/36/365206. MR2671718 (2011m:81266)

[40] C. Vafa and N. Warner, *Catastrophes and the classification of conformal theories*, Phys. Lett. B **218** (1989), no. 1, 51–58, DOI 10.1016/0370-2693(89)90473-5. MR983349 (90m:81135)

[41] E. Witten, *Phases of $N = 2$ theories in two dimensions*, Nuclear Phys. B **403** (1993), no. 1-2, 159–222, DOI 10.1016/0550-3213(93)90033-L. MR1232617 (95a:81261)

[42] E. Witten, *Fivebranes and Knots*, [arXiv:1101.3216].

SIMONS CENTER FOR GEOMETRY AND PHYSICS
E-mail address: `nils@carqueville.net`

MATHEMATICAL SCIENCES RESEARCH INSTITUTE
E-mail address: `dmurfet@msri.org`

Proceedings of Symposia in Pure Mathematics
Volume **90**, 2015
http://dx.doi.org/10.1090/pspum/090/01530

Grassmannian Twists, Derived Equivalences and Brane Transport

Will Donovan

Abstract. This note is based on a talk given at String-Math 2012 in Bonn, on a joint paper with Ed Segal. We exhibit derived equivalences corresponding to certain Grassmannian flops. The construction of these equivalences is inspired by work of Herbst–Hori–Page on brane transport for gauged linear σ-models: in particular, we define 'windows' corresponding to their grade restriction rules. We then show how composing our equivalences produces interesting autoequivalences, which we describe as twists and cotwists about certain spherical functors.

Our proofs use natural long exact sequences of bundles on Grassmannians known as twisted Lascoux complexes, or staircase complexes. We give a compact description of these. We also touch on some related developments, and work through some extended examples.

1. Introduction

It is conjectured [**18**, Conjecture 5.1] that for flops between smooth projective complex varieties there exist corresponding *derived equivalences*. In this note we discuss a particular set of examples, as studied in a joint paper of the author and Ed Segal [**8**]: local models for *Grassmannian flops*. These are obtained from geometric invariant theory (GIT) quotients of vector spaces by general linear group actions: the original variety corresponds to a certain stability condition, and the flopped variety to another.

The varieties occurring in our examples may be viewed as large-radius limits of particular gauged linear σ-models. Herbst, Hori and Page [**11**] studied *brane transport* between phases of such models, and produced a description in terms of grade-restriction rules. Inspired by this, we describe brane transport in our examples using 'windows' corresponding to grade-restriction rules. More precisely, we construct equivalences between the bounded derived categories of coherent sheaves on the respective GIT quotients: see Theorem A. By definition, these equivalences are *functorial*, and so may be interpreted as describing transport of branes, and also of the strings between them. We then describe *monodromy* for the brane transport, which turns out to be given by twists and cotwists about spherical functors: see Theorem B.

2010 *Mathematics Subject Classification.* Primary 14F05, 18E30; Secondary 14M15.

This note is intended to give a concise summary of the results in [**8**], concentrating on examples, and referring to the original paper for proofs and technical details.

1.1. Results. We now describe our examples, and give precise statements of our findings. We fix throughout a dimension d. Then we have:

DEFINITION 1.1. Taking V and $S^{(r)}$ to be complex vector spaces of dimension d and r respectively, we define a global quotient stack

$$\mathfrak{X}^{(r)} := \left[\frac{\operatorname{Hom}(S^{(r)}, V) \ \oplus \ \operatorname{Hom}(V, S^{(r)})}{\operatorname{GL}(S^{(r)})}\right].$$

The $\operatorname{GL}(S^{(r)})$-action here is the natural one, given by composition of linear maps.

We now consider two different GIT quotients, viewed as substacks of $\mathfrak{X}^{(r)}$:

DEFINITION 1.2. We define substacks $X_{\pm}^{(r)}$ of $\mathfrak{X}^{(r)}$ by requiring that the first map be injective, or the second surjective, as shown below:

$$X_{+}^{(r)} := \left[\frac{\operatorname{Hom}_{\hookrightarrow}(S^{(r)}, V) \ \oplus \ \operatorname{Hom}(V, S^{(r)})}{\operatorname{GL}(S^{(r)})}\right]$$
$$X_{-}^{(r)} := \left[\frac{\operatorname{Hom}(S^{(r)}, V) \ \oplus \ \operatorname{Hom}_{\twoheadrightarrow}(V, S^{(r)})}{\operatorname{GL}(S^{(r)})}\right]$$

REMARK 1.3. The geometric interpretation of these spaces $X_{\pm}^{(r)}$ is as follows:

(1) The GIT quotient $X_{+}^{(0)}$ is simply a point.
(2) The quotient $X_{+}^{(1)}$ is the total space of the bundle $l^{\oplus d}$ on $\mathbb{P}V$, where we take l to be tautological subspace bundle on $\mathbb{P}V$ (which is often denoted by $\mathcal{O}(-1)$).
(3) In general, the quotient $X_{+}^{(r)}$ is given by the total space of the bundle $\operatorname{Hom}(V, S^{(r)})$ on the Grassmannian $\mathbb{G}\mathrm{r}(r, V)$, where we reuse the notation $S^{(r)}$ for the tautological subspace bundle on the Grassmannian. After choosing a basis for V, this is of course just the bundle $S^{(r)\oplus d}$. A dual construction gives $X_{-}^{(r)}$: see [**8**, Section 3.1].

REMARK 1.4. The quotients $X_{\pm}^{(r)}$ should be thought of as corresponding to certain large-radius limits of a gauged linear σ-model determined by $\mathfrak{X}^{(r)}$. See Section 2.3 for some further remarks on this viewpoint.

We then have:

THEOREM A. [**8**, Theorem 3.7] *For each $k \in \mathbb{Z}$, there exists an equivalence of bounded derived categories*

$$\psi_k \colon \mathcal{D}(X_{+}^{(r)}) \xrightarrow{\sim} \mathcal{D}(X_{-}^{(r)}).$$

We consider now *autoequivalences* produced by composing the equivalences ψ_k and their inverses:

NOTATION. We write $\omega_{k,l} := \psi_k^{-1}\psi_l \in \operatorname{Aut}\left(\mathcal{D}(X_{+}^{(r)})\right)$, and refer to these $\omega_{k,l}$ as *window-shift autoequivalences.*

There is a well-developed theory associating twist autoequivalences of derived categories to *spherical objects* [**20**] or, more generally, to *spherical functors* (for definitions, see Section 2.1). These twists may be thought of as mirror to symplectic monodromies [**20, 21**]. The window-shift autoequivalences defined above can, at least heuristically, be seen as realising monodromy in the Stringy Kähler Moduli Space (SKMS) of our theory. Accordingly, the following result relates certain window-shift autoequivalences to twists of spherical functors. Other window-shifts are related to a dual notion of *cotwists*:

THEOREM B. [**8**, Theorems 3.12, 3.13] *We have that*

$$\begin{aligned}\omega_{0,+1} &= T_{F(r)},\\ \omega_{0,-1} &= C_{F(r+1)}[2(d-r)-1],\end{aligned}$$

where the twist $T_\bullet$ and the cotwist $C_\bullet$ are defined in Section 2.1, *and each $F(r)$ is a certain spherical functor*

$$F(r)\colon \mathcal{D}(X_+^{(r-1)}) \longrightarrow \mathcal{D}(X_+^{(r)})$$

defined in Section 5.2.

The functors involved in this theorem are illustrated in Figure 1.

REMARK 1.5. (1) When $r=1$, the functor $F(1)$ has domain $\mathcal{D}(\mathrm{pt})$ and we have $F(1)(-) = -\otimes\mathcal{O}_{\mathbb{P}V}$. Here $\mathcal{O}_{\mathbb{P}V}$ is the skyscraper sheaf on the zero section of the bundle $X_+^{(1)}$, where we use the description of $X_+^{(1)}$ given in Remark 1.3. We then have that $T_{F(1)} = T_{\mathcal{O}_{\mathbb{P}V}}$, the spherical twist about the spherical object $\mathcal{O}_{\mathbb{P}V}$ [**7**, Lemma 2.2].

(2) For $r \geq 2$, the twist $T_{F(r)}$ has a more complicated geometric description. Generically, it acts as a family spherical twist [**14**], but the behaviour on a certain closed locus is more elaborate. In the case $r=2$, this interesting locus is the zero section $\mathbb{G}\mathrm{r}(2,V)$ of $X_+^{(2)}$. See [**6**] for discussion.

Finally note that we obtain a description of all window-shift equivalences $\omega_{k,l}$ in our examples as a corollary of Theorem B: see Section 3.4 for details.

1.2. Outline.

- In Section 2, we explain some background, and give a brief discussion of the physical interpretation of our results.
- We construct our windows in Section 3, and indicate how they yield derived equivalences between GIT quotients (Theorem A).
- In Section 4, we give a compact guide to the *twisted Lascoux resolutions* which are used in our proofs to produce long exact sequences of bundles on Grassmannians: these may also be of independent interest.
- In Section 5, we explain the links between window-shift autoequivalences and twists (Theorem B).
- In Appendix A, we give explicit examples of the actions of window-shift autoequivalences.

ACKNOWLEDGEMENTS. This note is based on a joint paper with Ed Segal [**8**], which grew out of my thesis project suggested by Richard Thomas: their ideas and influence appear throughout, and this work would not have been possible without

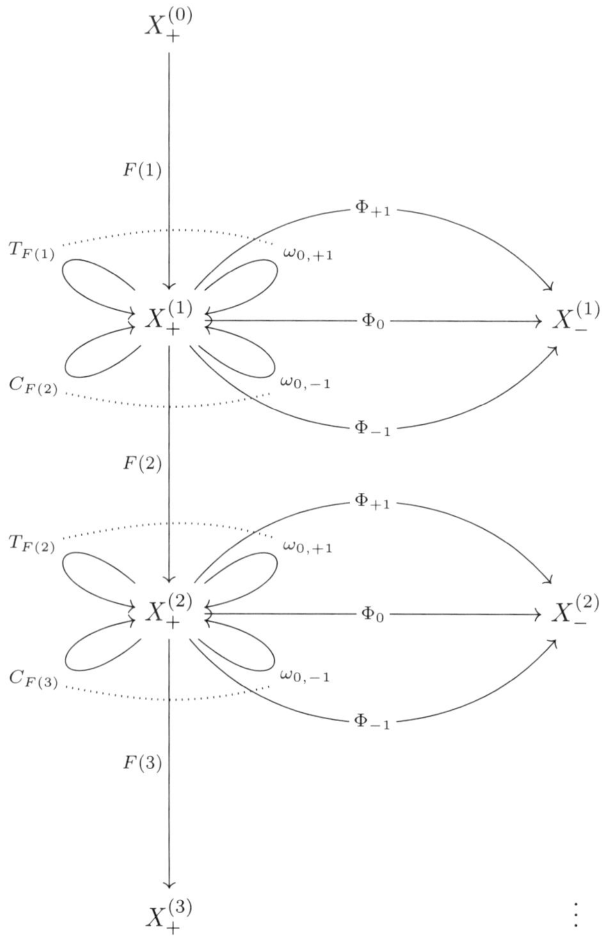

FIGURE 1. Schematic of the functors involved in Theorem B, with arrows denoting functors between the derived categories of the respective spaces $X_\pm^{(r)}$. Autoequivalences of the $\mathcal{D}(X_+^{(r)})$ which coincide (up to a shift) are connected by dotted lines.

their generous support. I also gratefully acknowledge the financial support of EPSRC during my doctoral work, and subsequently via grant EP/G007632 held in Edinburgh with Iain Gordon. Finally, I wish to thank the organisers of String-Math 2012 for an enjoyable and stimulating conference.

NOTATION. We write:

- $\mathbb{P}V$ for the projective space of lines in a vector space V, and $\mathbb{P}^\vee V$ for the dual space of 1-dimensional quotients;
- $\mathbb{Gr}(r, V)$ for the Grassmannian of r-dimensional subspaces of V, and $\mathbb{Gr}(V, r)$ for the dual Grassmannian of r-dimensional quotients;
- $\mathcal{D}(X)$ for the bounded derived category of coherent sheaves on a variety X;

- δ for a Young diagram, or the corresponding sequence of integers (see Definition 3.1);
- U^δ for a Schur power of the vector space, or bundle, U (see Definition 3.3);
- $\mathrm{Hom}_{\hookrightarrow}$ and $\mathrm{Hom}_{\twoheadrightarrow}$ for injective and surjective maps respectively.

2. Background

2.1. Spherical functors. Following [**1**, **2**] we have:

DEFINITION 2.1. Taking an integral functor $F\colon \mathcal{D}(Z) \to \mathcal{D}(X)$ with right adjoint R we have:

(1) a **twist** functor $T_F\colon \mathcal{D}(X) \to \mathcal{D}(X)$ defined so that

$$T_F(\mathcal{E}) := \mathrm{Cone}\,(FR(\mathcal{E}) \longrightarrow \mathcal{E})\,;$$

(2) a **cotwist** functor $C_F\colon \mathcal{D}(Z) \to \mathcal{D}(Z)$ such that

$$C_F(\mathcal{E}) := \mathrm{Cone}\,(\mathcal{E} \longrightarrow RF(\mathcal{E}))\,.$$

The morphisms here are provided by the (co)unit of the adjunction.

In all our examples, X and Z are Calabi–Yau (see [**7**, Section 3.2]). Under this assumption, T_F is an equivalence precisely when C_F is an equivalence. In this latter circumstance, we refer to F as *spherical.* Note that there exists a more general definition which relaxes the Calabi–Yau assumptions: see [**1**].

2.2. Related work concerning GIT. Halpern-Leistner [**12**] and Ballard, Favero, and Katzarkov [**4**], also employing ideas from [**11**] and [**19**], have developed a general theory of derived equivalences corresponding to certain variations of GIT: our equivalences ψ_k fit in this framework. Halpern-Leistner and Shipman [**13**] relate window-shift autoequivalences for general variations of GIT involving a single Hesselink stratum to twists of spherical functors. Our results for $r = 1$ follow from this general theory, so it is natural to ask whether the theory can be extended to cover the cases $r > 1$. We hope that this will be the subject of future work.

2.3. Links with gauged linear σ-models. As has been noted, the quotients $X_{\pm}^{(r)}$ should be thought of as corresponding to certain large-radius limits of a gauged linear σ-model determined by $\mathfrak{X}^{(r)}$, defined from the data of the underlying vector space of $\mathfrak{X}^{(r)}$ with its $\mathrm{GL}(r)$-action: see [**8**] for a more detailed account.

In physical language, our model has a $\mathrm{U}(r)$ gauge group coupled to d chiral multiplets Φ_i transforming in the fundamental representation $\mathbf{r}$, with another d fields transforming in the anti-fundamental representation $\bar{\mathbf{r}}$ (cf. [**15**, Section 2.1]): these latter fields ensure that the model is Calabi–Yau. Setting $d = 2$ and $r = 1$, we recover the brane transport analysis of [**11**, Section 8.4.1]. Note that each of our spherical functors $F(r)$ relates the categories of B-branes associated to a pair of GLSMs, with gauge groups $\mathrm{U}(r)$ and $\mathrm{U}(r + 1)$ respectively.

There has been some discussion [**3**, **13**, **16**] of how certain SKMS monodromies give rise to family spherical twists, or 'EZ twists' [**14**]. These twists may be described in the framework of Section 2.1: the associated cotwist is given by a composition of a shift, and tensoring by a line bundle. In our examples, by contrast, the cotwist is no longer of this form: for $r = 2$, it is itself a twist about a spherical object, whereas for $r > 2$ it is even more complicated [**8**]. It would be interesting to investigate how these exotic twists arise from monodromy. Furthermore, Theorem B suggests that, in our examples, SKMS monodromies yield both a twist and a *cotwist*: we do not know how this should be interpreted physically.

3. Windows

3.1. Schur powers.

NOTATION. We write $S^{(r)}$ for the tautological rank r bundle on the stack $\mathfrak{X}^{(r)}$, or simply S when the r is clear from context.

We describe how to obtain various other natural bundles on $\mathfrak{X}^{(r)}$ by the *Schur power* construction. These bundles will be used as generators for our windows. They are indexed by arrangements of boxes known as *Young diagrams*:

DEFINITION 3.1 (Young diagrams). Given a finite non-decreasing sequence $\delta = (\delta_1, \delta_2, \ldots, \delta_h)$ of non-negative integers δ_i there is a corresponding Young diagram. This is given by a stacked arrangement of boxes, with the sequence δ indicating the number of boxes contained in each successive layer (see Example 3.2 below).

NOTATION. We write $\mathrm{row}_i(\delta)$ and $\mathrm{col}_i(\delta)$ for the length of the ith row and i^{th} column of the diagram δ, respectively.

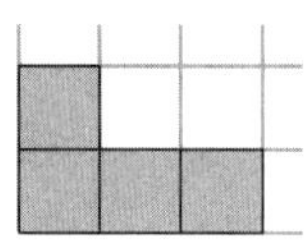

FIGURE 2. Young diagram $\delta = (3, 1)$.

EXAMPLE 3.2. The sequence $\delta = (3, 1)$ corresponds to the Young diagram in Figure 2. By definition, we have $\mathrm{row}_i(\delta) = \delta_i$. Observe also that

$$\mathrm{col}_i(\delta) = \begin{cases} 2 & i = 1, \\ 1 & i = 2, 3, \\ 0 & i \geq 4. \end{cases}$$

DEFINITION 3.3 (Schur powers). Given a vector space U, we write U^{δ} for the irreducible $\mathrm{GL}(U)$-representation with highest weight δ [**9**].

EXAMPLE 3.4. We have that

$$\begin{aligned} U^{(1,\ldots,1)} &\simeq \wedge^h U, \\ U^{(w)} &\simeq \mathrm{Sym}^w U. \end{aligned}$$

More general Schur powers have a more complicated description.

REMARK 3.5. Applying the same Schur power construction relative to a base, we may take Schur powers of *vector bundles* also. In particular, we obtain natural bundles $S^{\vee\delta}$ on the stacks $\mathfrak{X}^{(r)}$, which we use in the following Section 3.2.

3.2. Equivalences. We may now construct our windows:

DEFINITION 3.6 (windows). We define the window $\mathcal{W}_k$ for $k \in \mathbb{Z}$ to be the full subcategory of $\mathcal{D}(\mathfrak{X}^{(r)})$ split-generated by the following set of vector bundles:

$$W_k = \{ \ S^{\vee\delta} \otimes \det(S^\vee)^{\otimes k} \ | \ \mathrm{row}_\bullet(\delta) \leq d - r, \ \mathrm{col}_\bullet(\delta) \leq r \ \}.$$

REMARK 3.7. We choose the same representations as in Kapranov's exceptional collection for the Grassmannian [**17**]. As a consequence we find [**7**, Appendix C] that the restrictions of the bundles W_k, to either X_+ or X_-, sum to give a *tilting bundle*. This is the crucial point in the proof of the following lemma, which we omit:

LEMMA 3.8. [**8**, Proposition 3.6] *Writing $i_\pm$ for the inclusions $X_\pm \hookrightarrow \mathfrak{X}$, the (derived) restriction functors $\mathbb{L}i_\pm^*$ give equivalences*

$$\mathbb{L}i_\pm^* \colon \mathcal{W}_k \xrightarrow{\sim} \mathcal{D}(X_\pm).$$

Using this lemma we have a commutative diagram as follows, which defines the required equivalences ψ_k:

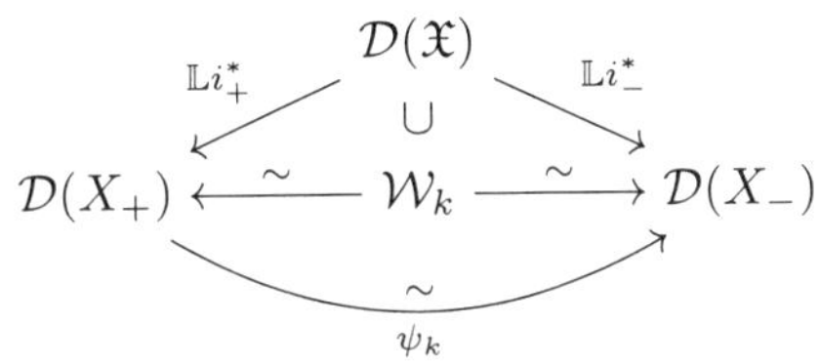

REMARK 3.9. The equivalences ψ_k are obtained in [**5**, Section 5] using a different method, which is characteristic-free.

3.3. Examples. It is an exercise [**9**] that if we take δ of length r then in fact

$$S^{\vee\delta} \otimes \det(S^\vee) \simeq S^{\vee\delta'}$$

where δ' is obtained by incrementing each element of the integer sequence δ. We deduce that windows $\mathcal{W}_k$ with consecutive k have a substantial overlap. Examples are as follows:

EXAMPLE 3.10. We illustrate the windows $\mathcal{W}_{+1}$ and $\mathcal{W}_0$ for $r = 1$ in Figure 3. Note that in this case $S = l$, say, a line bundle. We hence have simply that $S^{\vee(d)} = l^{\vee d}$, and so we recover the Beilinson tilting bundles on $X_\pm^{(1)}$.

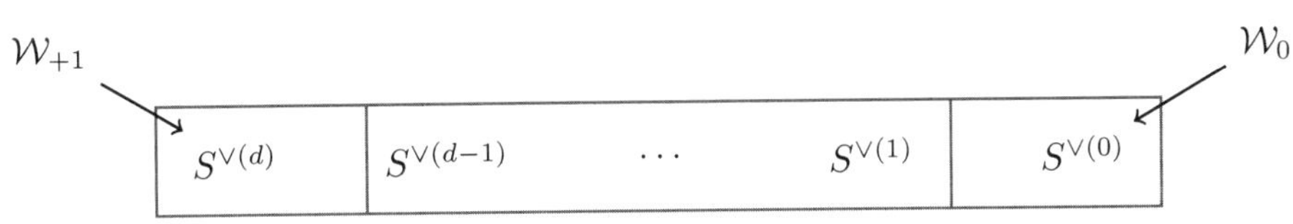

FIGURE 3. Windows for $r = 1$.

EXAMPLE 3.11. We give the windows $\mathcal{W}_{+1}$ and $\mathcal{W}_0$ for the Grassmannian example $d = 4$, $r = 2$, in Figure 4.

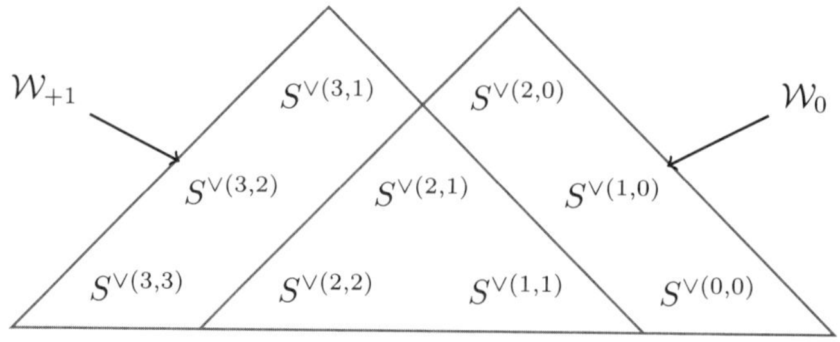

FIGURE 4. Windows for $d = 4$, $r = 2$.

3.4. Relations between window shifts. The following lemma, which follows immediately from the definitions, may be used to express a general window-shift autoequivalence $\omega_{k,l}$ in terms of those described in Theorem B:

LEMMA 3.12.

$$\omega_{m,k} \circ \omega_{k,l} = \omega_{m,l}$$

$$\omega_{k+m,l+m} = (- \otimes \det(S^\vee)^{\otimes m}) \circ \omega_{k,l} \circ (- \otimes \det(S)^{\otimes m})$$

4. Twisted Lascoux complexes

We explain a case of a certain generalised Koszul resolution known as the *twisted Lascoux resolution* [**22**, Section 6.1]. Weyman gives this resolution implicitly in *loc. cit.*: we perform a Borel–Weil–Bott calculation which makes it explicit in [**8**, Theorem A.7]. We summarise the results here, and give examples.

REMARK 4.1. A very closely-related construction appears in [**10**] under the name of *staircase complexes*.

We begin by defining a stack

$$\mathcal{T} := \left[\frac{\operatorname{Hom}(V, S^{(r)})}{\operatorname{GL}(S^{(r)})} \right].$$

Note that the dual Grassmannian $\mathbb{G}\mathrm{r}(V, r)$ arises as a substack of $\mathcal{T}$ by restricting to the locus of surjective Homs. The following Theorem 4.2 resolves certain natural torsion sheaves (4.2), supported on the complement of this locus. Corollary 4.3 then produces associated exact sequences on the Grassmannian.

We consider then the composition morphism j given as follows

$$j\colon \left[\frac{\operatorname{Hom}(V, S^{(r-1)}) \;\oplus\; \operatorname{Hom}_{\hookrightarrow}(S^{(r-1)}, S^{(r)})}{\operatorname{GL}(S^{(r-1)}) \;\times\; \operatorname{GL}(S^{(r)})} \right] \longrightarrow \mathcal{T},$$

and the direct images of certain Schur powers $(S^{(r-1)\vee})^\delta$ under j. We have:

THEOREM 4.2. [**8**, Theorem A.7] *Let δ be a Young diagram with*

$$\operatorname{row}_\bullet(\delta) \leq d - r + 1, \quad \operatorname{col}_\bullet(\delta) < r. \tag{4.1}$$

Then

$$j_*\big((S^{(r-1)\vee})^\delta\big) \tag{4.2}$$

may be resolved by a complex $\mathcal{E}_\bullet$ of bundles where

$$\mathcal{E}_0 := (S^{(r)\vee})^{\delta},$$
$$\mathcal{E}_k := (S^{(r)\vee})^{\delta_k} \otimes \wedge^{s_k} V.$$

The Young diagrams δ_k are defined, for $1 \leq k \leq d-r+1$, such that

$$\text{row}_i(\delta_k) = \begin{cases} \text{row}_i(\delta) & 1 \leq i < \text{col}_k(\delta)+1 \\ k & i = \text{col}_k(\delta)+1 \\ \text{row}_{i-1}(\delta)+1 & \text{col}_k(\delta)+1 < i \leq r \end{cases} \tag{4.3}$$

and $s_k := r + k - (\text{col}_k(\delta)+1)$.

The following corollary then yields exact sequences on $\mathbb{Gr}(V,r)$:

Corollary 4.3. *Take δ a Young diagram with shape as above in (4.1). Then we have an exact sequence on $\mathbb{Gr}(V,r)$, as follows:*

$$0 \longrightarrow S^{\vee\delta_K} \otimes \wedge^{s_K} V \longrightarrow \ \dots \ \longrightarrow S^{\vee\delta_1} \otimes \wedge^{s_1} V \longrightarrow S^{\vee\delta} \longrightarrow 0,$$

where

- *the δ_k are defined as previously by (4.3),*
- *$s_k := r + k - (\text{col}_k(\delta)+1)$ as above,*
- *$K := d - r + 1$, and*
- *S denotes the tautological quotient bundle on $\mathbb{Gr}(V,r)$.*

Proof. This follows immediately, as the support of any object in $\text{Im}(j_*)$, and in particular the sheaf (4.2), lies entirely in the locus which is removed from $\mathcal{T}$ to give the Grassmannian $\mathbb{Gr}(V,r)$. □

Example 4.4. We illustrate in Figure 5 the Young diagrams δ_k appearing in Theorem 4.2 and Corollary 4.3 above, as defined by (4.3), taking $\delta = (3,1)$, and $d = 7$, $r = 3$.

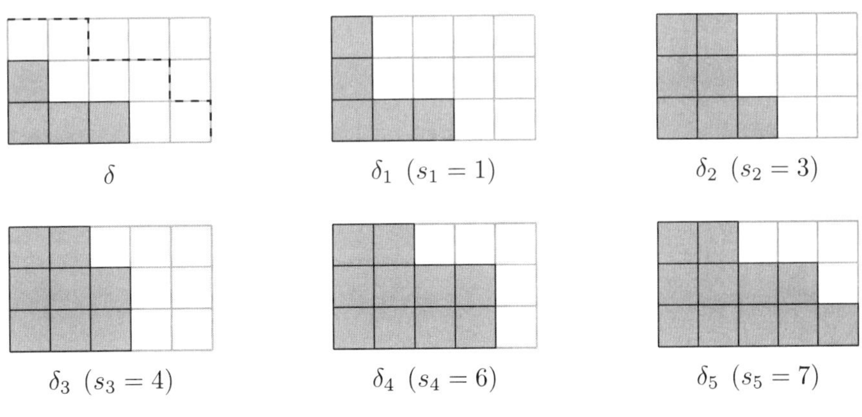

Figure 5. Young diagrams δ_k for $\delta = (3,1)$, with $d = 7$, $r = 3$. The dashed line, in the diagram for δ, indicates where boxes are added to produce each δ_k.

EXAMPLE 4.5. Taking $\delta = (0)$ with $r = 1$ we immediately see that $\delta_k = (k)$. Corollary 4.3 then recovers the following well-known exact sequences on the projective space of quotients $\mathbb{P}^\vee V$, where l denotes the tautological quotient bundle:

$$0 \longrightarrow l^{\vee d} \otimes \wedge^d V \longrightarrow \ \dots \ \longrightarrow l^\vee \otimes V \longrightarrow \mathcal{O} \longrightarrow 0$$

EXAMPLE 4.6. Taking $d = 4$ and $r = 2$, Corollary 4.3 gives exact sequences

$$0 \longrightarrow S^{\vee(3,1)} \otimes \wedge^4 V \longrightarrow S^{\vee(2,1)} \otimes \wedge^3 V \longrightarrow S^{\vee(1,1)} \otimes \wedge^2 V \longrightarrow S^{\vee(0,0)} \longrightarrow 0$$

$$0 \longrightarrow S^{\vee(3,2)} \otimes \wedge^4 V \longrightarrow S^{\vee(2,2)} \otimes \wedge^3 V \longrightarrow S^{\vee(1,1)} \otimes V \longrightarrow S^{\vee(1,0)} \longrightarrow 0$$

$$0 \longrightarrow S^{\vee(3,3)} \otimes \wedge^4 V \longrightarrow S^{\vee(2,2)} \otimes \wedge^2 V \longrightarrow S^{\vee(2,1)} \otimes V \longrightarrow S^{\vee(2,0)} \longrightarrow 0$$

with $\delta = (0,0)$, $(1,0)$ and $(2,0)$ respectively. Compare [**8**, Examples A.8, A.9].

REMARK 4.7. Notice that each of these sequences lies in two consecutive windows: all but the last term lie in $\mathcal{W}_{+1}$, and all but the first lie in $\mathcal{W}_0$. This observation will be key in what follows.

5. Windows and twists

5.1. Transfer functors.

DEFINITION 5.1 (transfer functor). A functor $\Phi \in \operatorname{End}(\mathcal{D}(\mathcal{X}))$ which

(1) restricts to a functor $\mathcal{W}_k \to \mathcal{W}_l$, and
(2) such that the following diagram commutes

$$\begin{array}{ccccc} \mathcal{D}(X^+) & \xleftarrow[\sim]{\mathbb{L}i_+^*} & \mathcal{W}_k & \xrightarrow[\sim]{\mathbb{L}i_-^*} & \mathcal{D}(X^-) \\ \downarrow{\scriptstyle\phi} & & \downarrow{\scriptstyle\Phi} & & \| \\ \mathcal{D}(X^+) & \xleftarrow{\mathbb{L}i_+^*} & \mathcal{D}(\mathcal{X}) & \xrightarrow{\mathbb{L}i_-^*} & \mathcal{D}(X^-) \end{array} \tag{5.1}$$

for some endofunctor $\phi \in \operatorname{End}(\mathcal{D}(X^+))$,

is referred to as a *transfer functor* for ϕ.

We think of Φ as transferring from window $\mathcal{W}_k$ to $\mathcal{W}_l$ in a manner compatible with the endofunctor ϕ on $\mathcal{D}(X^+)$. The following proposition then allows us to prove Theorem B, by constructing suitable transfer functors in our examples:

PROPOSITION 5.2. *Given a transfer functor Φ for ϕ as in Definition* 5.1 *above, we have*

$$\omega_{l,k} = \phi.$$

PROOF. This follows formally. See [**8**, Proposition 2.2]. □

5.2. Hecke correspondences. We now introduce natural correspondences which may be used to construct transfer functors in our examples. We put:

$$\mathcal{Z} := \left[\frac{\operatorname{Hom}(S^{(r)}, V) \ \oplus \ \operatorname{Hom}(V, S^{(r-1)}) \ \oplus \ \operatorname{Hom}_{\hookrightarrow}(S^{(r-1)}, S^{(r)})}{\operatorname{GL}(S^{(r-1)}) \ \times \ \operatorname{GL}(S^{(r)})} \right].$$

Observe that $\mathcal{Z}$ is equipped with natural morphisms to $\mathfrak{X}^{(r-1)}$ and $\mathfrak{X}^{(r)}$, given by composition of linear maps, which we denote as follows:

$$\mathfrak{X}^{(r-1)} \xleftarrow{\pi} \mathcal{Z} \xrightarrow{j} \mathfrak{X}^{(r)}.$$

By passing through the correspondence $\mathcal{Z}$, we then obtain a functor

$$\mathcal{F}(r)\colon \mathcal{D}(\mathfrak{X}^{(r-1)}) \longrightarrow \mathcal{D}(\mathfrak{X}^{(r)}),$$

defined by $\mathcal{F}(r) := \mathbb{R}j_* \circ \mathbb{L}\pi^*$.

If we replace $\operatorname{Hom}(S^{(r)}, V)$ in the definition of $\mathcal{Z}$ with $\operatorname{Hom}_{\hookrightarrow}(S^{(r)}, V)$, then we obtain a space, say Z_+, now with natural morphisms to $X_+^{(r-1)}$ and $X_+^{(r)}$. By passing through the correspondence Z_+, we obtain a functor

$$F(r)\colon \mathcal{D}(X_+^{(r-1)}) \longrightarrow \mathcal{D}(X_+^{(r)}).$$

Fixing now a particular r, and dropping this from the notation for simplicity, we have:

PROPOSITION 5.3. [**8**, Section 3.2.2] *$T_{\mathcal{F}}$ is a transfer functor for T_F mapping window $\mathcal{W}_{+1}$ to $\mathcal{W}_0$.*

We say a few words about the proof of Proposition 5.3 here:

(1) The statement that $T_{\mathcal{F}}$ maps window $\mathcal{W}_{+1}$ to $\mathcal{W}_0$ is [**8**, Lemma 3.16]. The core of the proof is that for a generator $\mathcal{E} \in W_{+1}$ (as in Example 3.11) we have

$$\mathcal{F}\mathcal{R}(\mathcal{E}) = \begin{cases} 0 & \mathcal{E} \in W_0, \\ \{\mathcal{E} \longrightarrow \ldots\} & \mathcal{E} \notin W_0, \end{cases}$$

where the omitted part of the complex is given by elements of W_0 (tensored by certain fixed vector spaces), and the generator $\mathcal{E}$ appears in homological degree 0. This all follows using the Lascoux resolutions of Section 4. We hence deduce that $T_{\mathcal{F}}(\mathcal{E}) \in \mathcal{W}_0$ as required.

(2) The commutativity of the left-hand square of (5.1) is [**8**, Lemma 3.17]: it follows from the formal similarity between $T_{\mathcal{F}}$ and T_F, and certain vanishing results. The commutativity of the right-hand square follows immediately from considering the support of objects in $\operatorname{Im}(\mathcal{F})$: by construction, they lie entirely in the locus which we remove to yield X_-.

REMARK 5.4. The cotwist proof is similar, though somewhat more delicate: see [**8**, Section 3.2.3].

Appendix A. Window shift examples

Finally, we give some examples of the action of window-shift autoequivalences, consistent with the twist action described in the previous Section 5. Complexes surrounded by braces denote objects of the derived category, with the left-most term in homological degree 0.

EXAMPLE A.1 ($r = 1$). Example 3.10 gives that the window $\mathcal{W}_{+1}$ has generators $\mathcal{E} = l^{\vee k}$ for $1 \leq k \leq d$. From the definitions we have that $\psi_{+1}\left(l^{\vee k}\right) = l^{\vee k}$. We thence obtain that $\omega_{0,+1}\left(l^{\vee k}\right) = l^{\vee k}$ for $k = 1, \ldots, d-1$, and that

$$\omega_{0,+1}\left(l^{\vee d}\right) = \{\ l^{\vee d-1} \otimes \wedge^{d-1} V \longrightarrow\ \ldots\ \longrightarrow l^{\vee} \otimes V \longrightarrow \mathcal{O}\ \}.$$

The only non-trivial part of this calculation is to use the exact sequence of Example 4.5 on X^- (we have to pull it up from $\mathbb{P}^\vee V$, and choose an isomorphism $\det V \simeq \mathbb{C}$) to express $l^{\vee d}$ as a complex of vector bundles in the window $\mathcal{W}_0$. See [**8**, Section 2.1.1] for the case $d = 2$.

EXAMPLE A.2 ($d = 4$, $r = 2$). Example 3.11 gives generators $\mathcal{E} = S^{\vee\delta}$ for the window $\mathcal{W}_{+1}$ with certain δ. For these δ we have $\omega_{0,+1}\left(S^{\vee\delta}\right) = S^{\vee\delta}$ for $\delta = (1,1)$, $(2,2)$, $(2,1)$, and

$$\omega_{0,+1}\left(S^{\vee\delta}\right) = \begin{cases} \{\ S^{\vee(2,2)} \otimes \wedge^2 V \longrightarrow S^{\vee(2,1)} \otimes V \longrightarrow S^{\vee(2,0)}\ \} & \delta = (3,3), \\ \{\ S^{\vee(2,2)} \otimes \wedge^3 V \longrightarrow S^{\vee(1,1)} \otimes V \longrightarrow S^{\vee(1,0)}\ \} & \delta = (3,2), \\ \{\ S^{\vee(2,1)} \otimes \wedge^3 V \longrightarrow S^{\vee(1,1)} \otimes \wedge^2 V \longrightarrow S^{\vee(0,0)}\ \} & \delta = (3,1). \end{cases}$$

The calculation proceeds as above, this time using the exact sequences in Example 4.6. Similar calculations (with a twist) may be found in [**8**, Section 2.1.2].

References

[1] R. Anno and T. Logvinenko. Orthogonally spherical objects and spherical fibrations. arXiv:1011.0707.

[2] R. Anno and T. Logvinenko. Spherical DG-functors. arXiv:1309.5035.

[3] P. S. Aspinwall, R. P. Horja, and R. L. Karp, *Massless D-branes on Calabi-Yau threefolds and monodromy*, Comm. Math. Phys. **259** (2005), no. 1, 45–69, DOI 10.1007/s00220-005-1378-6. MR2169967 (2006i:14012)

[4] Matthew Ballard, David Favero, and Ludmil Katzarkov. Variation of geometric invariant theory quotients and derived categories. arXiv:1203.6643.

[5] R. -O. Buchweitz, G. J. Leuschke, and M. Van den Bergh. Non-commutative desingularization of determinantal varieties, II: Arbitrary minors. http://arxiv.org/abs/1106.1833

[6] Will Donovan. *Grassmannian twists on derived categories of coherent sheaves.* PhD thesis, Imperial College London, July 2011.

[7] W. Donovan, *Grassmannian twists on the derived category via spherical functors*, Proc. Lond. Math. Soc. (3) **107** (2013), no. 5, 1053–1090, DOI 10.1112/plms/pdt008. MR3126391

[8] W. Donovan and E. Segal, *Window shifts, flop equivalences and Grassmannian twists*, Compos. Math. **150** (2014), no. 6, 942–978, DOI 10.1112/S0010437X13007641. MR3223878

[9] W. Fulton and J. Harris, *Representation theory: A first course*, Graduate Texts in Mathematics, vol. 129, Springer-Verlag, New York, 1991. MR1153249 (93a:20069)

[10] A. V. Fonarëv, *Minimal Lefschetz decompositions of the derived categories for Grassmannians* (Russian, with Russian summary), Izv. Ross. Akad. Nauk Ser. Mat. **77** (2013), no. 5, 203–224; English transl., Izv. Math. **77** (2013), no. 5, 1044–1065. MR3137200

[11] M. Herbst, K. Hori, and D. Page. Phases Of $\mathcal{N} = 2$ Theories In $1 + 1$ Dimensions With Boundary. arXiv:0803.2045.

[12] D. Halpern-Leistner, *The derived category of a GIT quotient*, J. Amer. Math. Soc. **28** (2015), no. 3, 871–912, DOI 10.1090/S0894-0347-2014-00815-8. MR3327537

[13] D. Halpern-Leistner and I. Shipman. Autoequivalences of derived categories via geometric invariant theory. arXiv:1303.5531.

[14] R. P. Horja, *Derived category automorphisms from mirror symmetry*, Duke Math. J. **127** (2005), no. 1, 1–34, DOI 10.1215/S0012-7094-04-12711-3. MR2126495 (2006a:14023)

[15] K. Hori and D. Tong, *Aspects of non-abelian gauge dynamics in two-dimensional $\mathcal{N} = (2,2)$ theories*, J. High Energy Phys. **5** (2007), 079, 41 pp. (electronic), DOI 10.1088/1126-6708/2007/05/079. MR2318130 (2009d:81351)

[16] M. Herbst and J. Walcher, *On the unipotence of autoequivalences of toric complete intersection Calabi-Yau categories*, Math. Ann. **353** (2012), no. 3, 783–802, DOI 10.1007/s00208-011-0704-x. MR2923950

[17] M. M. Kapranov, *On the derived categories of coherent sheaves on some homogeneous spaces*, Invent. Math. **92** (1988), no. 3, 479–508, DOI 10.1007/BF01393744. MR939472 (89g:18018)

[18] Y. Kawamata, *D-equivalence and K-equivalence*, J. Differential Geom. **61** (2002), no. 1, 147–171. MR1949787 (2004m:14025)
[19] E. Segal, *Equivalence between GIT quotients of Landau-Ginzburg B-models*, Comm. Math. Phys. **304** (2011), no. 2, 411–432, DOI 10.1007/s00220-011-1232-y. MR2795327 (2012f:81265)
[20] P. Seidel and R. Thomas, *Braid group actions on derived categories of coherent sheaves*, Duke Math. J. **108** (2001), no. 1, 37–108, DOI 10.1215/S0012-7094-01-10812-0. MR1831820 (2002e:14030)
[21] R. P. Thomas, *An exercise in mirror symmetry*, Proceedings of the International Congress of Mathematicians. Volume II, Hindustan Book Agency, New Delhi, 2010, pp. 624–651. MR2827812 (2012m:53197)
[22] J. Weyman, *Cohomology of vector bundles and syzygies*, Cambridge Tracts in Mathematics, vol. 149, Cambridge University Press, Cambridge, 2003. MR1988690 (2004d:13020)

The Maxwell Institute, School of Mathematics, University of Edinburgh, Edinburgh, EH9 3JZ, United Kingdom
E-mail address: Will.Donovan@ed.ac.uk

Proceedings of Symposia in Pure Mathematics
Volume **90**, 2015
http://dx.doi.org/10.1090/pspum/090/01526

Perturbative Terms of Kac-Moody-Eisenstein Series

Philipp Fleig and Axel Kleinschmidt

ABSTRACT. Supersymmetric theories of gravity can exhibit surprising hidden symmetries when considered on manifolds that include a torus. When the torus is of large dimension these symmetries can become infinite-dimensional and of Kac-Moody type. When taking quantum effects into account the symmetries become discrete and invariant functions under these symmetries should play an important role in quantum gravity. The new results here concern surprising simplifications in the constant terms of very particular Eisenstein series on these Kac-Moody groups. These are exactly the cases that are expected to arise in string theory.

1. Introduction

In string theory, discrete dualities have played a central role in the research of the last 15 years. These dualities can relate string theories on different backgrounds and with different matter content and are commonly referred to as U-dualities [**HT95**, **OP99**]. Their existence has led to the claim that there is a single M-theory underlying all string theories [**Wit95**].

A particular manifestation of this idea is given by type II superstring theory compactified on a $(d-1)$-dimensional torus, down to $D = 11 - d$ space-time dimensions. At low energies, the complete effective theory is maximal supergravity in D dimensions and possesses a continuous $E_{d(d)}(\mathbb{R})$ hidden symmetry group [**CJ78**, **Jul80**]. These are maximally split Lie groups that for $6 \leq d \leq 8$ are exceptional and for the other values of d are defined for our purposes in Table 1. We will write $E_d(\mathbb{R})$ instead of $E_{d(d)}(\mathbb{R})$ for ease of notation. For $d > 8$, the groups are infinite-dimensional and of Kac-Moody type. In general, the groups can be thought of as arising as the closure of the area preserving diffeomorphisms $SL(d, \mathbb{R})$ of the M-theory torus and the classical $SO(d-1, d-1, \mathbb{R})$ symmetry realising continuous T-duality [**OP99**]. The possible compactifications are labelled by the classical moduli space $\mathcal{M}_D^{(\mathrm{cl})} = E_d(\mathbb{R})/K(E_d(\mathbb{R}))$, where $K(E_d(\mathbb{R}))$ is the maximal compact subgroup of $E_d(\mathbb{R})$.

In string theory, these continuous symmetries are expected to be broken to the discrete $E_d(\mathbb{Z})$ U-duality group, via a Dirac-Schwinger-Zwanziger type quantisation condition related to the existence of charged states (branes) [**HT95**, **FILQ90**]. Table 1 shows a complete list of the U-duality groups. The effect of these discrete dualities is to identify classically inequivalent compactifications: The quantum moduli

2010 *Mathematics Subject Classification.* Primary 81T30, 83E30, 11F55, 11F03.

TABLE 1. List of the split real forms of the hidden symmetry groups $E_{d(d)}(\mathbb{R})$. We also list the corresponding maximal compact subgroups K and the last column contains the discrete U-duality versions that appear in string theory. The label $10B$ indicates that we are considering type IIB in ten dimensions rather than type IIA.

D	$E_{d+1}(\mathbb{R})$	$K(E_{d+1})$	$E_{d+1}(\mathbb{Z})$
$10B$	$SL(2,\mathbb{R})$	$SO(2)$	$SL(2,\mathbb{Z})$
9	$\mathbb{R}^+ \times SL(2,\mathbb{R})$	$SO(2)$	$SL(2,\mathbb{Z})$
8	$SL(2,\mathbb{R}) \times SL(3,\mathbb{R})$	$SO(3) \times SO(2)$	$SL(2,\mathbb{Z}) \times SL(3,\mathbb{Z})$
7	$SL(5,\mathbb{R})$	$SO(5)$	$SL(5,\mathbb{Z})$
6	$SO(5,5,\mathbb{R})$	$SO(5) \times SO(5)$	$SO(5,5,\mathbb{Z})$
5	$E_6(\mathbb{R})$	$USp(8)$	$E_6(\mathbb{Z})$
4	$E_7(\mathbb{R})$	$SU(8)/\mathbb{Z}_2$	$E_7(\mathbb{Z})$
3	$E_8(\mathbb{R})$	$Spin(16)/\mathbb{Z}_2$	$E_8(\mathbb{Z})$
2	$E_9(\mathbb{R})$	$K(E_9(\mathbb{R}))$	$E_9(\mathbb{Z})$
1	$E_{10}(\mathbb{R})$	$K(E_{10}(\mathbb{R}))$	$E_{10}(\mathbb{Z})$
0	$E_{11}(\mathbb{R})$	$K(E_{11}(\mathbb{R}))$	$E_{11}(\mathbb{Z})$

space of string compactifications to D dimensions is given by $(D = 11 - d)$

$$\mathcal{M}_D = E_d(\mathbb{Z})\backslash E_d(\mathbb{R})/K(E_d(\mathbb{R})). \tag{1}$$

A much studied example where U-duality is explicitly manifest is the type IIB superstring scattering amplitude of the four-graviton scattering process, see e.g. [**GG97, GRV10, GMRV10**]. More precisely, the amplitude of this process in D dimensions displays an invariance under the respective U-duality group discussed above. Instead of looking directly at the amplitude, one may also consider the corresponding low-energy effective action, where it is found that there is an infinite number of higher-order curvature corrections beyond the Einstein-Hilbert term, of the form

$$(\alpha')^{\frac{2-D}{2}} \sum_k \int \mathrm{d}^D x \, (\alpha')^{k+3} \mathcal{E}^D_{(p,q)}(\Phi) \nabla^{2k} R^4 \,. \tag{2}$$

These corrections constitute an expansion in orders of the Regge slope α' (of dimension (length)^2) and R^4 is given by a specific contraction of four Riemann tensors [**GW86**]. The first few terms in this expansion beyond the Einstein-Hilbert term occur for $k = 2p + 3q = 0, 2, 3, 4, \ldots$. The couplings $\mathcal{E}^D_{(p,q)}(\Phi)$ of these terms are functions of the moduli $\Phi \in \mathcal{M}_D$ of the classical moduli space (1).

The preservation of U-duality by the corrections in (2) puts strong constraints on the functions $\mathcal{E}^D_{(p,q)}$. Since the graviton is invariant under U-duality (in Einstein frame), each function $\mathcal{E}^D_{(p,q)}$ has to be a function on $E_d(\mathbb{R})/K(E_d(\mathbb{R}))$ that is invariant under $E_d(\mathbb{Z})$. Furthermore, consistency with string perturbation theory requires that $\mathcal{E}^D_{(p,q)}$ must have a well-defined 'weak coupling' expansion near the weak coupling cusp on $\mathcal{M}_D$. Finally, supersymmetry imposes differential equations on $\mathcal{E}^D_{(p,q)}$ [**GS99, GRV10**]; for the lowest order corrections $k = 0, 2$ these differential equations are homogeneous Laplace eigenvalue equations. Altogether, this means that $\mathcal{E}^D_{(p,q)}$ should be an $E_d(\mathbb{Z})$ automorphic function on $\mathcal{M}_D$.

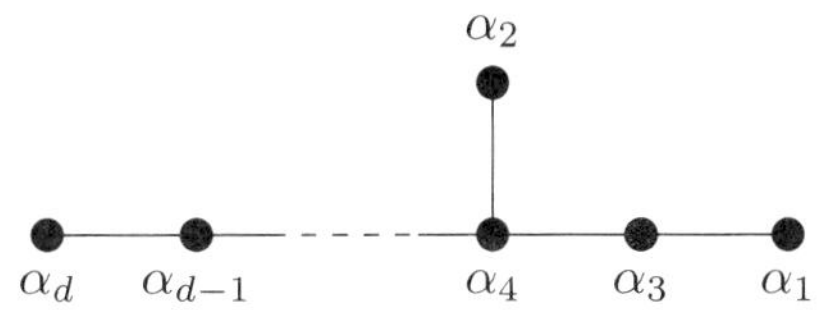

FIGURE 1. Dynkin diagram for E_d.

In some cases, these constraints are actually strong enough to identify $\mathcal{E}^D_{(p,q)}$ uniquely [**GG97**, **Pio98**, **GRV10**] as a non-holomorphic Eisenstein series and there is good evidence that for $k = 0, 2$ (corresponding to the so-called 1/2-BPS and 1/4-BPS couplings R^4 and $\partial^4 R^4$) the solution in any dimension $D \geq 3$ is given by such an Eisenstein series [**GRV10**, **Pio10**, **GMRV10**, **GMV11**]. For even higher derivative terms with $k > 2$, the Laplace equation becomes inhomogeneous and the automorphic function that is required is unlikely to be an Eisenstein series in general.

The purpose of the paper [**FK12**] —on which this contribution is based— was to extend the analysis of [**GMRV10**] to $D < 3$ for the cases $k = 0, 2$. This involves generalising the notion of Eisenstein series to Kac-Moody groups since for $D < 3$ the hidden symmetries E_{11-D} are of Kac-Moody type, see Table 1. Pioneering work for Eisenstein series over loop groups was carried out by Garland [**Gar01**]. We find surprising simplifications for the Kac–Moody Eisenstein series for $k = 0, 2$ and also more general Eisenstein series as will be shown below. The Dynkin diagram of E_d with our labelling conventions is shown in Figure 1.

These notes are structured as follows. In section 2 we will first discuss a general definition of an Eisenstein series, which applies both to the finite- and infinite-dimensional groups. In order to make this abstract definition more transparent, we will derive from it the explicit form of the series over $SL(2, \mathbb{R})$. In section 3 we discuss Fourier expansions of Eisenstein series and provide Langlands' formula for the zero-mode terms of such an expansion, which again is valid for the finite and more general Kac-Moody groups. In the same section we will show that particular Eisenstein series exhibit drastic simplifications in the structure of the constant terms. We will argue that these are the cases that are relevant in string theory and review some consistency checks on our claims. Finally we provide a short outlook. More details on many of the points discussed here can be found in [**FK12**].

2. Eisenstein Series on Kac-Moody groups

Eisenstein series are functions defined on a non-compact, semisimple real Lie group G and display invariance under a discrete subgroup $G(\mathbb{Z})$ of G, see for example [**Lan76**]. The invariance property is achieved by defining the series as a sum over orbits of $G(\mathbb{Z})$, typically quotienting by the stabiliser of a cusp to avoid overcounting.

To make this more transparent, consider the (non-holomorphic) $SL(2, \mathbb{R})$ Eisenstein series invariant under the discrete group $SL(2, \mathbb{Z})$. This series is normally defined as a sum over integers c and d which are co-prime

$$E_s^{SL(2,\mathbb{Z})}(\tau) = \frac{1}{2} \sum_{\gcd(c,d)=1} \frac{\tau_2^s}{|c\tau + d|^{2s}}. \tag{3}$$

Here s is a complex parameter and the sum is restricted, such that $(c,d) \neq (0,0)$. The argument of the series is $\tau = \tau_1 + i\tau_2 = \chi + ie^{-\phi}$, which lives in the upper-half of the complex plane. The variables ϕ and χ parameterise the classical moduli $\mathcal{M}_{10}^{(\mathrm{cl})}$ of uncompactified type IIB string theory and are identified as the dilaton and the axion field. The function (3) is clearly non-holomorphic due the appearance of the modulus in the denominator. The group $SL(2,\mathbb{Z})$ acts in the standard fractional linear fashion on τ

$$\tau \mapsto \frac{a\tau+b}{c\tau+d} \quad \text{for} \quad \begin{pmatrix} a & b \\ c & d \end{pmatrix} \in SL(2,\mathbb{Z}). \tag{4}$$

In order to make the definition of the Eisenstein series over $SL(2)$ given by (3) more easily generalizable, we now write it as a sum over $SL(2,\mathbb{Z})$ orbits. This is possible by realising that the summand in (3) can be written as

$$\frac{\tau_2^s}{|c\tau+d|^{2s}} = [\mathrm{Im}(\gamma\cdot\tau)]^s \quad \text{for} \quad \gamma = \begin{pmatrix} a & b \\ c & d \end{pmatrix} \in SL(2,\mathbb{Z}). \tag{5}$$

The matrix γ in this equation is not uniquely defined by c and d. But by invoking the modularity condition $ad - bc = 1$ all possible solutions for a and b can be obtained from a particular solution (a_0, b_0) through

$$\begin{pmatrix} a & b \\ c & d \end{pmatrix} = \begin{pmatrix} 1 & m \\ 0 & 1 \end{pmatrix} \begin{pmatrix} a_0 & b_0 \\ c & d \end{pmatrix} = \begin{pmatrix} a_0 + mc & b_0 + md \\ c & d \end{pmatrix} \tag{6}$$

with $m \in \mathbb{Z}$. The shift matrices form part of the Borel subgroup $B(\mathbb{Z})$ of $SL(2,\mathbb{Z})$:

$$B(\mathbb{Z}) = \left\{ \begin{pmatrix} \pm 1 & m \\ 0 & \pm 1 \end{pmatrix} : m \in \mathbb{Z} \right\} \tag{7}$$

and these matrices leave $\mathrm{Im}(\tau) = \tau_2$ invariant. Therefore the Eisenstein series (3) can be written equivalently as a sum over a coset

$$E_s^{SL(2,\mathbb{Z})}(\tau) = \sum_{\gamma \in B(\mathbb{Z})\backslash SL(2,\mathbb{Z})} [\mathrm{Im}(\gamma\cdot\tau)]^s . \tag{8}$$

Finally, we notice that $\mathrm{Im}(\tau)$ corresponds to the projection onto the abelian torus of a group element $g \in SL(2,\mathbb{R})$ written in Iwasawa decomposition

$$g = nak = \begin{pmatrix} 1 & \tau_1 \\ 0 & 1 \end{pmatrix} \begin{pmatrix} \tau_2^{1/2} & 0 \\ 0 & \tau_2^{-1/2} \end{pmatrix} k \tag{9}$$

with $k \in SO(2,\mathbb{R})$. Defining the projection to the Cartan subalgebra

$$H(g) = \log(a) = \frac{1}{2}\log(\tau_2)\cdot h_1, \quad \text{where} \quad h_1 = \begin{pmatrix} 1 & 0 \\ 0 & -1 \end{pmatrix} \tag{10}$$

is the standard $SL(2,\mathbb{R})$ Cartan generator, we see that as another equivalent form of (3) one obtains

$$E_s^{SL(2,\mathbb{Z})}(\tau) = E^{SL(2,\mathbb{Z})}(\lambda, g) = \sum_{\gamma \in B(\mathbb{Z})\backslash SL(2,\mathbb{Z})} e^{\langle \lambda+\rho | H(\gamma g)\rangle}, \tag{11}$$

where $\lambda = 2s\Lambda_1 - \rho$ if Λ_1 denotes the unique fundamental weight of $\mathfrak{sl}(2,\mathbb{R})$ and ρ the Weyl vector (equal to Λ_1 here). The angled brackets represent the action of a weight on the Cartan subalgebra element, using $\langle \Lambda_1 | h_1 \rangle = 1$. The addition and subtraction of the Weyl vector might seem a bit awkward here but it turns out that this is convenient for the general theory.

Expression (11) is the form of the non-holomorphic Eisenstein series that lends itself to a straightforward generalisation to other groups. This generalised definition for groups of arbitrary rank is

$$E^G(\lambda, g) \equiv \sum_{\gamma \in B(\mathbb{Z})\backslash G(\mathbb{Z})} e^{\langle \lambda+\rho | H(\gamma g)\rangle} . \tag{12}$$

and was given for finite-dimensional G by Langlands in [**Lan76**]. We will only be interested in cases when the weight λ appearing in the definition is given by $\lambda = 2s\Lambda_{i_*} - \rho$ with Λ_{i_*} the fundamental weight of node i_* of the Dynkin diagram of G. In that case we denote the Eisenstein series by

$$E^G_{i_*;s}(g) \equiv E^G(\lambda, g) \quad \text{for} \quad \lambda = 2s\Lambda_{i_*} - \rho. \tag{13}$$

Often we will leave out the argument $g \in G$ as well. The function $E^G_{i_*;s}$ will be referred to as a maximal parabolic Eisenstein series.

We note that the Eisenstein series of (12), satisfies the Laplace eigenvalue equation

$$\Delta^{G/K} E^G(\lambda, g) = \frac{1}{2}\left(\langle\lambda|\lambda\rangle - \langle\rho|\rho\rangle\right) E^G(\lambda, g) , \tag{14}$$

where $\Delta^{G/K}$ is a Laplacian defined on the fields which parameterise the coset G/K.

The definition of the Eisenstein series as given in (12) also applies in the case when G is a general Kac-Moody group and works in particular for affine E_9 [**Gar01**], the hyperbolic Kac-Moody group E_{10} and the group E_{11} [**FK12**]. In $D = 2$ space-time dimensions the situation is a bit special, since the corresponding U-duality group E_9 is an affine Kac-Moody group. The algebra of such a (non-twisted) affine group is constructed from the algebra of the underlying finite-dimensional algebra $\mathfrak{g}$ as

$$\hat{\mathfrak{g}} = \mathfrak{g}[[t, t^{-1}]] \oplus c\mathbb{R} \oplus d\mathbb{R} , \tag{15}$$

where the first summand represents the loop algebra of $\mathfrak{g}$, the second summand is associated with the central element and the last summand is the derivation which is counting the affine level. The corresponding Cartan subalgebra $\hat{\mathfrak{a}}$ has dimension $\dim(\mathfrak{a}) + 2$. The definition of an Eisenstein series over affine groups has been worked out rigorously by Garland in [**Gar01**] and convergence of the series was proven for sufficiently large real parts of the weight defining the Eisenstein series. The definition of the affine Eisenstein series differs subtly from the one of (12), in that one has to include a parameter v in the exponential, which parameterises the group associated with the derivation d. For the purpose of the presentation here, we will largely ignore this special case and refer the reader to [**FK12**] where its details are treated.

Let us also mention that Eisenstein over infinite-dimensional groups similarly satisfy the eigenvalue equation (14). Subtleties arise again for the case of E_9, which are linked to the appearance of scale invariance of gravity in $D = 2$ space-time dimensions [**FK12**].

We are now in the position to state the proposed automorphic functions $\mathcal{E}^D_{(p,q)}$ that appear in the four-graviton scattering process as discussed in the introduction. For the terms R^4 and $\partial^4 R^4$ and $D \geq 3$ these functions are given by [**GMRV10**,

OP00, Pio10]

$$\mathcal{E}^D_{(0,0)} = 2\zeta(3)E^G_{1;3/2}, \quad \text{and} \quad \mathcal{E}^D_{(1,0)} = \zeta(5)E^G_{1;5/2}\,. \tag{16}$$

This proposal has passed a variety of checks in the references just given. In [**FK12**] we propose that these expressions are also correct for $D < 3$ when the symmetry group G becomes infinite-dimensional. In $D = 2$, the proposal has to be modified slightly to accommodate properly the derivation d [**FK12**].

3. Fourier expansions of Eisenstein series

The physical information of the automorphic functions $\mathcal{E}^D_{(p,q)}$ is encoded in their Fourier expansion. In the present work we are only interested in the zero-mode Fourier terms of the expansion. Mathematically, these correspond to the constant terms of the automorphic functions; physically, they represent the perturbative contributions to the scattering process. Although referred to as the constant term the expressions do depend on the Cartan subalgebra degrees of freedom contained in A of the Iwasawa decomposition of $G = NAK$. The constant term is obtained by integrating out the degrees of freedom contained in the unipotent radical N. There exists a formula for the constant term due to Langlands [**Lan76**] given by

$$\int\limits_{N(\mathbb{Z})\backslash N(\mathbb{R})} E^G(\lambda, ng)\mathrm{d}n = \sum_{w\in\mathcal{W}} M(w,\lambda)e^{\langle w\lambda+\rho|H(g)\rangle}\,, \tag{17}$$

where the sum over the Weyl group $\mathcal{W}$ of G is due to the Bruhat decomposition of G. The factor $M(w, \lambda)$ is defined as

$$M(w,\lambda) = \prod_{\substack{\alpha\in\Delta_+\\ w\alpha\in\Delta_-}} \frac{\xi\left(\langle\lambda|\alpha\rangle\right)}{\xi\left(1+\langle\lambda|\alpha\rangle\right)} = \prod_{\substack{\alpha\in\Delta_+\\ w\alpha\in\Delta_-}} c\left(\langle\lambda|\alpha\rangle\right)\,. \tag{18}$$

The function ξ is the completed Riemann zeta function and its relation with the Riemann ζ function is $\xi(k) \equiv \pi^{-k/2}\Gamma\left(\frac{k}{2}\right)\zeta(k)$. The sets $\Delta_\pm$ represent the positive/negative roots of the Lie algebra of G. We will analyse the structure of the factor $M(w, \lambda)$ in some more detail in a moment. In particular we will see that its properties are responsible for drastic simplifications in the constant term of the Eisenstein series of (16). We refer to the type of expansion given in (17) as a minimal parabolic expansion of the constant term.

As an example, let us consider the $SL(2, \mathbb{Z})$ Eisenstein series (11). The Weyl group has two elements and the application of the constant term formula (17) gives

$$\int_{N(\mathbb{Z})\backslash N(\mathbb{R})} E^{SL(2,\mathbb{Z})}_s(ng)\mathrm{d}n = \int_0^1 \mathrm{d}\tau_1 E^{SL(2,\mathbb{Z})}_s(\tau) = \tau_2^s + \frac{\xi(s)}{\xi(s+1)}\tau_2^{1-s}. \tag{19}$$

The two terms have a very precise interpretation from string scattering calculations. For $s = 3/2$, the first term corresponds to the string tree level contribution and the second one to the string one-loop result [**GS82**]. There are no further perturbative corrections beyond one-loop due to supersymmetry and the numerical coefficients of string theory agree perfectly with those of the Eisenstein series [**GG97**].

There is also a second type of expansion which is possible that is commonly referred to as a maximal parabolic expansion, where one integrates out the degrees of freedom of the unipotent factor $N_{j_\circ}$ in a particular maximal parabolic subgroup

$P_{j_\circ} = N_{j_\circ} M_{j_\circ}$. Here, $j_\circ$ labels a choice of a simple root, with respect to which the maximal parabolic subgroup is defined [**GMRV10**, **FK12**]. The factor $M_{j_\circ}$ is called the Levi factor in the decomposition of $P_{j_\circ}$. The Levi factor itself can be written as

$$M_{j_\circ} = GL(1) \times G_{d-1}\,, \tag{20}$$

where G_{d-1} is the group with the Dynkin diagram that is left after deleting the $j_\circ$th node from the diagram of E_d. The $GL(1)$ factor in this product is parameterised by a single scalar $r \in \mathbb{R}^\times$. Langlands' formula for the constant term in a maximal parabolic expansion then becomes [**MW95**]

$$\int\limits_{N_{P_{j_\circ}}(\mathbb{Z})\backslash N_{P_{j_\circ}}(\mathbb{R})} E^G(\lambda, ng)\mathrm{d}n = \sum_{w\in\mathcal{W}_{j_\circ}\backslash\mathcal{W}} M(w,\lambda)e^{\langle (w\lambda+\rho)_{\|j_\circ}|H(g)\rangle} E^{G_d}\left((w\lambda)_{\perp j_\circ}, g\right). \tag{21}$$

Here the notation $(\lambda)_{\|j_\circ}$ denotes a projection operator on the component of λ which is proportional to the fundamental weight $\Lambda_{j_\circ}$ and $(\lambda)_{\perp j_\circ}$ is orthogonal to $\Lambda_{j_\circ}$.

As for the definition of the Eisenstein series, Langlands' formulæ above also apply in the case of the infinite-dimensional groups E_{10} and E_{11}. For the affine case slight modifications have to be made again, in order to account for the derivation d [**Gar01**]. Let us also mention that in this case, the Levi factor $M_{j_\circ} = GL(1) \times GL(1) \times G_{d-1}$. There are now two $GL(1)$ factors instead of only one as in (20). One factor corresponds to the central element c and the other to the derivation d. Hence we now have an additional parameter $v \in \mathbb{R}^\times$ appearing in the expression for the constant term besides r.

It is easy to see from (17) that the number of terms that make up the constant term is bounded from above by the order of the Weyl group $\mathcal{W}$. In particular, in the case of finite-dimensional groups, where the order of the Weyl group is also finite, this number is always finite and generically equal to the order of the Weyl group. For the particular choices $s = 3/2$ and $s = 5/2$ in (16), however, the number reduces drastically [**GMRV10**] in such a way that only very few non-zero terms are left. This is due to the structure of the coefficient $M(w,\lambda)$ of (18) and physically related to the BPS-ness of the R^4 and $\partial^4 R^4$ terms as was studied in detail for $D \geq 3$ in [**GMRV10**] and related to minimal and next-to-minimal automorphic representations in [**GRS97**, **Pio10**, **GMV11**].

When considering the infinite-dimensional symmetry groups $E_d(\mathbb{R})$ for $d > 8$ ($D < 3$) the situation is much less clear. An application of the formula (17) would lead generically to an infinite number of constant terms since the order of the Weyl group for indefinite Kac-Moody algebras is infinite. The remarkable result of our work [**FK12**] was that for the special values $s = 3/2$ and $s = 5/2$ this generic number reduces to a finite number as required by physical arguments.

Let us now explain the mechanism for this simplification along with a practical implementation [**FK12**], see also [**GMRV10**]. The function $M(w,\lambda)$ of (18) is of central importance. It satisfies the multiplicative property

$$M(w\tilde{w},\lambda) = M(w,\tilde{w}(\lambda))M(\tilde{w},\lambda). \tag{22}$$

The function $c(k)$ that appears in the factors that contribute to $M(w, \lambda)$ has special values only at arguments $k = \pm 1$, namely

$$c(-1) = 0\,, \qquad c(+1) = \infty \tag{23}$$

and $c(-1)c(1) = 1$. This means that if, for a particular Weyl word $w \in \mathcal{W}$, the product giving $M(w, \lambda)$ contains more $c(-1)$ than $c(1)$ factors, then $M(w, \lambda)$ will be zero. In addition, one can show by the multiplicative property (22) that if $M(\tilde{w}, \lambda) = 0$ for some $\tilde{w}$, then any longer Weyl word of the form $w\tilde{w}$ will also lead to $M(w\tilde{w}, \lambda) = 0$. Moreover, it is easy to see that any w that stabilises $\lambda + \rho$ will lead to $M(w, \lambda) = 0$. Restricting to the particular case $\lambda = 2s\Lambda_{i_*} - \rho$, the sum over Weyl words reduces therefore at least to the subset

$$\mathcal{S}_{i_*} \equiv \{w \in \mathcal{W} | w\alpha_i > 0 \text{ for all } i \neq i_*\} = \mathcal{W}/\mathcal{W}_{i_*}, \tag{24}$$

where $\mathcal{W}_{i_*}$ is the stabiliser of Λ_{i_*}. The set $\mathcal{S}_{i_*}$ is in bijection with the Weyl orbit of Λ_{i_*} and this also gives a convenient way of enumerating the set in a partially ordered manner. This was shown in [**FK12**]. Let us emphasise again that the number of Weyl words in $\mathcal{S}_{i_*}$ is only a 'small fraction' of the order of the whole Weyl group $\mathcal{W}$. Of course, this number is still infinite in the Kac-Moody case.

Representing $\mathcal{S}_{i_*}$ as a partially ordered set corresponding to the Weyl orbit of Λ_{i_*} also allows us to exploit the full power of (22). We can picture the partially ordered set of Weyl words as a tree rooted at the identity Weyl word. By parsing through the partially ordered set of Weyl words and computing $M(w, \lambda)$ along all branches of the tree, we know by (22) that we can terminate the investigation of a given branch if we reach a vertex $\tilde{w}$ of the tree where $M(\tilde{w}, \lambda)$ vanishes. In order to determine whether this happens we analyse the factors that contribute to the product (18). For this purpose, we define two different sets of roots $\Delta_s(\pm 1)$; one set for all positive roots producing $c(-1)$ factors and the other for roots producing $c(+1)$ factors in the product

$$\Delta_s(\pm 1) := \{\alpha \text{ contributing to } M(w, \lambda) \;:\; \langle \lambda | \alpha \rangle = \langle 2s\Lambda_{i_*} - \rho | \alpha \rangle = \pm 1\}\,. \tag{25}$$

These sets are well-defined as there are only finitely many α contributing to $M(w, \lambda)$ for a given $w \in \mathcal{W}$. By working out how many roots from each of the two sets will contribute in the product defining $M(w, \lambda)$, one can see that for specific choices of s and i_*, only a finite number of Weyl words in $\mathcal{S}_{i_*}$ will yield a non-zero $M(w, \lambda)$ factor. It turns out that such a specific choice is given by $s = 3/2$ and $5/2$ and $i_* = 1$. Therefore to summarise again, we can say that for special choices of maximal parabolic Eisenstein series the constant term, which was naïvely thought to contain an infinite number of terms, collapses to a finite sum of only a few terms.

The next step is to investigate the space of possible values of s across dimensions and in particular for $D < 3$. For this we have computed the number of terms in the constant term of a minimal parabolic expansion. Table 2 shows the result for $D \leq 5$ for a range of integer and half-integer values for s. For $D < 3$ there seem to be only a few values of s, amongst them of course $s = 3/2$ and $5/2$, for which one obtains the collapse of the infinite sum explained above. There is however a large number of values for which this collapse does not seem to happen. In particular for values of $s \geq 7/2$ the calculation of the constant term on a computer did not terminate within a reasonably short period of time (unlike it did for values of $s < 7/2$). This can be taken as a tentative indication that in these cases the number of Weyl words

TABLE 2. The table shows the number of Weyl words with non-vanishing coefficients $M(w,\lambda)$ in an expansion of $E^{E_d}_{1;s}$ in dimensions $0 \le D \le 5$ and for a range of values for the parameter s. An ellipsis signifies that the row is continued with the last number explicitly written out.

s	0	$\frac{1}{2}$	1	$\frac{3}{2}$	2	$\frac{5}{2}$	3	$\frac{7}{2}$	4	$\frac{9}{2}$	5	$\frac{11}{2}$	6	$\frac{13}{2}$
E_6	1	2	27	7	12	27	$\cdots$							
E_7	1	2	126	8	14	35	56	126	91	126	$\cdots$			
E_8	1	2	2160	9	16	44	72	408	534	1060	1460	1795	2160	$\cdots$
E_9	1	2	∞	10	18	54	90	∞	$\cdots$					
E_{10}	1	2	∞	11	20	65	110	∞	$\cdots$					
E_{11}	1	2	∞	12	22	77	132	∞	$\cdots$					

contributing to the sum in Langlands' formula is actually infinite and for this reason we put ∞ for the corresponding entries in Table 2. (Physically, this may be related to counterterms being unprotected by supersymmetry.)

Looking at Table 2 it is tempting to interpret it as a strong sign for the special properties associated with the small values of s in the set

$$s \in \{0, 1/2, 3/2, 2, 5/2, 3\}\ . \tag{26}$$

More precisely, by requiring the constant term to only encode a *finite* number of perturbative effects as required by supersymmetry, the range of possible values that s can take, gets reduced from a previously infinite set to a finite number of possible values. It would certainly be desirable to make these statements more precise and to prove them rigorously. In our paper [**FK12**] we not only compute the number of constant terms but also their precise form. Some of them develop logarithmic dependence on the Cartan subalgebra coordinates.

4. Remarks and outlook

The maximal parabolic Fourier expansion (21) can be used to check the consistency of the automorphic couplings $\mathcal{E}^{D}_{(p,q)}$ in the low-energy expansion. Namely, the functions (16) are subject to a number of strong consistency requirements [**GRV10**, **Pio10**] that arise from the interplay of string theory in various dimensions. The consistency conditions are typically phrased in terms of three (maximal parabolic) limits, corresponding to different combinations of the torus radii (in appropriate units) and the string coupling becoming large. The three standard limits correspond to

(i) decompactification from D to $D+1$ dimensions, where one torus circle becomes large,
(ii) string perturbation theory, where the D-dimensional string coupling is small, and
(iii) the M-theory limit, where the whole torus volume becomes large.

In terms of the E_d diagram this means singling out the nodes d, 1 or 2, respectively. Mathematically, these limits are tantamount to computing the constant terms of the Eisenstein series in different maximal parabolic expansions (21). We have performed these consistency checks for our Eisenstein series (16) in the Kac-Moody case. The functions (16) satisfy them for all $D < 3$. In the affine case, particular care must be taken due to the scale invariance of the two-dimensional gravity system. Naïve evaluations of the Laplace equations and decompactification limits will lead to nonsensical answers. A careful discussion of how to remedy this by the proper inclusion of the derivation and the central charge of the affine algebra can be found in [**FK12**].

Among the interesting future directions, we mention the question of the nonzero-mode Fourier coefficients. The abelian ones are expected to be related to instantons as in higher space-time dimensions [**Pio10**, **GMV11**], however, the study of instantons in low space-time dimensions bears its own subtleties since finite energy solutions are harder to construct than in higher space-time dimensions. This is due to the asymptotic behaviour of the Green functions of the Laplace operator. We anticipate that the study of the Fourier coefficients might shed some light on this question. Not unrelated is the issue of automorphic representations. The collapse of the constant term can be interpreted as resulting from $\mathcal{E}^D_{(p,q)}$ being associated with a small automorphic representation [**GRS97**, **Pio10**, **GMV11**]. Automorphic representations of Kac-Moody groups have not been studied to the best of our knowledge.

Yet another possible application of automorphic functions and of $E_{10}(\mathbb{Z})$ in particular is in the context of arithmetic quantum gravity as defined in [**Gan99**, **PW03**, **BGH04**, **KKN09**]. There it was argued that the wavefunction of the universe should be an automorphic function of $E_{10}(\mathbb{Z})$ with zero eigenvalue under the Laplacian. Whether such a function exists and also satisfies the additional boundary conditions is presently not known.

Acknowledgements. We would like the organisers of String Math 2012 for putting together a broad and interesting programme and an enjoyable conference. For [**FK12**], we benefitted from discussions with T. Damour, S. Fredenhagen, M. Green, H. Nicolai, S. Miller, B. Pioline, P. Vanhove and D. Persson who also made useful comments on the present manuscript.

References

[BGH04] J. Brown, O. J. Ganor, and C. Helfgott, *M-theory and E_{10}: billiards, branes, and imaginary roots*, J. High Energy Phys. **8** (2004), 063, 55 pp. (electronic), DOI 10.1088/1126-6708/2004/08/063. MR2109843 (2006d:81203)

[CJ78] E. Cremmer and B. Julia, *The N=8 Supergravity Theory. 1. The Lagrangian*, Phys.Lett. **B80** (1978), 48.

[FILQ90] A. Font, Luis E. Ibanez, D. Lust, and F. Quevedo, *Strong - weak coupling duality and nonperturbative effects in string theory*, Phys.Lett. **B249** (1990), 35–43.

[FK12] P. Fleig and A. Kleinschmidt, *Eisenstein series for infinite-dimensional U-duality groups*, J. High Energy Phys. **6** (2012), 054, front matter+47. MR3006875

[Gan99] O. J. Ganor, *Two conjectures on gauge theories, gravity, and infinite dimensional Kac-Moody groups*.

[Gar01] H. Garland, *Certain Eisenstein series on loop groups: convergence and the constant term*, Algebraic groups and arithmetic, Tata Inst. Fund. Res., Mumbai, 2004, pp. 275–319. MR2094114 (2005i:11118)

[GG97] M. B. Green and M. Gutperle, *Effects of D-instantons*, Nuclear Phys. B **498** (1997), no. 1-2, 195–227, DOI 10.1016/S0550-3213(97)00269-1. MR1459079 (99e:81193)

[GMRV10] M. B. Green, Stephen D. Miller, J. G. Russo, and P. Vanhove, *Eisenstein series for higher-rank groups and string theory amplitudes*, Commun. Num. Theor. Phys. **4** (2010) 551-596, arXiv:1004.0163.

[GMV11] M. B. Green, S. D. Miller, and P. Vanhove, *Small representations, string instantons, and Fourier modes of Eisenstein series (with an appendix by D. Ciubotaru and P. Trapa)*, Journal of Number Theory **146** (2015) 187-309, arXiv:1111.2983.

[GRS97] D. Ginzburg, S. Rallis, and D. Soudry, *On the automorphic theta representation for simply laced groups*, Israel J. Math. **100** (1997), 61–116, DOI 10.1007/BF02773635. MR1469105 (99c:11058)

[GRV10] M. B. Green, J. G. Russo, and P. Vanhove, *Automorphic properties of low energy string amplitudes in various dimensions*, Phys. Rev. D **81** (2010), no. 8, 086008, 49, DOI 10.1103/PhysRevD.81.086008. MR2653571 (2012f:81225)

[GS82] M. B. Green and J. H. Schwarz, *Supersymmetrical Dual String Theory. 3. Loops and Renormalization*, Nucl.Phys. **B198** (1982), 441–460.

[GS99] M. B. Green and S. Sethi, *Supersymmetry constraints on type IIB supergravity*, Phys. Rev. D (3) **59** (1999), no. 4, 046006, 15, DOI 10.1103/PhysRevD.59.046006. MR1683988 (2000c:83114)

[GW86] D. J. Gross and E. Witten, *Superstring modifications of Einstein's equations*, Nuclear Phys. B **277** (1986), no. 1, 1–10, DOI 10.1016/0550-3213(86)90429-3. MR858663 (88a:83070)

[HT95] C. M. Hull and P. K. Townsend, *Unity of superstring dualities*, Nuclear Phys. B **438** (1995), no. 1-2, 109–137, DOI 10.1016/0550-3213(94)00559-W. MR1321523 (96e:83069)

[Jul80] B. Julia, *Group disintegrations*, Superspace and Supergravity (S. Hawking and M. Rocek, eds.), vol. C8006162, Cambridge University Press, 1980, pp. 331–350.

[KKN09] A. Kleinschmidt, M. Koehn, and H. Nicolai, *Supersymmetric quantum cosmological billiards*, Phys. Rev. D **80** (2009), no. 6, 061701, 5, DOI 10.1103/PhysRevD.80.061701. MR2607652 (2011e:83132)

[Lan76] R. P. Langlands, *On the functional equations satisfied by Eisenstein series*, Lecture Notes in Mathematics, Vol. 544, Springer-Verlag, Berlin-New York, 1976. MR0579181 (58 #28319)

[MW95] C. Mœglin and J.-L. Waldspurger, *Spectral decomposition and Eisenstein series*, Cambridge Tracts in Mathematics, vol. 113, Cambridge University Press, Cambridge, 1995. Une paraphrase de l'Écriture [A paraphrase of Scripture]. MR1361168 (97d:11083)

[OP99] N. A. Obers and B. Pioline, *U-duality and M-theory*, Phys. Rep. **318** (1999), no. 4-5, 113–225, DOI 10.1016/S0370-1573(99)00004-6. MR1713479 (2001h:81208)

[OP00] N. A. Obers and B. Pioline, *Eisenstein series and string thresholds*, Comm. Math. Phys. **209** (2000), no. 2, 275–324, DOI 10.1007/s002200050022. MR1737986 (2001f:81159)

[Pio98] B. Pioline, *A note on non-perturbative R^4 couplings*, Phys. Lett. B **431** (1998), no. 1-2, 73–76, DOI 10.1016/S0370-2693(98)00554-1. MR1634961 (99c:81207)

[Pio10] B. Pioline, *R**4 couplings and automorphic unipotent representations*, JHEP **1003** (2010), 116.

[PW03] B. Pioline and A. Waldron, *Automorphic forms: A Physicist's survey*, arXiv:hep-th/0312068.

[Wit95] E. Witten, *String theory dynamics in various dimensions*, Nuclear Phys. B **443** (1995), no. 1-2, 85–126, DOI 10.1016/0550-3213(95)00158-O. MR1334520 (96f:81104)

Institut des Hautes Études Scientifiques, 91440 Bures-sur-Yvette, France
E-mail address: fleig@ihes.fr

Max-Planck-Institut für Gravitationsphysik (Albert-Einstein-Institut), Am Mühlenberg 1, DE-14476 Potsdam, Germany – and – International Solvay Institutes, ULB-Campus Plaine, C.P. 231, BE-1050 Brussels, Belgium
E-mail address: axel.kleinschmidt@aei.mpg.de

Proceedings of Symposia in Pure Mathematics
Volume **90**, 2015
http://dx.doi.org/10.1090/pspum/090/01528

Super-A-Polynomial

Hiroyuki Fuji and Piotr Sułkowski

Abstract. We review a construction of a new class of algebraic curves, called super-A-polynomials, and their quantum generalizations. The super-A-polynomial is a two-parameter deformation of the A-polynomial known from knot theory or Chern-Simons theory with $SL(2,\mathbb{C})$ gauge group. The two parameters of the super-A-polynomial encode, respectively, the t-deformation which leads to the "refined A-polynomial", and the Q-deformation which leads to the augmentation polynomial of knot contact homology. For a given knot, the super-A-polynomial encodes the asymptotics of the corresponding S^r-colored HOMFLY homology for large r, while the quantum super-A-polynomial provides recursion relations for such homology theories for each r. The super-A-polynomial also admits a simple physical interpretation as the defining equation for the space of SUSY vacua in a circle compactification of the effective 3d $\mathcal{N}=2$ theory associated to a given knot (complement). We discuss properties of super-A-polynomials and illustrate them in many examples.

1. Introduction

In past two decades remarkable relations between quantum field theory, string theory and knot theory have been found, following the seminal work by Witten [**43**]. Among the others, such important mathematical developments as polynomial knot invariants, volume conjectures, A-polynomials, homological knot invariants, and more, have been interpreted from the perspective of high energy physics. In this note we summarize a construction of a new object in this line of research, the so-called super-A-polynomial, introduced in [**15**]. The super-A-polynomial can be regarded as a two-parameter generalization of an ordinary A-polynomial [**9, 17, 19**]. One of these parameters encodes information about the t-deformation of knot invariants arising upon categorification. The one-parameter generalization of an A-polynomial, depending only on t, has been introduced in [**3**] as the so-called "refined A-polynomial". This parameter is related to the categorification of knot invariants and knot homologies, such as Khovanov homology [**28**], Khovanov-Rozansky homology [**29**], or HOMFLY homology [**13**]; more precisely, in this paper t appears upon taking the Poincaré characteristics of the (conjectural) colored HOMFLY homology [**21**]. The second parameter, denoted by a in this note, is related to Chern-Simons theory with $SU(N)$ gauge group. It also corresponds to the so-called Q-deformation of the A-polynomial introduced in [**2**], as well as the augmentation polynomial of

2010 *Mathematics Subject Classification.* Primary 81T45, 81T30.

knot contact homology [**35**]. The super-A-polynomial captures information about both a and t at once, and, among the others, it encodes the asymptotics of the corresponding S^r-colored HOMFLY homologies for large r. In addition, its quantum deformation, the so-called quantum super-A-polynomial, provides recursion relations for HOMFLY homology theories for each r. Further examples and properties of super-A-polynomials have been analyzed in [**16, 34**].

In this note we summarize the construction of the super-A-polynomial and illustrate it in several examples, following (mostly) [**3, 15**]. We start by recalling the original volume conjectures in section 2. In section 3 we generalize these conjectures in a way which leads to the super-A-polynomial and the quantum super-A-polynomial. Essential ingredients which make these new conjectures work are colored superpolynomials, introduced in section 4; we stress that developments of tools and techniques which enable to derive an explicit form of such colored superpolynomials is an important result, independent of the work reported here. In section 6 we discuss quantizability properties of super-A-polynomials, and in section 7 we present their interpretation in 3d, $\mathcal{N}=2$ SUSY gauge theories.

2. Volume conjectures

Originally the "volume conjecture" referred to the observation [**25**] that the so-called Kashaev invariant of a knot K defined at the n-th root of unity $q=e^{2\pi i/n}$ in the classical limit has a nice asymptotic behavior determined by the hyperbolic volume $\mathrm{Vol}(M)$ of the knot complement $M=S^3\setminus K$. Shortly after, it was realized [**33**] that the Kashaev invariant is equal to the n-colored Jones polynomial of a knot K evaluated at $q=e^{2\pi i/n}$, so that the volume conjecture could be stated simply as

$$\lim_{n\to\infty}\frac{2\pi\log|J_n(K;q=e^{2\pi i/n})|}{n}=\mathrm{Vol}(M)\,. \tag{2.1}$$

The physical interpretation of the volume conjecture was proposed in [**19**]. Besides explaining the original observation (2.1) it immediately led to a number of generalizations, in which the right-hand side is replaced by a function of various parameters (see [**11**] for a review).

2.1. Generalized volume conjecture.

Once the volume conjecture is put in the context of analytically continued Chern-Simons theory, it becomes clear that the right-hand side is simply the value of the classical $SL(2,\mathbb{C})$ Chern-Simons action functional on a knot complement M. Since classical solutions in Chern-Simons theory (*i.e.* flat connections on M) come in families, parametrized by the holonomy of the gauge connection on a small loop around the knot, this physical interpretation immediately leads to a "family version" of the volume conjecture [**19**]:

$$J_n(K;q=e^{\hbar})\ \overset{\substack{n\to\infty\\ \hbar\to 0}}{\sim}\ \exp\left(\frac{1}{\hbar}S_0(u)+\dots\right) \tag{2.2}$$

parametrized by a complex variable u. Here, the limit on the left-hand side is slightly more interesting than in (2.1) and, in particular, also depends on the value of the parameter u:

$$q=e^{\hbar}\to 1\,,\qquad n\to\infty\,,\qquad q^n=e^u\equiv x\ \text{(fixed)} \tag{2.3}$$

In fact, Chern-Simons theory predicts all of the subleading terms in the $\hbar$-expansion denoted by ellipsis in (2.2). These terms are the familiar perturbative coefficients of the $SL(2,\mathbb{C})$ Chern-Simons partition function on M.

2.2. Quantum volume conjecture (AJ-conjecture). Classical solutions in Chern-Simons theory (*i.e.* flat connections on M) are labeled by the holonomy eigenvalue $x = e^u$ or, to be more precise, by a point on the algebraic curve

$$\mathcal{C}: \quad \left\{ (x,y) \in \mathbb{C}^* \times \mathbb{C}^* \middle| A(x,y) \;=\; 0 \right\}, \tag{2.4}$$

defined by the zero locus of the A-polynomial, a certain classical invariant of a knot. In quantum theory, $A(x,y)$ becomes an operator $\hat{A}(\hat{x},\hat{y};q)$ and the classical condition (2.4) turns into a statement that the Chern-Simons partition function is annihilated by $\hat{A}(\hat{x},\hat{y};q)$. This statement applies equally well to Chern-Simons theory with the compact gauge group $SU(2)$ that computes the colored Jones polynomial $J_n(K;q)$ as well as to its analytic continuation that localizes on $SL(2,\mathbb{C})$ flat connections. In the former case, one arrives at the "quantum version" of the volume conjecture [**19**]:

$$\hat{A}\; J_*(K;q) \;\simeq\; 0\,, \tag{2.5}$$

which in the mathematical literature was independently proposed around the same time [**17**] and is known as the AJ-conjecture. The action of the operators $\hat{x}$ and $\hat{y}$ follows from quantization of Chern-Simons theory, and one finds that $\hat{x}$ acts as a multiplication by q^n, whereas $\hat{y}$ shifts the value of n:

$$\begin{aligned} \hat{x} J_n \;&=\; q^n J_n \\ \hat{y} J_n \;&=\; J_{n+1} \end{aligned} \tag{2.6}$$

In particular, one can easily verify that these operations obey the commutation relation

$$\hat{y}\hat{x} \;=\; q\hat{x}\hat{y} \tag{2.7}$$

that follows from the symplectic structure on the phase space of Chern-Simons theory. Therefore, upon quantization a classical polynomial relation of the form (2.4) becomes a q-difference equation for the colored Jones polynomial or Chern-Simons partition function. Further details, generalizations, and references can be found in [**11**].

3. New volume conjectures and the super-A-polynomial

Besides the "non-commutative" deformation (2.5), the A-polynomial also admits two commutative deformations that in a similar way encode the "color behavior" of two natural generalizations of the colored Jones polynomial: the t-deformation that corresponds to the categorification of colored Jones invariants [**3**] and Q-deformation that corresponds to extending $J_n(K;q)$ to higher rank knot polynomials [**2, 15**]. Our task in what follows is to combine these two deformations into a single unifying structure. In particular, this leads to a new unifying knot invariant. We call this invariant the *super-A-polynomial* since it describes how the S^{n-1}-colored superpolynomials $\mathcal{P}_n(a,q,t)$ depend on color, *i.e.* on the representation $R = S^{n-1}$, much in the same way as A-polynomial does it for the colored Jones polynomial.

We recall that, in the context of BPS states, the superpolynomial is defined as a generating function of refined open BPS invariants on a rigid Calabi-Yau 3-fold X in the presence of a Lagrangian brane supported on $L \subset X$:

$$\mathcal{P}(a,q,t) \ :=\ \mathrm{Tr}_{\mathcal{H}^{\mathrm{ref}}_{\mathrm{BPS}}}\, a^{\beta} q^{P} t^{F}\,, \qquad \beta \in H_2(X,L) \tag{3.1}$$

and, in application to knots, the superpolynomial $\mathcal{P}(K;a,q,t)$ is defined as a Poincaré polynomial of the triply-graded homology theory $\mathcal{H}(K)$ that categorifies the HOMFLY polynomial $P(K;a,q)$, see [**13**] for details. According to the conjecture of [**20**], these two definitions give the same result when X is the total space of the $\mathcal{O}(-1)\oplus\mathcal{O}(-1)$ bundle over $\mathbb{C}\mathbf{P}^1$ and L is the Lagrangian submanifold determined by the knot $K \subset \mathbf{S}^3$, *cf.* [**31, 36, 41**]. Lagrangian branes of multiplicity $r = n-1$ yield the so-called "n-colored" version of the superpolynomial which, in the context of knot homologies, was recently introduced in [**21**],

$$\mathcal{P}_n(K;a,q,t) \ :=\ \sum_{i,j,k} a^i q^j t^k \,\dim \mathcal{H}^{S^{n-1}}_{i,j,k}(K)\,, \tag{3.2}$$

as a Poincaré polynomial of a triply-graded homology theory categorifying the S^r-colored HOMFLY polynomial (see also [**3, 24, 32**]). For $t=-1$ the above expression reduces to the Euler characteristic and reproduces normalized (i.e. such that $P_n(\bigcirc;a,q)=1$ for the unknot $\bigcirc$) colored HOMFLY polynomial $P_n(K;a,q)$:

$$P_n(K;a,q) = \mathcal{P}_n(K;a,q,-1) = \sum_{i,j,k} a^i q^j (-1)^k \,\dim \mathcal{H}^{S^{n-1}}_{i,j,k}(K)\,. \tag{3.3}$$

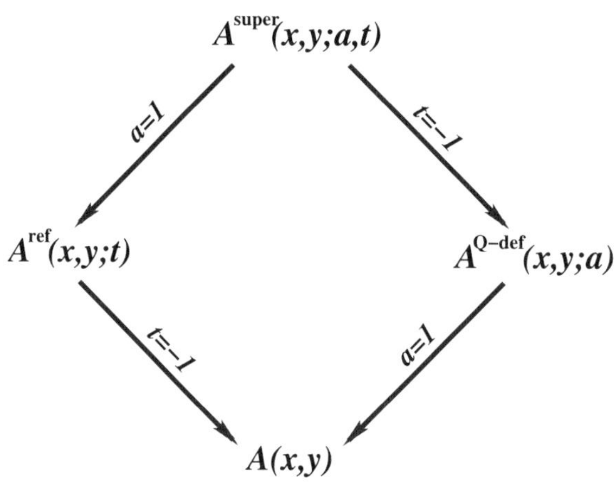

FIGURE 1. Various specializations of the super-A-polynomial. *(The color version of this graphic appears in the online version of the book.)*

We stress that superpolynomials were introduced in [**13**] based on arguments from physics. Rigorous mathematical construction of triply-graded homologies whose Euler characteristics reproduces HOMFLY polynomial was given in [**30**]. It is however unclear whether those triply graded homologies are related to sl_N homologies by the action of differentials, as postulated in [**13**] (such a construction for torus knots has been given e.g. in [**18**]; however, more generally, these two types of homologies could be related by spectral sequences that do not converge after the first page [**37**]). On the other hand, colored triply-graded homologies and corresponding superpolynomials, introduced in [**21**], have not been mathematically rigorously defined to date. Nonetheless, in this article we conjecture that these objects exist, and our main goal is to explain that S^{n-1}-colored superpolynomials $\mathcal{P}_n(K;a,q,t)$ depend on color (*i.e.* on the representation $R = S^{n-1}$) in a simple and controllable way, governed by the super-A-polynomial $A^{\text{super}}(x,y;a,t)$ and by its quantization $\hat{A}^{\text{super}}(\hat{x},\hat{y};a,q,t)$. Specifically, based on the physics arguments and the study of examples, we propose the following analog of the generalized volume conjecture [**19**] or its refined version [**3**]:

Conjecture 1: *In the limit*

$$q = e^{\hbar} \to 1\,, \qquad a = \text{fixed}\,, \qquad t = \text{fixed}\,, \qquad x = q^n = \text{fixed} \tag{3.4}$$

the n-colored superpolynomials $\mathcal{P}_n(K;a,q,t)$ exhibit the following "large color" behavior:

$$\mathcal{P}_n(K;a,q,t) \overset{\substack{n\to\infty \\ \hbar\to 0}}{\sim} \exp\left(\frac{1}{\hbar}\int \log y \frac{dx}{x} + \ldots\right) \tag{3.5}$$

where ellipsis stand for regular terms (as $\hbar \to 0$) and the leading term is given by the integral on the zero locus of the super-A-polynomial, cf. (2.4)*:*

$$A^{\text{super}}(x,y;a,t) \;=\; 0\,. \tag{3.6}$$

Moreover, just like the ordinary A-polynomial has its quantum analog (2.5), the super-A-polynomial is a characteristic polynomial of a quantum operator $\hat{A}^{\text{super}}(\hat{x},\hat{y};a,q,t)$ that combines commutative t- and a-deformations with the non-commutative q-deformation (2.7). We call this operator the *quantum super-A-polynomial.*

Conjecture 2: *For a given knot K, the colored superpolynomial $\mathcal{P}_n(K;a,q,t)$ satisfies a recurrence relation of the form* (2.5)*:*

$$a_k\,\mathcal{P}_{n+k}(K;a,q,t) + \ldots + a_1\,\mathcal{P}_{n+1}(K;a,q,t) + a_0\,\mathcal{P}_n(K;a,q,t) \;=\; 0 \tag{3.7}$$

where $\hat{x}$ and $\hat{y}$ act on $\mathcal{P}_n(K;a,q,t)$ as in (2.6)*, and where the rational functions $a_i \equiv a_i(\hat{x},a,q,t)$ are the coefficients of the "quantum super-A-polynomial"*

$$\hat{A}^{\text{super}}(\hat{x},\hat{y};a,q,t) \;=\; \sum_i a_i(\hat{x},a,q,t)\,\hat{y}^i\,, \tag{3.8}$$

whose characteristic polynomial is $A^{\text{super}}(x,y;a,t)$.

As in (2.5), sometimes we informally write (3.7) in the compact form

$$\hat{A}^{\text{super}}\;\mathcal{P}_*(K;a,q,t) \;=\; 0\,, \tag{3.9}$$

which is a quantum version of the classical curve (3.6).

The superpolynomial unifies many polynomial and homological invariants of knots that can be obtained from it via various specializations, applying differentials, *etc.* For example, for $\mathcal{H}$-thin knots the specialization to $a = q^2$ yields the Poincaré polynomial of the colored $sl(2)$ knot homology. Therefore, if K is a thin knot (*e.g.* if K is a two-bridge knot), in the limit $a = q^2$ we expect (3.5) and (3.7) to reproduce the corresponding versions of the refined volume conjectures proposed in [**3**]. In particular,

$$\hat{A}^{\text{super}}(\hat{x},\hat{y};a=q^2,q,t) \;=\; \hat{A}^{\text{ref}}(\hat{x},\hat{y};q,t)\,, \tag{3.10}$$

Quantum operator	provides recursion for	classical limit
$\hat{A}^{\text{super}}(\hat{x},\hat{y};a,q,t)$	colored superpolynomial	$A^{\text{super}}(x,y;a,t)$
$\hat{A}^{\text{ref}}(\hat{x},\hat{y};q,t)$	colored $sl(2)$ homology	$A^{\text{ref}}(x,y;t)$
$\hat{A}^{\text{Q-def}}(\hat{x},\hat{y};a,q)$	colored HOMFLY	$A^{\text{Q-def}}(x,y;a)$
$\hat{A}(\hat{x},\hat{y};q)$	colored Jones	$A(x,y)$

TABLE 1. Quantum super-A-polynomial and its specializations lead to recursion relations for various S^n-colored knot invariants.

and, via further specialization to the classical limit $q=1$,

$$A^{\text{super}}(x,y;a=1,t) \;=\; A^{\text{ref}}(x,y;t)\,. \tag{3.11}$$

Similarly, the specialization of the superpolynomial $\mathcal{P}_n(K;a,q,t)$ to $t=-1$ yields the HOMFLY polynomial or, in the problem at hand, the *colored* HOMFLY polynomial [**21**]. Therefore, at $t=-1$ the recursion relation (3.7) should reduce to the recursion relation for the S^{n-1}-colored HOMFLY polynomial, whose characteristic variety — called the Q-deformed A-polynomial in [**2**] — must be contained in $A^{\text{super}}(x,y;a,t=-1)$ as a factor. To avoid clutter, we include possible extra factors inherited from $A^{\text{super}}(x,y;a,t)$ in the definition of the Q-deformed A-polynomial, so that

$$A^{\text{super}}(x,y;a,t=-1) \;=\; A^{\text{Q-def}}(x,y;a)\,. \tag{3.12}$$

Moreover, the authors of [**2**] proposed an important conjecture that offers a new way of looking at this polynomial (that, in our Figure 1, occupies the right corner) and identifies it with the augmentation polynomial of knot contact homology [**35**]. In what follows we use the names "Q-deformed A-polynomial" and "augmentation polynomial" interchangeably. In fact, one justification for this comes from the fact (see [**35**, Proposition 5.9] for a proof) that the classical augmentation polynomial, when specialized further to $a=1$, reduces to the ordinary A-polynomial, possibly with some extra factors, which altogether we denote simply by $A(x,y)$:

$$A^{\text{super}}(x,y;a=1,t=-1) \;=\; A^{\text{Q-def}}(x,y;a=1) \;=\; A(x,y)\,, \tag{3.13}$$

as it should in order to fit perfectly in the diagram in Figure 1.

Therefore, our super-A-polynomial $A^{\text{super}}(x,y;a,t)$ can be viewed, on one hand, as a "refinement" of the augmentation polynomial $A^{\text{Q-def}}(x,y;a)$ and, on the other hand, as a "Q-deformation" of the refined A-polynomial $A^{\text{ref}}(x,y;t)$, see Figure 1. For important examples of super-A-polynomials (to be discussed in more detail in what follows) see table 2. Other interesting specializations of super-A-polynomials, *e.g.* involving setting $x=1$ or $q=1$, are discussed in [**15, 16**].

4. Essential ingredients: colored superpolynomials

The (quantum) super-A-polynomial arising from the conjectures presented in section 3 is intimately related to the colored superpolynomials introduced in (3.2). Indeed, the knowledge of S^{n-1}-colored superpolynomials $\mathcal{P}_n(K;a,q,t)$ for general

Knot	$A^{\text{super}}(x, y; a, t)$	
Unknot, $\bigcirc$	$(-a^{-1}t^{-3})^{1/2}(1 + at^3x) - (1 - x)y$	
Figure-eight, $\mathbf{4_1}$	$a^2t^5(x-1)^2x^2 + at^2x^2(1+at^3x)^2y^3+$ $+at(x-1)(1+t(1-t)x + 2at^3(t+1)x^2 - 2at^4(t+1)x^3$ $+a^2t^6(1-t)x^4 - a^2t^8x^5)y - (1+at^3x)(1+at(1-t)x + 2at^2(t+1)x^2$ $+2a^2t^4(t+1)x^3 + a^2t^5(t-1)x^4 + a^3t^7x^5)y^2$	
Trefoil, $\mathbf{3_1}$	$a^2t^4(x-1)x^3 - a\big(1 - t^2x + 2t^2(1+at)x^2 + at^5x^3 + a^2t^6x^4\big)y + (1+at^3x)y^2$	
$(2, 2p+1)$ torus knot	eliminate z_0 in $\begin{cases} 1 = \frac{(z_0-x)(t^2z_0-1)(1+at^3xz_0)}{t^{2+2p}z_0^{1+2p}(z_0-1)(atx+z_0)(t^2xz_0-1)} \\ y = \frac{a^pt^{2+2p}(x-1)x^{1+2p}(atx+z_0)(1+at^3xz_0)}{(1+at^3x)(x-z_0)(t^2xz_0-1)} \end{cases}$	see table 5

TABLE 2. Super-A-polynomials for simple knots.

n allows to determine the (quantum) super-A-polynomial, and this is how super-A-polynomials will be derived in all examples in what follows. Nonetheless, determining S^{n-1}-colored superpolynomials is itself a hard task, and these objects are not even defined mathematically in a rigorous and computable way. However it turns out that physics offers two possible ways to obtain (or, at least, to conjecture) the form of colored superpolynomials: either using the so-called refined Chern-Simons theory, or taking advantage of the structure of differentials in homological theories. For some knots both of these methods can be used, and then they lead to remarkable identities, which confirm validity of the physics approach. We need to get acquainted with these methods before we present the construction and examples of super-A-polynomials.

4.1. Refined Chern-Simons theory. As is well known [**43**], knot invariants are simply related to expectation values of Wilson loop operators $W_R(K)[A] := \mathrm{Tr}_R \mathrm{P}\exp\left[\oint_K A\right]$, supported on a knot K and decorated by a representation R, in Chern-Simons theory

$$Z_G^{\mathrm{CS}}(M, K_R; q) = \int [dA]\, W_R(K)[A]\, e^{ikS_{\mathrm{CS}}[A;M]}, \tag{4.1}$$

where the Chern-Simons action on a 3-manifold M reads

$$S_{\mathrm{CS}}[A; M] = \frac{1}{4\pi}\int_M \mathrm{Tr}_{\mathrm{adj}}\left(A \wedge dA + \frac{2}{3} A\wedge A\wedge A\right). \tag{4.2}$$

The quantum group invariants [**38**] are reproduced as the above expectation values normalized by that of the unknot (which we often denote as $\bigcirc$), and remarkably such expressions are simply polynomials in $q = e^{\frac{2\pi i}{k+h}}$ with integer coefficients (at least when $M = \S^3$)

$$J^{\mathfrak{g},R}(K;q) = \frac{Z_G^{\mathrm{CS}}(\S^3, K_R; q)}{Z_G^{\mathrm{CS}}(\S^3, \bigcirc_R; q)}. \tag{4.3}$$

In particular, for $\mathfrak{g} = sl(N)$ a dependence on N is very simple and $J^{sl(N),R}(K;q)$ turn out to be polynomials in q and $a = q^N$ which reproduce (normalized) colored HOMFLY polynomials $J^{sl(N),R}(K;q) = P^R(K; a = q^N, q)$. For $R = S^{n-1}$ they are denoted $P_n(K; a, q)$, and they already appeared above in (3.3).

Therefore our task is to introduce, from the perspective of the Chern-Simons theory, a dependence on the Poincaré variable t into the colored HOMFLY polynomial, so that Chern-Simons amplitudes given in (4.3) would be extended to t-dependent quantities, which should be identified with (3.2). Such a generalization has been introduced in [**1**] and is often referred to as refined Chern-Simons theory. More precisely, the fundamental formulation (in terms of a t-dependent action) of refined Chern-Simons theory is still not known. Nonetheless, the authors of [**1**] argued how a dependence on t should be introduced in a consistent manner in various quantities arising in the quantization of the original Chern-Simons theory. In particular they proposed a refined version of modular matrices S and T which satisfy the Verlinde formula. Similarly other objects in Chern-Simons theory become functions of q and t, or equivalently (as often arising in various calculations) q_1 and q_2 (such that $q=\frac{1}{q_2}, t=-\sqrt{\frac{q_2}{q_1}}$). For example, Schur polynomials that arise in original Chern-Simons theory (in particular as expectation values of the unknot) are replaced by Macdonald polynomials, and so on:

$$
\begin{array}{ccc}
\underline{\text{CS gauge theory}} & & \underline{\text{refined invariants}} \\
Z^{\rm CS}_{SU(N)}(\S^3, K_R; q) & \rightsquigarrow & Z^{\rm ref}_{SU(N)}(\S^3, K_R; q_1, q_2) \\
\dim_q R = s_R(q^\varrho) & \rightsquigarrow & M_R(q_2^\varrho; q_1, q_2) \\
q^{C_2(R)} & \rightsquigarrow & q_1^{\frac{1}{2}||R||^2} q_2^{-\frac{1}{2}||R^t||^2} q_2^{\frac{N}{2}|R|} q_1^{-\frac{1}{2N}|R|^2} \\
& \vdots &
\end{array}
\tag{4.4}
$$

Refined Chern-Simons theory is still quite mysterious; in particular explicit computations are possible only for some particular knots (i.e. those, whose complements are Seifert-fibered), and they involve various subtleties, related *e.g.* to the appearance of the so-called γ-factors (which are irrelevant for $t=-1$). Nonetheless, we are able to predict an explicit form of superpolynomials using refined Chern-Simons theory in various examples, such as the (unnormalized) unknot (see (5.1)), or $(2, 2p+1)$ torus knots (for arbitrary p, see (5.30)). Various aspects of refined Chern-Simons theory are discussed in detail in [**1, 3, 14**].

4.2. Differentials in knot homologies. Even though combinatorial definition of colored knot homologies (3.2) is, in general, not known, it turns out that various physics arguments predict how the action of various differentials in knot homologies should look like. Such differentials endow knot homologies with a very rich structure, which turns out to be very elegant and often so constraining that one can even compute colored superpolynomials $\mathcal{P}_n(K; a, q, t)$ based on this structure alone, with a minimal input. In particular, this is how nice formulas like (5.11), (5.21), or (5.31) can be produced. Referring the reader to [**16, 21**] for further details, here we merely state a simple rule of thumb: the factors of the form $(1+a^i q^j t^k)$ that we often see *e.g.* in (5.3), (5.11), (5.21), and (5.30) come from differentials of (a, q, t)-degree (i, j, k), *cf.* [**3**, eq. (3.54)]:

(4.5)

differentials	factors	(a, q, t) grading
$d_{N>0}$	$1+aq^{-N}t$	$(-1, N, -1)$
$d_{N<0}$	$1+aq^{-N}t^3$	$(-1, N, -3)$
$d_{\rm colored}$	$1+q$	$(0, 1, 0)$
	$1+at$	$(-1, 0, -1)$
	$\vdots$	

For example, notice that all terms with $k > 0$ in the expression (5.11) for the colored superpolynomial of the figure-eight knot manifestly contain a factor $(1 + aq^{n-1}t^3)$. Hence, the S^{n-1}-colored superpolynomial of the figure-eight knot has the following structure

$$\mathcal{P}_n(\mathbf{4_1}; a, q, t) \;=\; 1 + (1 + aq^{n-1}t^3)Q_n(a, q, t)\,, \tag{4.6}$$

which means that, when evaluated at $a = -q^{1-n}t^{-3}$, the sum (5.11) collapses to a single $k = 0$ term, $\mathcal{P}_n(\mathbf{4_1}; a = -q^{1-n}t^{-3}, q, t) = 1$. A proper interpretation of this fact is that a specialization to $N = 1 - n$ of the triply-graded S^{n-1}-colored HOMFLY homology, carried out by the action of the differential d_{1-n}, is trivial. In other words, the differential d_N with $N = 1 - n$ is canceling in a theory with $R = S^{n-1}$. A systematic implementation of such observations fully determines the form of colored superpolynomials – at least for some knots – as we will see in the examples in the next section.

5. Examples

In this section we illustrate ideas presented above in explicit examples of various knots. We start with the simplest example of the unknot, and then discuss a non-trivial example of a hyperbolic knot, *i.e.* the figure-eight knot, and the entire family of $(2, 2p+1)$ torus knots, with a special emphasis on the trefoil. In each case we start our considerations by providing explicit and general formulas for S^{n-1}-colored superpolynomials $\mathcal{P}_n(a, q, t)$, illustrating the power of two approaches described in section 4. Taking advantage of these representations, subsequently we derive classical and quantum super-A-polynomials for these knots and discuss their properties. For other examples of superpolynomials and super-A-polynomials see [**16**, **34**].

5.1. Unknot. Let us start with the simplest example of the unknot. Despite its simplicity, this is still an interesting and important example; as we will see, some objects associated to the unknot, which are trivial in the non-refined and non-super case, become rather non-trivial when t- or a-dependence is turned on.

We recall than in the unknot case we must consider unreduced (or "unnormalized") knot polynomials – in particular, unreduced colored superpolynomial $\bar{\mathcal{P}}_n(a, q, t)$ – since, by definition, reduced polynomials are normalized by the value of the unknot, so that $\mathcal{P}_n(\bigcirc; a, q, t) = 1$. From the viewpoint of the (refined) Chern-Simons theory the unreduced colored superpolynomial is defined as the ratio of partition functions on $\S^3$ in the presence and absence of a knot. In case of the unknot this ratio is given by the Macdonald polynomial, and after the change of variables $q = \frac{1}{q_2}, t = -\sqrt{\frac{q_2}{q_1}}$, we find that the S^{n-1}-colored superpolynomial reads

$$\begin{aligned}\bar{\mathcal{P}}_n(\bigcirc; a, q, t) \;&=\; \frac{Z^{\text{ref}}_{SU(N)}(\S^3, \bigcirc_{\Lambda^{n-1}}; q_1, q_2)}{Z^{\text{ref}}_{SU(N)}(\S^3; q_1, q_2)} = M_{\Lambda^{n-1}}(q_2^{\varrho}; q_1, q_2) \\ &=\; (-1)^{\frac{n-1}{2}} a^{-\frac{n-1}{2}} q^{\frac{n-1}{2}} t^{-\frac{3(n-1)}{2}} \frac{(-at^3; q)_{n-1}}{(q; q)_{n-1}}\,.\end{aligned} \tag{5.1}$$

We also recall that the q-Pochhammer symbol $(x,q)_n$, defined by a product formula, in the limit $q \to 1$ and q^n fixed ($c.f$ (3.4)), has the following asymptotics

$$(x,q)_n \equiv \prod_{i=0}^{n-1}(1-xq^i) \sim e^{\frac{1}{\hbar}(\mathrm{Li}_2(x)-\mathrm{Li}_2(xq^n))}. \tag{5.2}$$

Once the general expression for the colored superpolynomial is determined, we can find a recursion relation it satisfies. In particular, as the homological unknot invariant (5.1) has a product form, we can immediately write down the recursion relation it satisfies:

$$\bar{\mathcal{P}}_{n+1}(\bigcirc;a,q,t) \ = \ (-a^{-1}t^{-3}q)^{1/2}\frac{1+at^3q^{n-1}}{1-q^n}\,\bar{\mathcal{P}}_n(\bigcirc;q,t)\,. \tag{5.3}$$

This means that the quantum super-A-polynomial for the unknot reads

$$\hat{A}^{\mathrm{super}}(\hat{x},\hat{y};a,q,t) \ = \ (-a^{-1}t^{-3}q)^{1/2}(1+at^3q^{-1}\hat{x})-(1-\hat{x})\hat{y}\,. \tag{5.4}$$

In the classical limit $q \to 1$ this operator reduces to the classical super-A-polynomial

$$A^{\mathrm{super}}(x,y;a,t) \ = \ (-a^{-1}t^{-3})^{1/2}(1+at^3x)-(1-x)y\,. \tag{5.5}$$

The Newton polygon as well as the coefficients of monomials of this polynomial are shown in figure 2. In the unrefined limit $t=-1$ the relation (5.4) takes the form

$$\hat{A}^{\mathrm{Q\text{-}def}}(\hat{x},\hat{y};a,q) \ = \ (a^{-1}q)^{1/2}(1-aq^{-1}\hat{x})-(1-\hat{x})\hat{y}\,, \tag{5.6}$$

and specializing further to $q=1$ we get the augmentation polynomial

$$A^{\mathrm{Q\text{-}def}}(x,y;a) \ = \ a^{-1/2}(1-ax)-(1-x)y\,. \tag{5.7}$$

Interestingly, this polynomial does not factorize, and only in the limit of ordinary A-polynomial $a \to 1$ do we get a factorized form with $y-1$ factor representing the abelian connection

$$A(x,y) \ = \ (1-x)(1-y)\,. \tag{5.8}$$

It is instructive to show that the super-A-polynomial (5.5) can be also derived from the asymptotic analysis of (5.1). Indeed, using the asymptotics (5.2), in the limit (3.4) we can approximate (5.1) as

$$P_n(\bigcirc;a,q,t) = \exp\frac{1}{\hbar}\Big(\log x\,\log(-a^{-1}t^{-3})^{1/2} + \mathrm{Li}_2(x) - \mathrm{Li}_2(-at^3x) + \mathrm{Li}_2(-at^3) - \frac{\pi^2}{6} + \mathcal{O}(\hbar)\Big),$$

from which identify the potential $\widetilde{\mathcal{W}} = \int \log y \frac{dx}{x}$ in (3.5) as

$$\widetilde{\mathcal{W}} \ = \ \log x\,\log(-a^{-1}t^{-3})^{1/2} + \mathrm{Li}_2(x) - \mathrm{Li}_2(-at^3x) + \mathrm{Li}_2(-at^3) - \frac{\pi^2}{6}\,. \tag{5.9}$$

Differentiating it with respect to x, we now obtain

$$y = e^{x\partial_x \widetilde{\mathcal{W}}} = (-a^{-1}t^{-3})^{1/2}\frac{1+at^3x}{1-x}\,, \tag{5.10}$$

which reproduces the defining equation of the super-A-polynomial given in (5.5). We also note that for $a=-t=1$ the potential $\widetilde{\mathcal{W}}$ vanishes, which is related to the factorization occurring in (5.8) and can be attributed to the fact that the only $SL(2,\mathbb{C})$ flat connections on a solid torus (= complement of the unknot) are abelian flat connections. When $a \neq 1$ or $t \neq -1$, the potential $\widetilde{\mathcal{W}}$ is nonzero and presumably can be interpreted as a contribution of "deformed" abelian flat connections.

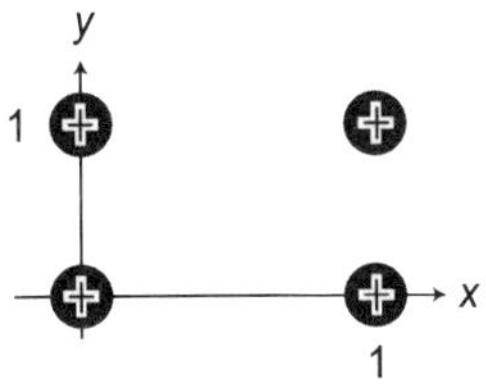

$$\begin{pmatrix} (-at^3)^{-1/2} & -1 \\ -(-at^3)^{1/2} & 1 \end{pmatrix}$$

FIGURE 2. Newton polygon for the super-A-polynomial of the unknot (left). Red circles denote monomials of the super-A-polynomial, and smaller yellow crosses denote monomials of its $a = -t = 1$ specialization. In this example both Newton polygons look the same, so that positions of all circles and crosses overlap. The coefficients of the super-A-polynomial are also shown in the matrix on the right. The role of rows and columns is exchanged in these two presentations: a monomial $a_{i,j}x^i y^j$ corresponds to a circle (resp. cross) at position (i, j) in the Newton polygon, while in the matrix on the right it is shown as the entry $a_{i,j}$ in the $(i+1)^{\text{th}}$ row and in the $(j+1)^{\text{th}}$ column. These conventions are the same as in [**3**, **15**, **16**]. *(The color version of this graphic appears in the online version of the book.)*

5.2. Figure-eight knot. In this section we consider the figure-eight knot, also denoted $\mathbf{4_1}$. This is a hyperbolic knot, and we stress that it provides a highly non-trivial example, for which many simplifications common in the realm of torus knots (to be discussed in the following sections) do not occur.

The colored superpolynomial (3.1) for figure-eight knot can be found using the highly constraining structure of differentials. This strategy has been employed in [**16**] and the resulting superpolynomial reads
(5.11)

$$\mathcal{P}_n(\mathbf{4_1}; a, q, t) = \sum_{k=0}^{\infty} (-1)^k a^{-k} t^{-2k} q^{-k(k-3)/2} \frac{(-atq^{-1}, q)_k}{(q, q)_k} (q^{1-n}, q)_k (-at^3 q^{n-1}, q)_k \,.$$

An independent, though not entirely unrelated derivation that gives the same result has been proposed in [**24**]. Explicit values of $\mathcal{P}_n(\mathbf{4_1}; a, q, t)$ for low values of n are given in table 3; note that they are all polynomials with positive coefficients, as necessarily expected from (3.2). We stress that (5.11) is in itself a very strong result, which illustrates the power of physics methods; to appreciate this fact and to confirm the validity of the above result we note that:

- for $a = q^2$ and $t = -1$, the formula (5.11) reduces to the familiar expression for the colored Jones polynomial studied *e.g.* in [**17**, **23**]:

$$J_n(\mathbf{4_1}; q) = \mathcal{P}_n(\mathbf{4_1}; q^2, q, -1) = \sum_{k=0}^{n-1} q^{nk} (q^{-n-1}, q^{-1})_k (q^{-n+1}, q)_k$$

- for $t = -1$, (5.11) agrees with the colored HOMFLY polynomial given in the unpublished work [**27**], which was also used in the analysis of [**2**] (for precise relation see [**15**]);

- for $n = 2$ the superpolynomial (5.11) agrees with the known result given *e.g.* in [**13**] (to match conventions we need to replace a and q in [**13**] respectively by $a^{1/2}$ and $q^{1/2}$);
- for $n = 3$ and $n = 4$ the expression (5.11) reproduces results given in [**21**];
- for $a = -q^j t^k$ the expression (5.11) correctly reproduces specializations predicted from the colored / canceling differentials with (a, q, t)-grading $(-1, j, k)$, see [**21**].

n	$\mathcal{P}_n(\mathbf{4_1}; a, q, t)$
1	1
2	$a^{-1}t^{-2} + t^{-1}q^{-1} + 1 + qt + at^2$
3	$a^{-2}q^{-2}t^{-4} + (a^{-1}q^{-3} + a^{-1}q^{-2})t^{-3} + (q^{-3} + a^{-1}q^{-1} + a^{-1})t^{-2} +$ $+(q^{-2} + q^{-1} + a^{-1} + a^{-1}q)t^{-1} + (q^{-1} + 3 + q) + (q^2 + q + a + aq^{-1})t +$ $+(q^3 + aq + a)t^2 + (aq^3 + aq^2)t^3 + a^2q^2t^4$
4	$1 + (1 + a^{-1}qt^{-1})(1 + a^{-1}t^{-1})(1 + a^{-1}q^{-1}t^{-1})\times$ $\times(1 + a^{-1}q^{-3}t^{-3})(1 + a^{-1}q^{-4}t^{-3})(1 + a^{-1}q^{-5}t^{-3})a^3q^6t^6 +$ $+(1 + q + q^2)(1 + a^{-1}qt^{-1})(1 + a^{-1}q^{-3}t^{-3})at^2 +$ $+(1 + q + q^2)(1 + a^{-1}qt^{-1})(1 + a^{-1}t^{-1})(1 + a^{-1}q^{-3}t^{-3})(1 + a^{-1}q^{-4}t^{-3})a^2q^2t^4$

TABLE 3. The colored superpolynomial of the $\mathbf{4_1}$ knot for $n = 1, 2, 3, 4$.

Recursion relations satisfied by (5.11) can be found using the Mathematica package qZeil.m developed by In the notation of (3.8) these recursions take form

$$\hat{A}^{\text{super}}(\hat{x}, \hat{y}; a, q, t) = a_0 + a_1\hat{y} + a_2\hat{y}^2 + a_3\hat{y}^3, \tag{5.12}$$

where

$$\begin{aligned}
a_0 &= \frac{at^3(1-\hat{x})(1-q\hat{x})(1+at^3q^2\hat{x}^2)(1+at^3q^3\hat{x}^2)}{q^3(1+at^3\hat{x})(1+at^3\hat{x}^2)(1+at^3q\hat{x})(1+at^3q^{-1}\hat{x}^2)} \\
a_1 &= -\frac{(1-q\hat{x})(1+at^3q^3\hat{x}^2)}{tq^3\hat{x}^2(1+at^3\hat{x})(1+at^3q\hat{x})(1+at^3q^{-1}\hat{x}^2)} \\
&\quad \times\Big(1 - t(t-1)q\hat{x} + at^3q^{-1}(1+q^3+qt+q^2t)\hat{x}^2 \\
&\quad -at^4(q+q^2+t+q^3t)\hat{x}^3 - a^2(t-1)t^6q\hat{x}^4 - a^2t^8q^2\hat{x}^5\Big) \\
a_2 &= -\frac{(1+at^3q^2\hat{x}^2)}{at^2q^2\hat{x}^2(1+at^3\hat{x}^2)(1+at^3q\hat{x})} \\
&\quad \times\Big(1 - at(t-1)\hat{x} + at^2(q+q^2+t+q^3t)\hat{x}^2 \\
&\quad +a^2t^4(1+q^3+qt+q^2t)\hat{x}^3 + a^2(t-1)t^5q^3\hat{x}^4 + a^3t^7q^3\hat{x}^5\Big) \\
a_3 &= 1
\end{aligned}$$

Taking the classical limit $q \to 1$ (and clearing the denominators), we find the following classical super-A-polynomial

$$
\begin{aligned}
A^{\text{super}}(x,y;a,t) \;&=\; a^2t^5(x-1)^2x^2 + at^2x^2(1+at^3x)^2y^3 \\
&+ at(x-1)(1+t(1-t)x + 2at^3(t+1)x^2 - 2at^4(t+1)x^3 \\
&+ a^2t^6(1-t)x^4 - a^2t^8x^5)y - (1+at^3x)(1+at(1-t)x + 2at^2(t+1)x^2 \\
&+ 2a^2t^4(t+1)x^3 + a^2t^5(t-1)x^4 + a^3t^7x^5)y^2 .
\end{aligned} \tag{5.13}
$$

The coefficients of the monomials in this polynomial are assembled into a matrix form presented in figure 3, and the corresponding Newton polygon is given in figure 4.

$$
\begin{pmatrix}
0 & -a\,t & -1 & 0 \\
0 & a\,t - a\,t^2 + a\,t^3 & -a\,t + a\,t^2 - a\,t^3 & 0 \\
a^2\,t^5 & a\,t^2 - a\,t^3 - 2\,a^2\,t^4 - 2\,a^2\,t^5 & -2\,a\,t^2 - 2\,a\,t^3 - a^2\,t^4 + a^2\,t^5 & a\,t^2 \\
-2\,a^2\,t^5 & 2\,a^2\,t^4 + 4\,a^2\,t^5 + 2\,a^2\,t^6 & -2\,a^2\,t^4 - 4\,a^2\,t^5 - 2\,a^2\,t^6 & 2\,a^2\,t^5 \\
a^2\,t^5 & -2\,a^2\,t^5 - 2\,a^2\,t^6 - a^3\,t^7 + a^3\,t^8 & a^2\,t^5 - a^2\,t^6 - 2\,a^3\,t^7 - 2\,a^3\,t^8 & a^3\,t^8 \\
0 & a^3\,t^7 - a^3\,t^8 + a^3\,t^9 & -a^3\,t^7 + a^3\,t^8 - a^3\,t^9 & 0 \\
0 & -a^3\,t^9 & -a^4\,t^{10} & 0
\end{pmatrix}
$$

FIGURE 3. Matrix form of the super-A-polynomial for the figure-eight knot. The conventions are the same as in the unknot example in figure 2.

According to the Conjecture 1, we should be able to reproduce the same polynomial from the asymptotic behavior of the colored superpolynomial (5.11). This is indeed the case. To show this, we introduce the variable $z = e^{\hbar k}$. Then, in the limit (3.4) with $z = \text{const}$ the sum over k in (5.11) can be approximated by the integral

$$
\mathcal{P}_n(\mathbf{4_1}; a,q,t) \;\sim\; \int dz\; e^{\frac{1}{\hbar}\left(\widetilde{\mathcal{W}}(\mathbf{4_1};z,x) + \mathcal{O}(\hbar)\right)} . \tag{5.14}
$$

The potential $\widetilde{\mathcal{W}}(\mathbf{4_1}; z, x)$ can be determined from the asymptotics (5.2):

$$
\begin{aligned}
\widetilde{\mathcal{W}}(\mathbf{4_1};z,x) &= \pi i \log z - \frac{\pi^2}{6} - (\log a + 2\log t)\log z - \frac{1}{2}(\log z)^2 \\
&+ \mathrm{Li}_2(x^{-1}) - \mathrm{Li}_2(x^{-1}z) + \mathrm{Li}_2(-at) - \mathrm{Li}_2(-atz) + Li_2(-axt^3) \\
&- \mathrm{Li}_2(-axt^3z) - \mathrm{Li}_2(z) .
\end{aligned} \tag{5.15}
$$

At the saddle point

$$
\left.\frac{\partial \widetilde{\mathcal{W}}(\mathbf{4_1}; z, x)}{\partial z}\right|_{z=z_0} = 0 \tag{5.16}
$$

it determines the leading asymptotic behavior (3.5), which at the same time is also computed by the integral along the curve (3.6), implying the key identity

$$
y = \exp\left(x \frac{\partial \widetilde{\mathcal{W}}(\mathbf{4_1}; z_0, x)}{\partial x} \right) . \tag{5.17}
$$

Plugging the expression (5.15) to the above two equations we obtain the following system

$$\left\{ \begin{array}{l} 1 = \frac{(x-z_0)(1+atz_0)(1+at^3xz_0)}{at^2xz_0(z_0-1)} \\ y = \frac{(x-1)(1+at^3xz_0)}{(1+at^3x)(x-z_0)} \end{array} \right. \tag{5.18}$$

Eliminating z_0 from these two equations we indeed reproduce the super-A-polynomial (5.13). Overall, the above statements verify the validity of the Conjecture 1 and Conjecture 2 for the figure-eight knot.

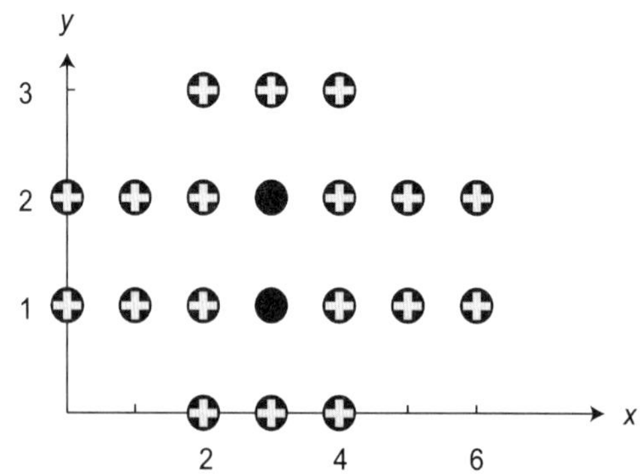

FIGURE 4. Newton polygon of the super-A-polynomial for the figure-eight knot and its $a = -t = 1$ limit. The conventions are the same as in figure 2. *(The color version of this graphic appears in the online version of the book.)*

Note that for $t = -1$ and $a = 1$, the expression (5.13) reduces to

$$A(x,y) = (x-1)^2(y-1)\Big(x^2(y^2+1) - (1-x-2x^2-x^3+x^4)y\Big), \tag{5.19}$$

which, apart from the $(x-1)^2$ factor, reproduces the A-polynomial of the $\mathbf{4_1}$ knot, including the $(y-1)$ factor representing the contribution of abelian flat connections. We stress that both the factorization and the explicit form of this abelian branch is seen only in the limit $a = -t = 1$ and is completely "mixed" with the other branches otherwise. In general the super-A-polynomial (5.13) does not factorize, as is also the case for the unknot and torus knots that will be discussed next.

More generally, after a simple change of variables

$$Q = a, \qquad \beta = x, \qquad \alpha = y\frac{1-\beta Q}{Q(1-\beta)}, \tag{5.20}$$

and for $t = -1$ we find that (5.13) becomes

$$\begin{aligned} A^{\text{Q-def}}(\alpha,\beta,Q) &= \frac{Q^2(1-\beta)^2}{\beta Q - 1}\Big((\beta^2 - Q\beta^3) + (2\beta - 2Q^2\beta^4 + Q^2\beta^5 - 1)\alpha + \\ &\quad +(1 - 2Q\beta + 2Q^2\beta^4 - Q^3\beta^5)\alpha^2 + Q^2(\beta-1)\beta^2\alpha^3\Big). \end{aligned}$$

Up to the first fraction, the expression in the big bracket reproduces the Q-deformed A-polynomial given in [**2**]. A related change of variables (for details see [**15**]) reveals the relation to the augmentation polynomial of [**35**].

5.3. Trefoil knot. In this section, we derive the classical and quantum super-A-polynomial for the trefoil knot (*i.e.* $(2,3)$ torus knot, also denoted $T^{(2,3)}$ or $\mathbf{3_1}$) and verify the validity of the Conjecture 1 and 2 for this knot. The analysis follows the same lines as in previous sections, and its starting point is the expression for the colored superpolynomial. We can provide such an expression from two sources. First, the colored superpolynomial for general $(2,2p+1)$ torus knot was derived in [**3**] from the perspective of the refined Chern-Simons theory. This superpolynomial is given in (5.30), as we will need it for the analysis of general torus knots in the next section. Even though in this section we only need $p=1$ specialization of (5.30), this is still quite an intricate expression. On the other hand, the analysis of constraints arising from the action of various differentials leads to the following expression

$$\mathcal{P}_n(\mathbf{3_1};a,q,t)=\sum_{k=0}^{n-1} a^{n-1}t^{2k}q^{n(k-1)+1}\frac{(q^{n-1},q^{-1})_k(-atq^{-1},q)_k}{(q,q)_k}\,, \tag{5.21}$$

and one can verify that this is equal to $p=1$ specialization of (5.30). Explicit values of $\mathcal{P}_n(\mathbf{3_1};a,q,t)$ following from (5.21) for low values of n are given in table 4. Again, note that they are polynomials with positive coefficients, as expected from (3.2).

n	$\mathcal{P}_n(\mathbf{3_1};a,q,t)$
1	1
2	$aq^{-1}+aqt^2+a^2t^3$
3	$a^2q^{-2}+a^2q(1+q)t^2+a^3(1+q)t^3+a^2q^4t^4+a^3q^3(1+q)t^5+a^4q^3t^6$
4	$a^3q^{-3}+a^3q(1+q+q^2)t^2+a^4(1+q+q^2)t^3+a^3q^5(1+q+q^2)t^4+$ $+a^4q^4(1+q)(1+q+q^2)t^5+a^3q^4(a^2+a^2q+a^2q^2+q^5)t^6+$ $+a^4q^8(1+q+q^2)t^7+a^5q^8(1+q+q^2)t^8+a^6q^9t^9$

TABLE 4. Colored superpolynomial of the $\mathbf{3_1}$ knot for $n=1,2,3,4$.

From the explicit form of the colored superpolynomial (5.21) we find the recursion relation it satisfies by using the Mathematica package qZeil.m, This recursion relation takes form

$$\hat{A}^{\text{super}}(\hat{x},\hat{y};a,q,t)=a_0+a_1\hat{y}+a_2\hat{y}^2\,, \tag{5.22}$$

where

$$\begin{aligned}
a_0 &= \frac{a^2t^4(\hat{x}-1)\hat{x}^3(1+aqt^3\hat{x}^2)}{q(1+at^3\hat{x})(1+at^3q^{-1}\hat{x}^2)}\\
a_1 &= -\frac{a(1+at^3\hat{x}^2)\big(q-q^2t^2\hat{x}+t^2(q^2+q^3+at+aq^2t)\hat{x}^2+aq^2t^5\hat{x}^3+a^2qt^6\hat{x}^4\big)}{q^2(1+at^3\hat{x})(1+at^3q^{-1}\hat{x}^2)}\\
a_2 &= 1
\end{aligned}$$

The classical super-A-polynomial for trefoil knot follows from the $q \to 1$ limit of $\hat{A}^{\rm super}$ and reads

$$
(5.23)\quad A^{\rm super}(x,y;a,t) \;=\; a^2t^4(x-1)x^3 + (1+at^3x)y^2 + \\ -a\big(1-t^2x+2t^2(1+at)x^2+at^5x^3+a^2t^6x^4\big)y\,.
$$

Matrix form of this polynomial is presented in figure 5, and its Newton polygon is shown in figure 6.

$$
\begin{pmatrix}
0 & -a & 1 \\
0 & a\,t^2 & a\,t^3 \\
0 & -2\,a\,t^2 - 2\,a^2\,t^3 & 0 \\
-a^2\,t^4 & -a^2\,t^5 & 0 \\
a^2\,t^4 & -a^3\,t^6 & 0
\end{pmatrix}
$$

FIGURE 5. Matrix form of the super-A-polynomial for the trefoil knot. The conventions are the same as in figure 2.

The same polynomial can be derived from the asymptotic behavior of the colored superpolynomial (5.21). Using integral representation as in (5.14),

$$
(5.24)\qquad \mathcal{P}_n(\mathbf{3_1};a,q,t) \sim \int dz\; e^{\frac{1}{\hbar}\left(\widetilde{\mathcal{W}}(\mathbf{3_1};z,x)+\mathcal{O}(\hbar)\right)},
$$

with the potential

$$
(5.25)\qquad \widetilde{\mathcal{W}}(\mathbf{3_1};z,x) = -\frac{\pi^2}{6} + \big(\log z + \log a\big)\log x + 2(\log t)(\log z) \\ +\mathrm{Li}_2(xz^{-1}) - \mathrm{Li}_2(x) + \mathrm{Li}_2(-at) - \mathrm{Li}_2(-atz) + \mathrm{Li}_2(z)\,,
$$

and in the limit (3.4) with $z = e^{\hbar k} = const$, we find that equations (5.16) and (5.17) take form

$$
(5.26)\qquad \begin{cases} 1 = \frac{t^2x(x-z_0)(1+atz_0)}{z_0(z_0-1)} \\ y = \frac{az_0^2(x-1)}{(x-z_0)} \end{cases}
$$

Eliminating z_0 from these two equations we reproduce the super-A-polynomial (5.23). We conclude that both Conjecture 1 and Conjecture 2 hold true for the trefoil knot.

We also note that for $t=-1$ and $a=1$ the super-A-polynomial (5.23) reduces to

$$
(5.27)\qquad A(x,y) = -(x-1)(y-1)(y+x^3)\,,
$$

which reproduces the well known A-polynomial for the trefoil knot, including the $(y-1)$ factor associated with abelian flat connections (and the overall immaterial factor $x-1$). More generally, under a change of variables

$$
(5.28)\qquad Q=a,\quad \beta = x,\quad \alpha = yQ^{-1}\beta^{-6}\frac{1-Q\beta}{1-\beta}\,,
$$

and in $t=-1$ limit, (5.23) reduces (up to an overall factor) to

$$
(5.29)\\ A(\alpha,\beta,Q) = (1-Q\beta) + (\beta^3-\beta^4+2\beta^5-2Q\beta^5-Q\beta^6+Q^2\beta^7)\alpha + (-\beta^9+\beta^{10})\alpha^2,
$$

which reproduces the Q-deformed or augmentation polynomial for the trefoil knot found in [**2**, **35**]. Relations between super-A-polynomial, Q-deformed A-polynomial and augmentation polynomial for torus knots are discussed in much more detail in [**15**].

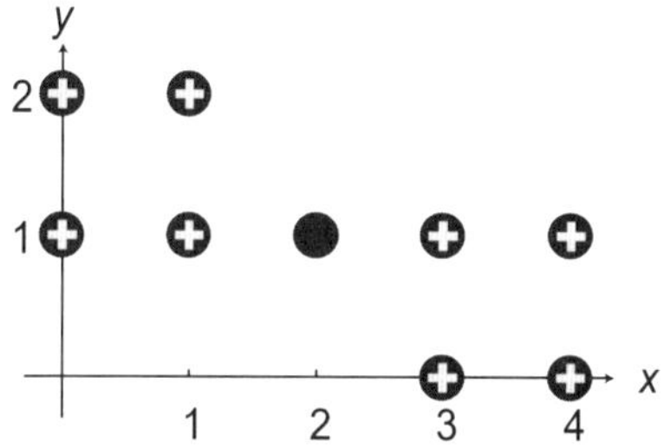

FIGURE 6. Newton polygon of the super-A-polynomial for the trefoil knot and its $a = -t = 1$ limit. The conventions are the same as in the unknot case in figure 2. *(The color version of this graphic appears in the online version of the book.)*

5.4. $(2, 2p+1)$ **torus knots.** As the last class of examples we discuss the entire family of $(2, 2p+1)$ torus knots, which are also denoted $T^{(2,2p+1)}$. The S^r-colored superpolynomials for this family can be found both from refined Chern-Simons theory, as well as from analysis of differentials. The former approach is described in detail in [**3**] and it leads (after taking into account appropriate γ-factors and other subtleties) to the expression for the reduced colored superpolynomial as the ratio of (refined) Chern-Simons partition functions in $\S^3$ in presence of a given knot and the unknot

$$\begin{aligned}
&\mathcal{P}^{S^r}(T^{(2,2p+1)}; a, q, t) = \\
&= (-1)^{pr}\left(\frac{q_1}{q_2}\right)^{\frac{pr}{2}} \frac{Z^{\mathrm{ref}}_{SU(N)}(\S^3, T^{(2,2p+1)}_{\Lambda^r}; q_1, q_2)}{Z^{\mathrm{ref}}_{SU(N)}(\S^3, \bigcirc_{\Lambda^r}; q_1, q_2)} \\
&= \sum_{\ell=0}^{r} \frac{(qt^2;q)_\ell(-at^3;q)_{r+\ell}(-aq^{-1}t;q)_{r-\ell}(q;q)_r}{(q;q)_\ell(q^2t^2;q)_{r+\ell}(q;q)_{r-\ell}(-at^3;q)_r} \frac{(1-q^{2\ell+1}t^2)}{(1-qt^2)} \\
&\quad \times(-1)^{n-1}a^{-\frac{r}{2}}q^{\frac{3(n-1)}{2}-\ell}t^{-(n-1)p-\ell+\frac{r}{2}}\left[(-1)^\ell a^{\frac{r}{2}} q^{\frac{r^2-\ell(\ell+1)}{2}} t^{\frac{3r}{2}-\ell}\right]^{2p+1}.
\end{aligned} \tag{5.30}$$

The second approach, based on analysis of differentials, has been employed in [**16**] and results in the following form of the superpolynomial

$$\begin{aligned}
\mathcal{P}^{S^r}(T^{2,2p+1}; a, q, t) &= a^{pr}q^{-pr} \sum_{0\le k_p\le\ldots\le k_2\le k_1\le r} \begin{bmatrix} r \\ k_1 \end{bmatrix}\begin{bmatrix} k_1 \\ k_2 \end{bmatrix}\cdots\begin{bmatrix} k_{p-1} \\ k_p \end{bmatrix} \times \\
&\times\ q^{(2r+1)(k_1+k_2+\ldots+k_p)-\sum_{i=1}^{p}k_{i-1}k_i}t^{2(k_1+k_2+\ldots+k_p)}\prod_{i=1}^{k_1}(1+aq^{i-2}t),
\end{aligned} \tag{5.31}$$

where $k_0 = r$. One can check that (5.30) and (5.31) agree up to relatively high values of r; it would be nice to find an analytic proof valid for all r. We are however convinced that both expressions are equal, and various consequences of this fact – in particular dualities between different UV descriptions of the corresponding $\mathcal{N}=2$, 3d SUSY theory (see section 7) – were presented in [**16**].

For the purpose of this presentation we will focus on the expression the colored superpolynomial in the form (5.30). In the asymptotic limit $\hbar \to 0$ limit we find

$$\mathcal{P}^{S^r}(T^{(2,2p+1)};a,q,t) \sim \int dz\, e^{\frac{1}{\hbar}\left(\widetilde{\mathcal{W}}(T^{(2,2p+1)};z,x)+\mathcal{O}(\hbar)\right)}, \tag{5.32}$$

with the potential

$$\begin{aligned}
&\widetilde{\mathcal{W}}(T^{(2,2p+1)};z,x) = \\
&= p\log(a)\cdot\log x - p\log(-t)\cdot\log x + (p+1)\pi i\log x + \log(x^{\frac{1}{2}}z^{-1})\cdot\log t \\
&\qquad +(2p+1)\left(\pi i\log z + \frac{1}{2}\left((\log x)^2-(\log z)^2\right) + \log(x^{\frac{3}{2}}z^{-1})\cdot\log t\right) \\
&\qquad +\mathrm{Li}_2(z)-\mathrm{Li}_2(x)-\mathrm{Li}_2(t^2z)+\mathrm{Li}_2(-at^3x)+\mathrm{Li}_2(t^2xz) \\
&\qquad -\mathrm{Li}_2(-at^3xz)+\mathrm{Li}_2(xz^{-1})-\mathrm{Li}_2(-atxz^{-1})+\mathrm{Li}_2(-at)-\mathrm{Li}_2(1),
\end{aligned} \tag{5.33}$$

where $z=q^\ell$.

For the above potential $\widetilde{\mathcal{W}}(T^{(2,2p+1)};z,x)$, the critical point condition can simply be expressed as $1=\exp\left(z\partial\widetilde{\mathcal{W}}/\partial z\right)|_{z=z_0}$:

$$1 = -\frac{t^{-2-2p}(x-z_0)z_0^{-1-2p}(-1+t^2z_0)(1+at^3xz_0)}{(-1+z_0)(atx+z_0)(-1+t^2xz_0)}. \tag{5.34}$$

and

$$\begin{aligned}
y(x,t,a) &= \exp\left(x\frac{\partial\widetilde{\mathcal{W}}(T^{(2,2p+1)};z_0,x)}{\partial x}\right) \\
&= \frac{a^pt^{2+2p}(-1+x)x^{1+2p}(atx+z_0)(1+at^3xz_0)}{(1+at^3x)(x-z_0)(-1+t^2xz_0)}.
\end{aligned} \tag{5.35}$$

Eliminating z_0 from the above equations, we find the super-A-polynomial $A^{\text{super}}(x,y;a,t)$ for any $(2,2p+1)$ torus knot. For small values of p, the resulting super-A-polynomials are listed in table 5 (where we omitted the extra factors which appear in the elimination, and picked up the factor that includes the non-abelian branch of the $SL(2)$ character variety). In particular, for $p=1$, we obtain (up to an inessential overall factor) the same super-A-polynomial for the trefoil (5.23), which was derived in the previous section starting from another expression for colored superpolynomials (5.21).

For $a=1$ the above super-A-polynomials reduce to the refined A-polynomials of [**3**]. On the other hand, for $t=-1$ we find the Q-deformed A-polynomial of [**2**] if the following identification of parameters is performed

$$Q=a,\quad \beta=x,\quad \alpha=yQ^{-p}\beta^{-4p-2}\frac{1-Q\beta}{1-\beta}, \tag{5.36}$$

and a related transformation reveals the form of the augmentation polynomial of [**35**]. Precise derivation of the above variable change, as well as explicit relations between super-A-polynomial, the augmentation polynomial and Q-deformed A-polynomial, are discussed in detail in [**15**].

Knot	$A_K^{\mathrm{super}}(x,y;a,t)$
$T^{(2,3)}$	$y^2+\frac{1}{1+at^3x}a(-1+t^2x-2t^2x^2-2at^3x^2-at^5x^3-a^2t^6x^4)y+\frac{(x-1)a^2t^4x^3}{1+at^3x}$
$T^{(2,5)}$	$y^3-\frac{a^2}{1+at^3x}\,(1-t^2x+2t^2x^2+2at^3x^2-2t^4x^3-2at^5x^3+3t^4x^4+4at^5x^4+a^2t^6x^4+at^7x^5$
	$-a^2t^8x^5+2a^2t^8x^6)y^2$
	$+\frac{a^4t^6(-1+x)x^5}{(1+at^3x)^2}\,(2-t^2x+at^3x+3t^2x^2+4at^3x^2+a^2t^4x^2+2at^5x^3+2a^2t^6x^3+2a^2t^6x^4$
	$+2a^3t^7x^4+a^3t^9x^5+a^4t^{10}x^6)y$
	$-\frac{a^6t^{12}(-1+x)^2x^{10}}{(1+at^3x)^2}$
$T^{(2,7)}$	$y^4-\frac{a^3}{1+at^3x}\,(1-t^2x+2t^2x^2+2at^3x^2-2t^4x^3-2at^5x^3+3t^4x^4+4at^5x^4+a^2t^6x^4-3t^6x^5$
	$-4at^7x^5-a^2t^8x^5+4t^6x^6+6at^7x^6+2a^2t^8x^6+at^9x^7-2a^2t^{10}x^7+3a^2t^{10}x^8)y^3$
	$+\frac{a^6t^8(-1+x)x^7}{(1+at^3x)^2}\,(3-2t^2x+at^3x+6t^2x^2+8at^3x^2+2a^2t^4x^2-3t^4x^3-2at^5x^3$
	$+a^2t^6x^3+6t^4x^4+12at^5x^4+10a^2t^6x^4+4a^3t^7x^4+3at^7x^5+2a^2t^8x^5-a^3t^9x^5$
	$+6a^2t^8x^6+8a^3t^9x^6+2a^4t^{10}x^6+2a^3t^{11}x^7-a^4t^{12}x^7+3a^4t^{12}x^8)y^2$
	$-\frac{a^9t^{16}(-1+x)^2x^{14}}{(1+at^3x)^3}\,(3-t^2x+2at^3x+4t^2x^2+6at^3x^2+2a^2t^4x^2+3at^5x^3+4a^2t^6x^3$
	$+a^3t^7x^3+3a^2t^6x^4+4a^3t^7x^4+a^4t^8x^4+2a^3t^9x^5+2a^4t^{10}x^5+2a^4t^{10}x^6$
	$+2a^5t^{11}x^6+a^5t^{13}x^7+a^6t^{14}x^8)y+\frac{a^{12}t^{24}(-1+x)^3x^{21}}{(1+t^3x)^3}$

TABLE 5. Super-A-polynomials for $(2,2p+1)$ torus knots with $p=1,2,3$.

6. Quantizability

In this section we discuss the super-A-polynomials that we found from the viewpoint of quantizability, by which we mean the following. For the Conjecture 1 to be formulated in a consistent way, we must ensure that the leading term $\int \log y \frac{dx}{x}$ in the integral (3.5) is well-defined, *i.e.* does not depend on the choice of the integration path on the algebraic curve (3.6). As explained in [**19, 22**], this requirement imposes the following constraints on the periods of the imaginary and real parts of $\log y \frac{dx}{x}$, respectively,

$$\oint_\gamma \Big(\log|x| d(\arg y) - \log|y| d(\arg x) \Big) \quad = \quad 0\,, \tag{6.1}$$

$$\frac{1}{4\pi^2}\oint_\gamma \Big(\log|x| d\log|y| + (\arg y) d(\arg x) \Big) \quad \in \quad \mathbb{Q}\,, \tag{6.2}$$

for all *closed* paths γ on the curve (3.6). It turns out that these conditions can be further reformulated and interpreted in a variety of ways. On one hand, it is amusing to observe that the integrand $\eta(x,y) = \log|x|d(\arg y) - \log|y|d(\arg x)$ in (6.1) is the image of the symbol $\{x,y\} \in K_2(\mathcal{C})$ under so-called regulator map, thereby constituting an immediate link to algebraic K-theory [**4–6**]. As discussed in [**22**], from this K-theory viewpoint the condition that the curve is quantizable can be rephrased simply as the requirement that $\{x,y\} \in K_2(\mathbb{C}(\mathcal{C}))$ is a torsion class. On the other hand, this more abstract condition also translates to the down-to-earth statement that quantizability of the curve requires its defining polynomial to be *tempered.*

By definition, a polynomial $A(x,y)$ is tempered if all roots of all face polynomials of its Newton polygon are roots of unity. Face polynomials are constructed as follows: we need to construct a Newton polygon corresponding to $A(x,y) = \sum_{i,j} a_{i,j} x^i y^j$, and to each point (i,j) of this polygon we associate the coefficient $a_{i,j}$. We label consecutive points along each face of the polygon by integers

$k = 1, 2, \dots$ and, for a given face, rename monomial coefficients associated to these points as a_k. Then, the face polynomial associated to a given face is defined to be $f(z) = \sum_k a_k z^k$. Therefore, the quantizability condition requires that all roots of $f(z)$ constructed for all faces of the Newton polygon must be roots of unity. In what follows we are going to examine super-A-polynomials which we found in examples in section 5 from this perspective.

Ordinary A-polynomials have numerical, integer coefficients [**9**], and therefore the above quantization condition imposes certain constraints on values of these coefficients. For example, the ordinary A-polynomial of the figure-eight knot given in (5.19) satisfies these constraints, while its close cousin with only slightly different coefficients, discussed *e.g.* in [**19, 22**], does not. Meeting these tight constraints might seem much less trivial in the case of t- or a-deformed curve, when coefficients of the defining polynomial depend on these extra parameters. Nonetheless, this is indeed possible and the outcome is very simple: the quantization condition implies that both a and t must be roots of unity. Therefore, even though such a and t cannot be completely arbitrary, they still take values in a dense set of points (on a unit circle). Below we verify that indeed all super-A-polynomials discussed in section 5 are tempered (and therefore quantizable) as long as both a and t are roots of unity. This condition very nicely fits with the fact that in specialization from colored superpolynomial or HOMFLY polynomial to $sl(N)$ quantum group invariant we substitute $a = q^N$ and in Chern-Simons theory with $SU(N)$ gauge group q is required to be a root of unity, so that $a = q^N$ is automatically a root of unity as well.

face	face polynomial
N	$z + at$
NE	$z + at^2$
E	$at^2(z + at^3)^2$
SE	$a^3t^8(z - at^2)$
S	$a^3t^9(z + at)$
SW	$a^2t^5(z - at^4)$
W	$a^2t^5(z - 1)^2$
NW	$at(z - at^4)$

TABLE 6. Face polynomials for the figure-eight knot, corresponding to faces of the octagonal shape formed by non-zero entries of the coefficient matrix in figure 3. Faces are labeled by compass directions (with N standing for North, *etc.*), with the first row $(0, -at, -1, 0)$ of the matrix in figure 3 located in the North.

Let us now illustrate the above claim in the examples of various knots discussed in section 5. For each of those knots we construct a Newton polygon and face polynomials of the corresponding super-A-polynomials. In order to construct face polynomials it is convenient to write down a matrix representation of the super-A-polynomials. For instance, for the unknot the Newton polygon and the corresponding matrix representation are shown in figure 2. In this case, it is clear that roots of face polynomials are all roots of unity if a and t are roots of unity. In fact, the unknot is so simple that even a weaker condition is sufficient to hold, namely that the combination at^3 is a root of unity.

The matrix coefficients and the Newton polygon for figure-eight knot are given, respectively, in figures 3 and 4, and the corresponding face polynomials are presented in table 6. The face polynomials are manifestly written as products of linear factors, and being tempered requires that both a and t are roots of unity. An analogous condition holds also for $(2, 2p+1)$ torus knots whose Newton polygons have hexagonal shape, and the corresponding face polynomials are given in table 7. Quantizability conditions are also met for other knots, as verified in [**16**]. To sum up, from all these examples we conclude that super-A-polynomials are quantizable if both a and t are roots of unity; we conjecture that this is the case for all knots.

face	face polynomial
first column	$-(at^2)^{p(p+1)}(z-1)^p$
last column	$(-1)^p(z+at^3)^p$
first row	za^p-1
last row	$-(at^2)^{p(p+1)}\big(z-(at^2)^p\big)$
lower diagonal	$(-1)^p\big(at^3)^p(z-a^{p+1}t^{2p+1}\big)^p$
upper diagonal	$(-1)^{p+1}a^p\big(z+a^pt^{2p+2}\big)^p$

TABLE 7. Face polynomials for $(2, 2p+1)$ torus knots, corresponding to faces of the hexagonal shape formed by non-zero entries of the coefficient matrices for $(2, 2p+1)$ torus knots, such as the matrix for the trefoil in figure 3.

7. Interpretation in 3d, $\mathcal{N} = 2$ theories

The objects we have considered so far, such as super-A-polynomials and colored superpolynomials, also have an interesting interpretation in 3d, $\mathcal{N} = 2$ SUSY gauge theories. We have already recalled that knot invariants can be described in terms of three-dimensional Chern-Simons theory, and from physics perspective the connection between these two classes of theories arises as a 3d-3d duality associated to complementary compactifications of M5-brane along appropriate three dimensions of its $3+3$ dimensional world-volume [**7, 10, 12, 42**]. In particular, important properties of both three-dimensional theories (i.e. Chern-Simons and $\mathcal{N} = 2$ gauge theory) are encoded in the same algebraic curve. From the perspective of $\mathcal{N} = 2$ theories this curve plays a role to some extent analogous to the Seiberg-Witten curves of four-dimensional gauge theories [**39, 40**]. In what follows we explain that the 3d-3d duality can be extended to incorporate dependence on a and t. On Chern-Simons side this corresponds to considering refined Chern-Simons theory with $SL(N)$ gauge group, and on $\mathcal{N} = 2$ side these parameters can be interpreted as twisted mass parameters for certain global symmetries $U(1)_Q$ and

$U(1)_F$. In this context, the algebraic curve mentioned above is precisely the super-A-polynomial, and so it carries important information about $\mathcal{N}=2$ theories with those symmetries.

To start with, we recall that the parameter t, responsible for the "refinement" or "categorification", can be interpreted [**3**] as a twisted mass parameter for the global symmetry $U(1)_F$ in the effective three-dimensional $\mathcal{N}=2$ theory T_M associated to the knot complement $M = \S^3 \setminus K$:

$$M \quad \rightsquigarrow \quad T_M \,. \tag{7.1}$$

Moreover, generically, every charged chiral multiplet in a theory T_M contributes to the effective twisted chiral superpotential a dilogarithm term:

$$\text{chiral field } \phi \qquad \longleftrightarrow \qquad \begin{array}{c}\text{twisted superpotential}\\ \Delta\widetilde{\mathcal{W}}(\vec{x};t) = \mathrm{Li}_2\Big((-t)^{n_F}\prod_i (x_i)^{n_i}\Big)\end{array} \tag{7.2}$$

where n_F is the charge of the chiral multiplet under the global R-symmetry $U(1)_F$ and $\{n_i\}$ is our (temporary) collective notation for all other charges of ϕ under symmetries $U(1)_i$, some of which may be global flavor symmetries and some of which may be dynamical gauge symmetries, depending on the problem at hand.[1] In particular, in the former case, the vev of the corresponding twisted chiral multiplet is usually called the twisted mass parameter $\tilde{m}_i = \log x_i$, of which $\tilde{m}_F = \log(-t)$ is a prominent example.

The second commutative deformation parameter a also admits a similar interpretation as a twisted mass parameter for a global symmetry that we denote $U(1)_Q$:

$$\log a \;=\; \tilde{m}_Q \,. \tag{7.3}$$

In fact, in the case of the a-deformation this interpretation is even more obvious and can be easily seen in the brane picture, where it corresponds to one of the Kähler moduli of the underlying Calabi-Yau geometry X. For example, the effective low-energy theory on a toric brane in the conifold geometry has two chiral multiplets that come from two open BPS states shown in blue and red in figure 7.

In this example, the symmetry $U(1)_Q$ responsible for the a-deformation comes from the 2-cycle in the conifold geometry X. (The corresponding gauge field A_μ comes from the Kaluza-Klein reduction of the RR 3-form field, $C \sim A \wedge \omega$, and becomes the starting point for the geometric engineering of $\mathcal{N}=2$ gauge theories in four dimensions [**26**].) In a basis of refined open BPS states shown in figure 7, one state is charged under the symmetry $U(1)_Q$, while the other state is neutral. Therefore, the effective twisted superpotential $\widetilde{\mathcal{W}}(x;a,t)$ of the corresponding model has two dilogarithm terms, one of which depends on a and the other does not.

Returning to the general theory T_M, now we are ready to explain the connection between the twisted superpotential in this theory and the algebraic curve (3.6) defined as the zero locus of the super-A-polynomial. Roughly speaking, the curve (3.6) describes the SUSY vacua in the $\mathcal{N}=2$ theory T_M. To make this more precise, we need to recall that among the parameters x_i in (7.2) some correspond to vevs of dynamical fields (and, therefore, need to be integrated out) and some are

[1] Below we shall return to the different role of gauge and global symmetries, but for now we wish to point out a simple rule of thumb that one can read off the matter content of the theory T_M by counting dilogarithm terms in the function $\widetilde{\mathcal{W}}(\vec{x};t)$.

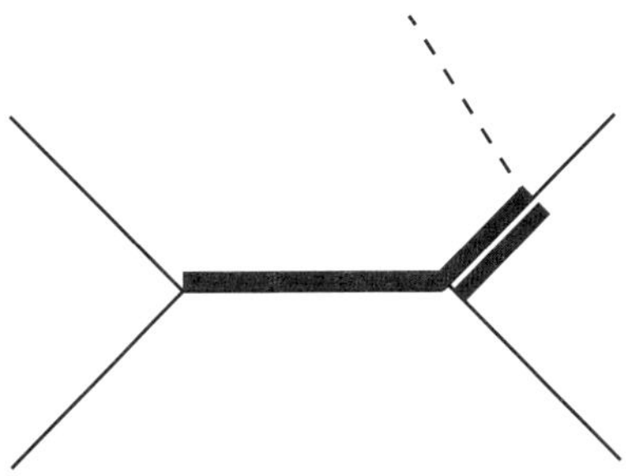

FIGURE 7. A toric Lagrangian brane in the conifold bounds two holomorphic disks (shown by red and blue intervals in the base of the toric geometry). *(The color version of this graphic appears in the online version of the book.)*

twisted masses for global flavor symmetries. To make the distinction clearer, let us denote the former by z_i (instead of x_i), so that the vevs of dynamical twisted chiral superfields are $\sigma_i = \log z_i$. Then, in order to find SUSY vacua of the theory T_M we need to extremize $\widetilde{\mathcal{W}}$ with respect to these dynamical fields,

$$\frac{\partial \widetilde{\mathcal{W}}}{\partial z_i} = 0\,. \tag{7.4}$$

This is exactly what we did *e.g.* in (5.16) when we extremized the potential function (5.15) for the figure-eight knot (*cf.* also (5.25) and (5.34) for the case of $(2, 2p+1)$ torus knots). Solving these equations for z_i and substituting the resulting values back into $\widetilde{\mathcal{W}}$ gives the effective twisted superpotential, $\widetilde{\mathcal{W}}_{\text{eff}}$, that depends only on twisted mass parameters associated with global symmetries of the $\mathcal{N}=2$ theory T_M.

Besides the symmetries $U(1)_F$ and $U(1)_Q$ which are responsible for t- and a-deformations, respectively, our $\mathcal{N}=2$ theories T_M come with additional global flavor symmetries, one for each component of the link K (or, more generally, one for every torus boundary of M). In particular, if K is a knot — which is what we assume throughout the present paper — then, in addition to $U(1)_F$ and $U(1)_Q$, there is only one extra global symmetry $U(1)_L$ with the corresponding twisted mass parameter that we simply denote $\tilde{m}$; it is $x = e^{\tilde{m}}$ that shortly will be identified with the variable by the same name in the super-A-polynomial. In the brane model,

$$\begin{array}{lccc} \text{space-time:} & \mathbb{R}^4 & \times & X \\ & \cup & & \cup \\ \text{D4-brane:} & \mathbb{R}^2 & \times & L \end{array} \tag{7.5}$$

this symmetry $U(1)_L$ can be identified with the gauge symmetry on the D4-brane supported on the Lagrangian submanifold $L \subset X$. The corresponding gauge field is dynamical when L has finite volume, while for non-compact L (of infinite volume) the symmetry $U(1)_L$ is a global symmetry. Moreover, the other global symmetry $U(1)_F$ that plays an important role in our discussion also can be identified in the brane setup (7.5): it corresponds to the rotation symmetry of the normal bundle of $\mathbb{R}^2 \subset \mathbb{R}^4$. We stress that this $U(1)_F$ symmetry is not simply an R-symmetry, but a certain modification thereof, as discussed in more detail in [**8**].

To summarize our discussion so far, we can incorporate $U(1)_Q$ and $U(1)_L$ charges in (7.2) and write the contribution of a chiral multiplet $\phi \in T_M$ to the twisted superpotential as

(7.6)

$$\text{chiral field } \phi \qquad \longleftrightarrow \qquad \begin{array}{l} \text{twisted superpotential} \\ \Delta\widetilde{\mathcal{W}}(x, z_i; a, t) = \mathrm{Li}_2\Big(a^{n_Q}(-t)^{n_F} x^{n_L} \prod_i (z_i)^{n_i}\Big) \end{array}$$

Using this dictionary and dilogarithm identities, such as the inversion formula $\mathrm{Li}_2(x) = -\mathrm{Li}_2\left(\frac{1}{x}\right) - \frac{\pi^2}{6} - \frac{1}{2}[\log(-x)]^2$, from (5.15) and (5.25) it is easy to read off the spectrum of the theory T_M for the trefoil knot and for the figure-eight knot:[2]

trefoil knot

	ϕ_1	ϕ_2	ϕ_3	ϕ_4	ϕ_5	parameter
$U(1)_{\text{gauge}}$	-1	0	0	-1	1	z
$U(1)_F$	0	0	1	-1	0	$-t$
$U(1)_Q$	0	0	1	-1	0	a
$U(1)_L$	1	-1	0	0	0	x

figure-eight knot

	ϕ_1	ϕ_2	ϕ_3	ϕ_4	ϕ_5	ϕ_6	ϕ_7
$U(1)_{\text{gauge}}$	0	-1	0	-1	0	-1	-1
$U(1)_F$	0	0	1	-1	3	-3	0
$U(1)_Q$	0	0	1	-1	1	-1	0
$U(1)_L$	-1	1	0	0	1	-1	0

TABLE 8. Spectrum of the $\mathcal{N} = 2$ theory T_M for the trefoil and figure-eight knots.

The terms of lower transcendentality degree, *i.e.* products of ordinary logarithms, also admit a simple interpretation in three-dimensional $\mathcal{N} = 2$ gauge theory T_M. Notice that, in the collective notations $\{x_i\}$ for global and gauge symmetries $U(1)_i$ used in (7.2), the dependence of the twisted superpotential $\widetilde{\mathcal{W}}$ on $\log x_i$ is always quadratic, see *e.g.* (5.15) and (5.25). Such terms correspond to supersymmetric Chern-Simons couplings for $U(1)$ gauge (resp. background flavor) fields:

$$(7.7) \qquad \frac{k_{ij}}{4\pi}\int A_i \wedge dA_j + \ldots \qquad \longleftrightarrow \qquad \begin{array}{l} \text{twisted superpotential} \\ \Delta\widetilde{\mathcal{W}}(\vec{x}; a, t) = \frac{k_{ij}}{2} \log x_i \cdot \log x_j \end{array}$$

At this point, we should remind the reader that a given $\mathcal{N} = 2$ theory T_M may admit many dual UV descriptions, with different number of gauge groups and charged matter fields [**10**]. However, all of these dual descriptions lead to the same space of supersymmetric moduli (twisted mass parameters) once all dynamical multiplets are integrated out, *i.e.* once the twisted superpotential is extremized (7.4) with respect to all z_i.

The resulting "effective" twisted superpotential $\widetilde{\mathcal{W}}_{\text{eff}}(x; a, t)$ depends only on the twisted mass parameters associated with the global symmetries $U(1)_L$, $U(1)_Q$, and $U(1)_F$. Then, the algebraic curve (3.6) defined as the zero locus of the super-A-polynomial is simply a graph of the function $x\frac{\partial \widetilde{\mathcal{W}}_{\text{eff}}}{\partial x}$, which in a circle compactification of the theory T_M is interpreted as the effective FI parameter:

$$(7.8) \qquad \boxed{\mathcal{M}_{\text{SUSY}}: \quad A^{\text{super}}(x, y; a, t) = 0 \qquad \Leftrightarrow \qquad \log y = x\partial_x \widetilde{\mathcal{W}}_{\text{eff}}(x; a, t)}$$

[2]To avoid any confusion, we stress that the descriptions of theories in table 8 are only schematic. In particular those theories may have many flavor symmetries other than $U(1)_F$, $U(1)_Q$, or $U(1)_L$; also superpotential couplings are not specified in this table.

We note that an amusing example of dualities between various UV descriptions of the same $\mathcal{N} = 2$ theory, associated to arbitrary $(2, 2p + 1)$ torus knot, has been analyzed in detail in [**16**]. In this case two distinct UV descriptions arise from two different colored superpolynomials given in (5.30) and (5.31).

Acknowledgments

We thank Sergei Gukov for a very nice and fruitful collaboration that led to the discovery of the super-A-polynomial and other results reviewed in this note. We also thank Hidetoshi Awata, Satashi Nawata and Marko Stošić for collaborations on these and related topics. The work of H.F. is supported by the Grant-in-Aid for Young Scientists (B) [# 21740179] from the Japan Ministry of Education, Culture, Sports, Science and Technology, and the Grant-in-Aid for Nagoya University Global COE Program, "Quest for Fundamental Principles in the Universe: from Particles to the Solar System and the Cosmos." The work of P.S. is supported by the Foundation for Polish Science and the ERC Starting Grant no. 335739 *"Quantum fields and knot homologies"*, funded by the European Research Council under the European Union's Seventh Framework Programme.

References

[1] Mina Aganagic and Shamil Shakirov, *Knot Homology from Refined Chern-Simons Theory* (2011), arXiv:1105.5117.

[2] Mina Aganagic and Cumrun Vafa, *Large N Duality, Mirror Symmetry, and a Q-deformed A-polynomial for Knots*, (2012), arXiv:1204.4709.

[3] Hiroyuki Fuji, Sergei Gukov, Piotr Sułkowski, and Hidetoshi Awata, *Volume conjecture: refined and categorified*, Adv. Theor. Math. Phys. **16** (2012), no. 6, 1669–1777. MR3065082

[4] A. A. Beĭlinson, *Higher regulators and values of L-functions of curves* (Russian), Funktsional. Anal. i Prilozhen. **14** (1980), no. 2, 46–47. MR575206 (81k:14020)

[5] Spencer Bloch, *The dilogarithm and extensions of Lie algebras*, Algebraic K-theory, Evanston 1980 (Proc. Conf., Northwestern Univ., Evanston, Ill., 1980), Lecture Notes in Math., vol. 854, Springer, Berlin-New York, 1981, pp. 1–23. MR618298 (83b:17010)

[6] D. Boyd, F. Rodriguez-Villegas, and N. Dunfield, *Mahler's measure and the dilogarithm (ii)* (2003), arXiv: math/0308041.

[7] Sergio Cecotti, Clay Cordova, and Cumrun Vafa, *Braids, walls, and mirrors* (2011), arXiv:1110.2115.

[8] Hee-Joong Chung, Tudor Dimofte, Sergei Gukov, and Piotr Sulkowski, *3d-3d Correspondence Revisited*, (2014), arXiv:1405.3663.

[9] D. Cooper, M. Culler, H. Gillet, D. D. Long, and P. B. Shalen, *Plane curves associated to character varieties of* 3*-manifolds*, Invent. Math. **118** (1994), no. 1, 47–84, DOI 10.1007/BF01231526. MR1288467 (95g:57029)

[10] Tudor Dimofte, Davide Gaiotto, and Sergei Gukov, *Gauge theories labelled by three-manifolds* (2011), arXiv:1108.4389.

[11] Tudor Dimofte and Sergei Gukov, *Quantum field theory and the volume conjecture*, Interactions between hyperbolic geometry, quantum topology and number theory, Contemp. Math., vol. 541, Amer. Math. Soc., Providence, RI, 2011, pp. 41–67, DOI 10.1090/conm/541/10678. MR2796627 (2012c:58037)

[12] Tudor Dimofte, Sergei Gukov, and Lotte Hollands, *Vortex counting and Lagrangian 3-manifolds*, Lett. Math. Phys. **98** (2011), no. 3, 225–287, DOI 10.1007/s11005-011-0531-8. MR2852983

[13] Nathan M. Dunfield, Sergei Gukov, and Jacob Rasmussen, *The superpolynomial for knot homologies*, Experiment. Math. **15** (2006), no. 2, 129–159. MR2253002 (2008c:57020)

[14] P. Dunin-Barkowski, A. Mironov, A. Morozov, A. Sleptsov, and A. Smirnov, *Superpolynomials for toric knots from evolution induced by cut-and-join operators* (2011), arXiv:1106.4305.

[15] Hiroyuki Fuji, Sergei Gukov, and Piotr Sułkowski, *Super-A-polynomial for knots and BPS states*, Nuclear Phys. B **867** (2013), no. 2, 506–546, DOI 10.1016/j.nuclphysb.2012.10.005. MR2992793

[16] Hiroyuki Fuji, Sergei Gukov, Piotr Sulkowski, and Marko Stosic, *3d analogs of Argyres-Douglas theories and knot homologies*, JHEP **01** (2013), 175.

[17] Stavros Garoufalidis, *On the characteristic and deformation varieties of a knot*, Proceedings of the Casson Fest, Geom. Topol. Monogr., vol. 7, Geom. Topol. Publ., Coventry, 2004, pp. 291–309 (electronic), DOI 10.2140/gtm.2004.7.291. MR2172488 (2006j:57028)

[18] Eugene Gorsky, Alexei Oblomkov, Jacob Rasmussen, and Vivek Shende, *Torus knots and the rational DAHA*, Duke Math. J. **163** (2014), no. 14, 2709–2794, DOI 10.1215/00127094-2827126. MR3273582

[19] Sergei Gukov, *Three-dimensional quantum gravity, Chern-Simons theory, and the A-polynomial*, Comm. Math. Phys. **255** (2005), no. 3, 577–627, DOI 10.1007/s00220-005-1312-y. MR2134725 (2006f:58029)

[20] Sergei Gukov, Albert Schwarz, and Cumrun Vafa, *Khovanov-Rozansky homology and topological strings*, Lett. Math. Phys. **74** (2005), no. 1, 53–74, DOI 10.1007/s11005-005-0008-8. MR2193547 (2007a:57014)

[21] Sergei Gukov and Marko Stošić, *Homological algebra of knots and BPS states*, Proceedings of the Freedman Fest, Geom. Topol. Monogr., vol. 18, Geom. Topol. Publ., Coventry, 2012, pp. 309–367, DOI 10.2140/gtm.2012.18.309. MR3084243

[22] Sergei Gukov and Piotr Sułkowski, *A-polynomial, B-model, and quantization*, J. High Energy Phys. **2** (2012), 070, front matter+56. MR2996110

[23] Kazuo Habiro, *On the quantum* sl_2 *invariants of knots and integral homology spheres*, Invariants of knots and 3-manifolds (Kyoto, 2001), Geom. Topol. Monogr., vol. 4, Geom. Topol. Publ., Coventry, 2002, pp. 55–68 (electronic), DOI 10.2140/gtm.2002.4.55. MR2002603 (2004g:57023)

[24] H. Itoyama, A. Mironov, A. Morozov, and An. Morozov, *HOMFLY and superpolynomials for figure eight knot in all symmetric and antisymmetric representations* (2012), arXiv:1203.5978.

[25] R. M. Kashaev, *The hyperbolic volume of knots from the quantum dilogarithm*, Lett. Math. Phys. **39** (1997), no. 3, 269–275, DOI 10.1023/A:1007364912784. MR1434238 (98b:57012)

[26] Sheldon Katz, Albrecht Klemm, and Cumrun Vafa, *Geometric engineering of quantum field theories*, Nuclear Phys. B **497** (1997), no. 1-2, 173–195, DOI 10.1016/S0550-3213(97)00282-4. MR1467889 (98h:81097)

[27] K. Kawagoe, *unpublished work.*

[28] Mikhail Khovanov, *A categorification of the Jones polynomial*, Duke Math. J. **101** (2000), no. 3, 359–426, DOI 10.1215/S0012-7094-00-10131-7. MR1740682 (2002j:57025)

[29] Mikhail Khovanov and Lev Rozansky, *Matrix factorizations and link homology*, arXiv: math/0401268.

[30] Mikhail Khovanov and Lev Rozansky, *Matrix factorizations and link homology II*, arXiv: math/0505056.

[31] Sergiy Koshkin, *Conormal bundles to knots and the Gopakumar-Vafa conjecture*, (2005), arXiv: math/0503248.

[32] A. Mironov, A. Morozov, Sh. Shakirov, and A. Sleptsov, *Interplay between MacDonald and Hall-Littlewood expansions of extended torus superpolynomials*, (2012), arXiv:1201.3339.

[33] Hitoshi Murakami and Jun Murakami, *The colored Jones polynomials and the simplicial volume of a knot*, Acta Math. **186** (2001), no. 1, 85–104, DOI 10.1007/BF02392716. MR1828373 (2002b:57005)

[34] Satoshi Nawata, P. Ramadevi, Zodinmawia, and Xinyu Sun, *Super-A-polynomials for Twist Knots*, arXiv:1209.1409.

[35] Lenhard Ng, *Framed knot contact homology*, Duke Math. J. **141** (2008), no. 2, 365–406, DOI 10.1215/S0012-7094-08-14125-0. MR2376818 (2008k:53202)

[36] Hirosi Ooguri and Cumrun Vafa, *Knot invariants and topological strings*, Nuclear Phys. B **577** (2000), no. 3, 419–438, DOI 10.1016/S0550-3213(00)00118-8. MR1765411 (2001i:81254)

[37] J. Rasmussen, *Some differentials on Khovanov-Rozansky homology*, arXiv: math/0607544.

[38] N. Reshetikhin and V. G. Turaev, *Invariants of 3-manifolds via link polynomials and quantum groups*, Invent. Math. **103** (1991), no. 3, 547–597, DOI 10.1007/BF01239527. MR1091619 (92b:57024)

[39] N. Seiberg and E. Witten, *Electric-magnetic duality, monopole condensation, and confinement in $N = 2$ supersymmetric Yang-Mills theory*, Nuclear Phys. B **426** (1994), no. 1, 19–52, DOI 10.1016/0550-3213(94)90124-4. MR1293681 (95m:81202a)

[40] N. Seiberg and E. Witten, *Monopoles, duality and chiral symmetry breaking in $N = 2$ supersymmetric QCD*, Nuclear Phys. B **431** (1994), no. 3, 484–550, DOI 10.1016/0550-3213(94)90214-3. MR1306869 (95m:81203)

[41] Clifford Henry Taubes, *Lagrangians for the Gopakumar-Vafa conjecture*, Adv. Theor. Math. Phys. **5** (2001), no. 1, 139–163. MR1894340 (2003f:53163)

[42] Yuji Terashima and Masahito Yamazaki, *Semiclassical Analysis of the 3d/3d Relation* (2011), arXiv:1106.3066.

[43] Edward Witten, *Quantum field theory and the Jones polynomial*, Comm. Math. Phys. **121** (1989), no. 3, 351–399. MR990772 (90h:57009)

California Institute of Technology, Pasadena, California 91125 – and – Nagoya University, Dept. of Physics, Graduate School of Science, Furo-cho, Chikusa-ku, Nagoya 464-8602, Japan

E-mail address: `fuji@th.phys.nagoya-u.ac.jp`

California Institute of Technology, Pasadena, California 91125 – and – Faculty of Physics, University of Warsaw, ul. Pasteura 5, 02-093 Warsaw, Poland

E-mail address: `psulkows@theory.caltech.edu`

Proceedings of Symposia in Pure Mathematics
Volume **90**, 2015
http://dx.doi.org/10.1090/pspum/090/01515

On Gauge Theory and Topological String in Nekrasov-Shatashvili Limit

Min-xin Huang

Dedicated to the Memory of Professor Friedrich Hirzebruch

ABSTRACT. We study the Nekrasov-Shatashvili limit of the $\mathcal{N}=2$ supersymmetric gauge theory and topological string theory on certain local toric Calabi-Yau manifolds. In this limit one of the two deformation parameters $\epsilon_{1,2}$ of the Ω background is set to zero and we study the perturbative expansion of the topological amplitudes around the remaining parameter. We derive differential equations from Seiberg-Witten curves and mirror geometries, which determine the higher genus topological amplitudes up to a constant. We show that the higher genus formulae previously obtained from holomorphic anomaly equations and boundary conditions satisfy these differential equations. We also provide a derivation of the holomorphic anomaly equations in the Nekrasov-Shatashvili limit from these differential equations.

1. Introduction

There has been much progress in the study of supersymmetric gauge theories since Seiberg and Witten discovered that the $\mathcal{N}=2$ supersymmetric gauge theories are exactly solvable [**SW1**, **SW2**]. The prepotential which characterizes the effective action can be determined by holomorphicity and monodromy in the moduli space. On the other hand, the instanton contributions in the prepotential can be directly computed by Nekrasov partition function [**Ne**]. The Nekrasov partition function is parametrized by two parameters ϵ_1 and ϵ_2 which deform the $\mathbb{R}^4$ space. It can be shown by saddle point method that the leading order contribution of Nekrasov function in small ϵ_1, ϵ_2 is equal to the Seiberg-Witten prepotential [**NO**, **NY**]. Furthermore, the higher order contributions in ϵ_1, ϵ_2 expansion of the Nekrasov function compute the gravitational coupling terms in the effective action, and is analogous to the higher genus amplitudes in topological string theory which can be computed by the method of holomorphic anomaly equation [**BCOV**] and gap conditions in the moduli space proposed in [**HK1**, **HK2**]. The two parameters ϵ_1, ϵ_2 correspond to a refinement of the string coupling in topological string theory, which was studied for certain toric Calabi-Yau manifolds in [**IKV**].

2010 *Mathematics Subject Classification.* Primary 81T60.

Key words and phrases. Topological string theory, gauge theory.

This is prepared for the proceedings of String-Math 2012 conference, a talk based on the paper [**Hu**] by the same author. The author thanks Amir-Kian Kashani-Poor and Albrecht Klemm for discussions and correspondences.

The homomorphic anomaly equation and gap conditions can be extended to the refined case and the higher order terms in $SU(2)$ Nekrasov function are solved exactly [**HK3**, **HKK**, **KW**]. The higher genus formulae are expressed in terms of quasi-modular forms such as Eisenstein series and Jacobi theta functions, and the formulae are exact in the sense that they sum up all instanton contributions at a fixed genus.

The Nekrasov partition function can be also related to the correlation function of 2d Liouville theory by the AGT (Alday-Gaiotto-Tachikawa) conjecture [**AGT**]. Recently there have been many works in this direction. We hope our works can provide some ideas for the AGT conjecture.

We will consider the so called Nekrasov-Shatashvili limit, also sometimes known as the chiral limit, of the Nekrasov function, which sets one of the deformation parameter $\epsilon_2 = 0$ and we expand the Nekrasov function for small $\epsilon \equiv \epsilon_1$. Nekrasov and Shatashvilli conjectures in this limit the $\mathcal{N} = 2$ gauge theories are described by certain quantum integrable systems [**NS**]. The quantum integrable systems provides another way to compute the Nekrasov function in the $\epsilon_2 = 0$ limit and has been considered in e.g. [**MM**, **ACDKV**]. In our previous paper [**HKK**] we showed that the formulae we derived from holomorphic anomaly equation satisfy the quantum equations in the sine-Gordon model for the pure $SU(2)$ Seiberg-Witten theory. Thus, if the quantum integrable system description of the Nekrasov-Shatashvili limit is correct, our higher genus formulae in this limit would be exactly proven.

In this talk we study the approach of using the saddle point method to compute Nekrasov function in the Nekrasov-Shatashvili limit. This is carried out quite explicitly in the papers [**FMPP**, **Po**], and seems to be on a more solid footing than the approach of using quantum integrable systems mentioned above. Furthermore, the saddle point method is readily applicable to the case of Seiberg-Witten gauge theory with matters and to higher rank gauge group. We will show that our $SU(2)$ higher genus formulae [**HK3**, **HKK**] in the Nekrasov-Shatashvili limit satisfy the saddle point equations in [**FMPP**, **Po**]. Since these equations uniquely fix the higher genus contributions (up to some constants, which can be easily checked), we would have proven our formulae exactly.

2. Review of the saddle point method

We will be interested in the small ϵ expansion of the logarithm of the Nekrasov partition function, which is called the free energy

$$\log Z(\epsilon_1, \epsilon_2, a_i) = \sum_{g,n=0}^{\infty} (\epsilon_1 + \epsilon_2)^{2n} (\epsilon_1\epsilon_2)^{g-1} F^{(n,g)}(a_i) \tag{2.1}$$

where a_i $(i = 1, 2, \cdots, N)$ are the periods or flat coordinates for the $SU(N)$ gauge theory, satisfying $\sum_{i=1}^{N} a_i = 0$. The leading term scales like $\frac{1}{\epsilon_1\epsilon_2}$ and is characteristic of the saddle point behavior in the small $\epsilon_{1,2}$ limit. The Nekrasov partition function are computed by sums over Young tableaux, and in the small $\epsilon_{1,2}$ limit its logarithm is dominated by the Young tableaux that have extremal contributions. It urns out the dominant Young tableaux have the number of boxes scaling as $\frac{1}{\epsilon_1\epsilon_2}$ in the $\epsilon_{1,2} \to 0$ limit. The leading term $F^{(0,0)}$ can be computed by finding the dominant Young tableau shapes, and it was shown by this saddle point method that the leading term $F^{(0,0)}$ is equal to the Seiberg-Witten prepotential [**NO**].

It turns out the saddle point method also works when we send only one of ϵ's, say ϵ_2 to zero. In this limit we consider the expansion around $\epsilon \equiv \epsilon_1$, and define the deformed prepotential $\mathcal{F}$ as

$$\mathcal{F}(a_i, \epsilon) = \sum_{n=0}^{\infty} \epsilon^{2n} F^{(n,0)}(a_i) \tag{2.2}$$

The deformed prepotential can be again computed by finding the extremal Young tableaux in the $\epsilon_2 \to 0$ limit [**Po**, **FMPP**]. Here we will not go into the details of the derivation but simply quote the results in [**Po**, **FMPP**]. For the case of $SU(N)$ theory with N_f fundamental matters, the saddle point equation is

$$qM(x-\epsilon)w(x)w(x-\epsilon) - w(x)P(x) + 1 = 0 \tag{2.3}$$

The explanation of the notations follows. Here q is a power of the dynamical scale for asymptotically free theories of $N_f < 2N$ and the gauge coupling $q = e^{2\pi i\tau}$ for the conformal theory $N_f = 2N$. The power of the q parameter counts the number of instanton in the contribution to the Nekrasov partition function. The $w(x)$ is a spectral function that encodes the dominant Young tableau configuration in the $\epsilon_2 \sim 0$ limit. The $P(x)$ is a degree N polynomial, and $M(x)$ is a degree N_f polynomial parametrized by the mass of fundamental matters

$$P(x) = \prod_{i=1}^{N}(x - b_i), \;\; M(x) = \prod_{i=1}^{N_f}(x + m_i) \tag{2.4}$$

Furthermore, the parameters b_i in $P(x)$ are related to the expectation value of the adjoint scalar field ϕ in the $\mathcal{N} = 2$ gauge multiplet

$$\langle \mathrm{tr}(\phi^J) \rangle = \sum_{i=1}^{N} b_i^J, \tag{2.5}$$

and the deformed periods $\tilde{a}_i$ can be computed by a residue formula

$$\tilde{a}_i = -\sum_{n=0}^{\infty} \mathrm{Res}_{x=b_i+n\epsilon}\, x\partial_x \log w(x), \tag{2.6}$$

where we use the tilde symbol to denote the period is deformed by ϵ parameter, as it turns out that it is different from the usual period a in Seiberg-Witten theory. The instanton parts of the deformed prepotential is computed by a generalized Matone relation [**Ma**]

$$2q\frac{d\mathcal{F}_{inst}(\tilde{a}_i, \epsilon, q)}{dq} = \sum_{i=1}^{N} \tilde{a}_i^2 - \langle \mathrm{tr}(\phi^2) \rangle = \sum_{i=1}^{N} (\tilde{a}_i^2 - b_i^2). \tag{2.7}$$

We will see that at low orders, the amplitudes $F^{(0,0)}$ and $F^{(1,0)}$ may also have some simple q-dependence in the classical and perturbative contributions, besides the main instanton contributions.

We consider the case of $SU(2)$ Seiberg-Witten theory whose Coulomb moduli space is described by a complex u-plane, where u is the expectation value $u = \frac{1}{2}\langle \mathrm{tr}(\phi^2) \rangle$. It turns out that in order to ensure the SU condition $\tilde{a}_1 = -\tilde{a}_2 \equiv \tilde{a}$, we can choose the parameters $b_1 = -b_2 \equiv b$. So the modulus can be written

$u = \frac{1}{2}(b_1^2 + b_2^2) = b^2$, and the polynomial $P(x) = x^2 - u$. The residue formula and the generalized Matone relation are

$$\tilde{a} = -\sum_{n=0}^{\infty} \text{Res}_{x=b+n\epsilon} x\partial_x \log w(x), \; q\frac{d\mathcal{F}_{inst}(\tilde{a},\epsilon,q)}{dq} = \tilde{a}^2 - u \tag{2.8}$$

The authors in [**Po**, **FMPP**] use the saddle point equation (2.3) and formulae (2.8) to solve the deformed prepotential $\mathcal{F}(\tilde{a},\epsilon,q)$ perturbatively in q parameter and the solution is exact in ϵ parameter. On the other hand, in order to make connection with the higher genus formulae in our paper [**HK3**, **HKK**], we need to instead solve the deformed prepotential exactly in q parameter and but perturbatively in ϵ parameter. We will do this in the following sections.

3. Pure $SU(2)$ theory

As an example we consider the simple case of pure $SU(2)$ theory without matter. In this case the polynomial $M(x) = 1$ and the saddle point equation becomes

$$qw(x)w(x-\epsilon) - w(x)P(x) + 1 = 0 \tag{3.1}$$

We write the $w(x)$ and the deformed period $\tilde{a}$ in small ϵ expansion as

$$w(x) = \sum_{n=0}^{\infty} w_n(x)\epsilon^n, \; \tilde{a} = \sum_{n=0}^{\infty} a_n\epsilon^n \tag{3.2}$$

We plug the expansion of $w(x)$ into the saddle point equation (3.1) and solve for $w_n(x)$'s to the few orders. With $P(x) = x^2 - u = x^2 - b^2$, we find

$$\begin{aligned} w_0(x) &= \frac{P(x) - \sqrt{P(x)^2 - 4q}}{2q}, \\ w_1(x) &= \frac{x(P(x) - \sqrt{P(x)^2 - 4q})^2}{2q(P(x)^2 - 4q)}, \\ w_2(x) &= \frac{1}{2q(P(x)^2 - 4q)^3}[P(x)^5(P(x) - \sqrt{P(x)^2 - 4q}) \\ &\quad -2qP(x)^2(12x^4 - 16ux^2 + 4u^2 + (3u - 11x^2)\sqrt{P(x)^2 - 4q}) \\ &\quad +8q^2(10x^4 - 12ux^2 + 2u^2 + (u - 4x^2)\sqrt{P(x)^2 - 4q})] \end{aligned} \tag{3.3}$$

At leading order $\epsilon = 0$, the equation for $w_0(x)$ is a simple quadratic equation. There are two solutions for $w_0(x)$ and we choose the one with minus sign in front of the quadratic discriminant. We will also use the sign convention $P(x) > 0$ when we expand the function perturbatively around $q \sim 0$. There are only rational functions of x in the perturbative series expansion around $q \sim 0$, so that the residue calculations are simple to do perturbatively. Our choice of convention for $w_0(x)$ and $P(x) > 0$ gives the correct sign for the leading period $a_0 = \sqrt{u} + \mathcal{O}(q)$.

Th deformed period $\tilde{a}$ can be computed perturbatively in ϵ parameter as residue around $b = \sqrt{u}$,

$$\tilde{a} = -\text{Res}_{x=b} x\partial_x[\log(w_0(x)) + \frac{w_1(x)}{w_0(x)}\epsilon + (\frac{w_2(x)}{w_0(x)} - \frac{w_1(x)^2}{2w_0(x)^2})\epsilon^2 + \mathcal{O}(\epsilon^3)] \tag{3.4}$$

Here in the $\epsilon \sim 0$ limit, all possible poles at $x = b + n\epsilon$ in (2.8) collapse to $x = b$, so we only need to compute the residue around $x = b$.

Now we consider the leading order period $a_0 = -\text{Res}_{x=b}\frac{xw_0'(x)}{w_0(x)}$, which can be computed perturbatively to the first few orders around $q \sim 0$. We assume $P(x) > 0$ and expand the expression for $w_0(x)$ around $q \sim 0$, and find

$$\begin{aligned} a_0 &= \text{Res}_{x=b}[\frac{2x^2}{P(x)} + \frac{4x^2}{P(x)^3}q + \frac{12x^2}{P(x)^5}q^2 + \mathcal{O}(q^3)] \\ &= \sqrt{u}(1 - \frac{q}{4u^2} - \frac{15q^2}{64u^4} + \mathcal{O}(q^3)) \end{aligned} \tag{3.5}$$

We realize the the leading order period is actually the conventional undeformed period $a \equiv a_0$ in Seiberg-Witten theory, which satisfies the Picard-Fuchs differential equation $4(4q - u^2)\partial_u^2 a = a$. This can be shown exactly

$$4(4q - u^2)\partial_u^2 a - a = \text{Res}_{x=b}\frac{d}{dx}[-\frac{2x^3(x^4 - 4ux^2 + 3u^2 + 4q)}{(P(x)^2 - 4q)^{\frac{3}{2}}}] = 0 \tag{3.6}$$

The residue vanishes since it can be written as a total derivative and there is no branch cut around the residue point $x = b$.

In general we find a contour integral or residue vanishes if the indefinite integral can be performed nicely, and the result is expressed a rational function of x and the square root $\sqrt{P(x)^2 - 4qM(x)}$, since there is usually no branch cut in the rational functions. We have to be a little more careful if the indefinite integral involving logarithm, but this case can be easily dealt with by taking account of the branch cut of the logarithm around the contour. Otherwise, if the indefinitely integral can not be done nicely, which implies that the integral is a generic elliptic integral, there will be branch cut around the contour and the residue will not vanish. In this case we will to relate the integral to other known integrals by adding some total derivatives of rational functions of x and $\sqrt{P(x)^2 - 4qM(x)}$, which have no branch cut around the contour.

We compute the deformed periods to the next few orders. We give the formulae

$$\begin{aligned} \tilde{a} &= a + a_2\epsilon^2 + a_4\epsilon^4 + \mathcal{O}(\epsilon^6) \\ &= a + \frac{\epsilon^2}{24}(\partial_u a + 2u\partial_u^2 a) + \frac{\epsilon^4}{5760}(75\partial_u^2 a + 120u\partial_u^3 a + 28u^2\partial_u^4 a) + \mathcal{O}(\epsilon^6) \end{aligned} \tag{3.7}$$

It turns out the deformed period (3.7) is the same as in the sine-Gordon quantum model studied in [**MM**, **HKK**]. One can probably prove the equivalence by some ingenious changes of variables. In [**MM**, **HKK**] the deformed dual period $\tilde{a}_D = \frac{\partial\mathcal{F}(\tilde{a})}{\partial\tilde{a}}$ is used to determine the equation for the deformed prepotential. Here we will follow a different procedure and use the generalized Matone relation which has been derived from the saddle point approach in [**Po**, **FMPP**].

We expand the generalized Matone relation by plugging the above equations in (2.2), and use (3.7)

$$\begin{aligned} & q\frac{d\mathcal{F}_{inst}(\tilde{a}, \epsilon, q)}{dq} - \tilde{a}^2 + u \\ = \ & \frac{1}{2}F^{(0,0)}(a) - \frac{1}{4}a\frac{\partial F^{(0,0)}(a)}{\partial a} + u + \frac{\epsilon^2}{4}[a_2(a_D + 2\pi i\tau a) - a\frac{\partial F^{(1,0)}(a)}{\partial a} + \frac{1}{6}] \\ & + \frac{\epsilon^4}{2}[-F^{(2,0)}(a) - \frac{a}{2}\frac{\partial F^{(2,0)}(a)}{\partial a} - \frac{a_2}{2}\partial_a(a\frac{\partial F^{(1,0)}(a)}{\partial a}) \\ & + \frac{a_4}{2}(a_D + 2\pi i\tau a) + \frac{a_2^2}{4}\partial_a(a_D + 2\pi i\tau a)] + \mathcal{O}(\epsilon^6) \end{aligned} \tag{3.8}$$

Here the second derivative of the prepotential is the gauge coupling $\frac{\partial^2 F^{(0,0)}(a)}{\partial^2 a} = -2\pi i\tau$, and we use the notation of the dual period $\frac{\partial F^{(0,0)}(a)}{\partial a} = a_D$. We note the definition of the parameter τ is the same as the elliptic parameter of the Seiberg-Witten curve, and is twice the convention used in [**HK1**, **HKK**]. The parameter q is the 4th power of the asymptotically free scale of the pure Seiberg-Witten gauge theory, and in the followings we will no longer need to compute the derivative of q, so for convenience we will set $q = 1$, which can always be easily recovered by dimensional analysis. The theory is then characterized by one independent parameter, the modulus parameter u on the complex plane, and the other parameters τ, a and a_D are functions of the modulus u. We will write down the functional relations between these parameters.

The leading order equation in (3.8) is the conventional Matone relation. We find differential equations for the higher order terms in the deformed prepotential at each order of ϵ expansion. The equations from the deformed Matone relation (3.8) at order ϵ^2 and ϵ^4 are

$$a\frac{\partial F^{(1,0)}(a)}{\partial a} - \frac{1}{6} = \frac{1}{24}(\partial_u a + 2u\partial_u^2 a)(a_D + 2\pi i\tau a) \tag{3.9}$$

$$\begin{aligned} &F^{(2,0)}(a) + \frac{a}{2}\frac{\partial F^{(2,0)}(a)}{\partial a} \\ = \ &-\frac{a_2}{2}\partial_a\Big(a\frac{\partial F^{(1,0)}(a)}{\partial a}\Big) + \frac{a_4}{2}(a_D + 2\pi i\tau a) + \frac{a_2^2}{4}\partial_a(a_D + 2\pi i\tau a) \end{aligned} \tag{3.10}$$

In [**HKK**] we derive higher genus formulae from holomorphic anomaly and gap conditions. The formulae for $F^{(1,0)}$ and $F^{(2,0)}$ are

$$\begin{aligned} F^{(1,0)}(a) &= \frac{1}{24}\log(u^2-4)\,, \\ F^{(2,0)}(a) &= -\frac{u(45uX + 4u^2 + 300)}{8640(u^2-4)^2}\,. \end{aligned} \tag{3.11}$$

where $X = \frac{E_2(\tau)E_4(\tau)}{E_6(\tau)}\frac{g_3(u)}{g_2(u)}$.

Now we can check our higher genus formulae (3.11) satisfy these equations (3.9, 3.10) derived from the saddle point method. The checks are straightforward but might become tedious if done manually, so one might resort to computer algebra manipulations. Thus we have proven these higher genus formulae for $F^{(1,0)}$ and $F^{(2,0)}$.

3.1. The deformed dual period. The deformed period $\tilde{a}$ is the residue for contour integral (2.8) of $-x\partial_x \log w(x)$ of the spectral function $w(x)$. We can define a deformed dual period $\tilde{a}_D$ as the contour integral of the same integrand but around a different B cycle

$$\tilde{a}_D = -\frac{1}{2\pi i}\oint_B x\partial_x \log w(x) \tag{3.12}$$

At leading order the dual period $a_{D0} \equiv a_D$ should satisfy the same Picard-Fuchs differential equation so it is the conventional dual period in Seiberg-Witten theory. The higher order contributions to the deformed dual period can be written as derivatives of leading dual period a_D, in the same way as the deformed period $\tilde{a}$, since the derivation of the formulae only depends on the integrand in the contour integral but not the contour. We find the same formulae as (3.7)

(3.13)

$$\begin{aligned}\tilde{a}_D &= a_D + a_{D2}\epsilon^2 + a_{D4}\epsilon^4 + \mathcal{O}(\epsilon^6)\\ &= a_D + \frac{\epsilon^2}{24}(\partial_u a_D + 2u\partial_u^2 a_D) + \frac{\epsilon^4}{5760}(75\partial_u^2 a_D + 120u\partial_u^3 a_D + 28u^2\partial_u^4 a_D) + \mathcal{O}(\epsilon^6)\end{aligned}$$

We shall show that the deformed prepotential satisfies the relation with dual deformed period

$$\frac{\partial \mathcal{F}(\tilde{a})}{\partial \tilde{a}} = \tilde{a}_D \tag{3.14}$$

This can be probably be done with arguments similar to those of Dijkgraaf and Vafa for showing the equivalence of the prepotential of topological string theory on a Calabi-Yau manifold with a corresponding matrix model in [**DV**]. Since this relation can also determine the higher order contributions of the deformed prepotential, we can prove our higher genus formulae for $F^{(n,0)}$ by showing they satisfy the relation (3.14). This is done for pure gauge theory in [**HKK**]. Here we show the formulae again for consistency of notation and prepare for the study for the case of Seiberg-Witten theory with matters.

We expand the relation (3.14) with the formulae for deformed period (3.7) and the dual deformed period (3.13)

$$\begin{aligned}\frac{\partial \mathcal{F}(\tilde{a})}{\partial \tilde{a}} - \tilde{a}_D \quad &= \quad \frac{\partial F^{(0,0)}(a)}{\partial a} - a_D + \epsilon^2(\partial_a F^{(1,0)}(a) - 2\pi i \tau a_2 - a_{D2})\\ &\quad +\epsilon^4[\partial_a F^{(2,0)}(a) + a_2\partial_a^2 F^{(1,0)}(a) - 2\pi i\tau a_4 - \pi i(\partial_a \tau)(a_2)^2 - a_{D4}]\\ &\quad +\mathcal{O}(\epsilon^6)\end{aligned} \tag{3.15}$$

The leading order is the well known Seiberg-Witten relation for the prepotential. We can again easily check that the higher genus formulae (3.11) satisfy the above equations at order ϵ^2 and order ϵ^4.

4. Derivation of the holomorphic anomaly equation

The generalized holomorphic anomaly equation is proposed in [**KW**, **HK3**, **HKK**] to solve the higher genus amplitudes of Seiberg-Witten gauge theory in general Ω background with generic ϵ_1, ϵ_2 parameters. In the chiral or Nekrasov-Shatashvili limit, the second derivative term in the generalized holomorphic anomaly equation vanishes and the equation is simplified as

$$\partial_{E_2} F^{(n,0)} = \frac{1}{24}\sum_{l=1}^{n-1} \partial_a F^{(l,0)}\partial_a F^{(n-l,0)} \tag{4.1}$$

Here the amplitude $F^{(n,0)}$ is a polynomial of $X = \frac{E_2(\tau)E_4(\tau)}{E_6(\tau)}\frac{g_3(u)}{g_2(u)}$, and the coefficients of the polynomial are rational functions of u. The partial derivative with respect to the second Eisenstein series E_2 is well defined, in the sense by regarding the other components E_4, E_6, and u in $F^{(n,0)}$ as constants under the partial derivative.

The second Eisenstein series $E_2(\tau)$ is holomorphic but not modular under $SL(2,\mathbb{Z})$ transformations. One can instead define a modular covariant but an-holomorphic quantity by a shift $\hat{E}_2(\tau) = E_2(\tau) - \frac{6i}{\pi(\tau-\bar{\tau})}$, which is called an almost holomorphic modular form. The holomorphic limit takes $\bar{\tau} \to \infty$ and we see $\hat{E}_2(\tau) \to E_2(\tau)$ in this limit. It is well known in the theory of modular forms that there is an isomorphism between the almost holomorphic modular forms and the holomorphic limit [**Za**].

It turns out that under certain simple assumptions, the holomorphic anomaly equation in the Nekrasov-Shatashvili limit can be derived from the equation $\frac{\partial \mathcal{F}(\tilde{a})}{\partial \tilde{a}} = \tilde{a}_D$ for deformed dual period, using the approach of induction. For more details see [**Hu**]. A related work is the derivation of holomorphic anomaly equation for the loop equations and topological recursion in matrix models in [**EO, EMO**].

In the saddle point method we are essentially working in the holomorphic limit where higher genus amplitude is holomorphic but not modular. The holomorphic anomaly appears when we use the isomorphism with almost holomorphic modular forms and replace expression in the holomorphic limit with the almost holomorphic modular counterpart. In our case, only E_2 is not modular covariant and needed to be replaced with the almost holomorphic modular form $\hat{E}_2$. The an-holomorphic derivative can be related to $\partial_{\hat{E}_2}$

$$\bar{\partial}_{\bar{\tau}} = (\bar{\partial}_{\bar{\tau}}\hat{E}_2)\partial_{\hat{E}_2} = \frac{6}{\pi i(\tau-\bar{\tau})^2}\partial_{\hat{E}_2}, \tag{4.2}$$

which is the origin of the appearance of ∂_{E_2} in holomorphic anomaly equation (4.1) in the holomorphic limit.

The free energy of $SU(2)$ Seiberg-Witten theory can be constructed by elements of a certain modular group, which is the monodromy group of the special points in the moduli space. For example, in the pure $SU(2)$ case, the modular group is a subgroup $\Gamma(2)$ of $SL(2,\mathbb{Z})$. The free energy can be written as polynomials of E_2 and the Jacobi theta functions $\theta_i(0,\tau)^4, i = 2,3,4$, up to a holomorphic modular factor. Here we will not need the details of the modular group, but only the property that all the elements are modular except E_2. It is more convenient to write them as rational functions of the modulus parameter u, and simply note that it has no modular anomaly.

We will need to derive some formulae involving the operator ∂_{E_2}. First we can work in the holomorphic limit for some simple formulae. The relations of various parameters τ, u, a in the Seiberg-Witten gauge theory are the followings

$$J(\tau) = \frac{E_4(\tau)^3}{E_4(\tau)^3 - E_6(\tau)^2} = \frac{g_2(u)^3}{g_2(u)^3 - 27g_3(u)^2}, \tag{4.3}$$

$$\frac{da}{du} = \sqrt{-\frac{1}{18}\frac{g_2(u)}{g_3(u)}\frac{E_6(\tau)}{E_4(\tau)}}, \tag{4.4}$$

It is easy to see from the above relations (4.3, 4.4) that the expressions for the following derivatives have only E_4, E_6, u but no E_2, so

$$\partial_{E_2}(\partial_a \tau) = 0, \ \partial_{E_2}(\partial_u a) = 0. \tag{4.5}$$

The E_2 series starts to appear when we take one more derivative. We can compute

$$\partial_{E_2}(\partial_u^2 a) = \frac{\pi i}{6}(\partial_u \tau)(\partial_u a) \tag{4.6}$$

We will assume the higher order contributions to the deformed period $\tilde{a} = \sum_{n=0}^{\infty} a_{2n}\epsilon^{2n}$ can be written as a linear combination of $\partial_u a$ and $\partial_u^2 a$. The dual deformed period has the same formula. For $n \geq 1$ we can write

$$\begin{aligned} a_{2n} &= g_{2n}(u)\partial_u a + f_{2n}(u)\partial_u^2 a, \\ a_{D2n} &= g_{2n}(u)\partial_u a_D + f_{2n}(u)\partial_u^2 a_D \end{aligned} \tag{4.7}$$

where the coefficients $g_{2n}(u)$ and $f_{2n}(u)$ are rational functions of u. We have checked the form (4.7)is valid up to second order, i.e. for $n \leq 2$ for all models studied in this paper. However, it seems difficult to find a rigorous proof of this assumption, and we leave for future works. Heuristically, we can argue the derivatives of the flat coordinate a form a linear complete basis for expanding the higher order periods a_{2n} with rational function coefficients, and since the Picard-Fuchs equations are third order differential equations, we know that the first and second derivatives suffice.

Using (4.5, 4.6) and $\partial_u a_D = -2\pi i\tau\partial_u a$, we find

$$\partial_{E_2}(a_{2n}) = \frac{\pi i}{6}(\partial_u\tau)(\partial_u a)f_{2n}(u) = -\frac{1}{12}(2\pi i\tau a_{2n} + a_{D2n}) \tag{4.8}$$

We obtain formulae for the higher order contributions in the deformed period

$$\partial_{E_2}(\tilde{a} - a) = -\frac{1}{12}[2\pi i\tau(\tilde{a} - a) + (\tilde{a}_D - a_D)], \tag{4.9}$$

$$\partial_{E_2}[2\pi i\tau(\tilde{a} - a) + (\tilde{a}_D - a_D)] = 0 \tag{4.10}$$

Our goal is to derive the holomorphic anomaly equation (4.1) from the equation for the deformed period. We can expand the equation for the deformed period

$$\begin{aligned} &\frac{\partial\mathcal{F}(\tilde{a},\epsilon)}{\partial\tilde{a}} - \tilde{a}_D = 0 \\ = \; &\sum_{n=1}^{\infty}\sum_{k=0}^{\infty}\partial_a^{k+1}F^{(n,0)}(a)\frac{(\tilde{a}-a)^k}{k!}\epsilon^{2n} + \sum_{k=0}^{\infty}\partial_a^{k+1}(-2\pi i\tau)\frac{(\tilde{a}-a)^{k+2}}{(k+2)!} \\ &-[2\pi i\tau(\tilde{a} - a) + (\tilde{a}_D - a_D)] \end{aligned} \tag{4.11}$$

where we have separated the prepotential $F^{(0,0)}$ and use the formulae $\partial_a F^{(0,0)} = a_D$ and $\partial_a^2 F^{(0,0)} = -2\pi i\tau$.

The order ϵ^2 and ϵ^4 equations in the above equation (4.11) can be written more explicitly as the followings

$$\begin{aligned} \partial_a F^{(1,0)}(a) &= 2\pi i\tau a_2 + a_{D2} \\ \partial_a F^{(2,0)}(a) &= -a_2\partial_a^2 F^{(1,0)}(a) + 2\pi i\tau a_4 + \pi i(\partial_a\tau)(a_2)^2 + a_{D4}. \end{aligned} \tag{4.12}$$

We can use the equations to compute $\partial_a F^{(n,0)}(a)$ recursively if we have the formulae for $g_{2n}(u)$ and $f_{2n}(u)$ in (4.7) for the higher order contributions in the deformed period. Furthermore, by dimensional analysis we know the asymptotic behavior of $F^{(n,0)} \sim a^{2-2n}$ for large a. So these equations determine $F^{(1,0)}$ up to a constant and completely fix $F^{(n,0)}$ for $n \geq 2$.

We would like to derive (4.1) recursively by induction. Taking the partial derivative ∂_{E_2} on both sides of (4.11), we find

$$\begin{aligned}
&\sum_{n=1}^{\infty}\partial_{E_2}\partial_a F^{(n,0)}\epsilon^{2n}\\
= &-\sum_{n=1}^{\infty}\sum_{k=1}^{\infty}\epsilon^{2n}(\partial_{E_2}\partial_a^{k+1}F^{(n,0)})\frac{(\tilde{a}-a)^k}{k!}-\sum_{n=1}^{\infty}\sum_{k=1}^{\infty}\epsilon^{2n}(\partial_a^{k+1}F^{(n,0)})\frac{(\tilde{a}-a)^{k-1}}{(k-1)!}\partial_{E_2}(\tilde{a}-a)\\
&+2\pi i\sum_{k=0}^{\infty}\partial_{E_2}\partial_a^{k+1}\tau\frac{(\tilde{a}-a)^{k+2}}{(k+2)!}+2\pi i\sum_{k=0}^{\infty}\partial_a^{k+1}\tau\frac{(\tilde{a}-a)^{k+1}}{(k+1)!}\partial_{E_2}(\tilde{a}-a),
\end{aligned}\tag{4.13}$$

where we have used the equation (4.10). At each order ϵ^{2n}, no $F^{(l,0)}$ with $l \geq n$ appears on the right hand side. So by induction we can use (4.1) to compute the right hand side, and we will complete the induction procedure by showing the left hand side also satisfies the holomorphic anomaly equation (4.1).

It is clear that in order to do the computations, it is crucial to understand how ∂_{E_2} and ∂_a commute with each others. This is mostly conveniently done in the almost holomorphic modular forms, instead of the holomorphic limit. To preserve the almost holomorphic modular structure, we need to use covariant derivatives with respect to the special Kahler metric of the moduli space. There are two contributions to the connection in covariant derivatives, one from the canonical line bundle and one from the Weil-Petersson metric. In our case, the moduli space of the Seiberg-Witten theory is similar to that of a one-parameter local Calabi-Yau space, and one can choose a gauge such that the contribution from the canonical line bundle vanishes. So we only need to include the connection from the Weil-Petersson metric. Furthermore, there is a flat coordinate a such that the connection for the flat coordinate vanishes in the holomorphic limit. The metric and connection in the flat coordinate a in Seiberg-Witten theory are well known, see e.g. [**HK2**],

$$G_{a\bar{a}}\sim(\tau-\bar{\tau}),\ \Gamma^a_{aa}=(G_{a\bar{a}})^{-1}\partial_a G_{a\bar{a}}=\frac{\partial_a\tau}{\tau-\bar{\tau}}\tag{4.14}$$

where we see the Christoffel connection indeed vanishes in the holomorphic limit $\bar{\tau}\to\infty$.

Suppose F_k is a tensor with k lower indices regarding to the metric of the moduli space in flat coordinate a, and it may has an-holomorphic dependence in terms of $\hat{E}_2$. The covariant derivative is then $D_aF_k = (\partial_a - k\Gamma^a_{aa})F_k$. We can compute the an-holomorphic derivative

$$\bar{\partial}_{\bar{\tau}}D_aF_k=(\partial_a-k\Gamma^a_{aa})\bar{\partial}_{\bar{\tau}}F_k-k(\bar{\partial}_{\bar{\tau}}\Gamma^a_{aa})F_k\tag{4.15}$$

We use (4.2) and then take the holomorphic limit to find the commutation relation

$$\partial_{E_2}\partial_aF_k=\partial_a\partial_{E_2}F_k-\frac{k\pi i}{6}(\partial_a\tau)F_k\tag{4.16}$$

The amplitude $F^{(n,0)}$ is a scalar in moduli space, and its derivative with ∂_a is a tensor with lower indices. We can compute the derivatives

$$\begin{aligned}
\partial_{E_2}\partial_a^{k+1}F^{(n,0)} &= \partial_a^{k+1}\partial_{E_2}F^{(n,0)}-\frac{\pi i}{6}\sum_{l=1}^{k}l\partial_a^{k-l}[\partial_a\tau\partial_a^lF^{(n,0)}]\\
&= \partial_a^{k+1}\partial_{E_2}F^{(n,0)}-\frac{\pi i}{6}\sum_{p=0}^{k-1}\binom{k+1}{p+2}(\partial_a^{p+1}\tau)(\partial_a^{k-p}F^{(n,0)}),
\end{aligned}\tag{4.17}$$

where we have used the binomial identity $\sum_{l=1}^{k-p}\binom{k-l}{p}l = \binom{k+1}{p+2}$. In particular, we note that in the case of $k=0$, the operators ∂_{E_2} and ∂_a commute when acting on $F^{(n,0)}$.

Similarly we derive the formula for $\tau = -\frac{1}{2\pi i}\partial_a^2 F^{(0,0)}$, using the first formula in (4.5)

$$\partial_{E_2}\partial_a^{k+2}\tau = -\frac{\pi i}{12}\sum_{p=0}^{k}\binom{k+4}{p+2}(\partial_a^{p+1}\tau)(\partial_a^{k+1-p}\tau) \tag{4.18}$$

Further using the equation for deformed dual period (4.11), the formula (4.9) can be written without the dual period as

$$\begin{aligned}\partial_{E_2}(\tilde{a}-a) &= -\frac{1}{12}\sum_{n=1}^{\infty}\sum_{k=0}^{\infty}\partial_a^{k+1}F^{(n,0)}(a)\frac{(\tilde{a}-a)^k}{k!}\epsilon^{2n} \\ &\quad +\frac{\pi i}{6}\sum_{k=0}^{\infty}(\partial_a^{k+1}\tau)\frac{(\tilde{a}-a)^{k+2}}{(k+2)!}\end{aligned} \tag{4.19}$$

We can now compute the right hand side of (4.13), by plugging the formulae (4.17, 4.18, 4.19) and then use (4.1) by induction. The calculation is quite lengthy, but surprisingly we encounter a lot of cancellations which drastically simplify the expression. In particular, the dependence on $(\tilde{a}-a)$ cancels out, so we don't need the specific formulae for $g_{2n}(u)$ and $f_{2n}(u)$ in (4.7). We keep the left hand side of (4.13) and write the final result of the calculations for the right hand side

$$\sum_{n=1}^{\infty}\partial_a\partial_{E_2}F^{(n,0)}\epsilon^{2n} = \frac{1}{12}\sum_{n=1}^{\infty}\epsilon^{2n}\sum_{l=1}^{n-1}\partial_a F^{(l,0)}\partial_a^2 F^{(n-l,0)} \tag{4.20}$$

It is easy to check $\partial_{E_2}F^{(1,0)}=0$, thus the above result proves the holomorphic anomaly equation (4.1) for $F^{(n,0)}$ with $n\geq 2$ up to an integration constant of a. From the asymptotic behavior $F^{(n,0)}\sim a^{2-2n}$ for large a, the constant must be zero, so we have proven (4.1) exactly by induction.

5. Conclusion

The analysis can be generalize to the case of $SU(2)$ Seiberg-Witten theory with matters, and also topological string theory on certain local toric Calabi-Yau manifolds. For more details see [**Hu**].

There are some questions for further study. The equation for the deformed dual period $\frac{\partial\mathcal{F}(\tilde{a})}{\partial\tilde{a}} = \tilde{a}_D$ should be derived more carefully, e.g. from the saddle point analysis for the Nekrasov function in Seiberg-Witten theory and from the refined topological vertex in toric Calabi-Yau models.

We have provided a derivation of the holomorphic anomaly equations in the Nekrasov-Shatashvili limit from the differential equations for the deformed dual period. It would be nice to also derive the gap boundary conditions from these differential equations.

References

[ACDKV] M. Aganagic, M. C. N. Cheng, R. Dijkgraaf, D. Krefl, and C. Vafa, *Quantum geometry of refined topological strings*, J. High Energy Phys. **11** (2012), 019, front matter + 52. MR3036500

[AGT] L. F. Alday, D. Gaiotto, and Y. Tachikawa, *Liouville correlation functions from four-dimensional gauge theories*, Lett. Math. Phys. **91** (2010), no. 2, 167–197, DOI 10.1007/s11005-010-0369-5. MR2586871 (2010k:81243)

[BCOV] M. Bershadsky, S. Cecotti, H. Ooguri, and C. Vafa, *Kodaira-Spencer theory of gravity and exact results for quantum string amplitudes*, Comm. Math. Phys. **165** (1994), no. 2, 311–427. MR1301851 (95f:32029)

[DV] R. Dijkgraaf and C. Vafa, *Matrix models, topological strings, and supersymmetric gauge theories*, Nuclear Phys. B **644** (2002), no. 1-2, 3–20, DOI 10.1016/S0550-3213(02)00766-6. MR1935579 (2003h:81166a)

[EMO] B. Eynard, N. Orantin, and M. Mariño, *Holomorphic anomaly and matrix models*, J. High Energy Phys. **6** (2007), 058, 20 pp. (electronic), DOI 10.1088/1126-6708/2007/06/058. MR2326593 (2009e:81219)

[EO] B. Eynard, N. Orantin, "Invariants of algebraic curves and topological expansion," [math-ph/0702045].

[FMPP] F. Fucito, J. F. Morales, R. Poghossian, and D. Ricci Pacifici, *Gauge theories on Ω-backgrounds from non commutative Seiberg-Witten curves*, J. High Energy Phys. **5** (2011), 098, 27, DOI 10.1007/JHEP05(2011)098. MR2834542

[Hu] M.-x. Huang, *On gauge theory and topological string in Nekrasov-Shatashvili limit*, J. High Energy Phys. **6** (2012), 152, front matter+39. MR3006836

[HK1] M.-x. Huang and A. Klemm, *Holomorphic anomaly in gauge theories and matrix models*, J. High Energy Phys. **9** (2007), 054, 33, DOI 10.1088/1126-6708/2007/09/054. MR2342377 (2009e:81225)

[HK2] M.-x. Huang and A. Klemm, *Holomorphicity and modularity in Seiberg-Witten theories with matter*, J. High Energy Phys. **7** (2010), 083, 49, DOI 10.1007/JHEP07(2010)083. MR2719968 (2011m:81274)

[HK3] M. x. Huang and A. Klemm, "Direct integration for general Omega backgrounds," arXiv:1009.1126 [hep-th].

[HKK] M.-x. Huang, A.-K. Kashani-Poor, and A. Klemm, *The Ω deformed B-model for rigid $N = 2$ theories*, Ann. Henri Poincaré **14** (2013), no. 3, 425–497, DOI 10.1007/s00023-012-0192-x. MR3035638

[IKV] A. Iqbal, C. Kozçaz, and C. Vafa, *The refined topological vertex*, J. High Energy Phys. **10** (2009), 069, 58, DOI 10.1088/1126-6708/2009/10/069. MR2607441 (2011d:81266)

[KW] D. Krefl and J. Walcher, *Extended holomorphic anomaly in gauge theory*, Lett. Math. Phys. **95** (2011), no. 1, 67–88, DOI 10.1007/s11005-010-0432-2. MR2764334 (2012d:81302)

[Ma] M. Matone, *Instantons and recursion relations in $N = 2$ SUSY gauge theory*, Phys. Lett. B **357** (1995), no. 3, 342–348, DOI 10.1016/0370-2693(95)00920-G. MR1352447 (96k:81260)

[MM] A. Mironov and A. Morosov, *Nekrasov functions and exact Bohr-Sommerfeld integrals*, J. High Energy Phys. **4** (2010), 040, 15, DOI 10.1007/JHEP04(2010)040. MR2673092 (2011j:81327)

[Ne] N. A. Nekrasov, *Seiberg-Witten prepotential from instanton counting*, Adv. Theor. Math. Phys. **7** (2003), no. 5, 831–864. MR2045303 (2005i:53109)

[NO] N. A. Nekrasov and A. Okounkov, *Seiberg-Witten theory and random partitions*, The unity of mathematics, Progr. Math., vol. 244, Birkhäuser Boston, Boston, MA, 2006, pp. 525–596, DOI 10.1007/0-8176-4467-9_15. MR2181816 (2008a:81227)

[NY] H. Nakajima and K. Yoshioka, *Lectures on instanton counting*, Algebraic structures and moduli spaces, CRM Proc. Lecture Notes, vol. 38, Amer. Math. Soc., Providence, RI, 2004, pp. 31–101. MR2095899 (2005m:14016)

[NS] N. A. Nekrasov and S. L. Shatashvili, *Quantization of integrable systems and four dimensional gauge theories*, XVIth International Congress on Mathematical Physics, World Sci. Publ., Hackensack, NJ, 2010, pp. 265–289, DOI 10.1142/9789814304634_0015. MR2730782 (2012b:81109)

[Po] R. Poghossian, *Deforming SW curve*, J. High Energy Phys. **4** (2011), 033, 12, DOI 10.1007/JHEP04(2011)033. MR2833280 (2012h:81272)

[SW1] N. Seiberg and E. Witten, *Erratum: "Electric-magnetic duality, monopole condensation, and confinement in $N = 2$ supersymmetric Yang-Mills theory"*, Nuclear Phys. B **430** (1994), no. 2, 485–486, DOI 10.1016/0550-3213(94)00449-8. MR1303306 (95m:81202b)

[SW2] N. Seiberg and E. Witten, *Monopoles, duality and chiral symmetry breaking in $N = 2$ supersymmetric QCD*, Nuclear Phys. B **431** (1994), no. 3, 484–550, DOI 10.1016/0550-3213(94)90214-3. MR1306869 (95m:81203)

[Za] D. Zagier, *Elliptic modular forms and their applications*, The 1-2-3 of modular forms, Universitext, Springer, Berlin, 2008, pp. 1–103, DOI 10.1007/978-3-540-74119-0_1. MR2409678 (2010b:11047)

Interdisciplinary Center for Theoretical Study, Department of Modern Physics, University of Science and Technology of China, 96 Jinzhai Road, Hefei, Anhui 230026, China

E-mail address: `minxin@ustc.edu.cn`

Proceedings of Symposia in Pure Mathematics
Volume **90**, 2015
http://dx.doi.org/10.1090/pspum/090/01524

AGT and the Topological String

Amir-Kian Kashani-Poor

Abstract. The AGT correspondence relates two dimensional conformal field theory and four dimensional $\mathcal{N} = 2$ gauge theory. The latter can be geometrically engineered within string theory. The worldsheet vantage point gives rise to the $\mathcal{N} = 2$ partition function in a genus expansion. The holomorphic anomaly equations provide modular expressions for the coefficients of this expansion. In this note, we interpret these expressions in terms of two dimensional conformal field theory, and study implications of two dimensional conformal field theory results for the genus expansion. In particular, we prove, in the case of $N_f = 4$ $SU(2)$ gauge theory, that the genus expansion does not converge.

1. Introduction

The AGT correspondence relates two dimensional conformal field theory to four dimensional $\mathcal{N} = 2$ supersymmetric gauge theory in an Ω-background. In its original formulation [**1**], evidence for the correspondence is provided by comparing gauge theory partition functions in an instanton expansion Z_{inst} [**2**] order by order in the instanton counting parameters to conformal blocks expanded in powers of complex structure parameters, which keep track of the weight of exchanged momenta. Gauge theory partition functions permit a second presentation,[1]

$$Z_{top} = \exp\left[\sum_{n,g} F^{(n,g)}(q) g_s^{2g-2} s^n\right] , \tag{1}$$

which permits the extraction of the prepotential $F^{(0,0)}$ of the $\mathcal{N} = 2$ theory as a leading term. The subleading terms have interpretations within gauge theory on a non-trivial gravitational background, but we prefer to think of them from the vantage point of topological string theory: upon geometric engineering of the gauge theory, they correspond to *generalized* topological string amplitudes in a field theory limit. In this language, this alternative expansion corresponds to the conventional genus expansion of string theory in terms of the coupling g_s, in conjunction with a less well understood expansion in terms of a parameter we have called s in (1), with $s \neq 0$ giving rise to the 'generalized' in 'generalized topological string'. We will subsume both the expansion in g_s and in s under the name of 'genus expansion'

2010 *Mathematics Subject Classification.* Primary 81T13, 81T40.

[1] As we argue further below, the expansion in g_s and s is not convergent. We hence prefer to distinguish notationally between Z_{inst} and Z_{top}.

for simplicity. In this note, based on [**3**], we explore some implications of the AGT correspondence in view of the genus expansion.

2. $N_f = 4$ gauge theory and the four-point conformal block

The example we will be focussing on is the instanton partition function Z_{inst} of $\mathcal{N} = 2$ $SU(2)$ gauge theory with $N_f = 4$ on an Ω-background [**2**]. The theory is specified by a coupling constant q_{inst} and a mass matrix $M = \sum_{i=1}^4 m_i H_i = \sum_i M_i \alpha_i \cdot H$. Here, $\{H_i\}$ specifies the standard basis of the Cartan of $\mathfrak{so}(8)$, while $\{\alpha_i \cdot H\}$ provides a basis of generators adapted to the inclusion $\mathfrak{su}(2)^4 \subset \mathfrak{so}(8)$. $\mathcal{N} = 2$ theories exhibit a moduli space of vacua, parametrized in the case of gauge group $SU(2)$ by a global coordinate u, or a local flat parameter a [**4**]. The Ω-background is characterized by two constants ϵ_1 and ϵ_2.

On the conformal field theory side, the central quantity in our discussion will be the four-point conformal block $F(c, h, h_i|x)$ on the sphere. This is a function of the central charge c of the theory, the complex structure x of the four punctured sphere, the weights h_i of the four insertions, and the exchanged momentum h.

The AGT dictionary states [**1**]

$$F(c, h, h_i|x) = Z_{inst}(\epsilon_1, \epsilon_2, a, m_i, q_{inst})(1-x)^{-\nu} . \tag{2}$$

The $U(1)$ factor $(1-x)^{-\nu}$ will not be important in our discussion, as it merely shifts the amplitudes $F^{(n,g)}$ at $n+g = 0, 1$ by a independent terms. With $Q = b + \frac{1}{b}$ such that $c = 1 + 6Q^2$,

$$b = \sqrt{\frac{\epsilon_1}{\epsilon_2}} , \quad h = \frac{Q^2}{4} - \frac{a^2}{\epsilon_1 \epsilon_2} , \quad h_i = \frac{Q^2}{4} - \frac{M_i^2}{\epsilon_1 \epsilon_2} , \tag{3}$$

$$x = q_{inst} . \tag{4}$$

2.1. S-duality, modularity, and the holomorphic anomaly equations.

$SU(2)$ Seiberg-Witten theory with four massless flavors is superconformal. Any coordinate on the space of exactly marginal deformations of the theory can serve as UV-coupling and corresponds to a different parametrization of the Seiberg-Witten curve. Two natural choices are τ and $\log q$ defined as follows:

$$y^2 = (x - u\, e_1(\tau))(x - u\, e_2(\tau))(x - u\, e_3(\tau)) \tag{5}$$

$$\Updownarrow$$

$$y^2 = x(x-u)(x - u\, q) , \quad q = \frac{e_3 - e_2}{e_1 - e_2}(\tau) = \frac{\vartheta_2^4(\tau)}{\vartheta_3^4(\tau)} . \tag{6}$$

The functions $e_i(\tau)$ appearing in equation (5) are obtained by evaluating the Weierstrass $\wp$-function at the three two-torsion points of the elliptic curve of complex structure τ. With the Seiberg-Witten curve as input, the holomorphic anomaly equations permit the recursive computation of the amplitudes $F^{(n,g)}$. Exact results in terms of quasi-modular forms were computed for the $N_f = 4$ theory in [**5**]. In fact, two instances of quasi-modularity arise: Quasi-modularity in the ultraviolet coupling τ is a reflection of S-duality. Quasi-modularity in the effective infrared coupling τ_{cplx} of the theory is built into the holomorphic anomaly equations in the form put forth in [**6**], and reflects the invariance of the theory under the monodromy group as singular points in moduli space are circled. In the massless limit, infrared and an appropriately chosen ultraviolet coupling coincide, and this distinction becomes blurred. Indeed, the curve (5) is parametrized in terms of its complex structure, $\tau = \tau_{cplx}$ (this is possible as its complex structure is independent of u). This is the parameter on which the S-duality group acts fractionally linearly. Upon computation, the parameter q in (6) is revealed to coincide with the instanton counting parameter q_{inst} of [**2**]. As evident from its explicit expression in terms of τ, it transforms as a cross-ratio under modular transformations.

The general structure of the amplitudes $F^{(n,g)}$ in the massless limit is [**5**]

$$F^{(n,g)} = \frac{1}{a^{2(n+g)-2}} p_{2(n+g)-2}(E_2(\tau), E_4(\tau), E_6(\tau)), \tag{7}$$

with p_n a weighted homogeneous polynomial of total weight n. The explicit expressions at $n + g = 2, 3$ are[2]

$$F^{(2,0)} = \frac{E_2(\tau)}{2^5 3a^2}, \quad F^{(1,1)} = -\frac{E_2(\tau)}{2^3 3a^2}, \quad F^{(0,2)} = \frac{E_2(\tau)}{2^5 a^2}, \tag{8}$$

$$\begin{aligned} F^{(3,0)} &= -\frac{1}{2^8 3^2 5a^4}(5E_2^2 + 13E_4)(\tau), \\ F^{(2,1)} &= \frac{1}{2^6 3^2 5a^4}(10E_2^2 + 17E_4)(\tau), \\ F^{(1,2)} &= -\frac{1}{2^8 3^2 5a^4}(95E_2^2 + 94E_4)(\tau), \\ F^{(0,3)} &= \frac{1}{2^7 3a^4}(2E_2^2 + E_4)(\tau). \end{aligned}$$

2.2. Recursion relations for conformal blocks. Which primary states occur in the spectrum of a given conformal field theory is a dynamical question. The occurrence of the corresponding descendants then follows by the representation theory of the Virasoro algebra. Conformal blocks capture the resulting universal contribution to correlators.

In the case of the four-point conformal block on the sphere, the corresponding decomposition takes the form

$$G_{1234}(x) = \sum_h C_{12h} C^h_{34} F^{23}_{14}(h|x) \bar{F}^{23}_{14}(h|\bar{x}), \tag{9}$$

where G_{1234} is related to the four-point function via

$$G_{1234}(x) = \langle h_1 | V_{h_2}(1) V_{h_3}(x) | h_4 \rangle, \tag{10}$$

with $x = \frac{(z_3-z_4)(z_2-z_1)}{(z_3-z_1)(z_2-z_4)}$ and $|h_4\rangle$ and $\langle h_1|$ defined with regard to the coordinate system $z'(z) = \frac{(z-z_4)(z_2-z_1)}{(z-z_1)(z_2-z_4)}$. In the decomposition (9), $F^{23}_{14}(h|x) = F(c, h, h_i|x)$ are the conformal blocks, and C_{ijk} encode the dynamical data of the CFT. The sum is over the primary fields in the spectrum of the theory of dimension h.

Conformal blocks exhibit poles at certain values of the central charge c and of the exchanged weight h. Considered as functions of c for fixed h, these lie at those values c_{mn} of the central charge c at which h corresponds to a degenerate weight of degree mn. Considered as functions of h for fixed c, they lie at degenerate weights $h_{mn}(c)$. The observation at the heart of the recursion relations [**7**] is that the residue of the poles is proportional to the conformal block itself, evaluated at the central charge or singular weight respectively.

[2]With the convention $Z_{top} \sim Z_{inst}$ (vs. $Z_{top} \sim Z_{pert} Z_{inst}$), we should strictly speaking subtract the constant piece in the Fourier expansion of the following amplitudes.

The genus expansion emerges from the presentation of the conformal blocks in which the poles in the exchanged momenta are explicit. It proves convenient to write the conformal block as a product of functions of the elliptic parameter q_2,

$$F(c,h,h_i|q_2) = f(c,h,h_i|q_2)H_h(c,h_i|q_2)\,. \tag{11}$$

Here, the parameter $q_2 = e^{\pi i\tau}$ and the cross ratio x are related via

$$x = \frac{e_3 - e_2}{e_1 - e_2}(\tau)\,, \tag{12}$$

with q_2 governing the leading h dependence of the conformal block in the $h \gg c \gg 1$ limit,[3] captured by the function $f(c,h,h_i|q_2)$. $f(c,h,h_i|q_2)$ was computed in [**8**][**9**] and is given by

$$\begin{aligned} f(c,h,h_i|q_2) &= (16q_2)^{h-\frac{c-1}{24}} x^{\frac{c-1}{24}-h_2-h_3}(1-x)^{\frac{c-1}{24}-h_2-h_1}\vartheta_3^{\frac{c-1}{2}-4(h_0+h_1+h_2+h_3)} \\ &= (16q_2)^{-\frac{a^2}{\epsilon_1\epsilon_2}}\left(\frac{\vartheta_3^2}{\vartheta_0\vartheta_2\vartheta_3}\right)^{4h_2}(\vartheta_0\vartheta_2\vartheta_3)^{Q^2}\prod_{i=1}^{3}(\vartheta_{i+1})^{-4h_{i+2}}\,. \end{aligned} \tag{13}$$

The second factor in the block (11) satisfies the relation

$$H_h(c,h_i|q_2) = 1 + \sum_{m,n}(16q_2)^{mn}\frac{R_{mn}(c,h_i)}{h - h_{mn}(c)}H_{h_{mn}+mn}(c,h_i|q_2)\,, \tag{14}$$

with the residue factor given by $R_{mn}(c,h_i) = A_{mn}\prod_{i=1}^{4} Y_{rs}(2m_i)$, and the functions A and Y by

(15)

$$A_{mn} = \frac{1}{2}\left(\prod_{r=1-m}^{m}\prod_{s=1-n}^{n}\right)'\frac{1}{rb+sb^{-1}}\,,\quad Y_{mn}(M) = \left(\prod_{r=1-m}^{m-1}\prod_{s=1-n}^{n-1}\right)''\left(\frac{M}{\sqrt{\epsilon_1\epsilon_2}} - \frac{rb+sb^{-1}}{2}\right).$$

The prime on the first product indicates that the factors $(r,s) = (0,0),(m,n)$ are to be omitted, and the double prime on the second product prescribes a product over pairs satisfying $(r,s) = (1-m,1-n)\bmod(2,2)$. The conformal dimensions h_{mn} of degenerate representations are given by

$$h_{mn} = \frac{Q^2}{4} - \frac{1}{4}\left(mb + \frac{n}{b}\right)^2\,. \tag{16}$$

The leading behavior q_2^h of the conformal block [**8**] gives rise to a recursion relation with regard to the order in the elliptic parameter q_2, providing a rationale for expressing the conformal block in terms of it [**9**].

3. AGT and the topological string

In the massless limit, the only massive parameter of the gauge theory aside from $\epsilon_{1,2}$ is a. The expansion in g_s and s hence coincides with an expansion in $\frac{1}{a}$, of the form

$$Z_{top} = \exp\sum_k \frac{\sum_{n+g=k}\mathcal{F}^{(n,g)}(q)g_s^{2g-2}s^n}{a^{2k-2}}\,. \tag{17}$$

[3]The notation q_2 follows the convention $q_n = \exp\frac{2\pi i}{n}$. It reflects the $\Gamma(2)$ modular symmetry of the massive theory.

Considering the a dependence of (11), we conclude that the coefficient of H_h, and the $U(1)$ factor in (2), contribute only to $F^{(n,g)}$ for $n+g \leq 1$. Furthermore, the a dependence on the RHS of (14) is contained entirely in the denominators

$$\frac{1}{h-h_{mn}(c)} = -\frac{\epsilon_1\epsilon_2}{a^2}\frac{1}{1-\frac{1}{4a^2}(m\epsilon_2+n\epsilon_1)^2}, \tag{18}$$

allowing us to conclude, formally,

$$H_h = \exp \sum_{n+g>1} F^{(n,g)} g_s^{2g-2} s^n . \tag{19}$$

We can use the existence of the recursion relation to rule out the convergence of the expansion (17). To this end, write the partition function in the form

$$\log Z'_{top} = \sum_{k=1}^{\infty}\left(\sum_{r=1}^{\infty}\mathcal{F}_r^{k+1}q^r\right)a^{-2k} . \tag{20}$$

We have introduced the notation $\sum_r \mathcal{F}_r^k q^r = \sum_{n+g=k}\mathcal{F}^{(n,g)}(q)g_s^{2g-2}s^n$, and dropped the irregular $g+n\leq 1$ contributions, as indicated by the prime. The sum in parentheses yields a finite sum of quasi-modular forms (minus the constant terms) for every value of k and is hence absolutely convergent in the upper half-plane. If the sum over a was also convergent, hence absolutely convergent within its radius of convergence, we could reverse the order of summation. This yields the partition function in the form in which it arises in the recursion relation for conformal blocks: In the $h \gg c \gg 1$ limit, $F(c,h,h_i|q_2)$ approaches $f(c,h,h_i|q_2)$, hence the sum over m and n in (14) becomes arbitrarily small, and $\log H_h$ can be expanded around 1. The equality (19) implies that the a dependence of $\sum_{k=1}^{\infty}\mathcal{F}_r^{k+1}a^{-2k}$ must equal a finite sum of products of rational functions (18). The formal expansion of (18) in $\frac{1}{a}$ fulfils this expectation. At given k, $\delta_k > 0$ exists such that this expansion is valid for $\frac{\epsilon_{1,2}}{a} < \delta_k$. Due to the dependence on m and n in (18), $\lim_{k\to\infty}\delta_k = 0$.

Nevertheless, performing the formal a expansion of (14) allows us to obtain correct all order in q expressions. The simplest such result arises upon isolating the $\frac{1}{a^2}$ contribution, and gives rise to the remarkable identity

$$\begin{aligned} -\frac{g_s^2}{a^2}\sum_{m,n}(16q)^{mn}R_{mn}(c,h_i)H_{h_{mn}+mn}(c,h_i) &= \left[\frac{s^2}{g_s^2}F^{(2,0)}+sF^{(1,1)}+g_s^2F^{(0,2)}\right]' \\ &= \left(\frac{s^2}{g_s^2}-4s+3g_s^2\right)\frac{[E_2(q)]'}{96a^2} . \end{aligned} \tag{21}$$

The notation $[\cdot]'$ indicates dropping the q independent term, see footnote 2. Below, we will use the notation $[\cdot]_n$ to indicate the n-th order term in q_2.

We can also reverse our reasoning to obtain all genus results for the topological string, at a given order in q. E.g., the first non-vanishing order in q gives rise to the equality

$$\begin{aligned} H_h(c,h_i|q_2) &= 1+(16q_2)^2\left(\frac{R_{12}}{h-h_{1,2}}+\frac{R_{21}}{h-h_{2,1}}\right)+\dots \\ &= 1+\left[\sum_{n+g>1}F^{(n,g)}g_s^{2g-2}s^n\right]_1 q+\dots \end{aligned} \tag{22}$$

with

$$\frac{R_{12}}{h-h_{1,2}}+\frac{R_{21}}{h-h_{2,1}}=\frac{\epsilon_2^5}{2^9\epsilon_1(\epsilon_1^2-\epsilon_2^2)a^2}\frac{1}{(1-\frac{(\epsilon_1+2\epsilon_2)^2}{4a^2})}+(\epsilon_1\leftrightarrow\epsilon_2)\,. \tag{23}$$

Returning to the a expansion, the recursion relation implies that at a given order in q_2, the increasing powers of a in (7) give rise to a geometric series, yielding an infinite number of poles in a. As we reviewed above, these have a very clear interpretation from the point of view of the Virasoro algebra. It would be interesting to find an interpretation in the context of gauge theory or the topological string. Note that the q independent contributions assembled in Z_{pert}, which can be extracted from the DOZZ three-point function of Liouville theory, yield zeros which exactly cancel these poles.

4. Conclusions

We have seen that studying the AGT correspondence from the vantage point of the genus expansion has interesting implications on both sides of the correspondence. For the topological string (in the field theory limit), it reveals an analytic partition function from which the genus expansion can be generated. For the conformal field theory, it exposes a hidden quasi-modular symmetry of the conformal blocks.

The analysis presented here clearly represents only a first step in exploiting the link provided by AGT between conformal field theory and the topological string (in the field theory limit). To remove the constant caveat *in the field theory limit*, one would need to explore a q-deformed version of the correspondence. Perhaps more pressing however is understanding the role of quasi-modularity: what are the signatures of this structure in the partition function proper, prior to its formal power series expansion?

References

[1] L. F. Alday, D. Gaiotto, and Y. Tachikawa, *Liouville correlation functions from four-dimensional gauge theories*, Lett. Math. Phys. **91** (2010), no. 2, 167–197, DOI 10.1007/s11005-010-0369-5. MR2586871 (2010k:81243)

[2] N. A. Nekrasov, *Seiberg-Witten prepotential from instanton counting*, Adv. Theor. Math. Phys. **7** (2003), no. 5, 831–864. MR2045303 (2005i:53109)

[3] A.-K. Kashani-Poor and J. Troost, *Transformations of spherical blocks*, J. High Energy Phys. **10** (2013), 009, front matter+29, DOI 10.1007/JHEP10(2013)009. MR3110708

[4] N. Seiberg and E. Witten, *Erratum: "Electric-magnetic duality, monopole condensation, and confinement in $N = 2$ supersymmetric Yang-Mills theory"*, Nuclear Phys. B **430** (1994), no. 2, 485–486, DOI 10.1016/0550-3213(94)00449-8. MR1303306 (95m:81202b)

[5] M.-x. Huang, A.-K. Kashani-Poor, and A. Klemm, *The Ω deformed B-model for rigid $N = 2$ theories*, Ann. Henri Poincaré **14** (2013), no. 3, 425–497, DOI 10.1007/s00023-012-0192-x. MR3035638

[6] M.-x. Huang and A. Klemm, *Holomorphicity and modularity in Seiberg-Witten theories with matter*, J. High Energy Phys. **7** (2010), 083, 49, DOI 10.1007/JHEP07(2010)083. MR2719968 (2011m:81274)

[7] A. B. Zamolodchikov and A. B. Zamolodchikov, "Conformal field theory and 2-D critical phenomena. 3. Conformal bootstrap and degenerate representations of conformal algebra," ITEP-90-31.

[8] Al. B. Zamolodchikov, *Two-dimensional conformal symmetry and critical four-spin correlation functions in the Ashkin-Teller model* (Russian), Zh. Èksper. Teoret. Fiz. **90** (1986), no. 5, 1808–1818; English transl., Soviet Phys. JETP **63** (1986), no. 5, 1061–1066. MR869395 (88d:82124)

[9] Al. B. Zamolodchikov, *Conformal symmetry in two-dimensional space: on a recurrent representation of the conformal block* (Russian, with English summary), Teoret. Mat. Fiz. **73** (1987), no. 1, 103–110. MR939798 (89d:81113)

LPTENS, École Normale Supérieure, 24 Rue Lhomond, 75005 Paris, France
E-mail address: kashani@lpt.ens.fr

Proceedings of Symposia in Pure Mathematics
Volume **90**, 2015
http://dx.doi.org/10.1090/pspum/090/01527

Generalized Chern-Simons Action and Maximally Supersymmetric Gauge Theories

M. V. Movshev and A. Schwarz

ABSTRACT. We study observables and deformations of generalized Chern-Simons action and show how to apply these results to maximally supersymmetric gauge theories. We describe a construction of large class of deformations based on some results on the cohomology of super Lie algebras proved in the Appendix.

1. Chern-Simons functional

One can construct Chern-Simons functional for every differential associative $\mathbf{Z}_2$-graded algebra $\mathcal{A}$ equipped with trace tr. By definition trace is a linear functional vanishing on supercommutators; we assume that the trace is closed (vanishes on elements of the form da). This linear functional can be odd or even; we are mostly interested in the case when it is it is odd.

We can define invariant inner product in terms of trace: $< a, b >= \mathrm{tr} ab$. Let us assume that this inner product is non-degenerate.

For every N we define the associative algebra $\mathcal{A}_N$ as tensor product $\mathcal{A} \otimes \mathrm{Mat}_N$ where Mat_N stands for the matrix algebra. (In other words, $\mathcal{A}_N$ is an algebra of $N \times N$ matrices with entries from $\mathcal{A}_N$.) Using the trace in $\mathcal{A}$ and the conventional matrix trace we define the trace and invariant inner product in $\mathcal{A}_N$.

Chern-Simons functional on $\Pi\mathcal{A}_N$ is defined by the formula

$$CS_N(A) = \frac{1}{2}\mathrm{tr} AdA + \frac{2}{3}\mathrm{tr} A^3 = \frac{1}{2}\mathrm{tr} AdA + \frac{1}{3}\mathrm{tr} A[A, A]$$

Here Π stands for the parity reversal. Chern-Simons functional is even if the trace has odd parity, as we will assume in what follows. We omit the index N in the notation CS_N.

Notice that in this definition we need really only the super Lie algebra structure and invariant inner product in the algebra $\mathcal{A}_N$:

$$CS(A) = \frac{1}{2} < A, dA > + \frac{1}{3} < A, [A, A] > .$$

The functional CS coincides with the standard Chern-Simons functional in the case when $\mathcal{A}$ is the algebra $\Omega(M)$ of differential forms on three-dimensional manifold M equipped with a trace $\mathrm{tr} C = \int_M C$.

2010 *Mathematics Subject Classification.* Primary 18G55.

The work of both authors was partially supported by NSF grants.

An odd non-degenerate inner product on vector space induces such a product on its dual that can be regarded as an odd symplectic form. This form $\omega = dz^a\omega_{ab}dz^b$ can be used to define an odd Poisson bracket on functions: $\{F, H\} = \overleftarrow{\partial}_a F\omega^{ab}\overrightarrow{\partial}_b H$, and to assign a vector field $\partial_b F\omega^{ab}$ to every function. (This vector field corresponds to the first order differential operator ξ_F defined by the formula $\xi_F(H) = \{F, H\}$.)

Applying this remark to $\mathcal{A}_N$ we see that the odd vector field Q corresponding to the functional CS can be written in the form

$$\delta_Q A = dA + \frac{1}{2}[A, A]. \tag{1.1}$$

This vector field obeys $[Q, Q] = 0$. [1].The relation $[Q, Q] = 0$ is equivalent to the BV master equation $\{CS, CS\} = 0$. Notice that that vector field Q does depend on super Lie algebra $L_N\mathcal{A} = Lie\mathcal{A}_N$ corresponding to the associative algebra $\mathcal{A}_N$, but it does not depend of inner product.

The construction of the functional CS can be generalized to the case when $\mathcal{A}$ is an A_∞-algebra equipped with invariant inner product. Recall that the structure of A_∞-algebra $\mathcal{A}$ on a $\mathbb{Z}_2$-graded space is specified by means of a sequence ${}^{(k)}m$ of operations, satisfying some relations that will be described later. The operation ${}^{(k)}m$ has k arguments, in a coordinate system it is specified by a tensor ${}^{(k)}m^a_{a_1,\dots,a_k}$ having one upper index and k lower indices. Having an inner product we can lower the upper index; invariance of inner product means that the tensor ${}^{(k)}\mu_{a_0,a_1,\dots,a_k} = \omega_{a_0a}m^a_{a_1,\dots,a_k}$ is cyclically symmetric (in graded sense). The Chern-Simons functional can be defined on $\mathcal{A} \otimes \mathrm{Mat}_N$ in the following way. In a basis $a_1, \dots, a_k$ for $\mathcal{A}$ the value of the functional on even $a = \sum a_i t^i, t^i \in \mathrm{Mat}_N$ is equal to

$$m(a) = \sum_k \pm m_{i_1\dots i_k}\mathrm{tr}(t^{i_1}\cdots t^{i_k})$$

with $\pm$ derived from the Koszul sign rules.

We assume that the inner product is odd; then the Chern-Simons functional generates an odd vector field Q on $\Pi\mathcal{A}_N$; the conditions that should be imposed on the operations in A_∞-algebra are equivalent to the condition $[Q, Q] = 0$ or , in other words, to the condition that Chern-Simons functional obeys the BV master equation.

Notice that two quasi-isomorphic A_∞-algebras are physically equivalent (i.e. corresponding Chern-Simons functionals lead to the same physical results).

A differential associative algebra can be considered as an A_∞-algebra where only operatious ${}^{(1)}m$ and ${}^{(2)}m$ do not vanish; in this case both definitions of Chern-Simons functional coincide.

2. Observables in Chern-Simons theory

Recall that in BV formalism a physical theory is specified by an action functional S defined on a space of fields considered as functions on odd symplectic manifold $\mathcal{E}$. A classical observable is defined as an even functional a obeying $\{S, a\} = 0$ or equivalently $\xi_S a = 0$ were S is the action functional obeying the master equa-

[1]An odd vector field Q obeying $[Q, Q] = 0$ is called homological vector field, because the corresponding first order differential operator $\hat{Q}$ obeys $\hat{Q}^2 = 0$ and therefore can be considered as a differential acting on the space of functions.

tion and ξ_S stands for the operator $\xi_S f = \{S, f\}$. (We restrict ourselves to the polynomial functionals.) Classical observables are related to infinitesimal deformations of the solution to the master equation. In what follows we will consider also odd functionals a obeying $\{S, a\} = 0$. These "odd observables" do not have a physical meaning of observables, but they correspond to odd infinitesimal deformations (deformations of the form $S + \epsilon a$ where ϵ is an odd parameter.) Trivial observables (observables of the form $a = \{S, b\}$) correspond to trivial deformations (deformations induced by infinitesimal change of variables). We will see that odd observables generate even symmetry transformations and even observables generate odd symmetry transformations.

Equivalently one can describe a theory by means of a homological vector field Q on $\mathcal{E}$ preserving the odd symplectic form; then the first order differential operator $\hat{Q}$ corresponding to Q can be represented in the form ξ_S. Notice that the set of of observables depends only on Q (the symplectic form and the action functional are irrelevant). The field Q determines the equations of motion (a solution of equation of motion can be interpreted as a point in the zero locus of Q.) If there exists a Q-invariant symplectic form these equations of motion come from action functional.

If Chern-Simons theory is constructed by means of associative graded differential algebra $\mathcal{A}$ with an odd inner product then every element of cyclic cohomology of $\mathcal{A}$ specifies an observable [**14**]. This fact follows from the statement that infinitesimal deformations of $\mathcal{A}$ into A_∞-algebra with inner product are labelled by cyclic cohomology $HC(\mathcal{A})$ of $\mathcal{A}$ [**13**]. Algebra $\mathcal{A}$ determines Chern-Simons theory for all N; the observables we are talking about are defined for every N.

The observables we consider can be constructed directly, without reference to [**13**]. By definition a cyclic cocycle on $\mathcal{A}$ is a polylinear map $\sigma : \mathcal{A}^n \to \mathbb{C}$ satisfying some conditions. Such a map can be used to construct a polylinear map $\sigma_N : \mathcal{A}_N^n \to \mathbb{C}$ specifying a cyclic cocycle of $\mathcal{A}_N$ .This follows from Morita invariance of cyclic cohomology([**10**]). It is easy to give an explicit formula for σ_N, namely

$$\sigma_N(A_1, \ldots, A_n) = \sum_{i_1,\ldots,i_n} \sigma(a_1^{i_1,i_2}, a_2^{i_2,i_3} \ldots, a_n^{i_n,i_1}), A_k = (a_k^{i,j})$$

An observable corresponding to the cyclic cocycle is the functional $\sigma_N(A, \ldots, A)$ defined on $\Pi\mathcal{A}_N$.

Obviously a product of observables is an observable, hence elements of $\mathrm{Sym}\Pi HC(\mathcal{A}) = \sum \mathrm{Sym}^k \Pi HC(\mathcal{A})$ specify observables defined for every N. Let us prove that all observables defined for every N are of this kind. To classify observables we should compute the cohomology of the first order differential operator $\hat{Q} = \xi_{CS}$ induced by the vector field Q in the space of polynomial functionals on $\mathcal{A}_N$. It is easy to check that up to parity reversion this operator can be identified with the differential in the definition of Lie algebra cohomology of $L_N\mathcal{A} = Lie\mathcal{A}_N$. It is well known [**10**] that for large N the cohomology of $\mathcal{A}$ with trivial differential is isomorphic to $\mathrm{Sym}[\Pi HC(\mathcal{A})]$.

For any dga $\mathcal{A}$ there exists a homomorphism of $\mathrm{Sym}[\Pi HC(\mathcal{A})]$ into the projective (inverse) limit of groups $H(L_N\mathcal{A})$; in the cases we are interested in this homomorphism is an isomorphism (we do not know whether this is true in general).

3. Ten-dimensional SUSY YM theory as generalized Chern-Simons theory

We have constructed Chern-Simons theory starting with differential associative algebra $\mathcal{A}$ equipped with closed trace tr that generates invariant inner product $< a, b >= \mathrm{tr}ab$. We have assumed that the inner product is non-degenerate; however, one can show that it suffices to assume the non-degeneracy of the induced inner product on homology. This remark allows us to consider ten- dimensional SUSY YM theory and its dimensional reductions as generalized Chern-Simons theory.

We define Berkovits algebra B as the algebra of polynomial functions of pure spinor λ, odd spinor ψ and $x = (x^1, \dots, x^{10}) \in \mathbb{R}^{10}$. Sometimes it is convenient to modify this definition considering an algebra B^∞ consisting of functions that are polynomial in λ and ψ but smooth as functions of $x \in \mathbb{R}^{10}$. The differential is defined as the derivation

$$d = \lambda^\alpha \left(\frac{\partial}{\partial \psi^\alpha} + \Gamma^i_{\alpha\beta} \psi^\beta \frac{\partial}{\partial x^i} \right). \tag{3.1}$$

The algebra B_d (Berkovits algebra reduced to d-dimensional space) is the algebra of functions depending on pure spinor λ, odd spinor ψ and $x = (x^1, \dots, x^d) \in \mathbb{R}^d$. The differential is defined by the same formula.

One can use (1.1) to define an odd vector field Q on $B_d \otimes Mat_N$. It is well known ([**2**], [**6**],[**3**]) that for every solution $A_\alpha(x, \theta)$ of equations of motion of ten-dimensional SUSY YM one can construct a point $\lambda^\alpha A_\alpha(x, \theta) \in B_{10} \otimes Mat_N$ belonging to the zero locus of Q. (We are working in superspace formalism, the superfield A_α takes values in $N \times N$ matrices.) Starting with a solution of equations of motion reduced to d-dimensional space we obtain a point of the zero locus of the vector field Q on $B_d \otimes Mat_N$.

For $d = 0$ we obtain the reduced Berkovits algebra related to ten-dimensional SUSY YM theory reduced to a point. Introducing an appropriate trace we can apply the techniques of generalized Chern-Simons theory to this algebra. To apply these techniques to the case $d > 0$ we should introduce the notions of local observables and local trace (see the next section).

4. Lie algebra of local observables in the classical BV formalism

Let $f : M \to N$ be a $\mathbb{C}$-linear map of graded R-modules where R is a graded commutative algebra over $\mathbb{C}$. We define $[x, f] : M \to N$ by the formula $xf(m) - (-1)^{|f||x|} f(xm), x \in R$. The map f is local if $[x_1, \dots, [x_n, f] \dots] \equiv 0$ for some n and all $x_1 \dots, x_n \in R$. If R is an algebra of smooth functions on $\mathbb{R}^{n|k}$ and M, N are the space of sections of finite rank vector bundles then f defines a differential operator between vector bundles.

If R, M, N are differential graded objects then the definition can be weakened by replacing $[x_1, \dots, [x_n, f] \dots] \equiv 0$ by $[x_1, \dots, [x_n, f] \dots] \equiv [d, g]$ for some $g : M \to N$.

Let A be an associative graded algebra over R and M is a graded A-bimodule. The graded space of Hochschild cochains $C^k(A, M) = \mathrm{Hom}_{\mathbb{C}}(A^{\otimes k}, M)$ contains a subspace $C^k_{R_{loc}}(A, M)$ of cochains $\mathcal{O}$ that are local with respect to all variables (i.e., $f(x) = \mathcal{O}(a_1, \dots, a_{i-1}, x, a_{i+1}, \dots, a_k)$ is a local map of R-bimodules for $i = 1, \dots, k$). We omit the standard definition of the Hochschild differential referring to [**1**] or to [**9**]. By definition multiplication and the differentials in A and M are local.

This is why $\prod_k C^k_{R_{loc}}(A,M)$ is a subcomplex in $\prod_k C^k(A,M)$. The cohomology of $\prod_k C^k_{R_{loc}}(A,M)$ is denoted by $\mathrm{HH}^k_{R_{loc}}(A,M)$. When no confusion is possible we will drop the R-dependence in the cohomology: $\mathrm{HH}^k_{loc}(A,M) = \mathrm{HH}^k_{R_{loc}}(A,M)$

When $A = R$ is an algebra of functions on a smooth or an affine algebraic manifold X and M is space of sections of a vector bundle over X then $\mathrm{HH}^k_{loc}(A,M) = \mathrm{HH}^k(A,M)$ (c.f. [**7**],Section 4.6.1.1.)

In the following we will assume that R is an algebra of functions on a smooth or an algebraic supermanifold X. This allows us to define the complex of integral forms $\{\Omega_R^{-i}\}$, equipped with the de Rham differential d_{dr}.

By definition a local trace is a series of graded local maps $\mathrm{tr}_i : A \to \Omega_R^{-i}$ that for $i \geq 0$ satisfy

$$\mathrm{tr}_i(d_A a) = -d_{dr}\mathrm{tr}_{i+1}(a), \mathrm{tr}_i([a,b]) = 0.$$

When $R = \mathbb{C}$ this becomes a definition of an ordinary trace that determines an inner product $tr(ab)$ on A. This inner product allows us to define a map of cohomology groups $\mathrm{HH}^i(A,A) \to \mathrm{HH}^i(A,A^*)$. Our next set of definitions is intended to create a setting which would accommodate a similar homomorphism in a local setting.

The complex $C^k(A,A^*)$ coincides with the space of graded maps $\mathrm{Hom}(A^{\otimes(k+1)}, \mathbb{C})$. The complex is equipped with the differential d_{hoch} (see [**1**], [**9**] for details) that computes Hochschild cohomology $\mathrm{HH}^i(A,A^*)$. By definition the complex $C^k_{loc}(A)$ is a subcomplex of the complex $\prod_{ij} \mathrm{Hom}_{\mathbb{C}}(A^{\otimes k+1}, \Omega_R^{-i})$ of R- local maps. The differential d is equal to the sum $d_{hoch} + d_{dr}$, where d_{dr} is acting in $\bigoplus_i \Omega_R^{-i}$. We denote the cohomology of $C^k_{loc}(A)$ by $\mathrm{HH}^k_{loc}(A)$. Note that when $R = \mathbb{C}$ the cohomology $\mathrm{HH}^k_{loc}(A)$ becomes equal to $\mathrm{HH}^k(A,A^*)$

The R-local cyclic cohomology are defined along the same line: the subspace $\prod_k CC^k(A)$ of $C^k_{loc}(A)$ consisting of cyclic cochains is a subcomplex. Corresponding cohomology is denoted by $\mathrm{HC}^n_{loc}(A)$.

The groups $\mathrm{HH}^n_{loc}(A,A)$, $\mathrm{HH}^n_{loc}(A)$ and $\mathrm{HC}^n_{loc}(A)$ have "multi-trace" generalizations. By definition the chains of $\mathrm{HH}^n_{loc,mt}(A,A)$ is a subspace of the R-local maps in $\mathrm{Hom}_{\mathbb{C}}(T(A) \otimes \mathrm{Sym}[\Pi CC(A)], A)$. The space of chains for $\mathrm{HH}^n_{loc,mt}(A)$ is a subspace of local maps in $\mathrm{Hom}_{\mathbb{C}}(T(A) \otimes A \otimes \mathrm{Sym}[\Pi CC(A)], \Omega_R)$, the space of chains for $\mathrm{HC}^n_{loc,mt}(A)$ is a subspace of local maps in $\mathrm{Hom}_{\mathbb{C}}(\mathrm{Sym}[\Pi CC(A)], \Omega_R)$. Keep in mind that such generalizations for the classical nonlocal theories reduce to the space of linear maps of $\bigoplus_k \mathrm{HH}^k(A,A)$ and $\bigoplus_k \mathrm{HH}^k(A,A^*)$ with $\mathrm{Sym}[\Pi\mathrm{HC}(A)]$.

The local version of the cyclic and Hochschild cohomology shares many common properties with their classical counterparts. In particular there is a long exact sequence

$$\cdots \to \mathrm{HH}^n_{loc}(A) \to \mathrm{HC}^{n-1}_{loc}(A) \to \mathrm{HC}^{n+1}_{loc}(A) \to \mathrm{HH}^{n+1}_{loc}(A) \to \cdots$$

There is a similar sequence for the multi-trace version of the theories.

A local trace functional $\{\mathrm{tr}_I\}$, which we defined above, is a cocycle in $\mathrm{HC}^0_{loc}(A)$. It defines maps $\mathrm{HH}^s_{loc}(A,A) \to \mathrm{HH}^s_{loc}(A), \mathrm{HH}^s_{loc,mt}(A,A) \to \mathrm{HH}^s_{loc,mt}(A)$ by the formula

$$c(a_1,\ldots,a_s) \to \sum_i \mathrm{tr}_i(c(a_1,\ldots,a_s)a_{s+1}).$$

We say that the trace $\mathrm{tr} = \{\mathrm{tr}_i\}$ is homologically nondegenerate if it induces an isomorphism $\mathrm{HH}^s_{loc}(A,A) \to \mathrm{HH}^s_{loc}(A)$.

Some of the discussion from Section 2 can be carried out in the local setting. In particular local vector fields, local functionals and local k-forms obviously make sense. There are homomorphisms

$$\mathrm{HH}^n_{loc,mt}(A, A) \to \mathrm{H}^n(L_N A, L_N A),$$

$$\mathrm{HC}^n_{loc,mt}(A) \to \mathrm{H}^n(L_N A),$$

$$\mathrm{HH}^n_{loc,mt}(A) \to \mathrm{H}^n(L_N A, L_N A^*)$$

The relative Hochschild cohomology group $\mathrm{HH}^n_R(A, A)$ maps to $\mathrm{HH}^n_{loc}(A, A)$. The relations of the relative groups $\mathrm{HH}^n_R(A, A^*)$ and $\mathrm{HC}^n_R(A)$ to $\mathrm{HH}^n_{loc}(A)$ and $\mathrm{HC}^n_{loc}(A)$ is obscure.

The last type of generalization is intended for a dga A without a unit over a commutative unital dga R. The definitions follow closely the above outline. The reader can consult on the details of cohomology theory of algebras without a unit in [**9**].

5. Symmetry preserving deformations

As we mentioned already infinitesimal deformations of the solution of master equation correspond to observables. Observables belonging to the same cohomology class specify equivalent deformations, i.e. deformations related by a change of variable (by a field redefinition). This means that in BV formalism the deformations of physical theory with action functional S are labeled by the cohomology of the differential ξ_S.

As we have seen an even element of $\Lambda HC(\mathcal{A})$ specifies an observable of Chern-Simons theory defined for every N; hence it determines an infinitesimal deformation of Chern-Simons theory defined for all N.

We will be interested in the deformations of Chern-Simons theory that are defined for every N and preserve the symmetry of original theory.

Let us make some general remarks about symmetries in BV formalism. If the equations of motion are specified by a homological vector field Q then every vector field q commuting with Q determines a symmetry of equations of motion. (The vector field q is tangent to the zero locus of Q.) The vector field q can be even or odd; in other words we can talk about super Lie algebra of symmetries. However, among these symmetries there are trivial symmetries, specified by vector fields of the form $[Q, a]$ where a is an arbitrary vector field. This means that the super Lie algebra $\mathcal{L}$ of non-trivial symmetries of EM can be described as homology of the space of all vector fields with respect to the differential defined as a commutator with Q. If we would like to consider only Lagrangian symmetries, i.e. symmetries corresponding to vector fields q having the form ξ_s we obtain a Lie algebra $\widetilde{\mathcal{L}}$ isomorphic to the homology of operator $\hat{Q}$. Notice, that both $\mathcal{L}$ and $\widetilde{\mathcal{L}}$ depend only on the vector field Q. However, the natural homomorphism $\widetilde{\mathcal{L}} \to \mathcal{L}$ does depend on the choice of odd symplectic structure on $\mathcal{E}$.

Let us calculate $\mathcal{L}$ and $\widetilde{\mathcal{L}}$ in the case of Chern-Simons theory restricting ourselves to the symmetry transformations, that can be applied for all N. The calculation of $\widetilde{\mathcal{L}}$ coincides with calculation of observables (up to parity reversal); we obtain $\widetilde{\mathcal{L}} = \Lambda HC(\mathcal{A})$.

To calculate $\mathcal{L}$ we should study the cohomology $H(L_N(\mathcal{A}), L_N(\mathcal{A}))$ (the cohomology of the Lie algebra $L_N(\mathcal{A})$ with the coefficients in adjoint representation).

In the case of algebra $\mathcal{A}$ with zero differential these cohomology are well known for large N ([**5**], [**10**]) ; they are equal to

$$\mathrm{Sym}[\Pi\mathrm{HC}(\mathcal{A})] \otimes \mathrm{HH}(\mathcal{A}, \mathcal{A}).$$

For dga algebras we have maps

$$\mathrm{H}(L_N\mathcal{A}, L_N\mathcal{A}) \leftarrow \mathrm{Sym}[\Pi\mathrm{HC}(\mathcal{A})] \otimes \mathrm{HH}(\mathcal{A}, \mathcal{A})$$

In interesting cases the LHS stabilizes for large N and becomes isomorphic to the RHS. It is not clear whether this situation is general.

We can say that $\mathfrak{g}$ is a symmetry Lie algebra if it is embedded into $\mathcal{L}$ (or into $\widetilde{\mathcal{L}}$ if we would like to consider only Lagrangian symmetries). Fixing some system of generators e_α in $\mathfrak{g}$ with structure constants $f^\gamma_{\alpha\beta}$ one can say that $\mathfrak{g}$ is a symmetry Lie algebra of BV theory with homological vector field Q if there exist symmetry transformations q_α satisfying commutation relations

$$[q_\alpha, q_\beta] = f^\gamma_{\alpha\beta} q_\gamma + [Q, q_{\alpha\beta}] \tag{5.1}$$

for some vector fields $q_{\alpha\beta}$. However, it is useful to accept a more restrictive definition of symmetry Lie algebra. We will say that $\mathfrak{g}$ is a symmetry algebra if we have an L_∞-homomorphism of $\mathfrak{g}$ into the differential Lie algebra of vector fields (this Lie algebra is equipped with a differential defined as a commutator with Q). L_∞-homomorphism of Lie algebra $\mathfrak{g}$ with generators e_α into differential Lie algebra $\mathcal{V}$ is defined as a sequence $q_\alpha, q_{\alpha_1,\alpha_2}, \dots \in \mathcal{V}$ obeying some relations, generalizing (5.1). (See [**8**] for details.)

Let us suppose that Lie algebra $\mathfrak{g}$ acts on the differential algebra $\mathcal{A}$. This means that we have fixed a homomorphism $\phi : \mathfrak{g} \to Der\mathcal{A}$ of $\mathfrak{g}$ into the Lie algebra of derivations of $\mathcal{A}$. (It is sufficient to assume that we have an L_∞action, i.e. an L_∞ homomorphism of $\mathfrak{g}$ into differential algebra $Der(\mathcal{A})$.) This action specifies $\mathfrak{g}$ as a Lie algebra of symmetries of Chern-Simons functional for every N. We are interested in infinitesimal deformations of this functional preserving these symmetries ($\mathfrak{g}$-invariant deformations). We identify two deformations related by the change of variables.

The space $C^\bullet(\mathcal{A}, \mathcal{A})$ of Hochschild cochains with coefficients in $\mathcal{A}$ has a natural L_∞ action of $\mathfrak{g}$, hence we can consider the cohomology of Lie algebra $\mathfrak{g}$ with coefficients in this module. We will denote this cohomology by $HH^\bullet_{\mathfrak{g}}(\mathcal{A}, \mathcal{A})$ and call it Lie -Hochschild cohomology with coefficients in $\mathcal{A}$. (For trivial $\mathfrak{g}$ it coincides with Hochschild cohomology of $\mathcal{A}$, for trivial $\mathcal{A}$ with Lie algebra cohomology of $\mathfrak{g}$.) Similarly we can define $HC^\bullet_{\mathfrak{g}}(\mathcal{A})$ (Lie-cyclic cohomology), $HH^\bullet_{\mathfrak{g}}(\mathcal{A}, \mathcal{A}^*)$. There are also multi-trace version of these groups. For example the multi-trace version of $HC^\bullet_{\mathfrak{g}}(\mathcal{A})$ uses uses the symmetric algebra of standard cyclic bicomplex $CC^i(\mathcal{A})$. The multi-trace cyclic cohomology group $HC^\bullet_{\mathfrak{g},mt}(\mathcal{A})$ is the cohomology of the bicomplex $C^\bullet(\mathfrak{g}, \mathrm{Sym}[CC^\bullet(\mathcal{A})])$. The multi-trace equivariant version of Hochschild cohomology is cohomology of the tri-complex $C^\bullet(\mathfrak{g}, \mathrm{Sym}[CC^\bullet(\mathcal{A})] \otimes C^\bullet(\mathcal{A}, \mathcal{A}))$ One can prove the following theorem:

The $\mathfrak{g}$-invariant deformations of Chern-Simons action functional $CS(\mathcal{A})$ that are defined for all N simultaneously are labelled by the elements of $\mathrm{HC}^\bullet_{\mathfrak{g},mt}(\mathcal{A})$.

We can use this theorem to study supersymmetric deformations of ten-dimensional SUSY YM theory represented as Chern-Simons theory corresponding to the Berkovits algebra.

There is a number of modifications of the Berkovits algebra B_{10} that depend on smoothness of its elements as functions on $\mathbb{R}^{10}$ and their asymptotics at infinity. Possible choices are polynomials functions , which are elements of $\mathbb{C}[x_1, \dots, x_{10}]$. This way we get B_{10}^{poly}. Similarly we can get an analytic modification B_{10}^{an} which contains the algebra of analytic functions $C^{an}(\mathbb{R}^{10})$ or the smooth version $B_{10}^{\infty} \supset C^{\infty}(\mathbb{R}^{10})$, with or without restriction on asymptotics at infinity. Our following computations don't depend on what pair of algebras

$$R = \mathbb{C}[x_1, \dots, x_{10}] \subset B_{10}^{poly} = A$$
$$R = C^{an}(\mathbb{R}^{10}) \subset B_{10}^{an} = A \text{ or}$$
$$R = C^{\infty}(\mathbb{R}^{10}) \subset B_{10}^{\infty} = A.$$

we choose for cohomology computations. This is why we will use B_{10} as a unifying notation for all modifications.

We can calculate groups $\mathrm{HH}^i_{loc\ \mathfrak{susy}}(B_{10}, B_{10})$. These linear spaces have an additional conformal grading by eigenvalues of the dilation operator scaled by the factor of two[2]: $\mathrm{HH}^i_{loc\ \mathfrak{susy}}(B_{10}, B_{10}) = \bigoplus_{k\in\mathbb{Z}} \mathrm{HH}^{i,k}_{loc\ \mathfrak{susy}}(B_{10}, B_{10})$.

They can be expressed in terms of the groups $\mathrm{H}^{s,t}(L, U(TYM))$ considered in [**11**]:

$$\mathrm{HH}^{i,k}_{loc\ \mathfrak{susy}}(B_{10}, B_{10}) = \mathrm{H}^{i+k,i}(L, U(TYM))$$

The groups $H^{k,t}(L, U(TYM))$ were calculated in [**11**] for $k = 2$. (See also [**12**].) Similar methods can be applied for other values of k.

6. Construction of deformations

One can construct some interesting symmetry preserving deformations starting with homology classes of symmetry Lie algebra $\mathfrak{g}$.

The application of the homology of $\mathfrak{g}$ to the analysis of deformations is based on the construction of the homomorphism

$$\psi : H_i(\mathfrak{g}, N) \to H^{s-i}(\mathfrak{g}, N) \tag{6.1}$$

for arbitrary differential module N with L_∞ action of $\mathfrak{g}$. Here s stands for the number of even generators of $\mathfrak{g}$. This homomorphism is described in the appendix.

We will apply the homomorphism ψ to the construction of symmetry preserving deformations.

Let $\mathcal{A}$ be a differential $\mathbb{Z}$-graded associative algebra.[3] Let us assume that A is equipped with L_∞ action of Lie algebra $\mathfrak{g}$. Then $C^\bullet(A, A^*)$ and $HH^\bullet(A, A^*)$ are L_∞ $\mathfrak{g}$- modules and we can talk about the homology and cohomology of $\mathfrak{g}$ with coefficients in these modules. (Recall that $C^k(A, A^*)$ stands for the module of Hochschild co-chains, i.e. of k-linear functionals on A with values in A^*. Notice, that these co-chains can be identified with $(k+1)$-linear functionals with values in $\mathbb{C}$.) The complex $C^\bullet(A, A^*)$ has an additional operation of degree minus one:

$$C^\bullet(A, A^*) \to C^{\bullet-1}(A, A^*)$$

—the Connes differential B . The map B is a composition of two maps αB_0, which look particularly simple if the degrees of all elements are even. The operator α

[2]We do this to avoid fractional gradation in spinor components.

[3]One can generalize our constructions to the case of A_∞ algebras using the fact that a $\mathbb{Z}$-graded A_∞ algebra is quasi-isomorphic to differential graded algebra,

is the operator of cyclic antisymmetrisation. The operator B_0 is defined by the formula

$$(B_0\psi)(a^0,\ldots,a^n) = \psi(1,a^0,\ldots,a^n) - (-1)^{n+1}\psi(a^0,\ldots,a^n,1)$$

The reader may consult [**4**], [**9**] for details. [4]

This operator induces map on $C^i_{\mathfrak{g}}(A,A^*)$ and $C^{\mathfrak{g}i}(A,A^*)$, denoted by the same symbol; it anticommutes with $d_{\mathfrak{g}}$ and d_c.

The Connes operator induces a differential on Hochschild cohomology. Let us assume in that the cohomology of B in $H^i(A,A^*)$ is trivial for $i>0$ and is one-dimensional for $i=0$. [5] This assumption permits us to construct an element of homology $H_\bullet(\mathfrak{g},C^\bullet(A,A^*))$ starting with any element $c_0 \in H_\bullet(\mathfrak{g},C^\bullet(A,A^*))$ obeying $Bd_{\mathfrak{g}}c_0 = 0$. The construction is based on the observation that due to triviality of the cohomology of B we can represent $d_{\mathfrak{g}}c_0$ as Bc_1. Applying $d_{\mathfrak{g}}$ to both parts of equation $d_{\mathfrak{g}}c_0 = Bc_1$ we obtain $Bd_{\mathfrak{g}}c_1 = 0$; this equality allows us to continue the process. The process will terminate when $d_{\mathfrak{g}}c_i = 0$. This must happen for some i because $d_{\mathfrak{g}}$ decreases the degree in $\mathrm{Sym}(\Pi\mathfrak{g})$ (the number of ghosts). The element c_i specifies the homology class we are interested in.

Let us describe the construction of elements of $H_\bullet(\mathfrak{g},C^\bullet(A,A))$, which uses homology classes of super Lie algebra $\mathfrak{g}$ with trivial coefficients and a $\mathfrak{g}$-equivariant trace tr as an input. (We assume that the trace specifies non-degenerate inner product on cohomology.) The trace determines a $\mathfrak{g}$-equivariant map $A^* \to A$ and therefore a homomorphism $H_\bullet(\mathfrak{g},C^\bullet(A,A^*)) \to H_\bullet(\mathfrak{g},C^\bullet(A,A))$. This means that it is sufficient to construct an element of $H_\bullet(\mathfrak{g},C^\bullet(A,A^*))$.

Let us take a representative $c \in \mathrm{Sym}(\Pi\mathfrak{g})$ of homology class of the Lie algebra $\mathfrak{g}$. Then we can define c_0 by the formula $c_0 = c\otimes\epsilon$, where ϵ stands for a homomorphism of the algebra A into a field. [6] It is easy to check that $B\epsilon = 0$ and that ϵ specifies a non-trivial class in the homology of B. We see that $Bd_{\mathfrak{g}}c_0 = -d_{\mathfrak{g}}Bc_0 = 0$. This means that we can apply the iterative construction described above to obtain a cycle c_l. The corresponding homology class $[c_l] \in H_\bullet(\mathfrak{g},C^\bullet(A,A^*))$ is the class we need.

Remark Recall that we have defined $\mathfrak{g}$-equivariant version of cyclic cohomology (Lie-cyclic cohomology) $HC^\bullet_{\mathfrak{g}}(\mathcal{A})$ as cohomology of the Lie algebra $\mathfrak{g}$ with coefficients in cyclic cochains considered as a differential $\mathfrak{g}$-module. It is rather straightforward to construct Connes long exact sequence. The main corollary of this construction is that classes c_l are images of classes in cyclic cohomology.

Remark The equivariant version of the package $\mathrm{HC}^\bullet(\mathcal{A})$, $\mathrm{HH}^\bullet(\mathcal{A},\mathcal{A})$, $\mathrm{HH}^\bullet(\mathcal{A})$ makes sense for both of local versions described in Section 4.

7. Appendix. Homology of super Lie algebras

7.1. Finite-dimensional super Lie algebras. Let us consider first the cohomology of finite-dimensional super Lie algebras. This cohomology is defined in terms of a differential

$$d = \frac{1}{2}(-1)^{|b^k|} f^m_{lk} b^k b^l c_m$$

[4]The operator B transforms Hochschild n-cochain into cyclic $(n-1)$-cochain; it generates a homomorphism $HH^n(A,A^*) \to HC^{n-1}(A)$ in Connes exact sequence.

[5]This is true, for example, if there exists an auxiliary grading by means of non-negative integers with one-dimensional grading zero component.

[6]If A is represented as a direct sum of one-dimensional subalgebra generated by the unit and ideal I (augmentation ideal) then ϵ is a projection of A on the first summand.

where f_{kl}^m are structure constants of super Lie algebra G in some basis t_k. The operators b^k and c_k correspond to elements of basis, but have parity opposite to the parity of elements of basis. They satisfy canonical (anti)commutation relations: $[c_k, b^l\} = \delta_k^l$; in other words they can be considered as generators of super Weyl algebra W_{rs} where r stands for the number of even generators and s stands for the number of odd generators .The differential acts in any representation of Weyl algebra (in any W_{rs}-module); the cohomology can be defined by means of any representation and depends on the choice of representation. We will assume that the representation $\mathcal{F}$ of Weyl algebra W_{rs} is graded in such a way that b^k raises grading by 1 and c_k decreases grading by 1, then the differential increases grading by 1. The cohomology is also graded in this case. One can define cohomology of super Lie algebra G with coefficients in G-module N by means of the differential $d + \sum T_k b^k$ on the space $\mathcal{F} \otimes N$ (Here T_k denotes the action of generator $t_k \in G$ on N.) We will use the notation $H^k(G, N|\mathcal{F})$ for k-dimensional cohomology of super Lie algebra G with coefficients in G-module N calculated by means of W_{rs}-module $\mathcal{F}$.

If G is a conventional Lie algebra then $r = 0$, the super Weyl algebra is a Clifford algebra with $2s$ generators. Irreducible representation in this case is unique; the representation space can be realized as Grassmann algebra (as algebra of functions of s anticommuting variables) where c_k and b^k act as derivatives and multiplication operators. We come to the standard notion of cohomology of Lie algebra.(However, in the case when G is an infinite-dimensional Lie algebra the irreducible representation of corresponding Clifford algebra is not unique; this remark leads to the notion of semi-infinite cohomology.) If $r > 0$ the super Weyl algebra is a tensor product of Weyl algebra W_r and Clifford algebra Cl_s; further W_r is a tensor product of r copies of Weyl algebra $W = W_1$. [7]

Let us consider first of all representations of the algebra W having generators b, c with relation $[c, b] = 1$. The simplest of these representations F_+ is realized in the space of polynomials $\mathbb{C}[t]$ where c acts as a derivation and b as a multiplication by t. The grading is given by the degree of polynomial. This representation can be described also as representation with cyclic vector Φ obeying $c\Phi = 0$ (Fock representation with vacuum vector Φ.) Another representation F_- can be constructed as a representation with cyclic vector Ψ obeying $b\Psi = 0$, $\deg \Psi = 0$. To relate these two representations we consider the representation F in the space $\mathbb{C}[t, t^{-1}]$ (polynomials of t and t^{-1}). The operators c and b again act as derivation and multiplication by t. It is easy to check that factorizing F with respect to subrepresentation F_+ we obtain a representation isomorphic to F_- (the polynomial t^{-1} plays the role of cyclic vector Ψ). Notice, however, that the grading in F_- does not coincide with the grading in F/F_+ (the degree of t^{-1} is equal to -1). One can say that as graded module F/F_+ is isomorphic to $F_-[-1]$ (to F_- with shifted grading).

Let us represent W_{rs} as a tensor product $W \otimes W_{r-1,s}$. For every representation E of second factor we can construct two representations of W_{rs} as tensor products $F_+ \otimes E$ and $F_- \otimes E$.The relation $F/F_+ = F_-[-1]$ permits us to construct a map

$$H^k(G, N|F_- \otimes E) \to H^k(G, N|F_+ \otimes E) \tag{7.1}$$

[7]Notice that we work in algebraic setting. It is well known that for correct definition of unitary representation Weyl algebra with finite number of generators has only one irreducible unitary representation. This statement cannot be applied to representations at hand.

This map is analogous to picture changing operator in BRST cohomology of superstring. It can be regarded as coboundary operator in exact cohomology sequence corresponding to short exact sequence

$$0 \to F_+ \otimes E \to F \otimes E \to F_- \otimes E \to 0.$$

Notice that coboundary operator raises degree by 1, but taking into account the the shift of grading in F_- we see that (7.1) does not change the degree.

We will consider irreducible representations $F_{\epsilon_1,\dots,\epsilon_r}$ of W_{rs} defined as tensor product of representations F_{ϵ_k} and irreducible representation of Clifford algebra. (Here $\epsilon_k = \pm$.)These representations can be defined also as Fock representations with vacuum vector Φ obeying $c_k\Phi = 0$ if $\epsilon_k = +$ and $b^k\Phi = 0$ if $\epsilon_k = 0$. The grading is determined by the condition $\deg\Phi = 0$. The cohomology corresponding to representation with all $\epsilon_k = +$ coincides with standard cohomology of super Lie algebra:

$$H^k(G, N|F_{+\cdots+}) = H^k(G, N).$$

The cohomology corresponding to representation with all $\epsilon_k = -$ is closely related to homology of super Lie algebra. If $f_{kl}^l = 0$ we have

$$H^k(G, N|F_{-\dots-}) = H_{s-k}(G, N). \tag{7.2}$$

To check (7.2) we notice that homology can be defined by means of differential

$$\partial = \frac{1}{2}(-1)^{|\gamma_k|} f_{lk}^m \gamma_m \frac{\partial^2}{\partial\gamma_k\partial\gamma_l} + T_k \frac{\partial}{\partial\gamma_k}$$

acting in the space of polynomial functions of variables γ_k (ghost variables). (Here as earlier T_k denotes the action of $t_k \in G$ on G-module N. The ghost variables have the parity opposite to the parity of t_k.) We can rewrite this differential in the form

$$\partial = \frac{1}{2}(-1)^{|b_k|} f_{lk}^m c_m b^k b^l + T_k b^k$$

where c_k, b^l satisfy canonical (anti)commutation relations. If $f_{kl}^l = 0$ the differential takes the form of cohomology differential acting in the space $F_{-\dots-}$. (The constant polynomial is a cyclic vector Φ obeying $b_k\Phi = 0$.) However, the grading is different: in the space of polynomial functions of γ_k the operator c_k increases degree by 1 (instead of decreasing it by 1 in cohomological grading). The grading of the cyclic vector Φ (of the Fock vacuum) is also different in homological and cohomological setting (0 versus s). We obtain the formula (7.2).

Applying r times the homomorphism (7.1) we obtain a homomorphism from homology into cohomology. More precisely, if $f_{kl}^l = 0$ we obtain a homomorphism

$$H_i(G, N) \to H^{s-i}(G, N)$$

.

7.2. Infinite-dimensional super Lie algebras. Recall that the cohomology of finite-dimensional super Lie algebra G with coefficients in G-module N were defined by means of differential

$$d = \frac{1}{2}(-1)^{|b_k|} f_{lk}^m b^k b^l c_m + T_k b^k \tag{7.3}$$

acting on the tensor product $\mathcal{F} \otimes N$. Here $\mathcal{F}$ is a representation of super Weyl algebra with generators b^k, c_l , the symbol f_{kl}^m denotes structure constants of G in the basis t_k and T_k stands for the operator in N corresponding to t_k, The space $\mathcal{F}$ can be considered as a G-module; the elements $t_k \in G$ act as operators

$$\tau_k = f_{kl}^m b^l c_m. \tag{7.4}$$

These operators obey relations

$$[\tau_k, \tau_l\} = f_{kl}^m \tau_m, \tag{7.5}$$

$$[\tau_k, b^m\} = f_{kl}^m b^l, \tag{7.6}$$

$$[\tau_k, c_l\} = f_{kl}^m c_m. \tag{7.7}$$

The differential d obeys

$$[d, b^m\} = \frac{1}{2} f_{kl}^m b^k b^l, \tag{7.8}$$

$$[d, c_l\} = \tau_l + T_l. \tag{7.9}$$

Let us consider now the case when G is an infinite-dimensional super Lie algebra and N is a projective representation of G (=a module over central extension of G). We will keep the notation t_k for the elements of basis of G and f_{kl}^m for structure constants. *We will assume that for fixed indices k, l there exists only finite number of indices m such that $f_{kl}^m \neq 0$. Similarly, if indices k, m are fixed then $f_{kl}^m \neq 0$ only for finite number of indices l.* The formulas (7.3) and (7.4) in general do not make sense in this situation. However, the RHS of (7.6) and (7.7) is well defined. We will assume $\mathcal{F}$ is an irreducible representation of Weyl algebra; then these formulas specify τ_k uniquely up to an additive constant. If the solution for τ_k does exist it specifies a projective representation of G:

$$[\tau_k, \tau_l\} = f_{kl}^m \tau_m + \gamma_{kl}.$$

The constants γ_{kl} determine a two-dimensional cocycle of G; in physics it is related to central charge.

We assume that the two-dimensional cohomology class of G corresponding to the projective module N is opposite to the cohomology class of γ. This means that for appropriate choice τ_l the expression $\tau_l + T_l$ (the RHS of (7.9)) specifies a genuine representation of G.

We will consider the case when $\mathcal{F} = \mathbb{F}_I$ is a Fock module (a module with a cyclic vector Φ obeying $b^k\Phi = 0$ for $k \in I$, $c_l\Phi = 0$ if $l \in J$ where J denotes the complement to I). Here I stands for some set of indices;*we assume that there exists only a finite set of triples (k, l, m) with $f_{kl}^m \neq 0$ obeying $k, l \in J, m \in I$.* [8] Then τ_k obeying equations (7.6) and (7.7) can be written in terms of normal product

$$\tau_k = f_{kl}^m : b^l c_m : . \tag{7.10}$$

Under our assumptions the RHS of (7.8) and (7.9) specifies a well defined operator on $\mathcal{F}$. Considering these formulas as equations for d we see that they determine d

[8] In many interesting situations G as a vector space can be represented as a direct sum of two subalgebras; the representation of the set of indices as a a disjoint union of I and J is related to this decomposition. In this case the cohomology we are interested in is called semi-infinite cohomology.

up to an additive constant. Requiring $d^2 = 0$ we obtain the following expression for d:

$$d = \frac{1}{2}(-1)^{|\gamma_k|} f_{lk}^m : b^k b^l c_m : + T_k b^k. \tag{7.11}$$

One defines the cohomology of G with coefficients in N by means of differential d. The cohomology in general depends on the choice of set I (on the choice of picture). One can introduce grading in $\mathcal{F}$ assuming that b^k increases degree by 1 ,c_l decreases degree by 1 and that $\deg \Phi = 0$. Using this grading (and grading in N) one can define grading on cohomology. We will use the notation $H^n(G, N; I)$ for k-dimensional cohomology.

We would like to study relation between $H^n(G, N; I)$ and $H^n(G, N; I')$ (the dependence of cohomology on the choice of the picture). We will analyze the case when I' is obtained from I by deleting one index k. Let us notice first of all that in the case when t_k is an even generator (corresponding ghosts are odd) $H^n(G, N; I) = H^n(G, N; I')$; this follows from the fact that $\mathbb{F}_I$ can be identified with $\mathcal{F}'$. (If a vector Φ obeys $b^k\Phi = 0$ for $k \in I$, $c_l\Phi = 0$ for $l \in J$ the vector $\Phi' = c_k\Phi \in \mathcal{F}_I$ obeys $b^k\Phi' = 0$ for $k \in I'$, $c_l\Phi = 0$ if $l \in J'$ where J' stands for the complement of I'.) If the generator t_k is odd (corresponding ghosts are even) then repeating the arguments used for finite-dimensional Lie algebras we can construct a homomorphism

$$H^n(G, N; I) \to H^n(G, n; I') \tag{7.12}$$

(picture changing operator).

This homomorphism is not an isomorphism in general. However, it is an isomorphism in cases relevant for string theory (when G is a superanalog of Virasoro algebra).

References

[1] H. Abbaspour. On algebraic structures of Hochschild complex. arXiv:1302.6534.

[2] E. Abdalla, M. Forger, and M. Jacques, *Higher conservation laws for ten-dimensional supersymmetric Yang-Mills theories*, Nuclear Phys. B **307** (1988), no. 1, 198–220, DOI 10.1016/0550-3213(88)90529-9. MR958496 (89h:81096)

[3] N. Berkovits, *Covariant quantization of the superparticle using pure spinors*, J. High Energy Phys. **9** (2001), Paper 16, 17, DOI 10.1088/1126-6708/2001/09/016. MR1867182 (2002m:81238)

[4] A. Connes, *Noncommutative geometry*, Academic Press, Inc., San Diego, CA, 1994. MR1303779 (95j:46063)

[5] T. G. Goodwillie, *On the general linear group and Hochschild homology*, Ann. of Math. (2) **121** (1985), no. 2, 383–407, DOI 10.2307/1971179. MR786354 (86i:18013)

[6] P. S. Howe, *Pure spinors, function superspaces and supergravity theories in ten and eleven dimensions*, Phys. Lett. B **273** (1991), no. 1-2, 90–94, DOI 10.1016/0370-2693(91)90558-8. MR1140168 (92h:83070)

[7] M. Kontsevich, *Deformation quantization of Poisson manifolds*, Lett. Math. Phys. **66** (2003), no. 3, 157–216, DOI 10.1023/B:MATH.0000027508.00421.bf. MR2062626 (2005i:53122)

[8] T. Lada and J. Stasheff, *Introduction to SH Lie algebras for physicists*, Internat. J. Theoret. Phys. **32** (1993), no. 7, 1087–1103, DOI 10.1007/BF00671791. MR1235010 (94g:17059)

[9] J.-L. Loday, *Cyclic homology*, 2nd ed., Grundlehren der Mathematischen Wissenschaften [Fundamental Principles of Mathematical Sciences], vol. 301, Springer-Verlag, Berlin, 1998. Appendix E by María O. Ronco; Chapter 13 by the author in collaboration with Teimuraz Pirashvili. MR1600246 (98h:16014)

[10] J.-L. Loday and D. Quillen, *Cyclic homology and the Lie algebra homology of matrices*, Comment. Math. Helv. **59** (1984), no. 4, 569–591, DOI 10.1007/BF02566367. MR780077 (86i:17003)

[11] M.V. Movshev. Deformation of maximally supersymmetric Yang-Mills theory in dimensions 10. An algebraic approach, 2006. arXiv:hep-th/0601010v1.

[12] M. V. Movshev and A. Schwarz, *Supersymmetric deformations of maximally supersymmetric gauge theories*, J. High Energy Phys. **9** (2012), 136, front matter+76. MR3044913

[13] M. Penkava and A. Schwarz, *A_∞ algebras and the cohomology of moduli spaces*, Lie groups and Lie algebras: E. B. Dynkin's Seminar, Amer. Math. Soc. Transl. Ser. 2, vol. 169, Amer. Math. Soc., Providence, RI, 1995, pp. 91–107. MR1364455 (97b:16015)

[14] A. Schwarz, *A-model and generalized Chern-Simons theory*, Phys. Lett. B **620** (2005), no. 3-4, 180–186, DOI 10.1016/j.physletb.2005.06.030. MR2149784 (2006j:81159)

Department of Mathematics, Stony Brook University,Stony Brook, New York 11794-3651

E-mail address: `mmovshev@math.sunysb.edu`

Department of Mathematics, University of California, Davis, California 95616

E-mail address: `schwarz@math.ucdavis.edu`

Selected Published Titles in This Series

90 **Ron Donagi, Sheldon Katz, Albrecht Klemm, and David R. Morrison, Editors,** String-Math 2012, 2015

89 **D. Dolgopyat, Y. Pesin, M. Pollicott, and L. Stoyanov, Editors,** Hyperbolic Dynamics, Fluctuations and Large Deviations, 2015

88 **Ron Donagi, Michael R. Douglas, Ljudmila Kamenova, and Martin Rocek, Editors,** String-Math 2013, 2014

87 **Helge Holden, Barry Simon, and Gerald Teschl, Editors,** Spectral Analysis, Differential Equations and Mathematical Physics: A Festschrift in Honor of Fritz Gesztesy's 60th Birthday, 2013

86 **Kailash C. Misra, Daniel K. Nakano, and Brian J. Parshall, Editors,** Recent Developments in Lie Algebras, Groups and Representation Theory, 2012

85 **Jonathan Block, Jacques Distler, Ron Donagi, and Eric Sharpe, Editors,** String-Math 2011, 2012

84 **Alex H. Barnett, Carolyn S. Gordon, Peter A. Perry, and Alejandro Uribe, Editors,** Spectral Geometry, 2012

83 **Hisham Sati and Urs Schreiber, Editors,** Mathematical Foundations of Quantum Field Theory and Perturbative String Theory, 2011

82 **Michael Usher, Editor,** Low-dimensional and Symplectic Topology, 2011

81 **Robert S. Doran, Greg Friedman, and Jonathan Rosenberg, Editors,** Superstrings, Geometry, Topology, and C^*-algebras, 2010

80 **D. Abramovich, A. Bertram, L. Katzarkov, R. Pandharipande, and M. Thaddeus, Editors,** Algebraic Geometry, 2009

79 **Dorina Mitrea and Marius Mitrea, Editors,** Perspectives in Partial Differential Equations, Harmonic Analysis and Applications, 2008

78 **Ron Y. Donagi and Katrin Wendland, Editors,** From Hodge Theory to Integrability and TQFT, 2008

77 **Pavel Exner, Jonathan P. Keating, Peter Kuchment, Toshikazu Sunada, and Alexander Teplyaev, Editors,** Analysis on Graphs and Its Applications, 2008

76 **Fritz Gesztesy, Percy Deift, Cherie Galvez, Peter Perry, and Wilhelm Schlag, Editors,** Spectral Theory and Mathematical Physics: A Festschrift in Honor of Barry Simon's 60th Birthday, 2007

75 **Solomon Friedberg, Daniel Bump, Dorian Goldfeld, and Jeffrey Hoffstein, Editors,** Multiple Dirichlet Series, Automorphic Forms, and Analytic Number Theory, 2006

74 **Benson Farb, Editor,** Problems on Mapping Class Groups and Related Topics, 2006

73 **Mikhail Lyubich and Leon Takhtajan, Editors,** Graphs and Patterns in Mathematics and Theoretical Physics, 2005

72 **Michel L. Lapidus and Machiel van Frankenhuijsen, Editors,** Fractal Geometry and Applications: A Jubilee of Benoît Mandelbrot, 2004

71 **Gordana Matić and Clint McCrory, Editors,** Topology and Geometry of Manifolds, 2003

70 **Michael D. Fried and Yasutaka Ihara, Editors,** Arithmetic Fundamental Groups and Noncommutative Algebra, 2002

69 **Anatole Katok, Rafael de la Llave, Yakov Pesin, and Howard Weiss, Editors,** Smooth Ergodic Theory and Its Applications, 2001

68 **Robert S. Doran and V. S. Varadarajan, Editors,** The Mathematical Legacy of Harish-Chandra, 2000

67 **Wayne Raskind and Charles Weibel, Editors,** Algebraic K-Theory, 1999

For a complete list of titles in this series, visit the
AMS Bookstore at **www.ams.org/bookstore/pspumseries/**.

ISBN 978-0-8218-9495-8